ENDOTOXIN AND THE LUNGS

LUNG BIOLOGY IN HEALTH AND DISEASE

Executive Editor

Claude Lenfant
Director, National Heart, Lung and Blood Institute
National Institutes of Health
Bethesda, Maryland

1. Immunologic and Infectious Reactions in the Lung, *edited by Charles H. Kirkpatrick and Herbert Y. Reynolds*
2. The Biochemical Basis of Pulmonary Function, *edited by Ronald G. Crystal*
3. Bioengineering Aspects of the Lung, *edited by John B. West*
4. Metabolic Functions of the Lung, *edited by Y. S. Bakhle and John R. Vane*
5. Respiratory Defense Mechanisms (in two parts), *edited by Joseph D. Brain, Donald F. Proctor, and Lynne M. Reid*
6. Development of the Lung, *edited by W. Alan Hodson*
7. Lung Water and Solute Exchange, *edited by Norman C. Staub*
8. Extrapulmonary Manifestations of Respiratory Disease, *edited by Eugene Debs Robin*
9. Chronic Obstructive Pulmonary Disease, *edited by Thomas L. Petty*
10. Pathogenesis and Therapy of Lung Cancer, *edited by Curtis C. Harris*
11. Genetic Determinants of Pulmonary Disease, *edited by Stephen D. Litwin*
12. The Lung in the Transition Between Health and Disease, *edited by Peter T. Macklem and Solbert Permutt*
13. Evolution of Respiratory Processes: A Comparative Approach, *edited by Stephen C. Wood and Claude Lenfant*
14. Pulmonary Vascular Diseases, *edited by Kenneth M. Moser*
15. Physiology and Pharmacology of the Airways, *edited by Jay A. Nadel*
16. Diagnostic Techniques in Pulmonary Disease (in two parts), *edited by Marvin A. Sackner*
17. Regulation of Breathing (in two parts), *edited by Thomas F. Hornbein*
18. Occupational Lung Diseases: Research Approaches and Methods, *edited by Hans Weill and Margaret Turner-Warwick*
19. Immunopharmacology of the Lung, *edited by Harold H. Newball*
20. Sarcoidosis and Other Granulomatous Diseases of the Lung, *edited by Barry L. Fanburg*
21. Sleep and Breathing, *edited by Nicholas A. Saunders and Colin E. Sullivan*
22. *Pneumocystis carinii* Pneumonia: Pathogenesis, Diagnosis, and Treatment, *edited by Lowell S. Young*

23. Pulmonary Nuclear Medicine: Techniques in Diagnosis of Lung Disease, *edited by Harold L. Atkins*
24. Acute Respiratory Failure, *edited by Warren M. Zapol and Konrad J. Falke*
25. Gas Mixing and Distribution in the Lung, *edited by Ludwig A. Engel and Manuel Paiva*
26. High-Frequency Ventilation in Intensive Care and During Surgery, *edited by Graziano Carlon and William S. Howland*
27. Pulmonary Development: Transition from Intrauterine to Extrauterine Life, *edited by George H. Nelson*
28. Chronic Obstructive Pulmonary Disease: Second Edition, Revised and Expanded, *edited by Thomas L. Petty*
29. The Thorax (in two parts), *edited by Charis Roussos and Peter T. Macklem*
30. The Pleura in Health and Disease, *edited by Jacques Chrétien, Jean Bignon, and Albert Hirsch*
31. Drug Therapy for Asthma: Research and Clinical Practice, *edited by John W. Jenne and Shirley Murphy*
32. Pulmonary Endothelium in Health and Disease, *edited by Una S. Ryan*
33. The Airways: Neural Control in Health and Disease, *edited by Michael A. Kaliner and Peter J. Barnes*
34. Pathophysiology and Treatment of Inhalation Injuries, *edited by Jacob Loke*
35. Respiratory Function of the Upper Airway, *edited by Oommen P. Mathew and Giuseppe Sant'Ambrogio*
36. Chronic Obstructive Pulmonary Disease: A Behavioral Perspective, *edited by A. John McSweeny and Igor Grant*
37. Biology of Lung Cancer: Diagnosis and Treatment, *edited by Steven T. Rosen, James L. Mulshine, Frank Cuttitta, and Paul G. Abrams*
38. Pulmonary Vascular Physiology and Pathophysiology, *edited by E. Kenneth Weir and John T. Reeves*
39. Comparative Pulmonary Physiology: Current Concepts, *edited by Stephen C. Wood*
40. Respiratory Physiology: An Analytical Approach, *edited by H. K. Chang and Manuel Paiva*
41. Lung Cell Biology, *edited by Donald Massaro*
42. Heart–Lung Interactions in Health and Disease, *edited by Steven M. Scharf and Sharon S. Cassidy*
43. Clinical Epidemiology of Chronic Obstructive Pulmonary Disease, *edited by Michael J. Hensley and Nicholas A. Saunders*
44. Surgical Pathology of Lung Neoplasms, *edited by Alberto M. Marchevsky*
45. The Lung in Rheumatic Diseases, *edited by Grant W. Cannon and Guy A. Zimmerman*
46. Diagnostic Imaging of the Lung, *edited by Charles E. Putman*
47. Models of Lung Disease: Microscopy and Structural Methods, *edited by Joan Gil*

48. Electron Microscopy of the Lung, *edited by Dean E. Schraufnagel*
49. Asthma: Its Pathology and Treatment, *edited by Michael A. Kaliner, Peter J. Barnes, and Carl G. A. Persson*
50. Acute Respiratory Failure: Second Edition, *edited by Warren M. Zapol and Francois Lemaire*
51. Lung Disease in the Tropics, *edited by Om P. Sharma*
52. Exercise: Pulmonary Physiology and Pathophysiology, *edited by Brian J. Whipp and Karlman Wasserman*
53. Developmental Neurobiology of Breathing, *edited by Gabriel G. Haddad and Jay P. Farber*
54. Mediators of Pulmonary Inflammation, *edited by Michael A. Bray and Wayne H. Anderson*
55. The Airway Epithelium, *edited by Stephen G. Farmer and Douglas Hay*
56. Physiological Adaptations in Vertebrates: Respiration, Circulation, and Metabolism, *edited by Stephen C. Wood, Roy E. Weber, Alan R. Hargens, and Ronald W. Millard*
57. The Bronchial Circulation, *edited by John Butler*
58. Lung Cancer Differentiation: Implications for Diagnosis and Treatment, *edited by Samuel D. Bernal and Paul J. Hesketh*
59. Pulmonary Complications of Systemic Disease, *edited by John F. Murray*
60. Lung Vascular Injury: Molecular and Cellular Response, *edited by Arnold Johnson and Thomas J. Ferro*
61. Cytokines of the Lung, *edited by Jason Kelley*
62. The Mast Cell in Health and Disease, *edited by Michael A. Kaliner and Dean D. Metcalfe*
63. Pulmonary Disease in the Elderly Patient, *edited by Donald A. Mahler*
64. Cystic Fibrosis, *edited by Pamela B. Davis*
65. Signal Transduction in Lung Cells, *edited by Jerome S. Brody, David M. Center, and Vsevolod A. Tkachuk*
66. Tuberculosis: A Comprehensive International Approach, *edited by Lee B. Reichman and Earl S. Hershfield*
67. Pharmacology of the Respiratory Tract: Experimental and Clinical Research, *edited by K. Fan Chung and Peter J. Barnes*
68. Prevention of Respiratory Diseases, *edited by Albert Hirsch, Marcel Goldberg, Jean-Pierre Martin, and Roland Masse*
69. *Pneumocystis carinii* Pneumonia: Second Edition, Revised and Expanded, *edited by Peter D. Walzer*
70. Fluid and Solute Transport in the Airspaces of the Lungs, *edited by Richard M. Effros and H. K. Chang*
71. Sleep and Breathing: Second Edition, Revised and Expanded, *edited by Nicholas A. Saunders and Colin E. Sullivan*
72. Airway Secretion: Physiological Bases for the Control of Mucous Hypersecretion, *edited by Tamotsu Takishima and Sanae Shimura*
73. Sarcoidosis and Other Granulomatous Disorders, *edited by D. Geraint James*
74. Epidemiology of Lung Cancer, *edited by Jonathan M. Samet*

75. Pulmonary Embolism, *edited by Mario Morpurgo*
76. Sports and Exercise Medicine, *edited by Stephen C. Wood and Robert C. Roach*
77. Endotoxin and the Lungs, *edited by Kenneth L. Brigham*
78. The Mesothelial Cell and Mesothelioma, *edited by Jean Bignon and Marie-Claude Jaurand*

ADDITIONAL VOLUMES IN PREPARATION

Regulation of Breathing: Second Edition, Revised and Expanded, *edited by Jerome A. Dempsey and Allan I. Pack*

Pulmonary Fibrosis, *edited by Sem H. Phan and Roger S. Thrall*

ENDOTOXIN AND THE LUNGS

Edited by

Kenneth L. Brigham

Vanderbilt University School of Medicine
Nashville, Tennessee

Marcel Dekker, Inc. **New York • Basel • Hong Kong**

Library of Congress Cataloging-in-Publication Data

Endotoxin and the lungs / edited by Kenneth L. Brigham.
p. cm. — (Lung biology in health and disease; v. 77)
Includes bibliographical references and indexes.
ISBN 0-8247-9222-X (alk. paper)
1. Respiratory infections—Pathophysiology. 2. Respiratory infections—Molecular aspects. 3. Endotoxins—Pathophysiology. I. Brigham, Kenneth L. II. Series.
[DNLM: 1. Lung Diseases—etiology. 2. Lung Diseases—therapy. 3. Lung—physiopathology. 4. Endotoxins—adverse effects. W1 LU62 v. 77 1994 / WF 600 E565 1994]
RC740.E53 1994
616.2'4—dc20
DNLM/DLC
For Library of Congress 94-20079
CIP

The publisher offers discounts on this book when ordered in bulk quantities. For more information, write to Special Sales/Professional Marketing at the address below.

This book is printed on acid-free paper.

MARCEL DEKKER, INC.
270 Madison Avenue, New York, New York 10016

Current printing (last digit):
10 9 8 7 6 5 4 3 2 1

Printed in the United States of America

INTRODUCTION

Will Hell ever open? Of course, no one would venture to answer this question, suggested in the Preface, as the risks are great. Years of research and an "enormous literature on endotoxin" have not yet solved an important and old public health problem. Bacteremia and endotoxemia have been, and remain, the main cause of the lung injury called acute respiratory distress syndrome, which kills more than half of its victims.

Endotoxemia is not new; it can be traced back to the most ancient texts on medicine. However, in the nineteenth century major and consequential observations began to emerge from the work of such great researchers as von Behring, Robert Koch, Louis Pasteur, and Emile Roux. In 1901, von Behring received the Nobel Prize in Medicine and Physiology for his landmark work on serum therapy.

Few fields of research have been as fertile, and yet as frustrating. Despite years of effort and immense progress the bottom line remains, as Dr. Brigham says in his Preface, "Endotoxin kills people."

This volume opens the door on new directions. Undoubtedly, it will challenge investigators and clinicians as well. From a "philosophical" point of view, it goes steps further, as it represents a paradigm of what basic research can contribute to clinical investigations. Indeed, it displays a spectrum of information in chapters that range from the most fundamental to the practical. The last chapter, which discusses the potential for gene therapy, is a vibrant illustration of what is today referred to as molecular medicine. Will genetic transfection become a cure? Will genetic material be used as a drug? Certainly, current evidence and experimental work lead us to answer in the affirmative. But much remains to be done so that "do no harm" remains the primary goal.

This book is a major contribution to a rapidly advancing field. Dr. Brigham and his contributors deserve much commendation for sharing their thoughts and their expectations. This effort may constitute only a "keyhole," but let us not forget that without a keyhole the door cannot be unlocked.

Claude Lenfant, M.D.
Bethesda, Maryland

PREFACE

> Endotoxin and mycoplasma are Nature's darkest secrets. If they are ever solved, Hell itself will open.
>
> *Lewis Thomas*

Lewis Thomas spent much of his life studying endotoxin and reflecting literately on what he learned from that and myriad other experiences. I was a fellow, already seduced by the subject of endotoxin, when I heard him make the preceding statement on receiving the Kober Medal from the Association of American Physicians. I was tempted to rethink my research direction.

It does not take a book this big to say that endotoxin pathobiology is complex; the goal is bigger than that. The reader will judge, but I think the serious efforts of a daunting list of expert authors have distilled from a vast scientific spectrum the critical core of information relevant to endotoxin as a clinical problem. The exciting thing is that, although complex, the pathobiology may be manipulable to the lungs' (and possibly to other organs') advantage. That is the good news.

The bad news is that, as Professor Thomas intimated, understanding how to manipulate the pathobiology to do more good than harm is not easy. This volume also makes that point. The structure of endotoxin and how that structure relates to its bioactivity is complex. Endotoxin has fascinating (and complex) interactions with proteins, cell surface molecules and signal transduction apparatuses. Cellular and humoral components of the inflammatory process and their networking are at endotoxin's beck and call. Somehow, endotoxin orchestrates all of that complexity to mischievous ends. Endotoxin kills.

Endotoxin kills peoples' lungs and therefore kills people. Such mischief lends urgency to applying even incomplete knowledge about mechanisms to designing therapy. Several chapters in this volume discuss current, experimental and possible future therapies. There are lessons. One can be too vulnerable to the urgency and make mistakes. But waiting for complete understanding of endotoxin pathobiology is simply not possible. Endotoxin kills people.

There is an enormous literature on endotoxin. As a subject for investigation it has always been irresistible—a somewhat morbid attraction. In the references supplied at the ends of the chapters, we have tried to direct the reader to the relevant literature, but if you go that route, be forewarned.

The hope is that this volume casts new light on the core of information about endotoxin relevant to clinical medicine by putting it together in a different way and that this will be useful to scientists and clinicians and everyone in between. The book is probably not close enough to solving the endotoxin riddle to risk unlocking the Gates of Hell. But it gives a pretty revealing look through the keyhole.

Kenneth L. Brigham

CONTRIBUTORS

Kenneth L. Brigham, M.D. Director, Center for Lung Research, and Professor, Department of Medicine, Vanderbilt University School of Medicine, Nashville, Tennessee

Angelo E. Canonico, M.D. Instructor in Medicine, Department of Medicine, Vanderbilt University School of Medicine, Nashville, Tennessee

Tai-Ying Chen, Ph.D. Research Assistant Professor, Department of Microbiology, Molecular Genetics, and Immunology, University of Kansas Medical Center, Kansas City, Kansas

Stephen W. Chensue, Ph.D., M.D. Associate Professor, Department of Pathology, University of Michigan Medical School, Ann Arbor, Michigan

James A. Cook, Ph.D. Professor, Department of Physiology, Medical University of South Carolina, Charleston, South Carolina

Jon T. Conary, Ph.D. Research Assistant Professor, Departments of Medicine and Biochemistry, Vanderbilt University School of Medicine, Nashville, Tennessee

Dennis E. Doherty, M.D. Associate Professor of Medicine, National Jewish Center for Immunology and Respiratory Medicine, Associate Professor of Medicine, Denver Veterans Affairs Medical Center, and Associate Professor of Medicine, Division of Pulmonary and Critical Care Medicine, University of Colorado Health Sciences Center, Denver, Colorado

Mary Durando, D.V.M. Department of Physiology, Medical University of South Carolina, Charleston, South Carolina

H. Dieter Grimmecke Forschungsinstitut Borstel, Institut für Experimentelle Biologie und Medizin, Borstel, Germany

Perry V. Halushka, M.D., Ph.D. Professor and Director, Division of Clinical Pharmacology, Department of Pharmacology, Medical University of South Carolina, Charleston, South Carolina

John M. Harlan, M.D. Head, Division of Hematology, and Professor, Department of Medicine, University of Washington, Seattle, Washington

Jacek Hawiger, M.D., Ph.D. Oswald T. Avery Professor and Chairman, Department of Microbiology and Immunology, Vanderbilt University School of Medicine, Nashville, Tennessee

Peter M. Henson, Ph.D., D.V.M. Executive Vice President, Academic Affairs, National Jewish Center for Immunology and Respiratory Medicine, and Professor of Medicine and Pathology, University of Colorado Health Sciences Center, Denver, Colorado

Leonard D. Hudson, M.D. Professor of Medicine and Head, Division of Pulmonary and Critical Care Medicine, Department of Medicine, University of Washington, Seattle, Washington

Teruo Kirikae, M.D., Ph.D. Senior Research Associate, Department of Microbiology, Jichi Medical School, Tochigi-Ken, Japan

Steven L. Kunkel, Ph.D. Department of Pathology, University of Michigan Medical School, Ann Arbor, Michigan

Mei-Guey Lei, Ph.D. Research Assistant Professor, Department of Microbiology, Molecular Genetics, and Immunology, University of Kansas Medical Center, Kansas City, Kansas

Michel Makhlouf Department of Physiology, Medical University of South Carolina, Charleston, South Carolina

Thomas R. Martin, M.D. Professor of Medicine, Department of Pulmonary and Critical Care Medicine, University of Washington School of Medicine, and Seattle Veterans Affairs Medical Center, Seattle, Washington

John C. Mathison, Ph.D. Scientific Associate, Department of Immunology, The Scripps Research Institute, La Jolla, California

James E. McFeely, M.D. Acting Instructor, Division of Pulmonary and Critical Care Medicine, Department of Medicine, University of Washington, Seattle, Washington

Barbara O. Meyrick, Ph.D. Professor of Pathology and Medicine, Departments of Pathology and Medicine, Vanderbilt University School of Medicine, Nashville, Tennessee

David C. Morrison, Ph.D. Associate Director, Cancer Center, University of Kansas Medical Center, Kansas City, Kansas

John H. Newman, M.D. Elsa S. Hanigan Professor of Medicine, Center for Lung Research, Vanderbilt University School of Medicine, Nashville, Tennessee

Polly E. Parsons, M.D. Associate Professor, Pulmonary Division, University of Colorado Health Sciences Center, Denver, Colorado

Jonathan D. Plitman, M.D. Assistant Professor of Medicine, Center for Lung Research, Vanderbilt University School of Medicine, Nashville, Tennessee

Margaret A. Read, Ph.D. Vascular Research Division, Department of Pathology, Brigham and Women's Hospital, Boston, Massachusetts

John E. Repine, M.D. President and Director, Webb-Waring Institute for Biomedical Research, and Professor of Medicine, University of Colorado Health Sciences Center, Denver, Colorado

David W. H. Riches, Ph.D. Associate Faculty Member, Division of Basic Sciences, Department of Pediatrics, National Jewish Center for Immunology and Respiratory Medicine, and Associate Professor, Department of Biochemistry, Biophysics, and Genetics, University of Colorado Health Sciences Center, Denver, Colorado

Ernst Th. Rietschel, Ph.D., M.D. Professor and Head, Department of Immunochemistry and Biochemical Microbiology, Forschungsinstitut Borstel, Institut für Experimentelle Biologie und Medicin, Borstel, Germany

Hans Schreier, Ph.D. Associate Professor of Medicine, Department of Medicine, Vanderbilt University School of Medicine, Nashville, Tennessee

Sam R. Sharar, M.D. Assistant Professor of Medicine, Department of Pediatric Anesthesia and Critical Care, Department of Anesthesiology, University of Washington, Seattle, Washington

Arthur S. Slutsky, M.D. Director, Division of Respiratory Medicine, University of Toronto and Mount Sinai Hospital, Toronto, Ontario, Canada

Robert M. Smith, M.D. Staff Physician, Department of Medicine, Pulmonary Section, San Diego VA Medical Center, San Diego, California

James R. Snapper, M.D. Professor of Medicine and Senior Investigator, Center for Lung Research, Vanderbilt University School of Medicine, Nashville, Tennessee

H. R. Anders Sonesson Research Associate, Center for Environmental Biotechnology, University of Tennessee, Knoxville, Tennessee

Roger G. Spragg, M.D. Chief of Medical Service, San Diego VA Medical Center, San Diego, California

Theodore J. Standiford, M.D. Assistant Professor, Department of Internal Medicine, University of Michigan Medical School, Ann Arbor, Michigan

Robert M. Strieter, M.D. Associate Professor, Department of Internal Medicine, University of Michigan Medical School, Ann Arbor, Michigan

Fletcher B. Taylor, Jr., M.D. Cardiovascular Biology Research, Oklahoma Medical Research Foundation, Oklahoma City, Oklahoma

Peter S. Tobias, Ph.D. Associate Member, Department of Immunology, The Scripps Research Institute, La Jolla, California

Richard J. Ulevitch, Ph.D. Member and Chairman, Department of Immunology, The Scripps Research Institute, La Jolla, California

Jesus Villar, M.D., Ph.D., F.C.C.M. Critical Care Scientist, Department of Medicine, Mount Sinai Hospital, Toronto, Ontario, Canada

Otto Westphal Forschungsinstitut Borstel, Institut für Experimentelle Biologie und Medizin, Borstel, Germany

Arthur P. Wheeler, M.D. Assistant Professor, Division of Pulmonary and Critical Care Medicine, Vanderbilt University Medical Center, Nashville, Tennessee

G. Scott Worthen, M.D. Head, Pulmonary Division, Department of Medicine, National Jewish Center for Immunology and Respiratory Medicine, and Associate Professor of Medicine, University of Colorado Health Sciences Center, Denver, Colorado

Robert K. Winn, Ph.D. Research Professor, Departments of Surgery, Physiology and Biophysics, University of Washington, Seattle, Washington

Ulrich Zähringer Head, Immunochemistry Division, Department of Immunochemistry and Biochemical Microbiology, Forschungsinstitut Borstel, Institut für Experimentelle Biologie und Medizin, Borstel, Germany

CONTENTS

1

Bacterial Endotoxin
Chemical Structure and Biological Activity

H. R. ANDERS SONESSON

Center for Environmental Biotechnology
University of Tennessee
Knoxville, Tennessee

ULRICH ZÄHRINGER, H. DIETER GRIMMECKE, OTTO WESTPHAL, and ERNST Th. RIETSCHEL

Forschungsinstitut Borstel
Institut für Experimentelle Biologie und Medizin
Borstel, Germany

I. Introduction

Gram-negative bacteria, with a few exceptions, share certain fundamental properties of cell-wall assembly: the presence of a structurally and functionally complex outer membrane, an inner cytoplasmic membrane, and an interjecting thin layer of peptidoglycan (1). The bilayered outer membrane is organized asymmetrically. In the outermost layer, pointing to the environment, certain proteins and in particular the *lipopolysaccharides* (LPS, *endotoxin*) are the dominating constituents, whereas the inner layer is composed of proteins and phospholipids (2).

LPS, being unique to gram-negative bacteria, is anchored by its lipid portion into the hydrophobic phase of the outer membrane, with the LPS polysaccharide part directed out from the surface (3,4). As integral components, LPS participate in a number of membrane functions. LPS are important for maintaining the integrity of the outer membrane and the cell shape, and thus are essential to cell growth and survival (1,3,4). LPS are the primary heat-stable antigens of the bacterial cell, and their general antigenic properties have long been utilized in diagnostics as the *O-antigens* for the classification (serotyping) of gram-negative species. The host's immune system responds to such exposed surface antigens during an infection, and antibodies are produced that are directed against epitopes located primarily in the polysaccharide region of LPS (4–6).

Injection of LPS into experimental animals causes a wide spectrum of biological (i.e., endotoxic) effects, including pathophysiological reactions such as fever, hypotension, changes in white blood cell counts, disseminated intravascular coagulation, or—with higher doses—irreversible shock (7). Some of these effects are seen as manifestations of gram-negative bacterial infections, and experimental data support the role of LPS as an important pathogenic factor of gram-negative sepsis (8). Other biological effects of LPS include the polyclonal stimulation of B-lymphocytes to differentiate, proliferate, and secrete immunoglobulins; and the activation of macrophages to enhance cytotoxicity, phagocytosis, and secretion of endogenous peptide and lipid mediators such as tumor necrosis factor α (TNFα), Interleukin-1 (IL-1) and IL-6 as well as thromboxane, leukotriens (LTs), and platelet aggregating factor (PAF).

Purified LPS or its endotoxically active center, free lipid A, biologically speaking, are artifacts which in nature during infection, or in continuous contact with higher animals and humans, probably do not occur as such. In viable bacteria, endotoxin is firmly bound to the outer cell membrane and released only during multiplication or disintegration of microbial organisms. Under these conditions, the high-molecular-weight "endotoxic complex," consisting of polysaccharide, lipid, and protein, will be subject to stepwise degradation. Intermediate products may differ in the spectrum of their endotoxic activities and in the induction of tolerance and other phenomena, such as conditioning of the body to standard endotoxic activities as we observe them in puristic animal experiments.

Another point of importance is the mode of application of endotoxin. Intravenous injection appears to be the most efficient mode of administration. But it is also known, and may be especially relevant here, that purified endotoxin or endotoxin released from bacteria will, under certain conditions, be inhaled (9–11) or absorbed through mucous membranes, such as in the lungs or along the digestive tract. Inhalation of endotoxin may lead to serious pathophysiologic disorders (9–11) but also may prevent the development of cancer (12). On the other hand, the continuous exposure of higher animals and humans to small amounts of endotoxin, to which we are adapted during evolution, may have a life-important influence, due to continuous stimulation. Interestingly, it seems to be technically impossible to create experimental conditions of endotoxin-free life (not to be mistaken for germfree life).

In the present volume many related questions are touched upon by a variety of authors. The elucidation of the biologically active structures of endotoxin certainly contributes to an ever-better understanding of the complex relationships between the higher animals and endotoxins in disease and normal life. Thus work with highly purified and standardized LPSs have allowed us to study and to understand more clearly the various responses to their action by a given individual. In this chapter we summarize some aspects of the chemical structure of LPS in relation to its biological activity and principal mechanisms involved in the endotoxic activity of LPS.

II. General Architecture of Lipopolysaccharide

Although a great structural diversity is found on comparison of different bacterial serotypes. LPS from various gram-negative bacteria share a common architecture in that they all consist of a polysaccharide portion covalently linked to a lipid component, termed lipid A (Fig. 1). The chemical composition of LPS has been found to be useful in taxonomic studies of gram-negative bacteria (13). Monomeric glycosyl units and types of linkages vary considerably between taxa (5,14). As pointed out earlier (15,16), the lipid A composition, especially, has proven to be of value for phylogenetic studies. Observed structural variations of lipid A have been shown to correlate well with the phylogenetic grouping of bacteria based on studies of ribosomal RNA sequence analysis. The taxonomic tree proposed on the basis of 16S

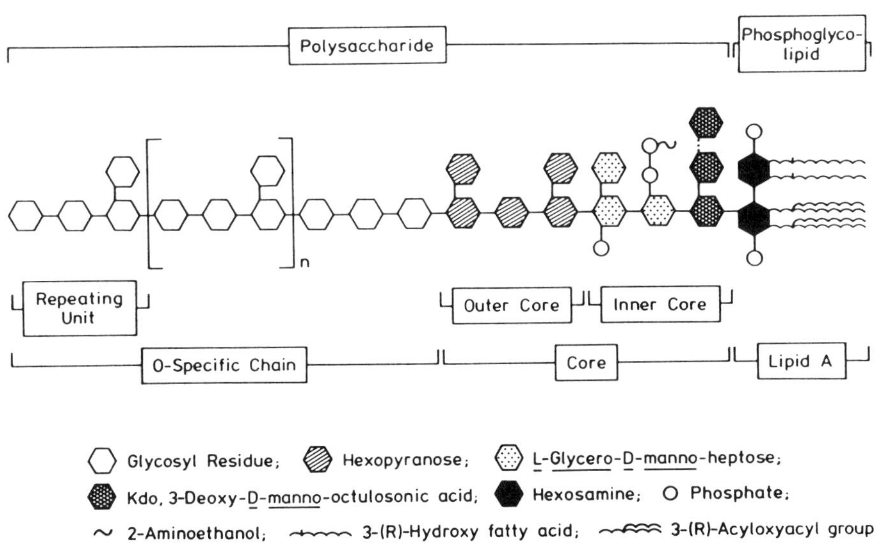

Figure 1 Schematic structure of enterobacterial wild-type (S-form) lipopolysaccharide.

and 5S rRNA homology studies has been well supported by the chemical composition analysis of lipid A components (17). Hence lipid A is considered a valuable chemotaxonomic marker for ascertaining the relationships of various bacteria.

LPS of various *Enterobacteriaceae* have been investigated most thoroughly and usually serve as a standard to which structures of LPS from other genera are compared (Fig. 1). The polysaccharide component of enterobacterial LPS comprises two distinct portions that are under separate genetic control: the core oligosaccharide and a polysaccharide chain of several repeating units of oligosaccharides, termed O-specific chain.

A. Polysaccharide Component

In general, the repeating units building up the O-specific chain are composed of two to eight distinct glycosyl constituents, and they form chains of different lengths. Homopolymeric O-chains occur in a few species (e.g., *Escherichia coli* 09, *Klebsiella pneumoniae* 01 and 02). Monosaccharides of neutral, basic, and acidic nature, and rare glycosyl residues such as 3,6-dideoxyhexoses and pseudaminic acid have been found to be constituents of the repeating units. From species to species the number of glycosyl residues present varies, as do their sequence and type of linkage (18).

Even when isolated from a single culture, an LPS preparation shows considerable heterogeneity regarding the lengths of the individual O-chains. This diversity can be visualized electrophoretically by polyacrylamide gel electrophoresis (19). Thus if a typical enterobacterial LPS preparation is analyzed, a ladder-like banding pattern is observed (Fig. 2, lanes 1,2). The distance between adjacent bands corresponds to the molecular weight difference of one repeating unit of the O-chain. Using this technique, it can be demonstrated that molecules with only 1 and up to 50 repeating units are present. Sometimes it is difficult to visualize a well-resolved banding pattern. This is true, for example, in the case of monosaccharidic repeating units (Fig. 2, lane 7) and LPS having more than one, and different O-side chains (Fig. 2, lane 3) which may overlap and produce a diffuse banding pattern. LPS with O-specific chains are termed smooth (S)-LPS, owing to the colonial morphology of the strain on nutrient-rich solid medium (20). Mutants of enterobacterial species showing a rough (R)-colony morphology have defects in their O-chain (or core) biosynthesis, resulting in the synthesis of LPS with a polysaccharide part lacking O-chains or parts of the core (21). Polyacrylamide electrophoresis of R-form LPS preparations shows only one or a few bands in the low-molecular-weight regions (Fig. 2).

In enterobacterial LPS, the core portion consists of a heterooligosaccharide which can formally be subdivided into an outer and an inner core region. The core structures are more uniform, structural diversity being found

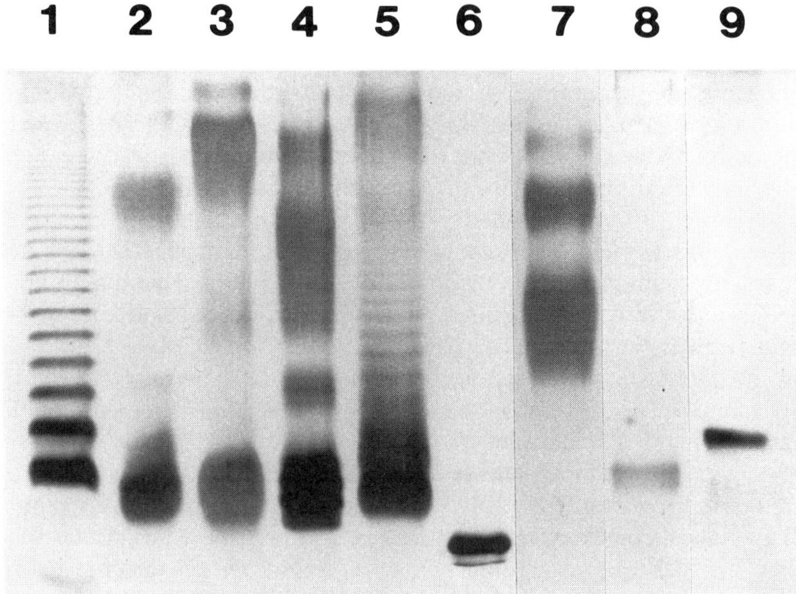

Figure 2 SDS-gelelectrophoresis pattern of enterobacterial S- and R-form and nonenterobacterial R-type lipopolysaccharides. Separation of LPS in 18% polyacrylamide allows both demonstration of O-side chain heterogeneity and analysis of core type. Distances of bands in the ladder pattern correspond to the size of the repeating units (RU) of the O-side chain. Lane 1: *Escherichia coli* 0111. The LPS consists of complete core, core with one RU (SR), and more RU up to the high-molecular-weight region. The O-specific side chain is built up of a pentasaccharide RU. Lane 2: *E. coli* 09. The LPS consists of large amounts of complete core, and a high-molecular-weight fraction of about 10 to 15 pentasaccharide RU. An RU contains mannose as the only glycosyl constituent. Lane 3: *Klebsiella* K5 (01). The LPS consists of complete core and high-molecular-weight LPS having two distinct O-side chains of disaccharide RU. Lane 4: *Klebsiella* K2 (02a, 02b). The LPS consists of different core oligosaccharides and a polydisperse O-side chain of a disaccharide RU. Lane 5: *Pseudomonas aeruginosa* Fischer immunotype 2. The LPS consists of complete core, core having one and more trisaccharide RU up to the high-molecular-weight region. Lane 6: *Haemophilus influenzae* I-69 Rd^-/b^+. The LPS consists only of lipid A and one Kdo phosphate residue (29). Lane 7: *Legionella pneumophila* serogroup 1, strain Philadelphia 1. The LPS consists of at least four populations of molecules of moderate and high-molecular-weight O-side chains. No free-core LPS is seen. The narrow banding pattern indicates a monosaccharide RU. Lane 8: *Salmonella minnesota* R60. The LPS consists of the complete core, lacking an O-side chain (Ra LPS). Lane 9: *Salmonella helsinki* 777. The LPS consists of the complete core and one RU (SR LPS).

primarily in the outer core. A single core type is present in all *Salmonella* serotypes; five different core types have been recognized for *E. coli* serotypes, the same number for *Proteus*, and three for *Citrobacter* (22). The outer core usually contains D-glucose (Glc), D-galactose (Gal), *N*-acetyl-D-glucosamine (GlcNAc), and *N*-acetyl-D-galactosamine (GalNAc), whereas the inner core is composed of the characteristic and LPS-specific components heptose (Hep), primarily in the L-*glycero*-D-*manno* and the D-*glycero*-D-*manno* configuration, and 3-deoxy-D-*manno*-octulosonic acid [or 2-keto-3-deoxyoctanoic (Kdo)]. The Hep and Kdo residues are, in general, substituted by charged groups, such as phosphate, pyrophosphate, 2-aminoethylphosphate, and 2-aminoethylpyrophosphate, leading to an agglomeration of charged residues in this inner part of the core region (4,18).

A distinct LPS type is found, for example, in *Neisseria, Haemophilus,* and *Bordetella* (23–25). In these cases the polysaccharide region lacks an O-chain part, but its chain length is somewhat larger than the core structures of enterobacterial S- and R-form LPS. LPS of this type may preferably be termed R-type LPS (20). The innermost part of the core region appears to be well conserved among taxa, as Kdo forms the linkage to lipid A in almost all LPS, an exception being *Acinetobacter calcoaceticus*, where D-*glycero*-D-*talo*-octulosonic acid (Ko) positionally replaces Kdo (26).

B. Lipid A Component

The ketosidic linkage between the reducing Kdo unit of the core and lipid A is particularly susceptible to acidic hydrolysis. Hence free lipid A can be prepared from LPS under mildly acidic conditions. In contrast, the Ko-lipid A linkage present in *A. calcoaceticus* LPS is highly resistant to acid hydrolysis (26).

Lipid A has been considered a highly conserved structure among gram-negative taxa (27,28). However, newer chemotaxonomic studies covering a phylogenetically wider range of organisms have revealed that the lipid A structure is less homogeneous than previously believed (4,17,28). Nevertheless, it constitutes, together with Kdo, the structurally most conserved region of LPS (28). As exemplified by *Haemophilus influenzae* (29) (Fig. 3), the primary structure of some nonenterobacterial lipid A and all the enterobacterial species investigated so far, including *E. coli* (30), *Salmonella* (31), and *Proteus* (32), consists of a β-(1,6)-linked D-GlcN disaccharide carrying two phosphoryl groups, one α-linked in position 1 and one in position 4′ (Fig. 3). The hydrophilic lipid A backbone carries, in general, 4 mol of (*R*)-3-hydroxy-fatty acids [14:0(3-OH) in the case of *Haemophilus* and *Enterobacteriaceae*], of which two are amide bound and two are ester linked.

These primary hydroxylated fatty acids may carry secondary (nonhydroxylated) ester-linked acyl groups, as is the case in *H. influenzae*, where

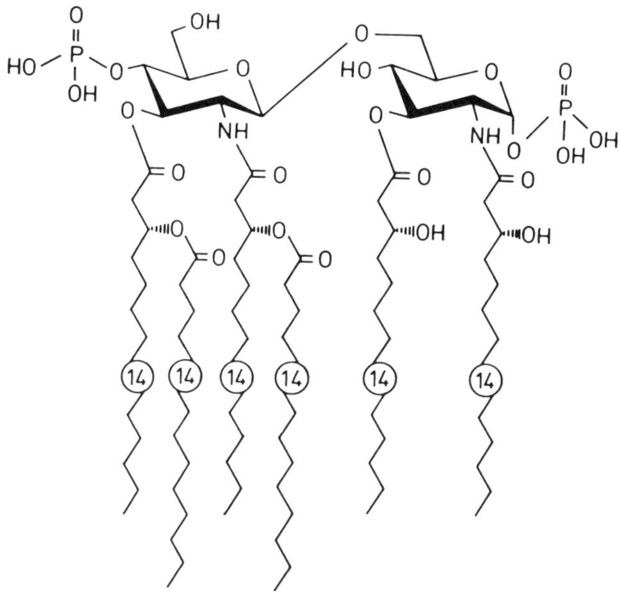

Figure 3 Chemical structure of the lipid A component of *Haemophilus influenzae*. (From Ref. 29.)

tetradecanoic acid (14:0) represents the only type of secondary fatty acid (Fig. 3). In other lipid A (30), the phosphoryl groups may be substituted by polar headgroups such as phosphate, 2-aminoethanol, 2-aminoethylphosphate, and 4-amino-4-deoxy-L-arabinopyranose (Ara4N). In free lipid A, positions 4 and 6′ are nonsubstituted, the latter hydroxyl group constituting the attachment site for the reducing Kdo group of the polysaccharide portion. Within a lipid A preparation, a considerable structural heterogeneity is observed. This is due primarily to nonstoichiometric presence of fatty acids and the substitution of phosphate groups (for literature, see Refs. 4, 17, 18, 22, and 30).

Lipid A structures of some nonenterobacterial LPS have been found to differ from *Haemophilus* and enterobacterial species in several parameters. For example, in lipid A of *Chromobacterium violaceum* (33,34), *Neisseria meningitidis* (35), *Pseudomonas aeruginosa* (36), and *Bacteroides fragilis* (37), the nature, number, chain length, and location of fatty acids are different. In contrast to the asymmetric acylation pattern of fatty acids over the GlcN disaccharide in lipid A of *Haemophilus* and *E. coli*, a symmetric distribution is present in *C. violaceum* and *N. meningitidis* lipid A (33,34). The 3-hydroxy fatty acid chain lengths are smaller in lipid A of *C. violaceum, N. meningitidis,* and *P. aeruginosa* than in *Haemophilus* and enterobacterial lipid A. The major species of *P. aeruginosa* lipid A contains only five fatty acids, as the primary acyl residue at position 3 of the reducing GlcN residue is lacking (36). The same number of fatty acids, but comparatively long-chain, in part iso-branched, fatty acids are present in the major component of *B. fragilis* lipid A. Also deviating in this structure is the absence of the nonglycosidic phosphate group at the nonreducing GlcN unit (37).

Lipid A GlcN disaccharide structures completely lacking phosphate residues have been encountered in *Rhodomicrobium vannielii* and *Rhodopseudomonas acidophila.* These structures and several other phosphate-devoid lipid A structures have been found to carry D-mannopyranosyl residues at position 4′ (17,38). A growing number of species are found to contain a lipid A backbone harboring a 2,3-diamino-2,3-dideoxy-D-glucose (GlcN3N) disaccharide. This amino sugar is present in at least 25 species belonging to 12 different genera including the important pathogen *Brucella abortus* (17). The diamino sugar GlcN3N has so far been encountered in nature only as a backbone sugar of lipid A. Hybrid lipid A disaccharide backbones of GlcN and GlcN3N have also been detected. As an example, lipid A of *Campylobacter jejuni* (39) consists of a β-(1,6)-linked D-*gluco*-hexosamine disaccharide containing GlcN3N and GlcN and three amide-linked 3-hydroxy fatty acids, in contrast to only two found in GlcN disaccharide structures (Fig. 3). Lipid A structures containing only a monosaccharidic GlcN3N backbone were bound, for example, in *Rhodopseudomonas viridis* (16).

Several lipid A variants with distinct fatty acid constituents and patterns clearly deviating from the enterobacterial fatty acid composition have been found among taxa belonging to *Proteobacteria* (17,40). In *Rhodobacter sphaeroides* and all species of the α-3 branch of the phylogenetic tree, the amide-linked fatty acids of the GlcN disaccharide constitute 3-oxotetradecanoic acid. A number of these species have also been found to contain unsaturated fatty acids as lipid A constituents (e.g., delta-7-tetradecenoic acid in *Rh. sphaeroides* and delta-14-docosenoic acid in *Rhodomicrobium vanniellii*) (17,38). Most species belonging to the α_2 subclass were recently shown to contain long-chain $n - 2$ hydroxylated fatty acids, regardless of the lipid A backbone type (40). This unusual hydroxylated fatty acid carries the functional hydroxyl group at the penultimate position of the hydrocarbon chain and not at position 3. The most prominent acid, 27-hydroxyoctacosanoic acid [28:0(27-OH)], possesses the double length of the usual 14:0(3-OH) acid and may stretch through the entire outer membrane (41). It has been suggested that this may result in a more stable anchoring of LPS and an increased stability of the bilayered outer membrane (40).

Two intracellular parasites belonging to the γ-branch, *Coxiella burnetii* (causative agent of Q-fever) and *Legionella pneumophila* (main *Legionella* species causing legionnaires' disease), have been shown to contain an unusually complex fatty acid pattern. LPS of *C. burnetii* contains nine different 3-hydroxy fatty acids (methyl branched and nonbranched) in the range 13 to 17 carbon atoms, and 15 different nonhydroxy fatty acids (42). The fatty acid composition of *L. pneumophila* lipid A is even more heterogeneous, nineteen 3-hydroxy fatty acids (methyl-branched and linear) in the range of 13 to 23 carbon atoms, and six nonhydroxy fatty acids have been identified (43–45). This complicated fatty acid composition is most likely a common feature of the presently recognized *Legionella* species as judged by the cellular fatty acid composition data recorded (46). In *Chlamydia trachomatis*, and *C. pneumoniae* also a complex lipid A fatty acid composition is present comprising 12 fatty acids of the long-chain type, the longest being 22:0(3-OH) (47). In LPS of *Alteromonas putrefaciens*, a complex composition of 3-hydroxy fatty acids with shorter chain lengths (i.e., 11 different branched and linear fatty acids with 11 to 15 carbon atoms) was found (48).

Dihydroxy fatty acids have been identified in lipid A of a few species. Lipid A of *Pseudomonas vesicularis* and *P. diminuta* contains 5,9-dihydroxytetradecanoic acid (49). The lipid A of *L. pneumophila* contains three different 2,3-dihydroxy fatty acids: of the isotetradecanoic, tetradecanoic, and anteisopentadecanoic chain types (43). These dihydroxy fatty acids have also been identified in lipid A of other *Legionella* species, such as *L. israelensis, L. maceachernii*, and *L. micdadei* (50).

III. Lipid A: The Toxic Principle of LPS

In the early 1950s it was postulated that the lipid A compound constitutes the endotoxically active structure of LPS (51). Chemical isolation of lipid A and convertion to a water-soluble form (triethylammonium salt) confirmed that lipid A was the active domain responsible for the induction of all known pathophysiological LPS effects (5,51,52). The successful chemical synthesis of complete *E. coli* lipid A and partial structures have proven this concept to be correct and have offered the possibility to establish relationships between chemical structure and biological activity at the molecular level (4,28, 53). Thus the chemical and biological properties of natural and synthetic *E. coli* lipid A (compound 506) were shown to be identical (52–55). The results of a great number of biological experiments show that for full expression of typical in vivo manifestations, such as fever and skin necrosis, the simultaneous presence of a bisphosphorylated β-(1,6)-linked D-*gluco*-hexosamine disaccharide carrying six acyl groups—4 mol of (*R*)-3-hydroxy fatty acids (C_{10} to C_{16}) and two secondary fatty acids in the form of two (*R*)-3-acyloxyacyl groups in a defined structural arrangement (as in *E. coli*)—is a prerequisite (54). Slight modification of any site of the lipid A architecture results in a significant reduction of biological activity. It therefore appears that endotoxicity is determined by a unique molecular (and supramolecular) conformation of lipid A, termed endotoxic conformation (55).

Chemically synthesized lipid A partial structures lacking one constituent or having a different distribution of constituents are endotoxically less active or inactive in the same endotoxin test system. The structural elucidation of naturally occurring lipid A with deviating architectures and, as a consequence, deviating biological activities was and still is of great help in gaining further insight into the chemical and physical prerequisites governing endotoxicity. Thus LPS containing a different biphosphorylated β-(1,6)-linked-D-*gluco*-hexosamine structure [e.g., *C. jejuni* (possessing a hybrid disaccharide of GlcN and GlcN3N) and *P. diminuta* (GlcN3N disaccharide)] show pronounced endotoxic activity (39,56,57). LPS of *P. aeruginosa* is significantly less toxic than enterobacterial LPS, and the presence of only five acyl residues in lipid A may account for the low endotoxic activity (36,58). The absence of a 4′-*O*-phosphoryl group in lipid A of *B. fragilis* and the presence of only five fatty acids is probably the reason for the low endotoxic activity of *B. fragilis* LPS (37,59). Lipid A of *Rh. sphaeroides* is virtually free of toxicity, presumably due to the presence of only five fatty acids. The other major change in the hydrophobic part of the molecule is caused by the 3-oxotetradecanoic and unsaturated fatty acids, probably leading to a disturbed packing arrangement of acyl groups (60,61). LPS containing a highly complex fatty acid composition (i.e., LPS of *C. burnetti, Ch. trachomatis*, and *L. pneumophila*) are all of low endotoxicity (62–64).

IV. Endotoxic Activity of LPS

For a long time, interest in the biological activity of endotoxin has focused on its effect after parenteral administration to mammals. Striking parallels exist between the biological activities of LPS in experimental animals and those observed in patients with gram-negative sepsis. In fact, LPS have been recognized as playing an important role in the pathogenesis and manifestations of severe gram-negative infection, in general, and in septic shock, in particular (3–7).

A most remarkable feature of LPS is its almost ubiquitous capacity to elicit the entire spectrum of host effector molecules. With regard to its effects in the mammalian body, LPS exhibits two salient features: (1) a highly diversified pattern of immediate pathophysiological activities, including fever, hypotension, changes in leukocyte counts, and disseminated intravascular coagulation, and (2) more delayed effects, such as the induction of antibodies being specific for defined LPS structures but also for other antigens (18). LPS is known to interact with both cellular and humoral systems of the host. In serum, LPS binds to specific proteins (i.e., high- and low-density lipoproteins) and, in particular, to LPS-binding protein (LBP) (65), which appears to mediate the transport of LPS to organs involved in LPS accumulation (i.e., the liver and, to some extent, the spleen and lung). At these body sites and in the circulation, LPS interacts with cells, and this event is believed to play a decisive role in the mechanism of endotoxin action. Cell types that are responsive to endotoxic LPS include macrophages, monocytes, endothelial cells, platelets, mast cells, polymorphonuclear leukocytes, lymphocytes, and fibroblasts.

The interaction of LPS with cells of the mononuclear phagocyte system is perhaps the central event that triggers the reaction cascade, resulting in endotoxic effects (66). On the one hand, LPS activates mononuclear phagocytes both in vivo and in vitro to enhanced functional capacity. Thus both cytotoxic activity and phagocytic capacity of macrophages are significantly enhanced on exposure to biologically active LPS (67,68). On the other hand, and as shown schematically in Fig. 4, LPS induces the production and release of inflammatory mediators and immunoregulatory cytokines, including prostaglandins (PGs), leukotrienes (LTs), platelet activating factor (PAF), superoxide anion, hydrogen peroxide, nitric oxide, and interleukines (ILs) as well as tumor necrosis factor α (TNFα). These secondary, hormone-like mediators possess potent intrinsic bioactivity and contribute to the overall manifestations of endotoxemia (69). In this sense, the LPS molecule is not toxic in itself. As recognized by L. Thomas, it is the response of the host organism to LPS (i.e., the overproduction and disregulation of endogenous mediators) that makes endotoxin poisonous (70).

V. Specificity of Lipid A–Target Cell Interaction

As stated, the key event in endotoxin action consists of the interaction of the lipid A component of LPS with host monocytes/macrophages. It had been widely believed that this interaction was dominated by the hydrophobic region of lipid A. Thus models were envisaged implicating the nonspecific

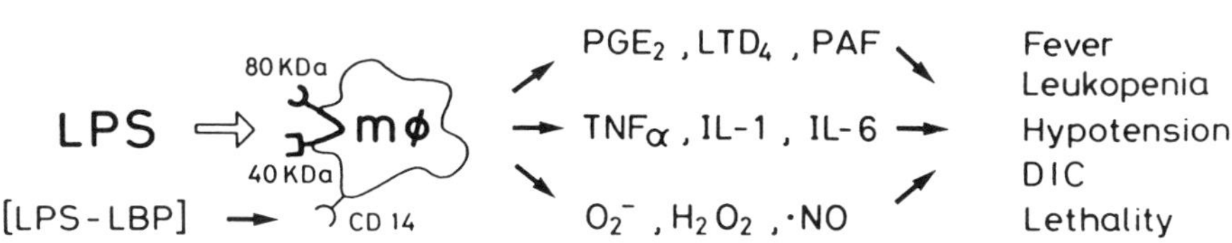

Figure 4 Schematic representation of the interaction of endotoxin with macrophages and the subsequent release of bioactive lipid, peptide, and oxygen radical mediators.

integration of lipid A fatty acid into the macrophage membrane, resulting in cellular activation. It has only been during recent years that the concept of the existence of specific cellular binding protein or receptors for LPS was developed (71–73). Today, several proteins, present on the surface of LPS-reactive cells, are known to interact specifically with LPS. These molecules include proteins of 73, 40, and 28 kDa, the CD11/18 family of adhesins, the 95-kDa scavenger receptor, and in particular, the CD14 monocyte surface marker. The latter appears to be essential and functional in serum-dependent cell activation if low doses of LPS are employed (72). The serum effect is due to the presence of LBP, which possesses high affinity to LPS, and it is the LBP/LPS complex that is recognized by the CD14 molecule. Evidence has been provided that the 73-kDa protein may serve as a functional receptor (71). Thus there is little doubt that host cells dispose of one or more functional receptors capable of engaging specifically with the lipid A component of LPS. It is reasonable to assume, then, that it is a characteristic structure of lipid A which is involved in receptor binding. In fact, as hypothesized above, it is likely to be unique—endotoxic—conformation of lipid A that is essential in the interaction with the host recognition system.

In very recent studies we have addressed the question of which portions of lipid A are involved in the binding to and activation of macrophages/monocytes. The binding capacity of lipid A partial structures and analogs to cells was estimated by analyzing their capacity to compete in a specific LPS binding assay (74,75). The ability of compounds to trigger endotoxic effects was estimated by their capacity to induce the production and release of TNF and IL-1 from mouse macrophage-like cells (J774.1) or human peripheral monocytes (71–75). The compounds used for the competition experiments all shared the β-(1,6)-linked D-GlcN backbone of lipid A, but differed in the anomeric configuration of the reducing GlcN, the phosphorylation pattern, and the nature, number, and distribution of fatty acids.

The results of studies concerning the specific binding of LPS to cells and the inhibition of this binding can be summarized as follows:

1. *E. coli*-type hexaacyl lipid A is a good competitor.
2. A hexaacyl compound, having, however, a symmetrical distribution of acyl chains (chain length 14 carbon atoms) largely lacks competitive activity.
3. A compound in which the reducing GlcN residues possesses the β-configuration, is a poor competitor.
4. A tetraacyl compound containing only the two ester- and two amide-linked primary 14:0(3-OH) residues (called precursor Ia or 406) constitutes the most potent competitor, whereas a *N,N*-bisacyl compound is only weakly active.

5. Completely O- and N-deacylated LPS shows no inhibitory activity of LPS binding.
6. Monophosphoryl *E. coli*-type lipid A lacking either the 1 or 4′-phosphate group are weak competitors of LPS binding.

In summary, these data show that the bisphosphorylated D-GlcN disaccharide plays a crucial role in specific binding of LPS or lipid A to cells. Consistent with this idea is that monophosphoryl partial structures and analogs with the β-configurated GlcN residue are poor binding competitors. Since deacylated LPS preparations (containing the regular backbone) do not bind to cells, it is clear that the presence of a minimal number of fatty acids is required for the interaction of lipid A with the cellular recognition site. This minimal number of fatty acids is probably two, as the bisacyl compound exhibits low but significant binding (i.e., competitive activity).

Further analyses were conducted as to whether the tendency to bind specifically to cells correlated with the ability of compounds to stimulate mouse macrophages or human monocytes to secrete bioactive peptide mediators (TNF and IL-1). These studies revealed the following results:

1. *E. coli*-type hexaacyl lipid A represents a potent monokine inducer.
2. The hexaacyl compound with the symmetrical acyl distributions is only weakly active.
3. Very low stimulatory activity was seen with compounds possessing a β-configurated reducing GlcN group.
4. The tetraacyl compound completely lacked stimulatory activity in the human system and was only weakly active in the mouse macrophage cell line J774.1.
5. Monophosphoryl (tetraacyl) partial structures were (in the mouse system) of even lower activity.
6. O- and N-deacylated preparations exhibited no mediator-inducing activity.

These data permit the conclusion that it is the number, nature, and distribution of fatty acids (i.e., the acylation pattern) that determine the mediator-inducing (i.e., cell activation) capacity of compounds. It is also evident that for the activation of cells the binding of lipid A is a necessary but not sufficient event. An additional receptor-mediated triggering reaction is required for which the acylation pattern of lipid A plays a most critical role and which provides the signal for cellular stimulation.

These data also provide a basis for understanding why certain lipid A compounds are capable of antagonizing LPS-induced cytokine induction in macrophages/monocytes (75–80). Previously it has been shown that the tetraacyl compound (precursor Ia, 406) is an excellent inhibitor of LPS or lipid A activity in vitro. This is probably due to the fact that this compound binds strongly to the LPS binding site, thereby preventing access of LPS or mature lipid A to its receptor. Studies along these lines are followed at present with great intensity since lipid A receptor antagonists may prove to represent potent suppressors of endotoxic activities in vivo and therefore may constitute promising suitable antiendotoxin drugs.

VI. Concluding Remarks

The significance of endotoxin in bacteria-induced inflammatory processes is now widely appreciated, as are the mechanisms leading to endotoxic manifestations involving the production of endogenous, monocyte-derived bioactive peptide and lipid mediators. Also, great progress has been made in our understanding of the molecular structure of endotoxin. Lipid A has been proven to represent the toxic center of endotoxin, its primary structure has been elucidated, it has been chemically synthesized, and the principal structural requirements for its biological activities are known. As a result of these studies, the concept emerged that as compared with other phosphoglycolipids, the constitution of lipid A is unique and that because of its unique primary structure, lipid A adopts a peculiar conformation that is responsible for its characteristic (endotoxic) biological activities. In view of the unique structure of bioactive lipid A and the demonstration of particular humoral and cellular lipid A–binding proteins, we presently favor the concept that a peculiar conformation (i.e., a three-dimensional structure) interacts with a defined cellular recognition molecule and that this interaction represents the key event in endotoxin action. To understand it at the molecular level, all efforts should be directed at the characterization of the conformation of lipid A, of the cellular-binding proteins, and of the interaction between them. It is to be expected that once this interaction is elucidated in all details, advantage will be taken of the knowledge acquired to develop new immunological and pharmacological strategies to control endotoxin-induced local and general inflammatory reactions.

Acknowledgments

The financial support of the Fonds der Chemie and the SFB 367 (EThR) is gratefully acknowledged. H.-D. Grimmecke received a fellowship of the Alexander von Humboldt Foundation. We thank Mrs. I. Bendt for excellent secretarial assistance and Mrs. M. Lohs, Mrs. F. Richter, Mrs. G. Stegelmann-Müller, and Mrs. B. Köhler for preparing illustrations and photographs.

References

1. Nikaido H, Nakae T. The outer membrane of gram-negative bacteria. In: Rose AH, and Morris, JG, eds. Advances in Microbial Physiology. Vol. 20. London: Academic Press, 1979:163–250.
2. Lugtenberg B, van Alphen L. Molecular architecture and functioning of the outer membrane of *Escherichia coli* and other gram-negative bacteria. Biochim Biophys Acta 1983; 737:51–115.
3. Rietschel ETh, Brade H. Bacterial endotoxins. Sci Am 1992; 267:54–61.
4. Rietschel ETh, Brade L, Schade FU, et al. Bacterial endotoxins: properties and structure of biologically active domains. In: Schrinner E, Richmond MH, Seibert G, Schwarz U, eds. Surface structures of microorganisms and their interactions with the mammalian host. Weinheim: VCH, 1988:1–41.
5. Galanos C, Lüderitz O, Rietschel ETh, Westphal O. Newer aspects of the chemistry and biology of bacterial lipopolysaccharides, with special reference to their lipid A component. In: Goodwin TW, ed. Biochemistry of Lipids II. International Review of Biochemistry. Vol. 14. Baltimore: University Park Press, 1977: 239–335.
6. Kronborg G, Fomsgaard A, Galanos C, Freudenberg MA, Hoiby N. Antibody responses to lipid A, core, and O sugars of the *Pseudomonas aeruginosa* lipopolysaccharide in chronically infected cystic fibrosis patients. J Clin Microbiol 1992; 30:1848–1855.
7. Morrison DC, Ryan JL. Endotoxins and disease mechanisms. Ann Rev Med 1987; 38:417–432.
8. Kadis S, Weinbaum G, Ajl SJ, eds. Microbial Toxins. Vol. 5. New York: Academic Press, 1971.
9. Rylander R, Bake B, Fischer JJ, Helander IM. Pulmonary function and symptoms after inhalation of endotoxin. Am Rev Respir Dis 1989; 140:981–986.
10. de Rochemonteix-Galve B, Marchat-Amoruso B, Dayer J-M, Rylander R. Tumor necrosis factor and interleukin-1 activities in free lung cells after single and repeated inhalation of bacterial endotoxin. Infect Immun 1991; 59:3646–3650.
11. Fogelmark B, Goto H, Yuasa K, Marchat B, Rylander R. Acute pulmonary toxicity of inhaled β-1,3-glucan and endotoxin. Agents Actions 1992; 35:50–56.
12. Enterline PE, Sykora JL, Keleti G, Lange JH. Endotoxins, cotton dust, and cancer. Lancet 1985; 934–935.
13. Jantzen E, Bryn K. Whole-cell and lipopolysaccharide fatty acids and sugars of gram-negative bacteria. In: Goodfellow M, Minnikin DE, eds. Chemical Methods in Bacterial Systematics. The Society for Applied Bacteriology Technical Series No. 20. London: Academic Press, 1985:145–171.
14. Wilkinson SG. Composition and structure of bacterial lipopolysaccharides. In: Sutherland I, ed. Surface Carbohydrates of the Prokaryotic Cell. London: Academic Press, 1977:97–175.
15. Mayer H. Significance of lipopolysaccharide structure for questions of taxonomy and phylogenetical relatedness of gram-negative bacteria. In: Haber E, ed. The Cell Membrane: Its Role in Interaction with the Outside World. New York: Plenum Press, 1984:71–83.

16. Weckesser J, Mayer H. Different lipid A types in lipopolysaccharides of phototrophic and related non-phototrophic bacteria. FEMS Microbiol Rev 1988; 54: 143–154.
17. Mayer H, Krauss JH, Yokota A, Weckesser J. Natural variants of lipid A. In: Friedman H, Klein TW, Nakano M, Nowotny A, eds. Advances in Experimental Medicine and Biology. Vol. 256. New York: Plenum Press, 1990:45–70.
18. Lüderitz O, Freudenberg MA, Galanos C, Lehmann V, Rietschel ETh, Shaw DH. Lipopolysaccharides of gram-negative bacteria. In: Razin S, Rottem S, eds. Microbial Membrane Lipids. Vol. 17. Current Topics in Membranes and Transport. New York: Academic Press, 1982: 79–151.
19. Jann B, Reske K, Jann K. Heterogeneity of lipopolysaccharides. Analysis of polysaccharide chain lengths by sodium dodecylsulfate–polyacrylamide gel electrophoresis. Eur J Biochem 1975; 60:239–246.
20. Hitchcock PJ, Leive L, Mäkelä PH, Rietschel ETh, Strittmatter W, Morrison DC. Lipopolysaccharide nomenclature: past, present and future. J Bacteriol 1986; 166:699–705.
21. Schmidt G, Schlecht S, Lüderitz O, Westphal O. Studies of the typing of *Salmonella* R-forms. I. Communication: microbiological and serological studies on *Salmonella minnesota* mutants. Zentralbl Bakteriol Parasitenkd Infektionskr Hyg Abt I, Orig Reihe A 1968; 209:483–496.
22. Holst O, Brade H. Chemical structure of the core region of lipopolysaccharides. In: Morrison DC, Ryan JL, eds. Bacterial Endotoxic Lipopolysaccharides. Vol. 1. Boca Raton, FL: CRC Press, 1992:135–170.
23. Inzama TJ. Electrophoretic heterogeneity and interstrain variation of the lipopolysaccharide of *Haemophilus influenzae.* J Infect Dis 1983; 148:492–499.
24. Jennings HJ, Bhattacherjee AK, Kenne L, Kenny CP, Lukas G. The R-type lipopolysaccharides of *Neisseria meningitidis.* Can J Biochem 1980; 58:128–136.
25. Le Dur A, Chaby R, Szabo L. Isolation of two protein-free and chemically different lipopolysaccharides from *Bordetella pertussis* phenol-extracted endotoxin. J Bacteriol 1980; 143:78–88.
26. Kawahara K, Brade H, Rietschel ETh, Zähringer U. Studies on the chemical structure of the core–lipid A region of the lipopolysaccharide of *Acinetobacter calcoaceticus* NCTC 10305. Eur J Biochem 1987; 163:489–495.
27. Lüderitz O, Galanos C, Lehmann V, Mayer H, Rietschel ETh, Weckesser J. Chemical structure and biological activities of lipid A's from various bacterial families. Naturwissenschaften 1978; 65:578–585.
28. Rietschel ETh, Seydel U, Zähringer U, et al. Bacterial endotoxin: molecular relationships between structure and activity. In: Young LS, Glauser MP, eds. Infectious Disease Clinic of North America. Gram-Negative Septicemia and Septic Shock. Philadelphia: W.B. Saunders, 1991:753–779.
29. Helander IM, Lindner B, Brade H, et al. Chemical structure of the lipopolysaccharide of *Haemophilus influenzae* strain I-69 Rd^{-}/b^{+}. Eur J Biochem 1988; 177:483–492.
30. Rietschel ETh, Wollenweber H-W, Brade H, et al. Structure and conformation of the lipid A component of lipopolysaccharides. In: Proctor R, ed. Handbook of Endotoxin. Vol. 1. Rietschel ETh, ed. Chemistry of Endotoxin. Amsterdam: Elsevier, 1984:187–220.

31. Takayama K, Quereshi N, Mascagni P. Complete structure of lipid A obtained from the lipopolysaccharides of the heptoseless mutant of *Salmonella typhimurium*. J Biol Chem 1983; 258:12801–12803.
32. Sidorczyk Z, Zähringer U, Rietschel ETh. Chemical structure of the lipid A component of the lipopolysaccharide from a *Proteus mirabilis* Re-mutant. Eur J Biochem 1983; 137:15–22.
33. Wollenweber H-W, Seydel U, Lindner B, Lüderitz O, Rietschel ETh. Nature and location of amide-bound (*R*)-3-acyloxyacyl groups in lipid A of lipopolysaccharides from various gram-negative bacteria. Eur J Biochem 1984; 145: 265–272.
34. Hase S, Rietschel ETh. The chemical structure of the lipid A component of lipopolysaccharides from *Chromobacterium violaceum* NCTC 9694. Eur J Biochem 1977; 75:23–34.
35. Kulshin VA, Zähringer U, Lindner B, et al. Structural characterization of the lipid A component of pathogenic *Neisseria meningitidis*. J Bacteriol 1992; 174: 1793–1800.
36. Kulshin VA, Zähringer U, Lindner B, Jäger K-E, Dmitriev BA, Rietschel ETh. Structural characterization of the lipid A component of *Pseudomonas aeruginosa* wild-type and rough mutant lipopolysaccharides. Eur J Biochem 1991; 198:697–704.
37. Weintraub A, Zähringer U, Wollenweber H-W, Seydel U, Rietschel ETh. Structural characterization of the lipid A component of *Bacteroides fragilis* strain NCTC 9343 lipopolysaccharide. Eur J Biochem 1989; 183:425–431.
38. Holst O, Borowiak D, Weckesser J, Mayer H. Structural studies on the phosphate-free lipid A of *Rhodomicrobium vannielii* ATCC 17100. Eur J Biochem 1983; 137:325–332.
39. Moran AP, Zähringer U, Seydel U, Scholz D, Stütz P, Rietschel ETh. Structural analysis of the lipid A component of *Campylobacter jejuni* CCUG 10936 (serotype O:2) lipopolysaccharide. Description of a lipid A containing a hybrid backbone of 2-amino-2-deoxy-D-glucose and 2,3-diamino-2,3-dideoxy-D-glucose. Eur J Biochem 1991; 198:459–469.
40. Bhat UR, Carlson RW, Busch M, Mayer H. Distribution and phylogenetic significance of 27-hydroxy-octacosanoic acid in lipopolysaccharides from bacteria belonging to the alpha-2 subgroup of *Proteobacteria*. Int J Syst Bacteriol 1991; 41:213–217.
41. Bhat UR, Mayer H, Yokota A, Hollingsworth RI, Carlson RW. Occurrence of lipid A variants with 27-hydroxyoctacosanoic acid in lipopolysaccharides from members of the family *Rhizobiaceae*. J Bacteriol 1991; 173:2155–2159.
42. Wollenweber H-W, Schramek S, Moll H, Rietschel ETh. Nature and linkage type of fatty acids present in lipopolysaccharides of phase I and phase II *Coxiella burnetii*. Arch Microbiol 1985; 142:6–11.
43. Sonesson A, Jantzen E, Bryn K, Larsson L, Eng J. Chemical composition of a lipopolysaccharide from *Legionella pneumophila*. Arch Microbiol 1989; 153: 72–78.
44. Sonesson A, Moll H, Jantzen E, Zähringer U. Long-chain α-hydroxy-(ω-1)-oxo fatty acids and α-hydroxyl-(1,ω)-dioic fatty acids are cell wall constituents of *Legionella jordanis, L. macechernii*, and *L. micdadei*. FEMS Microbiol Lett 1993; 106:315–320.

45. Moll H, Sonesson A, Jantzen E, Marre R, Zähringer U. Identification of 27-oxo-octacosanoic and heptacosane-1,27-dioic acid in *Legionella pneumophila*. FEMS Microbiol Lett 1992; 97:1–6.
46. Mayberry WR. Monohydroxy and dihydroxy fatty acid composition of *Legionella* species. Int J Syst Bacteriol 1984; 34:321–326.
47. Nurminen M, Rietschel ETh, Brade H. Chemical characterization of *Chlamydia trachomatis* lipopolysaccharide. Infect Immun 1985; 48:573–575.
48. Moule AL, Wilkinson SG. Composition of lipopolysaccharides from *Alteromonas putrefaciens* (*Shewanella putrefaciens*). J Gen Microbiol 1989; 135:163–173.
49. Arata S, Hirayama T, Kasai N, Itoh T, Ohsawa A. Isolation of 9-hydroxy-δ-tetradecalactone from lipid A of *Pseudomonas diminuta* and *Pseudomonas vesicularis*. FEMS Microbiol Lett 1989; 60:219–222.
50. Sonesson A, Jantzen E. Chemical composition of lipopolysaccharide from 12 *Legionella* species (abstr). International Symposium on Legionella, Jan. 26–29, 1992, Orlando, FL.
51. Westphal O, Lüderitz O. Chemische Erforschung von Lipopolysacchariden gram-negativer Bakterien. Angew Chem 1954; 66:407–417.
52. Galanos C, Lüderitz O, Rietschel ETh, et al. Synthetic and natural *Escherichia coli* free lipid A express identical endotoxic activities. Eur J Biochem 1985; 148: 1–5.
53. Takada H, Kotani S. Structural requirements of lipid A for endotoxicity and other biological activities. CRC Crit Rev Microbiol 1989; 16:477–523.
54. Rietschel ETh, Brade L, Brandenburg K, et al. Chemical structure and biological activity of bacterial and synthetic lipid A. Rev Infect Dis 1987; 9:S527–S536.
55. Rietschel ETh, Brade H, Brade L, et al. Lipid A, the endotoxic center of bacterial lipopolysaccharides: relation of chemical structure to biological activity. Prog Clin Biol Res 1987; 231:25–53.
56. Brangiano MR, Alviano CS, Ricciardi ID. Chemical composition and biological action of lipopolysaccharide (LPS) of *Campylobacter fetus* ss. *jejuni*. Rev Microbiol (Sao Paulo) 1983; 14:90–96.
57. Kasai N, Arata S, Mashimo JI, et al. *Pseudomonas diminuta* LPS with a new endotoxic lipid A structure. Biochem Biophys Res Commun 1987; 142:972–978.
58. Kropinski AM, Jewell B, Kuzio J, Milazzo F, Berry D. Structure and functions of *Pseudomonas aeruginosa* lipopolysaccharide. Antibiot Chemother 1985; 36:58–73.
59. Kasper DL. Chemical and biological characterization of the lipopolysaccharide of *Bacteroides fragilis* subspecies *fragilis*. J Infect Dis 1976; 134:59–66.
60. Salimath PV, Weckesser J, Strittmatter W, Mayer H. Structural studies on the non-toxic lipid A from *Rhodopseudomonas sphaeroides* ATCC 17023. Eur J Biochem 1983; 136:195–200.
61. Qureshi N, Takayama K, Meyer KC, et al. Chemical reduction of 3-oxo and unsaturated groups in fatty acids of diphosphoryl lipid A from the lipopolysaccharide of *Rhodopseudomonas sphaeroides*. J Biol Chem 1991; 266:6532–6538.

62. Hackstadt T, Peacock MG, Hitchcock PJ, Cole RL. Lipopolysaccharide variation in *Coxiella burnetti*: intrastrain heterogeneity in structure and antigenicity. Infect Immun 1985; 48:359–365.
63. Brade L, Schramek S, Schade FU, Brade H. Chemical, biological, and immunochemical properties of the lipopolysaccharide of *Chlamydia psittaci*. Infect Immun 1986; 54:568–574.
64. Wong KH, Moss CW, Hochstein DH, Arko RJ, Schalla WO. "Endotoxicity" of the Legionnaires' disease bacterium. Ann Intern Med 1979; 90:624–627.
65. Schumann RR, Leong SR, Flaggs GW, et al. Structure and function of lipopolysaccharide binding protein. Science 1990; 249:1429–1431.
66. Freudenberg MA, Keppler D, Galanos C. Requirement for lipopolysaccharide-responsive macrophages in galactosamine-induced sensitization to endotoxin. Infect Immun 1986; 51:891–895.
67. Schade FU. Involvement of lipoxygenase in the activation of mouse macrophages by endotoxin. Biochem Biophys Res Commun 1986; 138:842–849.
68. Schade FU, Burmeister I, Elekes E, Engel R, Wolter DT. Mononuclear phagocytes and eicosanoids: aspects of their synthesis and biological activities. Blut 1989; 59:475–485.
69. Vogel SN, Hogan MM. The role of cytokines in endotoxin-mediated host responses. In: Oppenheim JJ, Shevack EM, eds. Immunopharmacology: The Role of Cells and Cytokines in Immunity and Inflammation. New York: Oxford University Press, 1990:238–258.
70. Thomas L. The Lives of a Cell. New York: Viking Press, 1974:75–80.
71. Morrison DC. The case for specific lipopolysaccharide receptors expressed on mammalian cells. Microb Pathogen 1989; 7:389–398.
72. Wright SD, Ramos RA, Tobias PS, Ulevitch RJ, Mathison JC. CD14, a receptor for complexes of lipopolysaccharide (LPS) and LPS binding protein. Science 1990; 249:1431–1433.
73. Kirikae T, Kirikae F, Schade FU, et al. Detection of lipopolysaccharide-binding proteins on membranes of murine lymphocyte and macrophage-like cell lines. FEMS Microbiol Immunol 1991; 76:327–336.
74. Kirikae T, Schade FU, Zähringer U, et al. Inhibition of lipopolysaccharide (LPS)-binding to J774.1 mouse macrophage-like cells and cytokine release by defined lipopolysaccharide (LPS) partial structures. FEMS Microbiol Immunol 1994; 8:13–26.
75. Ulmer AJ, Feist W, Heine H, et al. Modulation of endotoxin-induced monokine release in human monocytes by lipid A partial structures that inhibit binding of ^{125}I-lipopolysaccharide. Infect Immun 1992; 60:5145–5152.
76. Loppnow H, Brade H, Dürrbaum I, et al. Interleukin 1 induction-capacity of defined lipopolysaccharide partial structures. J Immunol 1989; 142:3229–3238.
77. Feist W, Ulmer AJ, Wang M-H, et al. Modulation of lipopolysaccharide-induced production of tumor necrosis factor, interleukin 1, and interleukin 6 by synthetic precursor Ia of lipid A. FEMS Microbiol Immunol 1992; 89:73–90.
78. Wang M-H, Feist W, Brade H, et al. Suppressive effect of lipid A partial structures on lipopolysaccharide or lipid A-induced release of interleukin 1 by human monocytes. FEMS Microbiol Immunol 1990; 64:179–186.

79. Ulmer AJ, Heine H, Feist W, et al. Biological activity of synthetic phosphonooxyethyl analogs of lipid A and lipid A partial structures. Infect Immun 1992; 60:3309–3314.
80. Wang M-H, Flad H-D, Feist W, et al. Inhibition of endotoxin-induced interleukin-6 production by synthetic lipid A partial structures in human peripheral blood mononuclear cells. Infect Immun 1991; 59:4655–4664.

2

Interactions of Endotoxin with Membranes and Membrane Receptors

TERUO KIRIKAE

Jichi Medical School
Tochigi-Ken, Japan

MEI-GUEY LEI, TAI-YING CHEN, and DAVID C. MORRISON

University of Kansas Medical Center
Kansas City, Kansas

I. Introduction

Bacterial lipopolysaccharides (LPSs) have been recognized for many years as one of the most potent microbial products in the induction of host inflammatory responses in both experimental animals and in the human population. As amply documented by other contributors to this volume, the ability of LPS to function as inducer of inflammation results from its almost ubiquitous capacity to interact, via its conserved lipid A region, with a variety of host target cells, resulting in the production of a broad spectrum of proinflammatory immunologic and pharmacologic mediators. Such mediator molecules include, but are certainly not limited to cytokines [interleukin-1 (IL-1), IL-6, IL-8, tumor necrosis factor (TNF)-α, interferon (IFN)-β, IFN-γ], metabolic products of lipid metabolism (prostaglandins, leukotrienes, and platelet activating factor), and others (procoagulant activity, reactive oxygen intermediates, enzymes, etc.) (1). The self-amplifying effects of these mediators on host tissues and cells result in the establishment of a generalized inflammatory response, termed the sepsis syndrome, characterized by hypotension, disseminated intravascular coagulation, metabolic acidosis, multiple-organ failure, and ultimately death (2). As the various contributors to this volume review, the lung is among the critical organs that are affected profoundly by the presence of bacterial LPS.

An understanding of the pathogenesis of the sepsis syndrome begins with a detailed delineation of the biochemical mechanisms by which LPS interactions with host target cells results in the production of the plethora of mediators that define the inflammatory response. In most (but not all) instances, and in contrast to the effects of most bacterial (exo)toxins, the effects of bacterial endotoxins are noncytotoxic. With many host cells, the in vitro activation for production of mediators can be demonstrated at extremely low concentrations (e.g., nanograms per milliliter). Although the aggregated state of LPS precludes precise calculations of molar concentrations, such observations suggest that effective stimulatory concentrations of the order of hundreds of molecules per cell may be effective. In many instances, as discussed below, serum proteins serve as either essential cofactors or potentiators of the LPS response.

The list of host cells that can respond directly to LPS includes many of the cells important to normal lung function, specifically the macrophage and the endothelial cell. The capacity of LPS to induce the production of mediators in these cells both in vitro and in vivo has been well established (3). In addition, however, LPS effects on polymorphonuclear leukocytes, platelets, lymphocytes (4), and even blood basophils (5) have also been reported. Thus there would appear to be stong experimental evidence to support the concept that responsiveness to LPS is a general property of host cells that are involved with either host defense or the vasculature.

As pointed out above, and as discussed in detail in Chapter 1, there is unequivocal evidence for a dominant role for the lipid A region of LPS in the activation of host cells. The use of synthetic lipid A, prepared by total organic synthesis, has precluded the possibility for alternative LPS structures as being necessary for LPS/lipid A–dependent cell activation (6). There is, however, a growing body of evidence that other LPS structures, particularly those associated with the inner core, may contribute to activation of host cells and may, under some circumstances, be required for optimal activity of LPS (7,8).

The use of synthetic lipid A structures has also provided evidence of a relatively strong requirement for specific lipid A substructures for effective binding to, and activation of, host cells (9,10). These studies suggest that the diglucosamine backbone structure of lipid A and both acyl groups and substituents at the 1 and 4 positions are important for full activity of lipid A. Other evidence for specificity has derived both from studies using an unusual lipid A obtained from *Rhodopseudomonas sphaeroides* (11) and a modified lipid A obtained following deacylation with an acyloxyacyl hydrolase, an LPS-detoxifying enzyme found in the granule compartment of polymorphonuclear leukocytes (12).

While the collective studies summarized above would provide support for the concept of a relatively high degree of specficity for lipid A in interaction with host cells, such conclusions are limited by the potential confounding influence of limited lipid A solubility on expression of activity, due to the relatively hydrophobic nature of lipid A and the pronounced tendency of LPS/lipid A to form high-molecular-weight micellar structures. Indeed, published studies from our laboratory using hybrid LPS structures, created by detergent-mediated disassociation–reassociation of S-LPS and R-LPS preparations, have provided strong support for the concept that lipid A content, per se, cannot be used to predict the relative cellular activating potential of a given LPS preparation (13). Rather, the aggregate structure of LPS and the presence of large amounts of O-antigen, either covalently or noncovalently associated with biologically active lipid A, can restrict the activity of lipid A within the LPS aggregate structure.

In addition to the concerns about LPS solubility in influencing stimulating activity, it is also relevant to consider the contribution of LPS intrinsic affinity for biological membranes. It is well recognized that in its native state in the bacterial cell surface, LPS is localized to the outermost component of the outer membrane. Thus in its natural state, LPS is a natural membrane structure, and numerous studies have confirmed the affinity of LPS and lipd A for mammalian cell membranes in general (reviewed in Ref. 14). The recently published experiments of Daemen et al. (15), showing markedly different cellular activating properties of LPS encapsulated in liposomes depending on the ability of the liposomes to fuse with cells, provide provocative evidence for the potential contribution of hydrophobic membrane interactions in LPS-dependent host cell responses.

Although all of the factors cited above would be expected to contribute to structure–function relationships in cellular responses to LPS, the concept that specific receptors for these potent bacterial products exist has intrigued LPS researchers since the earliest studies to investigate LPS-dependent cell responses in vitro were first undertaken. The pioneering studies of Neter (16), who studied binding of LPS to erythrocytes, and Springer (17), who characterized biochemically a receptor for LPS on erythrocytes, provided much of the early conceptual framework underlying the search for LPS receptors. These efforts were fueled by Sultzer's discovery of the C3H/HeJ mutant mouse strain and the realization that immune cells from this animal manifest a defective response to LPS in vitro (18), thus providing a potentially testable model that these cells may lack a specific receptor for LPS. Although early studies to provide support for this hypothesis were promising, it is now clear that the C3H/HeJ defect is not so simple (reviewed in

Ref. 19), and as discussed below, this mutation is unlikely to be manifest strictly at the LPS receptor level. In fact, the precise nature of the defect in the C3H/HeJ mouse responsible for its altered response to LPS has remained remarkably elusive to date, and biochemical factors that might provide specific clues about this genetic defect have been identified only recently (20).

As a consequence, efforts to identify specific LPS receptors have taken more traditional pathways using a variety of approaches, including generation and characterization of mutant cell lines, monoclonal antibodies with specificity for defined surface structures, immunoblotting with specific LPS probes, and photocross-linking techniques (reviewed in Ref. 21). During the past decade, therefore, a wealth of new information concerning candidate LPS receptors has emerged and a number of surface structures relevant to LPS–cellular interactions have been defined. Although the existing knowledge base has not yet allowed a unifying hypothesis with respect to LPS–host cell interactions, there are now a number of important findings that seem relatively solid. It is possible, and perhaps may be likely, that multiple receptors for LPS exist on the various cell types studied and that, depending on particular experimental circumstances, interaction with one type of receptor may be favored. There is growing experimental evidence that such may be the case in serum-dependent versus serum-independent LPS interactions with macrophages and endothelial cells, when, as reviewed below, specific requirements for both serum factors and specific membrane structural components have been identified.

Such differences may, in fact, serve as a useful differentiation point in the consideration of receptors for LPS. As reviewed below, studies intiated in our own laboratory have allowed identification of a membrane glycoprotein of 73-kDa molecular mass which binds to LPS and would appear to function as a specific receptor for LPS (21). More recent studies from other laboratories have confirmed the identity of this glycoprotein on a variety of host cells and tissues, and much information about this unusual protein is now available. The ability of a monoclonal antibody with specificity for this surface glycoprotein to serve as a weak agonist in activating mouse macrophages has provided additional experimental evidence for a functional role for this protein in cellular activation events (22).

Using an alternative experimental approach, Wright, Ulevitch, and their co-workers have identified the monocyte differentiation antigen CD14, a phosphatidylinositol–linked glycoprotein expressed on monocytes, macrophages, and to a lesser extent, on polymorphonuclear leukocytes, as an important surfact structure for binding of molecular complexes of LPS and lipopolysaccharide-binding protein (LBP) (23). The latter has been shown to function as an acute-phase protein whose concentration is markedly elevated in the circulation following the administration of an inflammatory stimulus to experimental animals (24). The LBP protein has been cloned and sequenced (25) and shown to have remarkable sequence indentity with a polymorphonuclear leukocyte granule protein, first identified by Weiss et al. (26) as bacterial permeability-inducing protein. As will be described, the markedly enhanced sensitivity of a mouse B-cell line to respond to LPS following transfection with a cDNA encoding the gene for mouse CD14 clearly underscores the relevance of this surface structure to LPS activation events (27). Althought, as reviewed below, recent published data have provided some evidence that factors in serum other than LBP may also be important in LPS-dependent activation events, and although the precise role of surface-localized CD14 in these cell responses is not yet entirely clear, there is no question that an important role for serum factors in activation of LPS-responsive cells has been documented. On the other hand, the fact that host cells which do not express CD14 can easily be shown to respond to LPS (including monocytes from patients with paroxysmal nocturnal hemoglobulinuria) (28), as well as the fact that a portion of the activation event in CD14-expressing cells can be shown to be CD14 independent, together clearly implicate CD14-independent events in LPS stimulation of host cells. In the following paragraphs we review the existing literature with respect to both CD14-independent and CD14-dependent candidate LPS-mediated receptor triggering of host cell responses.

II. CD14-Independent LPS Receptor Pathway

During the past decade, efforts form a number of laboratories, including our own, have endeavored to identify surface structures on macrophages and other host inflammatory mediator cells that can interact directly with LPS/lipid A and initiate transmembrane signaling. Using a variety of experimental approaches, LPS-binding proteins of 95 to 96 kDa (29,30), 70 to 80 kDa (31,32), 38 to 40 kDa (33,34), and 18 kDa (35) were identified initially as candidate LPS receptors. The majority of the available information to date has addressed the functional relevance of the 70 to 80 kDa, and this section of the review will thereore focus evidence in support of a contribution of this protein to LPS-dependent activation events.

The first report of an LPS-binding protein with molecular mass 70 to 80 kDa derived from studies carried out in our laboratory using a photoactivatable, radioiodinated, disulfide-reducible derivative of *Escherichia coli* LPS (36) to probe LPS-binding substances on mouse splenocytes (31,32). Fractionation of photocross-linked, solubilized cellular extracts by two-dimensional gel electrophoresis followed by autoradiography resulted in the identification of a dominant LPS binding protein of approximately 80 kDa molecular mass and pI value of about 6.5. Experiments demonstrating inhibition of binding of LPS to this 80-kDa protein all suggested that this LPS-binding protein might served as a specific candidate receptor for LPS.

Of interest was the observation that expression of this LPS-binding protein could be demonstrated on a variety of mouse lymphoid cell subpopulations and host inflammatory cells, including B-lymphocytes, T-lymphocytes, macrophages, polymorphonuclear leukocytes, and platelets, as well as some nonlymphoid cells, particularly fibroblasts (31,37). A variety of cells lines were also demonstrated to express this 80-kDa binding protein; notably, however, neither erythrocytes nor an undifferentiated myeloma cell were positive. Extension of these studies to peripheral blood cells of a variety of species confirmed that, in fact, expression of the 73-kDa LPS-binding protein was ubiquitous in mammals, including humans (38,39).

Since publication of these studies, a number of investigators have used similar approaches to confirm the existence of this, or a very similar, structure as a dominant LPS-binding moiety on a mouse macrophage cell line (40), mouse B-cells and macrophages (41), human peripheral blood monocytes (42), a human T-cell line (G. D. Armstrong, personal communication), and a human hepatoma cell line (Blake et al., personal communication). In addition, using a sensitive immunoblotting procedure in which solubilized cell extracts of mouse macrophages were fractionated by polyacrylamide gel electrophoresis, blotted onto nitrocellulose, and probed with LPS followed by monoclonal anti-LPS antibody, Kirikae et al. (43) have independently confirmed the existence of LPS-binding proteins with an approximate molecular mass of 80 kDa. Thus there would appear to be strong experimental evidence form a variety of laboratories for the existence of a dominant LPS-binding protein of 70 to 80 kDa expressed on the surface of many mammalian cells.

Although most investigators who have studied this 70 to 80-kDa LPS-binding protein have demonstrated that the addition of excess underivatized LPS or lipid A will reduce binding of the photoprobe-directed radioiodination signigicantly, thus implicating a binding specificity for lipid A and/or LPS, several recent studies have raised questions about the extent of the specificity of binding for LPS per se. In the first of these studies, Dziarski, in a comprehensive set of experiments designed to investigate petidoglycan-binding proteins on mouse lymphoreticular cells by photocross-linking techniques similar to those described above for LPS, demonstrated specific and saturable binding of peptidoglycan to a 70-kDa protein (44). Competitive inhibition studies using either soluble peptidoglycan or LPS provided evidence in support of the concept that both bacterial cell-wall constituents were binding to the same or similar structures on mouse splenocyte subpopulations (41). The observation that reciprocal inhibition profiles could be generated using photocross-linking–derivatized LPS provided additional data in support of this conclusion. Finally, Dziarski demonstrated that proteolytic fragments of solubilized cell extracts radiolabeled by either cross-linking with peptidoglycan or LPS fractionated by SDS-PAGE were identical. Collectively, these experiments have provided data that implicate the 74-kDa protein as a binding site for both LPS and peptidoglycan. Dziarski speculated that the shared *N*-acetylglucosamine found in peptidoglycan and in the backbone of lipid A might provide a common binding motif for 73-kDa receptor's specificity.

In contrast to these results, data from our own laboratory using a variety of peptidoglycan and peptidoglycan–polysaccharide preparations were unable to demonstrate significant inhibition of LPS binding (37). Although arguments to suggest that such studies were flawed because the molar concentrations of the competitive inhibitors used were inappropriate (45), the highly aggregated and polymeric nature of both peptidoglycan and LPS suspensions in aqueous solution would preclude even order-of-magnitude estimates of molar concentrations of these reagents (46). Nevertheless, more recent competitive inhibitor studies in our laboratory using soluble peptidoglycan preparations generated by 3-h sonication (generously provided by Dr. Dziarski) have confirmed the earlier reports that soluble peptidoglycan will, in fact, inhibit LPS binding to mouse splenocyte subpopulations (M. G. Lei and D. C. Morrison, unpublished results). Similar findings have very recently been published in the human monocyte system (N. H. Teng, personal communication). Thus it would be fair to conclude that at least one preparation of soluble peptidoglycan can inhibit LPS interaction with the 73-kDa LPS-binding protein. The relevance of this finding to either LPS receptor or peptidoglycan receptor research remains to be determined.

A second and perhaps more curious finding has recently been reported by both our laboratory and independently by Clark and Armstrong (47). Both studies address the potential relationship between the 73-kDa LPS-binding protein first identified by our laboratory (31) and a 70-kDa pertussis toxin–binding protein identified by Clark and Armstrong (47). Our own studies were prompted by the observation that treatment of mouse macrophages with pertussis toxin, an A-B type of exotoxin with the capacity to ADP-ribosylate regulatory G-proteins associated with receptors in mammalian cells, would significantly reduce the capacity of these cells to respond to LPS by production of the reactive oxygen intermediate nitric oxide while enhancing their capacity to secret TNF-α (48). Given these significant effects of pertussis toxin (PT) on LPS responses and the fact that the pertusis toxin receptor was suggested also to exist as a 70-kDa surface protein, experiments were carried out to investigate the effect of pertussis toxin (PT) on binding of LPS to the 73-kDa LPS-binding protein. The results of these studies indicated, rather unexpectedly, that PT pretreatment of mouse splenocytes would, in fact, reduce LPS binding to the 73-kDa binding protein, coincident with photolabeling of the S1, S2, and S3 subunits of PT (49). Control experiments established unequivocally that these findings were not the result of a direct interaction of LPS with PT, even though other published studies from our laboratory have shown that LPS, will, in the fluid phase, bind selectively to the S2 subunit of PT (50). We have tentatively concluded from these studies that PT binds to cells at a site that is topologically closely related to, and is perhaps identical with, the 73-kDa LPS-binding protein. Further, we have suggested that the reason that LPS photocross-linking of the 73-kDa LPS-binding protein is reduced is because the PT serves as an alternative acceptor target for the LPS photocross-linking group. These results are entirely consistent with data obtained by Armstrong using photocross-linking derivatized PT (personal communication), and collaborative studies are under way between our two laboratories to confirm and extend these observations.

Of particular note in the studies which have characterized the 70-kDa PT-binding protein is the fact that the binding interaction is specific for carbohydrate determinants and may readily be inhibited by a variety of lectins as well as by prebinding of PT to carbohydrate-rich glycoproteins (G. D. Armstrong et al., personal communication). Earlier experiments from our laboratory have also suggested that the 73-kDa LPS-binding protein is glycosylated (51). It is unlikely that LPS/lipid A binding specificity to the 73-kDa protein is mediated via carbohydrate specific determinants and, as a consequence, it is attractive to consider this glycoprotein as having potential duel specificities (i.e., both LPS- and PT-binding sites). Alternatively, it is possible that the 73-kDa protein is not glycosylated but is closely associated with a glycolipid for which PT has high binding affinity. Given the strong evidence for exotoxin binding to specific gangliosides, such a hypothesis is particularly attractive, and preliminary experiments from our laboratory would provide additional support for this concept. Using mouse splenocytes pretreated with PT and photocross-linked with LPS so as to radioiodinate both the 73-kDa protein and the PT subunits, we have been able to release some of the 73-kDa protein with octylglucoside while retaining radiolabeled PT as cell associated (M. G. Lei and D. C. Morrison, unpublished results). These preliminary findings support the concept that the PT-binding site and the LPS-binding site, while closely associated at the cell surface, may not be covalently associated with each other.

Whatever the precise relationship between the 73-kDa LPS-binding protein and its capacity to interact with peptidoglycan and, either directly or indirectly, pertussis toxin, it would appear reasonable to conclude that the 73-kDa LPS-binding protein exists on the cell membrane of many cells and can readily be detected as specifically binding to LPS and lipid A. Of some interest, studies to date have not detected any significant difference in the properties of the 73-kDa protein on cells from the endotoxin-hyporesponsive C3H/HeJ mouse in comparison with its histocompatible C3Heb/FeJ (22,31) or C3H/OuJ (52) normal mouse strain. Since, however, other studies in our laboratory have shown that the defect on cells from the C3H/HeJ mouse is primarily a qualitative one, and that LPS responsiveness can be elicited with LPS derived from some rough LPS chemotypes (53,54), such a result may not be totally unexpected. In the latter respect, there is a growing body of evidence that the mutational defect on the C3H/HeJ mouse may well be one of the intracellular triggering mechanisms rather than a receptor defect per se (20,52).

It is important that functional distinctions be delineated between the capacity of a membrane-localized structure to serve as a binding site versus its capacity to serve as a specific receptor. One important distinction between the two would be the capacity of the latter interaction to initiate a transmembrane signal. Evidence that the 73-kDa LPS-binding protein can function as a specific receptor for LPS has derived from studies using monoclonal antibodies with specificity for the 73-kDa binding protein. One IgM monoclonal antibody, termed MAb5D3, generated by immunization of Armenian hamsters with partially purified 73-kDa protein and subsequent fusion of splenocytes from immunized animals with a mouse myeloma cell line, has proven of interest (51). This monoclonal antiboy binds specifically to mouse lymphoreticular cells which express the 73-kDa LPS-binding protein, and binding both inhibits, and is inhibited by, LPS binding. Although efforts to establish binding specificity of this monoclonal antibody by standard Western blot procedures have, to date, not been successful, as assessed by ELISA, the antibody does bind to highly purified 73-kDa protein prepared by elution from nitrocellulose blots following two-dimensional electrophoretic separation of solubilized splenocyte extracts.

Studies to assess the functional activity of MAb5D3 have indicated that this monoclonal antibody can serve as an agonist on mouse bone marrow–derived macrophages in activating these cells for tumor cell killing (22) as well as peritoneal macrophages for production of nitric oxide and tumor necrosis factor (55). As with LPS, the activity of MAb5D3 is potentiated by IFN-γ, and no activity can be demonstrated using bone marrow–derived macrophages from the C3H/HeJ mouse strain. Finally, pretreatment of mice with MAb5D3 provides protection against subsequent lethality due to LPS challenge in the D-galactosamine sensitization model, probably by desensitizing macrophages to LPS (56). Similar desensitization effects can also be shown in vitro (Chang, T. Chen, and D. C. Morrison, unpublished observations). Collectively, these studies would support the concept that the 73-kDa protein can serve as a functional receptor for LPS.

It is clear, however, that studies using monoclonal antibodies to mimic LPS effects must rigorously exclude potential effect that might be caused by minute levels of contaminating endotoxin introduced during the in vitro culture and subsequent purification steps. Although we believe that it would be virtually impossible to rule out this possibility completely, a number of experimental approaches have provided evidence in support of the conclusion that MAb5D3-mediated effects are not the result of endotoxin contamination. First, as assessed by elution profiles following column chromatography of MAb5D3 on both ion-exchange columns and gel permeation columns, both antibody binding to splenocytes and macrophage-stimulating activity comigrate together (T. Chen and D. C. Morrison, unpublished observations). Second, direct assays using the *Limulus* amoebocyte lysate assay indicate that endotoxin levels are at least an order of magnitude lower than those required to elicit detectable LPS responses in vitro or in vivo. Third, unlike LPS, whose biological activity is retained after heating to 100°C, MAb5D3 activity is destroyed by heating to 100°C. Fourth, quantitative estimates of polymyxin B–mediated inhibition of LPS versus MAb5D3 activation of mouse macrophages indicate that the former is four to five orders of magnitude more sensitive by weight than the latter for inhibition of activity. Fifth, in recent studies, we have shown that unlike LPS, whose ability to activate macrophages in vitro for nitric oxide (NO) production is enhanced significantly by the presence of fetal bovine serum (FBS), the capacity of MAb5D3 to effect macrophage activation is markedly suppressed (T. Chen and D. C. Morrison, unpublished observations). Interestingly, the latter suppressive effect of FBS on MAb5D3 activation can be observed only when serum is added either before or within the first 2 h of MAb5D3 stimulation of macrophages, even though NO production is not measured until 18 h. Collectively, therefore, these results would not support the conclusion that MAb5D3 activity is the result of endotoxin contamination, and thus that the 73-kDa LPS-binding protein, is, indeed, a functional receptor for LPS.

At present, the precise molecular identity of this LPS receptor has not been established; however, active investigation of this protein is under way in several laboratories, including our own, and it is anticipated that the molecular cloning and sequencing of the cDNA that encodes this unique cell surface constituent will be witnessed within the next 12 months. The elucidation of the structure of this LPS receptor and its potential to share properties with previously identified LPS-binding proteins will shed considerable additional light on the role of this protein in LPS-dependent cellular responses to LPS.

In addition to the dominant 73-kDa LPS receptor described above, several additional LPS-binding proteins have been tentatively identified and merit specific comment. Perhaps the first candidate surface structure recognized to interact with LPS was the CD11/18 heterodimeric adhesin complex found on mononuclear phagocytes and polymorphonuclear leukocytes. This particular class of adhesin molecules consist of a conserved β chain (CD18) in association with one of several α chains (CD11a,b,c). As first reported by Wright and Jong (57), antibodies with specificity for the α subunit would inhibit the capacity of erthrocytes, to which LPS had been absorbed, to adhere to monolayers of human monocytes. The authors suggested from these studies that this family of inflammatory cell adhesins may serve as functional receptors for LPS. More recently published studies by these investigators, however, have demonstrated that leukocytes from patients genetically deficient in CD11/18 expression maintain the capacity to respond in vitro to LPS as assessed by the production of cytokines (58). Thus, while it is possible and, in fact, may be probable that adhesins contribute in the host pathogenic response to LPS and/or gram-negative bacteria, perhaps by direct association with bacterial-associated LPS (for which LPS-coated erythrocytes would serve as a prototype experimental model), it is not likely that these surface receptors function as the pivotal recognition structures essential for host responsiveness to LPS

A second surface structure present on mononuclear phagocytes that has been implicated in lipid A–dependent responses has been the scavenger receptor for low-density lipoproteins. This protein was first identified by Raetz and his colleagues as a 95-kDa lipid A binding protein, detected by probing nitrocellulose blots of solubilized and SDS-PAGE fractionated extracts of the murine macrophage-like cell line RAW264.1 with a high-specific-activity radiolabeled lipid A precursor IVa (59). Of interest in these studies was the fact that binding of the lipid A precursor was inhibited by both intact lipid A and an Re-chemotype LPS but not by polysaccharide-containing S-LPS. An extension of these studies provided data to support an identity between the 95-kDa LPS binding protein and the previously characterized scavenger receptor (60). It is of interest to speculate that this receptor may well prove of critical importance in clearance of lipid A/LPS following host-dependent destruction of gram-negative bacteria. It is, however, not likely that the scavenger receptor is responsible for the LPS–receptor interaction that provides the initial triggering event leading to production and release of immunologic and pharmacologic mediators (61).

Finally, there are several additional reports of LPS-binding proteins which have been reported in preliminary studies as having the potential to function as candidate LPS receptor molecules. The first is a protein of 38 to 40 kDa identified on mouse lymphoreticular cells first reported by our laboratory (33). In contrast to the 73-kDa LPS receptor, however, the available evidence would suggest the interesting possibility that specificity of binding may be localized to Kdo rather than lipid A determinants (34). A second relatively low molecular weight protein of approximately 18 kDa has been tentatively identified on the mouse 70Z/3 pre-B-cell line (35). It might be noted that the latter findings do not confirm the published results from our own laboratory (37), although differences in conditions of electrophoresis may, at least in part, account for these differences. Preliminary studies in our laboratory using the conditions defined by Kirkland et al. have, in fact, confirmed those observations (unpublished data). Finally, preliminary studies reported by Parent, also using photocross-linking technology, have demonstrated the existence of a 47-kDa LPS-binding protein on hepatocytes which also may manifest binding specificity for LPS core oligosaccharide determinants (62). Although evidence in support of a functional role for these LPS-binding proteins is not available at present, their potential to bind LPS should be considered in overall analyses to define LPS host-cell interactions.

III. CD14-Dependent LPS Receptor Pathway

Identification of the role of the surface protein CD14, long recognized as a specific differentiation antigen for mononuclear phagocytes and, to a lesser extent, polymorphonuclear leukocytes (64), in LPS-dependent cellular recognition has proven to be one of the most significant findings in the field in recent years. Research in this area had as its genesis the discovery by Ulevitch and his colleagues of an acute-phase protein with the capacity to bind specifically to lipid A region of LPS (24). This protein, termed lipopolysaccharide-binding protein (LBP), although present in low concentrations in normal serum, was shown to be markedly elevated in the circulation in response to inflammatory stimuli (64).

Biochemical characterization of LPS has shown this protein to be approximately 60 kDa in molecular mass and to share a high degree of sequence homology between the human and rabbit protein (25,65). A rather striking sequence similarity was also reported between LBP and a protein termed bacterial permeability-inducing factor (BPI), a polymorphonuclear leukocyte granule constituent first recognized by Elsbach and Weiss for its capacity to bind to, and effect the killing of, gram-negative bacteria (26). There is also a report that both LBP and BPI share sequence similarities with an additional serum protein, termed cholesterol ester transport protein (CETP)(67). Ulevitch et al. have suggested that LBP may be prototypic of an entire family of mammalian-drived LPS-binding proteins (65).

In contrast to the effects of BPI on LPS-mediated activation of host inflammatory cells, where dose-dependent inhibition is observed (67), the primary effect of LBP is to potentiate LPS-dependent responses. As shown by Wright el al. (68), LBP can markedly potentiate the attachment of LPS-coated erythrocytes to macrophages, and control experiments have provided strong evidence that this potentiation requires recognition by the phagocytes of molecular complexes of LPS and LBP. Further, LBP will enhance LPS priming of polymorphonuclear leukocytes for enchanced superoxide production (69), as well as increasing the capacity of LPS to stimulate TNF-α production in rabbit macrophages as much as two orders of magnitude (25). The observations in the latter studies that isolated LBP by itself had no effects on macrophages and did not potentiate the production of TNF-α by *Staphylococcus aureus*, a gram-positive stimulus, again confirmed the requirement for LPS–LBP complexes.

Thus there is now strong evidence to suggest an important role for LBP present in normal serum and, to a greater extent, in acute-phase serum in the enhancement of LPS-dependent host cell responses. It is of interest to note that potentiation of LPS responses by serum factors has, in fact, been recognized for more than 30 years, since the early studies by Cohn and Morse (70) showing that the presence of serum was required for LPS stimulation of polymorphonuclear leukocyte bactericidal activity. Similarly, serum factors have been implicated in LPS-induced up-regulation of CD18 expression on polymorphonuclear leukocytes (71), LPS-dependent priming for an oxidative burst (72), and LPS-induced cytotoxicity of bovine endothelial cells (73).

Whether LBP is the only factor in serum that can influence LPS acivity, or even in every situation is the predominant serum factor is less clear. In this respect it has been recognized for a number of years that high-density lipoprotein (HDL) can also bind LPS and may, under some circumstances, be the primary serum factor interacting with LPS (74). Interestingly, the major effects HDL appear to be inhibition of cellular responses (75). Finally, other well-recognized serum proteins can also, under the appropriate circumstances, interact with LPS, including lysozyme (76), serum complement components (77,78), and even albumin (36).

More recently, Wright et al. (79) have reported the existence of normal human plasma factors termed septins, which can also mediate LPS-dependent monocyte activation. These studies have provided convincing evidence that the observed activity of septin is distinct from that of LBP, in that activity is not blocked by neutralizing antibody to LBP. The fact that activity was, however, blocked by protease inhibitors is strongly suggestive of an enzymatic component for septin.

It would appear, therefore, that there are no lack of candidate serum factors that can interact with, and modify, the capacity of LPS to trigger host inflammatory cells. As suggested above, the consequences of such LPS–serum factor interactions may serve to either enhance or inhibit LPS responses. Indeed, recent data from our own laboratory suggest further that the extent of the effect of serum factors on LPS-dependent responses may also depend, to a significant extent, on the chemotype of the LPS employed in the in vitro studies (80). In those experiments, the presence of as little as 0.02% of fetal bovine serum (FBS) would enhance smooth-type (S)-LPS-dependent production of either TNF-α or nitric oxide from the macrophage-like J774.1 cell line.

Optimal potentiation of responsiveness, as well as an increase in the kinetics of the responses, was observed with approximately 2% FBS, and was also noted with normal mouse peritoneal macrophages. Interestingly, the extent of the serum-dependent enhancement of LPS macrophage responsiveness observed was approximately 200- to 500-fold with S-LPS but was only 5- to 10-fold with R-chemotype LPS or purified lipid A. The precise identity of the factor in FBS that is responsible for this serum-dependent enhancement has not yet been identified. Nevertheless, it is likely that the serum factor functions by an interaction with LPS since, under identical conditions where LPS–macrophage responses are enhanced, the MAb5D3-mediated macrophage responses, which presumably function through the same stimulatory pathway, are significantly inhibited (T. Chen and D. C. Morrison, unpublished observations). It may well be that the differential capacity of serum to effect different LPS chemotypes may contribute to our understanding of the role of surface structures in LPS activation.

Although, there is clearly no question that serum can effect LPS responses, it would appear reasonable to conclude that LPS can interact with at least some host inflammatory mediator and/or immune cells and cells lines in the absence of serum factors. As summarized above, the addition of purified factors isolated from serum can manifest predictable and reproducible enhancing or inhibiting effects on the cellular responses. The addition of whole serum/plasma to LPS may, however, serve to have a net consequence of enhancement or inhibition, depending on the source of the serum and its final concentration, the chemotype of the LPS, and the particular system under study. Examples of each can, as has been pointed out above, be identified readily in the scientific literature.

An important milestone in the elucidation of specific target molecules on the surface of human monocytes that function by recognition of LPS complexed with one serum factor, LBP, was the report of Wright et al. in 1991, demonstrating that monoclonal antibody to CD14 would inhibit the ability of LPS–LBP complexes on erythrocytes to bind to monolayers of human monocytes and also the capacity of LPS–LBP to stimulate TNF-α production by these cells (23). Other experiments demonstrated that the molecular recognition by cell-bound CD14 was for the molecular complex of LPS and LBP, and that not all monoclonals against CD14 were effective in inhibiting activity. These data thus provided the first strong evidence that CD14 may be an important surface structure in the interaction of LPS with human monocytes, particularly in the presence of LBP.

As a follow-up to these studies, Ulevitch and his colleagues (27) have provided direct proof for a role for CD14 in LPS cell responses. These investigators transfected the cDNA encoding human CD14 into the murine pre-B-cell line 70Z/3, a cell line that does not normally express CD14 antigen. These cells will, nevertheless, respond to LPS in vitro by synthesis and expression of surface IgM, indicating that CD14 is not essential for manifestation of cellular LPS responsiveness. Importantly, however, as shown in this study, transfection with cDNA for CD14 markedly potentiated the capacity of these cells to respond to LPS, as assessed by surface IgM expression. In contrast, there was no detectable enhancement of the IFN-γ-mediated 70Z/3 cell response, supporting the specificity of the CD14 in enhancement of LPS-dependent responses. Of additional interest in these studies was the observation that some enhancement of activity was noted in LPS-stimulated CD14 transfected cells, even in the absence of exogenous LBP, suggesting that the presence of the latter may not be an absolute requirement for CD14-mediated potentiation of LPS responsiveness. That the presence of membrane CD14 was actually enhancing LPS-dependent signal transduction was suggested by a coordinate up-regulation of the transcription factor NFKB.

Independent confirmation of a functional role for CD14 in macrophage responses to LPS has derived from studies in which mutant cell lines, selectively unresponsive to LPS, have been derived from the J774 macrophage-like cell line. The first such cell line, termed LR9, was described by Hara-Kuge et al. (40). This cell line was suggested to have less cellular responsiveness to LPS than that of the parent J774 cell line. Analysis of the LPS-binding defect in the LR9 cells by LPS-photocross-linking technology suggested an absence of LPS-specific binding proteins of approximately 55 and 65 kDa (40). More recent studies by these investigators have reported that the 55-kDa LPS-binding protein, which is altered in LR9 cells, is probably identical to the murine CD14 antigen (81).

In similar studies carried out in our own laboratory, experiments were designed to isolate mutants of the macrophage-like cell line J774.1 by selection of cells with reduced fetal calf serum-dependent binding to Re-LPS. One mutant cell line, termed J7.DEF3, has been characterized extensively (82). This cell line manifests reduced binding to LPS and, as assessed by either TNF-α or nitric oxide production, has less LPS responsiveness than that exhibited by the parent cells. Of importance, this differential responsiveness is manifest only in the presence of fetal calf serum. In the absence of serum, both parent and mutant cells reflect identical dose-response profiles to LPS. The fact that parent and mutant cells respond to other macrophage stimuli (zymogen and heat-killed *Listeria*) provides support for the concept that the defect is LPS-specific. Extensive dose-response studies have documented, however, that the differential activity of the mutant relative to the parent is highly dependent on the LPS chemotype, similar to what we have observed with serum-dependent and serum-independent LPS responses in the parent (see above). Thus, in response to S-LPS, the differences between parent and mutant are approximately two to three orders of magnitude. In contrast, differences are reduced to only 10-fold when Re-LPS or lipid A are used as stimuli. Preliminary studies to define the biochemical nature of the defect in the J7.DEF3 cells have indicated that these cells, as assessed by photocross-linking with LPS, manifest LPS binding profiles identical to those of the parent cells, with a dominance of LPS-binding proteins at approximately 80 and 40 kDa, similar to what has been described earlier (T. Kirikae et al., unpublished observations). However, when solubilized extracts of cells fractionated by SDS-PAGE and blotted onto nitrocellulose are probed with Re-LPS in the presence of serum, followed by monoclonal antibody to LPS, an additional LPS binding protein of 55 kDa can readily be detected in the parent cells but not in the mutant cells. Analysis of parent and mutant cells by flow cytometry using monoclonal antibody specific for mouse CD14 developed by Goyert has demonstrated that the J7.DEF3 cells manifest significantly reduced levels of surface CD14 (83). Collectively, these data would be consistent with the concept that the defect in these cells which renders them hyporesponsive to serum-dependent LPS responses is an altered CD14 antigen.

Direct evidence for a role for surface-bound CD14 in mediating the pathophysiological effects of LPS has derived from the recently published studies of Goyert and her colleagues, who have created transgenic mice expressing cell membrane–localized human CD14 (84). Of interest, compared to nontransgenic controls, these experiments documented an approximate threefold increase in sensitivity to the lethal effects of LPS. It was also shown that the human CD14 was highly expressed in macrophages and polymorphonuclear leukocytes but was also found on Thy-1 + cells. The potential role of the latter cells in mediating the effects of endotoxin is, at present, unclear. Nevertheless, these studies clearly implicate cell surface CD14 as an important molecular target for LPS-dependent cellular interactions.

The precise functional role of CD14 in contributing to cell activation remains, however, to be fully defined, and several points may be of relevance. First, as discussed earlier, one of the major effects of CD14 appears to be to potentiate LPS responses, and this potentiation effect can be as much as three to four orders of magnitude in the case of 70Z/3 cells (27). Nevertheless, the fact that these (and other) cells, which do not normally express CD14, can still respond to LPS would support the concept that the presence of CD14 is not essential for LPS-dependent cell activation. Second, the extent of the enhancement of LPS activation would appear, at least in part, to depend on the type of LPS employed, the type of cell under investigation, and the presence of specific serum factors (82). Third, a very recent report from Ulevitch's laboratory has provided evidence to suggest that the biochemical nature of the membrane association of CD14 with the cell, normally a phosphatidylinositol-linked membrane protein, is not critical for the manifestation of CD14-dependent LPS effects (85). In this very elegant series of studies, Lee et al. showed that construction of a modified CD14 with a traditional transmembrane domain and cytoplasmic tail was equally as effective as the native CD14 in enhancing LPS responses when transfected with 70Z/3 cells. These results would support the concept that the phosphatidylinositol linkage of CD14 is not critical for manifestation of LPS effects. Finally, a recent report by Millham et al. (86) has suggested that the presence of CD14 alone does not necessarily confer LPS responsiveness on cells. Using techniques similar to those described by Lee et al. (85), these investigators transfected CD14 into CHO cells. When successful transfectants were subsequently analyzed for binding of lipid IVa, no specific binding could be detected either in the presence or the absence of serum, suggesting that other factors in addition to membrane CD14 are essential for effective LPS cell interactions.

Indeed, several very recent studies have suggested that a facilitating role for CD14 in mediating LPS responses is not limited to membrane-bound CD14 but may extend to soluble CD14 as well. It is known that CD14 can readily be detected in normal serum and may be, in part, generated by shedding of CD14 from the surface of mononuclear cells. Evidence has been presented to implicate soluble CD14 as an important cofactor in the activation of human endothelial cells (87) and an epithelial cell line (88) by LPS. These recent findings help to explain earlier studies showing a serum dependence for LPS-mediated endothelial cell activation (86) and its inhibition by antibody directed against CD14. In addition, it would be of interest to speculate whether soluble CD14 could inhibit the capacity of LPS to mediate LPS-dependent activation of monocytes. The precise role of soluble versus membrane-bound CD14 in modulating LPS-dependent effects will continue to prove a fascinating story for future investigators.

IV. Summary and Conclusions

In the preceding paragraphs we have summarized a substantial body of new evidence that has accumulated during the past decade which addresses the question of the identity and molecular and functional characterization of specific receptors for LPS on host inflammatory mediator, vascular, and immune cells. While a number of candidate LPS-binding target molecules on these cells have tentatively been shown to interact with LPS, at the present time there are two molecules that appear likely to serve as important effector molecules for LPS signaling pathways in mammalian host cells. The first of these is the 73-kDa LPS-binding protein, which is expressed on a variety of host cells and has been shown to manifest all the properties necessary to characterize it as a specific receptor for LPS. The second is the mononuclear cell/polymorphonuclear cell-specific phosphatidylinositol-linked CD14 antigen, which has been documented to bind to molecular complexes of LPS and LBP and to potentiate cellular responses to LPS.

To date, there is no evidence that would necessarily link these two LPS interactive proteins at the cell surface, nor is there yet a compelling reason to suggest that they must function in a coordinate or cooperative fashion. Earlier we speculated that the function of CD14 may be to alter the primary physicochemical structure of the LPS (and/or LPS–LBP) macromolecular complex so as to expose more readily the normally hydrophobic lipid A ligand determinants necessary for effective receptor interactions (82). It is noteworthy in the latter respect that as assessed by binding of monoclonal antibodies with a specificity for lipid A, only weak binding to S-LPS is usually detected in aqueous environments. Since lipid A is essential for manifestation of LPS activity, a requirement for molecular events that make lipid A epitopes accessible for binding to lipid A–specific receptors is not unreasonable. Coupled with the fact that serum (LBP or other) potentiation of LPS responsiveness (via CD14) is dependent on LPS chemotype with less potentiation observed for Re-LPS and lipid A, a functional role for CD14 in "presenting" LPS to the 73-kDa receptor would not be an unreasonable hypothesis.

On the other hand, it may well be that these two LPS recognition systems function independently of one another. In the latter respect, it is equally likely that multiple mechanisms may exist for recognition of LPS and that different signaling pathways exist as well. Given the remarkable degree of redundance in the immune signaling and intercellular communications pathways, and the postulated importance of appropriate recognition by the immune system of LPS as a harbinger of gram-negative bacteria (89), the existence of multiple receptors for LPS would certainly help to ensure survival of the organism.

Although a number of important questions remain to be answered before the mechanism(s) by which LPS triggers host cellular responses is fully understood, the remarkable progress that has been made in the past several years instills confidence that these molecular mechanisms will, in the next several years, become clear. This knowledge will be of considerable value in defining and implementing appropriate therapeutic strategies for the treatment of gram-negative septic shock.

Acknowledgments

This research was supported by NIH grants R37 AI-23447 and PO1 CA-54474. The authors would also like to acknowledge Ms. Kathy Rode for preparation of this manuscript.

References

1. Morrison DC, Ryan JL. Endotoxins and disease mechanisms. In: Annual Review of Medicine. Palo Alto, CA: W.P. Creger Annual Reviews, 1987:417–432.
2. Strieter RM, Lynch JP, Basha MA, Standiford TJ, Kasahara K, Kunkel SL. Host responses in mediating sepsis and adult respiratory distress syndrome. Semin Respir Infect 1990; 5:233–247.
3. Schlag G, Redl H, Hallstrom S. The cell in shock: the origin of multiple organ failure. Resuscitation 1991; 21:137–180.
4. Morrison DC, Ulevitch RJ. The effects of bacterial endotoxin on host mediation systems. Am J Pathol 1976; 93:527–617.
5. Smith T, Alveolet M, Morrison DC. The effect of bacterial lipopolysaccharide on histamine release from human basophils. I. Enhancement of immunologic release by *S. minnesota* R595 LPS. Clin Immunol Immunopathol 1985; 34: 355–365.
6. Kotani S, Takada H, Tsujimoto M, et al. Synthetic lipid A with endotoxic and related biological activities comparable to those of a natural lipid A from an *Escherichia coli* Re-mutant. Infect Immun 1985; 49:225–237.
7. Loppnow H, Brade H, Durrbaum I, et al. IL-1 induction-capacity of defined lipopolysaccharide partial structures. J Immunol 1989; 142:3229–3237.
8. Haeffner-Cavaillon N, Caroff M, Cavaillon JM. Interleukin-1 induction by lipopolysaccharide: structure requirements of the 3-deoxy-D-manno-2-octulosonic acid (KDO). Mol Immunol 1989; 26:485–494.
9. Wang MH, Flad HD, Feist W, et al. Inhibition of endotoxin-induced interleukin-6 production by synthetic lipid A partial structures in human peripheral blood mononuclear cells. Infect Immun 1991; 59:4655–4664.
10. Ulmer AJ, Heine H, Feist W, et al. Biological activity of synthetic phosphonooxyethyl analogs of lipid A and lipid A partial structures. Infect Immun 1992; 60:3309–3314.
11. Qureshi N, Takayama K, Meyer KC, et al. Chemical reduction of 3-oxo and unsaturated groups in fatty acids of diphosphoryl lipid A from the lipopolysaccharide of *Rhodopseudomonas sphaeroides*. J Biol Chem 1991; 266:6532–6538.
12. Munford RS, Hall CL. Purification of actyloxyacyl hydrolase, a leukocyte enzyme that removes secondary acyl chains from bacterial lipopolysaccharides. J Biol Chem 1989; 264:15613.
13. Vukajlovich SW, Morrison DC. Activation of murine spleen cells by lipid A: negative modulation of lipid A mitogenic activity by O-antigen polysaccharide. J Immunol 1985; 135:2546–2550.
14. Morrison DC. Non-specific interactions of bacterial lipopolysaccharides with membranes and membrane components. In: Berry LJ, ed. Handbook of Endotoxin. Vol. 3. New York: Elsevier, 1985:25.
15. Daemen T, Veninga A, Dijkstra J, Scherphof G. Differential effects of liposome-incorporation on liver macrophage activating potencies of rough lipopolysaccharide, lipid A, and muramyl dipeptide. Differences in susceptibility to lysosomal enzymes. J Immunol 1989; 142:2469–2474.

16. Neter E, Zalewski J, Zak DA. Inhibition by lectin and cholesterol of bacterial (*Escherichia coli*) hemagglutination and hemolysis. J Immunol 1953; 145–151.
17. Springer GF, Adye JC, Bezkorovainy A, Jirgensons B. Properties and activity of the lipopolysaccharide-receptor from human erythrocytes. Biochemistry 1974; 13:1379–1389.
18. Sultzer BM. Genetic control of leukocyte responses to endotoxin. Nature (London) 1968; 219:1252–1254.
19. Morrison DC. C3H/HeJ mouse strain: role in elucidation of host response to bacterial endotoxins. In: Schlessinger D, ed. Microbiology. Washington, DC: ASM Publications, 1986:23–28.
20. Nakano M, Shinomiya H. The *Lps* mutational defect in C3H/HeJ mice. In: Morrison DC, Ryan JL, eds. Bacterial Endotoxic Lipopolysaccharides. Vol. I. Boca Raton, FL: CRC Press, 1992:213–237.
21. Morrison DC, Lei MG, Kirikae T, Chen TY. Endotoxin receptors on mammalian cells. Immunobiology 1993. In press.
22. Chen TY, Bright SW, Pace JL, Russell SW, Morrison DC. Induction of macrophage mediated tumor cytotoxicity by a hamster monoclonal antibody with specificity for LPS receptor. J Immunol 1990; 145:8–12.
23. Wright SD, Ramos RA, Tobias PS, Ulevitch RJ, Mathison JC. CD14, a receptor for complexes of lipopolysaccharide (LPS) and LPS binding protein. Science 1990; 249:1431–1433.
24. Tobias PS, Soldau K, Ulevitch RJ. Isolation of lipopolysaccharide-binding acute phase reactant from rabbit serum. J Exp Med 1986; 164:777–793.
25. Schumann RR, Leong SR, Flaggs GW, et al. Structure and function of lipopolysaccharide binding protein. Science 1990; 249:1429–1431.
26. Weiss J, Muello K, Victor M, Elsbach P. The role of lipopolysaccharides in the action of the bactericidal/permeability-increasing neutrophil protein on the bacterial envelope. J Immunol 1984; 132:3109–3115.
27. Lee JD, Kato K, Tobias PS, Kirkland TN, Ulevitch RJ. Transfection of CD14 into 70Z/3 cells dramatically enhances the sensitivity to complexes of lipopolysaccharide (LPS) and LPS binding protein. J Exp Med 1992; 175:1697–1705.
28. Cavaillon JM, Munoz C, Fitting C, Couturier C, Haeffner-Cavaillon N. Signals involved in interleukin-1 production induced by endotoxins. In: Nowotny A, Spitzer JJ, Ziegler EJ, eds. Cellular and Molecular Aspects of Endotoxin Reactions. New York: Elsevier, 1990:257.
29. Hampton RY, Golenbock DT, Raetz RHC. Lipid A binding sites in membranes of macrophage tumor cells. J Biol Chem 1988; 263:14802–14807.
30. Kirikae T, Inada K, Hirata M, Yoshida M, Kondo H, Hisatsune K. Identification of Re lipopolysaccharide-binding protein on murine erythrocyte membrane. Microbiol Immunol 1988; 32:33–44.
31. Lei MG, Morrison DC. Specific endotoxin lipopolysaccharide binding protein on murine splenocytes. I. Detection of LPS binding sites on splenocytes and splenocyte subpopulation. J Immunol 1988; 141:996–1005.
32. Lei MG, Morrison DC. Specific endotoxic lipopolysaccharide binding protein on murine splenocytes. II. Membrane localization and binding characteristics. J Immunol 1988; 141:1006–1011.

33. Lei MG, Morrison DC. Identification of an LPS binding protein with specificity for inner core region (KDO) determinant. FASEB J 1991; 5:A1363.
34. Lei MG, Qureshi N, Morrison DC. Lipopolysaccharide (LPS) binding to 73 kDa and 38 kDa surface proteins on lymphoreticular cells: preferential inhibition of LPS binding to the former by *Rhodopseudomonas sphaeroides* lipid A. Immunol Lett. In press.
35. Kirkland TN, Virca GD, Kuus-Reichel T, et al. Identification of lipopolysaccharide-binding proteins in 70Z/3 cells by photoaffinity crosslinking. J Biol Chem 1990; 265:9520–9525.
36. Wollenweber HW, Morrison DC. Synthesis and biochemical characterization of photoactivatable, iodinatable, clevable bacterial lipopolysaccharide derivative. J Biol Chem 1985; 260:15068–15074.
37. Lei MG, Stimpson SA, Morrison DC. Specific endotoxic lipopolysaccharide-binding receptors on murine splenocytes. III. Binding specificity and characterization. J Immunol 1991; 147:1925–1932.
38. Roeder D, Lei M-G, Morrison DC. Specific endotoxic lipopolysaccharide (LPS) binding proteins on lymphoid cells of various animal species: correlation with endotoxin susceptibility. Infect Immun 1989; 57:1054–1058.
39. Halling JL, Hamill D, Lei MG, Morrison DC. Identification and characterization of lipopolysaccharide-binding proteins on human peripheral blood cell populations. Infect Immun 1992; 60:845–852.
40. Hara-Kuge S, Amano F, Nishijima M, Akamatsu Y. Isolation of a lipopolysaccharide (LPS)-resistant mutant, with defective LPS binding, of cultured macrophage-like cells. J Biol Chem 1990; 265:6606–6610.
41. Dziarski R. Peptidoglycan and lipopolysaccharide bind to the same binding site on lymphocytes. J Biol Chem 1991; 266:4719.
42. Rabin RL, Bieber MM, Teng NH. Identification of specific LPS binding proteins on human peripheral blood monocytes. J Cell Biochem 1992 (Abstr Suppl 16C):165.
43. Kirikae T, Kirikae F, Schade FU, et al. Detection of lipopolysaccharide-binding proteins on membranes of murine lymphocyte and macrophage-like cell lines. FEMS Microbiol Immunol 1992; 76:327–336.
44. Dziarski R. Demonstratiion of peptidoglycan binding sites on lymphocytes and macrophages by photoaffinity crosslinking. J Biol Chem 1991; 266:4713.
45. Dziarski R. Letter to the editor. J Immunol 1992; 148:1590–1591.
46. Lei MG, Stimpson S, Morrison DC. Letter to the editor. J Immunol 1992; 148: 1591.
47. Clark CG, Armstrong GD. Lymphocyte receptors for pertussis toxin. Infect Immun 1990; 58:3840–3846.
48. Zhang X, Morrison DC. A pertussis toxin-sensitive factor differentially regulates LPS-induced tumor necrosis factor-α and nitric oxide production in mouse peritoneal macrophages. J Immunol 1993; 150:1011–1018.
49. Lei MG, Morrison DC. Evidence that lipopolysaccharide and pertussis toxin bind to different domains on the same p73 receptor on murine splenocytes. Infect Immun 1993; 61:1359–1364.

50. Lei MG, Morrison DC. Lipopolysaccharide interaction with S2 subunit of pertussis toxin. J Biol Chem 1993; 268:1488–1493.
51. Bright SW, Chen TY, Flebbe LM, Lei MG, Morrison DC. Generation and characterization of hamster-mouse hybridomas secreting monoclonal antibodies with specificity for lipopolysaccharide receptor. J Immunol 1990; 145:1–7.
52. Perera PY, Chen TY, Morrison DC, Vogel SN. Detection and analysis of the 73 kD LPS receptor in macrophages derived from LPS^n and LPS^d mice. J Leukoc Biol 1992; 51:501–506.
53. Flebbe LM, Vukajlovich SW, Morrison DC. Immunostimulation of C3H/HeJ lymphoid cells by R-chemotype LPS preparations. J Immunol 1989; 142:642–652.
54. Flebbe LM, Chapes SK, Morrison DC. Activation of C3H/HeJ macrophage tumoricidal activity and cytokine release by R-chemotype lipopolysaccharide preparations. Differential effects of IFN-γ. J Immunol 1990; 145:1505–1511.
55. Green TS, Chen TY, Crawford MR, Nacy C, Morrison DC, Melzer MS. Cytotoxic activity and production of toxic nitrogen oxides by macrophages treated with IFN-γ and monoclonal antibodies against the 73 kDa lipopolysaccharide receptor. J Immunol 1992; 149:2069–2075.
56. Morrison DC, Silverstein R, Bright SW, Chen TY, Flebbe LM, Lei MG. Monoclonal antibody to mouse lipopolysaccharide receptor protects mice against the lethal effects of endotoxin. J Infect Dis 1990; 162:1063–1068.
57. Wright SD, Jong MTC. Adhesion-promoting receptor on human macrophages recognize *Escherichia coli* by binding to lipopolysaccharide. J Exp Med 1986; 164:1876–1888.
58. Wright SD, Detmers PA, Aida Y, et al. CD18-deficient cells respond to lipopolysaccharide in vitro. J Immunol 1990; 144:2566–2571.
59. Golenbock DT, Hampton RY, Raetz CRH, Wright SD. Human phagocytes have multiple lipid A-binding sites. Infect Immun 1990; 58:4069–4075.
60. Hampton RY, Golenbock DT, Penman M, Krieger M, Raetz CRH. Recognition and plasma clearance of endotoxin by scavenger receptors. Nature 1991; 352:342–334.
61. Hampton RY, Raetz CRH. Macrophage catabolism of lipid A is regulated by endotoxin stimulation. J Biol Chem 1991; 266:19499–19509.
62. Parent JB. Core specific receptor for lipopolysaccharide on hepatocytes. Circ Shock 1989; 27:341–342.
63. Goyert SM, Ferrero E, Rettig W, Yenamandra AK, Obata F, LeBeau MM. The CD14 monocyte differentiation antigen maps to a region encoding growth factors and receptors. Science 1988; 239:497–500.
64. Ramadori G, Meyer zum Buschenfelde KH, Tobias PS, Mathison JC, Ulevitch RJ. Biosynthesis of lipopolysaccharide-binding protein in rabbit hepatocyte. Pathobiology 1990; 58:89–94.
65. Tobias PS, Soldau K, Ulevitch RJ. Identification of a lipid A binding site in the acute phase reactant lipopolysaccharide binding protein. J Biol Chem 1989; 264:10867–10871.
66. Drayna D, Jarnagin AS, McLean J, et al. Cloning and sequencing of human cholesterol ester transfer protein cDNA. Nature 1987; 327:632–634.

67. Marra MN, Wilde CG, Griffith JE, Snable JL, Scott RW. Bactericidal/permeability-increasing protein has endotoxin-neutralizing activity. J Immunol 1990; 144:662–666.
68. Wright SD, Tobias PS, Ulevitch RJ, Ramos RA. Lipopolysaccharide (LPS) binding protein opsonizes LPS-bearing particles for recognition by a novel receptor on macrophages. J Exp Med 1989; 170:1231–1241.
69. Wright SD, Ramos RA, Hermanowsky-Vosatka A, Rockwell P, Detmers PA. Activation of the adhesive capacity of CR3 on neutrophils by endotoxin: dependence on lipopolysaccharide binding protein and CD14. J Exp Med 1991; 173:1281–1286.
70. Cohn ZA, Morse SI. Functional and metabolic properties of polymorphonuclear leukocytes. II. The influence of a lipopolysaccharide endotoxin. J Exp Med 1960; 111:689–704.
71. Lynn WA, Reatz CRH, Qureshi N, Golenbock DT. Lipopolysaccharide-induced stimulation of CD11b/CD18 expression on neutrophils. Evidence of specific receptor-based response and inhibition by lipid A-based antagonists. J Immunol 1991; 147:3072–3079.
72. Aida Y, Pabst MJ. Priming of neutrophils by lipopolysaccharide for enhanced release of superoxide. Requirement for plasma but not tumor necrosis factor-α. J Immunol 1990; 145:3017–3025.
73. Patrick D, Betts J, Frey EA, Prameya R, Dorovini-Zis K, Finlay BB. *Haemophilus influenzae* lipopolysaccharide disrupts confluent monolayers of bovine brain endothelial cells via a serum-dependent cytotoxic pathway. J Infect Dis 1992; 165:865–872.
74. Freudenberg MA, Bog-Hansen TC, Back U, Galanos C. Interaction of lipopolysaccharides with plasma high density lipoproteins in rats. Infect Immun 1980; 28:373–380.
75. Munford RS, Dietschy JM. Effects of specific antibodies, hormones, and lipoproteins on bacterial lipopolysaccharides injected into the rat. J Infect Dis 1985; 152:177–184.
76. Ohno N, Morrison DC. Lipopolysaccharide interaction with lysozyme. Binding of lipopolysaccharide to lysozyme and inhibition of lysozyme enzymatic activity. J Biol Chem 1989; 264:4434–4441.
77. Vukajlovich SW. Interaction of LPS with serum complement. In: Morrison DC, Ryan JL, eds. Bacterial Endotoxic Lipopolysaccharides. Boca Raton, FL: CRC Press, 1992:213–237.
78. Kinoshita T. Biology of complement: the overture. Immunol Today 1991; 12: 291–295.
79. Wright SD, Ramos RA, Patel M, Miller DS. Septin: a factor in plasma that opsonizes lipopolysaccharide-bearing particles for recognition by CD14 on phagocytes. J Exp Med 1992; 176:719–727.
80. Kirikae T, Kirikae F, Morrison DC. Serum-dependent enhancement of S-form lipopolysaccharide-induced tumor necrosis factor production by macrophage-like J774.1 cells. In preparation.
81. Nishijima M, Hara-Kuge S, Amano F, Akamatsu Y. Genetic and biochemical approaches to understanding macrophage LPS receptor. J Cell Biochem Suppl 1992; 16C:152.

82. Kirikae T, Schade FU, Kirikae F, Rietschel ET, Morrison DC. Isolation of a macrophage-like cell line defective in binding of lipopolysaccharide: influence of serum and lipopolysaccharide chain length on macrophage activation. J Immunol 1993; 151:2742–2752.
83. Kirikae T, Goyert SM, Kirikae F, Morrison DC. Identification of a defective molecule in LPS-binding mutant cells isolated from macrophage-like J774.1 cells. Abstract 159. 2nd International Endotoxin Conference, Vienna, Austria, 1993:100.
84. Ferrero E, Jiao D, Tsuberi BZ, et al. Transgenic mice expressing human CD14 are hypersensitive to LPS. Proc Natl Acad Sci USA 1993; 90:2380–2384.
85. Lee JD, Krauchenko V, Kirkland TN, et al. GPI-anchor or integral membrane CD14 mediate identical responses to LPS. FASEB J 1993; 7:A598.
86. Millham F, Lynn W, Liu Y, Golenbock D. CD14 and serum-independent pathways of cellular activation by LPS. Abstract. 2nd Conference of the International Endotoxin Society, Vienna, Austria, 1992:100.
87. Frey EA, Miller DS, Jahr TG, et al. Soluble CD14 participates in the response of cells to lipopolysaccharide. J Exp Med 1992; 176:1665–1671.
88. Pugin J, Schurer-Maly CC, Leturco D, Moriarty A, Ulevitch RJ, Tobias PS. Soluble CD14 and LPS binding protein mediate epithelial cell responses to LPS. FASEB J 1993; 7:A142.
89. Morrison DC, Lei M-G, Chen TY, Flebbe LM, Halling J, Field S. Identification and characterization of mammalian cell membrane receptors for LPS-endotoxin. In: Friedman H, Klein T, Yamaguchi H, eds. Microbial Infections: Role of Biological Response Modifiers. New York: Plenum Press, 1992:23–29.

3

Interactions Between Endotoxin and Endotoxin-Binding Protein

THOMAS R. MARTIN

University of Washington School of Medicine
and Seattle Veterans Affairs Medical Center
Seattle, Washington

PETER S. TOBIAS, JOHN C. MATHISON, and RICHARD J. ULEVITCH

The Scripps Research Institute
La Jolla, California

I. Introduction

Endotoxin is an important mediator of the biological responses to gram-negative bacteria. Although endotoxins by themselves have no intrinsic toxicity, endotoxin stimulates mononuclear phagocytes to produce an array of cytokines that stimulate many aspects of the inflammatory response (1). A comparison of the effects of endotoxin in vivo and in vitro reveals a striking "endotoxin paradox." When studied in vitro in the absence of serum, relatively high concentrations of endotoxin (e.g., 1 μg/mL or more) usually are required to activate macrophages. In contrast, endotoxin stimulates cytokine production in whole blood at concentrations of 1 ng/mL or less (2), similar to the serum concentrations that have been found in patients with sepsis (3,4), and substantially lower than the concentrations required to activate macrophages in vitro in the absence of serum. In addition, there has not been a consistent relationship between circulating endotoxin levels and the clinical course of illness. These observations indicate that in biological systems the pathways by which endotoxin activates inflammatory responses are more complex than the simple direct activation of macrophages that is often modeled in vitro. Therefore, discovering how endotoxin activates cells in biological systems has been of major interest. Developing effective strategies to inhibit the interaction of endotoxin with target cells should be important in modifying early inflammatory responses. Such strategies have the potential to prevent the production of cascades of mediators that initiate and intensify inflammatory reactions.

II. Discovery of Endotoxin-Binding Protein

It has been known for a number of years that endotoxin binds to circulating proteins. Initially, endotoxin was found to activate proteins involved in the coagulation pathways (reviewed in Ref. 1). Subsequently, endotoxin was also found to bind to plasma lipoproteins, including high-density lipoproteins (5). Endotoxin bound to high-density lipoproteins is less effective in stimulating cytokine production by mononuclear cells (6). In this respect, the binding of endotoxin to lipoproteins provides a mechanism to sequester or detoxify endotoxin, thereby preventing it from activating cellular inflammatory responses.

In the early 1980s, studies by Tobias and Ulevitch led to the discovery of an alternative endotoxin binding system in plasma (7–9). These studies established that the binding of lipopolysaccharide (LPS) to plasma proteins differed in normal and acute-phase sera in several important ways. First, the binding of LPS to high-density lipoproteins is rapid in normal serum but much slower in acute-phase serum (7). Second, when the binding of [^{3}H]LPS to serum proteins was studied using density-gradient centrifugation, it was found that LPS was present in a higher-density protein fraction in acute-phase serum than in normal serum. This led to the suggestion that an acute-phase reacting protein accounted for the differences in LPS–protein interactions in serum. Fractionation of normal and acute-phase serum using ion-exchange chromatography revealed that acute-phase serum contains a prominent protein peak that binds LPS (Fig. 1). This peak was absent in normal serum. Purification of the endotoxin binding activity in this peak led to the isolation of a 60-kDa glycoprotein that was named lipopolysaccharide-binding protein (LBP) (9). Immunodepletion of acute-phase rabbit serum with polyclonal antisera raised against purified LBP eliminated the high-density endotoxin-binding activity in acute-phase serum and converted the endotoxin-binding profile of acute-phase serum to that of normal rabbit serum (Fig. 2). This demonstrated that during acute-phase responses, a plasma protein appears that binds endotoxin. Initially, this protein was thought to provide another mechanism by which endotoxin in plasma might be bound and prevented from activating target cells, as is the case with high-density

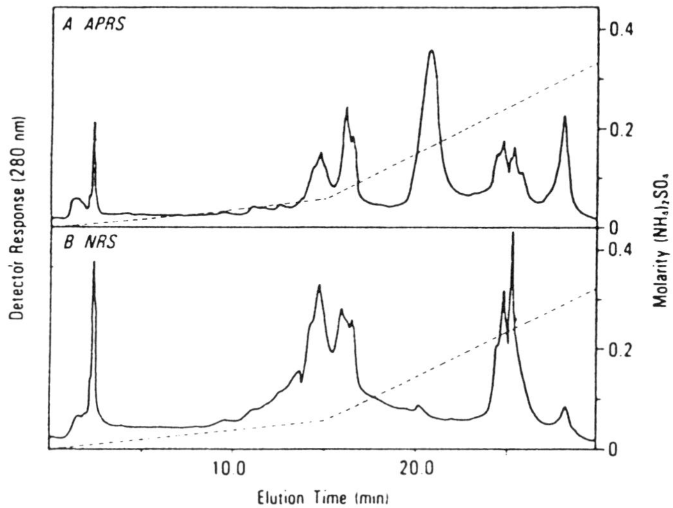

Figure 1 Ion-exchange chromatography of acute-phase rabbit serum (A, APRS) and normal rabbit serum (B, NRS). The APRS contained a major peak eluting at 20 min that was shown to have LPS-binding activity using [^{3}H]LPS. This peak was not present in NRS. (From Ref. 9.)

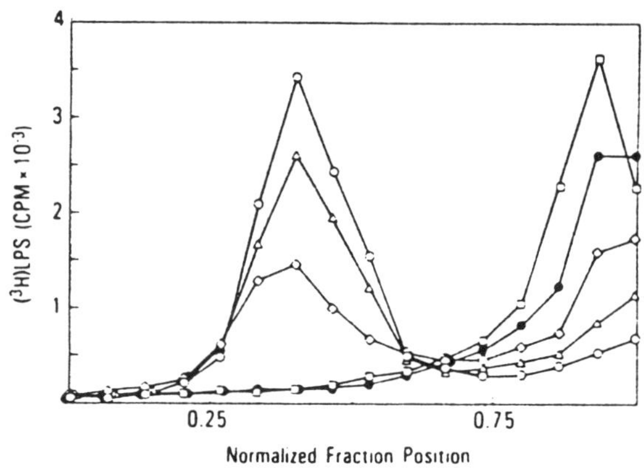

Figure 2 CsCl density-gradient analysis of LPS-binding activity in the immunoprecipitate supernatants obtained from mixtures of acute-phase rabbit serum (APRS), or normal rabbit serum (NRS) with rat anti-LBP antiserum. APRS, no antiserum (○); APRS, 1.7% (vol/vol) antiserum (△); APRS, 4.6% antiserum ( ); APRS, 14% antiserum (□); NRS, 14% antiserum (●). The anti-LBP serum abolished the LPS-binding activity of the acute-phase rabbit sterum. (From Ref. 9.)

lipoproteins (5). However, it was soon found that depleting LBP from rabbit blood substantially reduced the production of TNF-α in response to endotoxin (10). When endotoxin was added to rabbit blood ex vivo, TNF-α production was detectable at very low concentrations of endotoxin (e.g., 1 to 10 ng/mL). However, when rabbit blood was immunodepleted of LBP prior to adding endotoxin, the endotoxin dose response for TNF-α production was shifted dramatically to the right, so that low concentrations of endotoxin no longer stimulated TNF-α production. The dose response of the LBP-depleted rabbit blood was similar to the dose response for TNF-α production by mononuclear cells stimulated in vitro in the absence of serum, where concentrations of 1 μg/mL or more are often required to stimulate maximal TNF-α production. These findings suggested that LBP was an important component of a pathway by which endotoxin activates mononuclear cells in vivo.

Experiments using rabbit peritoneal exudate macrophages also demonstrated that LBP enhances the endotoxin-dose response for isolated exudate macrophages (10). LBP shifted the dose response for endotoxin-induced TNF-α production to the left, resulting in TNF-α production at 1 ng/mL concentrations of endotoxin, similar to that seen in whole blood. The increased TNF-α production was associated with an increase in steady-state levels of mRNA for TNF-α. This provided further support for the interpretation that LBP is an important mediator of macrophage responses to endotoxin. The data also suggested that LBP might account for the differences between endotoxin dose responses in vivo and in vitro.

The purification and N-terminal sequencing of purified rabbit LBP led to the cloning of a complete cDNA for rabbit, and subsequently human, LBP and the prediction of the entire amino acid sequences (10). The amino acid sequences of human and rabbit LBP are shown in Fig. 3. Human and rabbit LBP are composed of 456 amino acids, 69% of which are identical. LBP is a glycoprotein, with five N-linked glycosylation sites in human LBP and three in rabbit LBP. Signal peptide sequences of 25 and 26 residues are present in the human and rabbit LBP proteins, respectively. Comparison of the amino acid sequence of LBP with other known proteins indicates that LBP is 45% homologous with a neutrophil granule protein called bactericidal/permeability increasing protein (BPI) (11,12). LBP and BPI both bind endotoxin and share antigenic reactivity (11). The physiological role of BPI may be to enhance bacterial killing by neutrophils, but its function in endotoxemia has not been defined.

III. Tissue Distribution of LBP

LBP appears to be produced mainly in the liver. By Western analysis, LBP protein was detected in the liver of rabbits undergoing acute-phase responses

```
LBP-HUMAN     MGALARA-LP SILLALLLTS TPEALGANPG LVARITDKGL QYAAQEGLLA   49
LBP-RABBIT    MGTWARALLG SILLSLLLAA APGALGINPG LITRITDKGL EYAAREGLLA   50
Consensus     MG..ARA.L. S.LL.LLL.. .P.ALG.NPG L..RITDKGL .YAA.EGLLA   50

LBP-HUMAN     LQSELLRITL PDFTGDLRIP HVGRGRYEFH SLNIHSCELL HSALRPVPGQ   99
LBP-RABBIT    LQRKLLEVTL PDSDGDHRIK HHGRAQYKFY SLKIPRHELL RGTLRPLPGQ  100
Consensus     LQ..LL..TL PD..GD.RI. H.GR..Y.F. SL.I...ELL ...LRP.PGQ  100

LBP-HUMAN     GLSLSISDSS IRVQGRWKVR KSFFKLQGSF DVSVKGISIS VNLLLGSESS  149
LBP-RABBIT    GLSLDISDAY IHVRGSWKVR KAFLRLKNSF DLYVKGLTIS VHLMLGSESS  150
Consensus     GLSL.ISD.. I.V.G.WKVR K.F..L..SF D..VKG..IS V.L.LGSESS  150

LBP-HUMAN     GRPTGYCLSC SSDIADVEVD MSGDSGWLLN LFHNQIESKF QKVLESRICE  199
LBP-RABBIT    GRPTVTTSSC SSDIQNVELD IEGDLEELLN LLQSQIDARL REVLESKICR  200
Consensus     GRPT....SC SSDI..VE.D ..GD...LLN L...QI.... ..VLES.IC.  200

LBP-HUMAN     MIQKSVSSDL QPYLQTLPVT TEIDSFADID YSLVEAPRAT AQMLEVMFKG  249
LBP-RABBIT    QIEEAVTAHL QPYLQTLPVT TQIDSFAGID YSLMEAPRAT AGMLDVMFKG  250
Consensus     .I...V...L QPYLQTLPVT T.IDSFA.ID YSL.EAPRAT A.ML.VMFKG  250

LBP-HUMAN     EIFHRNHRSP VTLLAAAEE- ---HNKMVYF AISDYVFNTA SLVYHEEGYL  295
LBP-RABBIT    EIFPLDHRSP VDFLAHAMNL PEAHSRMVYF SISDYVFNTA SLAYHKSGYW  300
Consensus     EIF...HRSP V..LA.A... ...H..MVYF .ISDYVFNTA SL.YH..GY.  300

LBP-HUMAN     NFSITDDMIP PDSNIRLTTK SFRPFVPRLA RLYPNMNLEL QGSVPSAPLL  345
LBP-RABBIT    NFSITDAMVP ADLNIRRTTK SFRPFVPLLA NLYPNMNLEL QGIVNSEQLV  350
Consensus     NFSITD.M.P .D.NIR.TTK SFRPFVP.LA .LYPNMNLEL QG.V.S..L.  350

LBP-HUMAN     NFSPGNLSVD PYMEIDAFVL LPSSSKEPVF RLSVATNVSA TLTFNTSKIT  395
LBP-RABBIT    NLSTENLLEE PHMDIHALVV LPSSAREPVF RLGVATNVSA TLTLNTRKIT  400
Consensus     N.S..NL... P.M.I.A.V. LPSS..EPVF RL.VATNVSA TLT.NT.KIT  400

LBP-HUMAN     GFLKPGKVKV ELKESKVGLF NAELLEALLN YYILNTLYPK FNDKLAEGFP  445
LBP-RABBIT    GFLKPGRLQV ELKESKVGGF NVELLEALLN YYILNNLYPK VNEKLAHRFP  450
Consensus     GFLKPG...V ELKESKVG.F N.ELLEALLN YYILN.LYPK .N.KLA..FP  450

LBP-HUMAN     LPLLKRVQLY DLGLQIHKDF LFLGANVQYM RV                     477
LBP-RABBIT    LPLLRHIQLY DLLLQIHENF LLVGANIQYR RV                     482
Consensus     LPLL...QLY DL.LQ.H..F L..GAN.QY. RV                     482
```

Figure 3 Comparison of the cDNA-derived amino acid sequences of human and rabbit LBP. The arrow denotes the beginning of the mature secreted peptides. Rabbit and human LBP have conserved N-linked glycosylation sites at residues 301, 351, and 387 in the consensus sequence. Rabbit LBP has a single plasmin (or trypsin) cleavage site at Arg (125)–Leu (126). Cleavage at this site destroys LPS-binding activity.

(13). Using the sensitive polymerase chain reaction (PCR), we were able to amplify mRNA signals for LBP in total cellular RNA from rabbit liver but not in RNA extracted from whole lungs, normal alveolar macrophages, alveolar macrophages stimulated with endotoxin, or alveolar macrophages from rabbits undergoing acute-phase responses (14). Pulse-chase labeling studies with rabbit hepatocytes showed that LBP is produced in the liver as a 50-kDa intracellular product that is later found extracellularly as a 58- and 60-kDa doublet (13). Treatment of hepatocytes with tunicamycin, which prevents glycosylation, resulted in the intracellular accumulation of the 50-kDa LBP, suggesting that LBP is synthesized as a nonglycosylated precursor, then is posttranslationally modified and secreted as a mature glycoprotein.

IV. Biochemical Interactions Between LPS and LBP

Studies with different bacterial endotoxin structures have shown that LBP binds to a wide variety of endotoxins that contain differing O-chain structures, as well as to lipid A, which lacks an O-chain (15). However, LBP binds only weakly to partial lipid A structures such as lipid X, which contains only one glucosamine and two fatty acid side chains. These observations suggested that the interaction between endotoxin and LBP is not dependent on the specific sugar moieties in the O-chain of endotoxin, and focused attention on the lipid A portion of endotoxin as the critical site of interaction with LBP. Endotoxin binds to LBP with high affinity, as the apparent dissociation constant for complexes of LBP and *Salmonella minnesota* Re595 LPS is in the nanomolar range. Studies with peritoneal macrophages have shown that the ability of LBP to bind to LPS structures is paralleled by the ability of LBP to augment LPS-stimulated cytokine production (16). LBP does not enhance cytokine production in response to gram-positive organisms, as TNF-α production is normal in LBP-immunodepleted blood when *Staphylococcus aureus* is used as the stimulus (10). LBP does not enhance TNF-α production by peritoneal macrophages in response to gram-positive cell-wall peptidoglycans, or the protein kinase C–activator phorbol myristyl acetate (PMA) (16).

V. Identification of CD14 as a Receptor for LPS–LBP Complexes

The recognition that LPS–LBP complexes activate macrophages at low concentrations of LPS suggested that LPS–LBP complexes interact with a specific receptor on macrophages. Using LPS-coated erythrocytes (RBCs), it was shown that LBP could opsonize LPS-coated RBCs for attachment to human monocyte–derived macrophages (17). Studies with a panel of monoclonal antibodies demonstrated that antibodies to CD14, but not other membrane epitopes, blocked the binding of the LPS-coated, LBP-opsonized RBCs to macrophages. When anti-CD14 antibodies were added to human blood, TNF-α production was diminished in response to LPS. These experiments suggested that CD14 was an important component of the recognition of LPS–LBP complexes by mononuclear cells in vivo. Although LPS can bind to a number of different cell surface proteins (18), several monoclonal antibodies against CD14 consistently inhibit the activation of cells by LPS or LPS–LBP complexes (14,16,17). Studies with LPS that contains a photoactivatable group have demonstrated that LBP promotes the cross-linking of LPS to CD14, and not to other proteins on the surface of mononuclear cells (19).

CD14 was originally identified as a differentiation antigen on the surface of mononuclear cells (20). It is a member of a family of surface proteins that are anchored in the cell membrane by phosphatidylinositol linkages (21,22). Although CD14 is expressed primarily on blood monocytes and tissue macrophages, it is also detectable in lesser amounts on neutrophils, and its expression increases with neutrophil activation (23). Some B-cells also have been found to react with the anti-CD14 monoclonal antibody MY4 but not with other anti-CD14 antibodiesm such as LeuM3, that react strongly with CD14 on monocytes and macrophages (24). The relationship between this antigenic reactivity on B-cells and the CD14 molecule on monocytes remains to be determined.

Studies using the THP-1 mononuclear cell line have established that the maturation of LPS-driven cytokine responses in mononuclear phagocytes correlates with the appearance of CD14 on the cell surface (25). The THP-1 cell line was originally recovered from a young boy with myelomonocytic leukemia (26,27). THP-1 cells have characteristics of monocytes, including surface Fc receptors, nonspecific esterase staining, and the ability to phagocytose particles such as neutral red. The parental THP-1 cell line expresses minimal amounts of CD14 on the surface and shows only a minimal response to LPS. Treatment with the maturational agent 1,25$(OH)_2$-vitamin D_3 (D_3) induces THP-1 cells to increase their expression of CD14, as detected by flow cytometry (19,25,28) (Fig. 4). The D_3 treatment induces THP-1 cells to become responsive to LPS, although activation requires high concentrations of LPS (e.g., greater than 1 μg/mL). However, D_3 treatment causes THP-1 cells to become very sensitive to LPS–LBP complexes (Fig. 5), and this response is completely inhibitable by monoclonal antibodies to CD14. Furthermore, as THP-1 cells mature in the presence of D_3, they become responsive to the effect of serum in the media. In the presence of 1% human serum, low concentrations of LPS (e.g., 1 to 10 ng/mL) trigger the release of large amounts

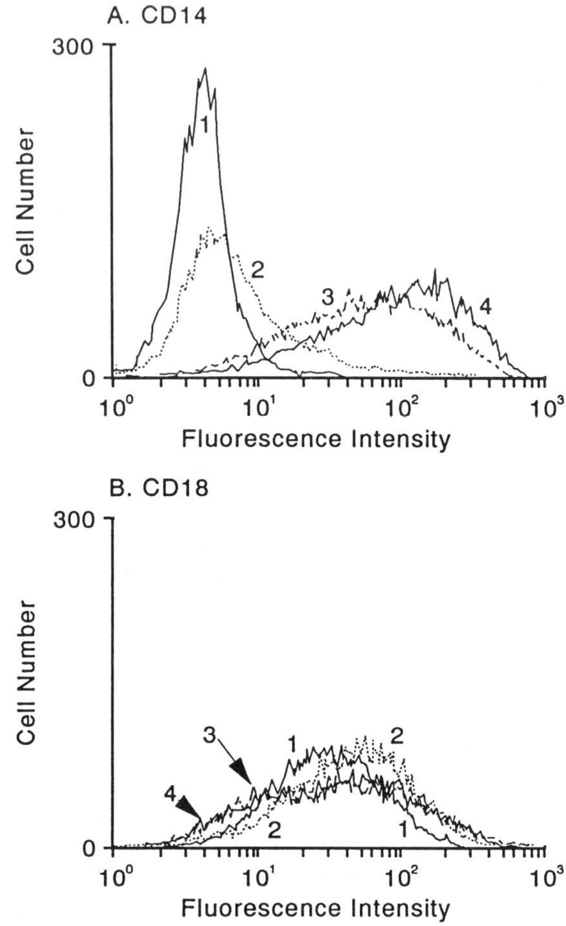

Figure 4 Dose-dependent expression of CD14 on THP-1 cells treated with increasing concentrations of D_3. THP-1 were incubated for 48 h with the indicated concentrations of D_3, then labeled by indirect immunofluorescence using murine 3c10 (anti-CD14), or murine IB4 (anti-CD18). Panel A, CD14; panel B, CD18. 1, media only; 2, 10^{-9} *M* D_3; 3, 10^{-9} *M* D_3; 4, 10^{-7} *M* D_3. The fluorescence pattern of cells labeled with an isotype control antibody was similar to curve 1 in panel A. The D_3 treatment caused a dose-dependent increase in the expression of CD14 (A), whereas the change in expression of CD18 was minimal.

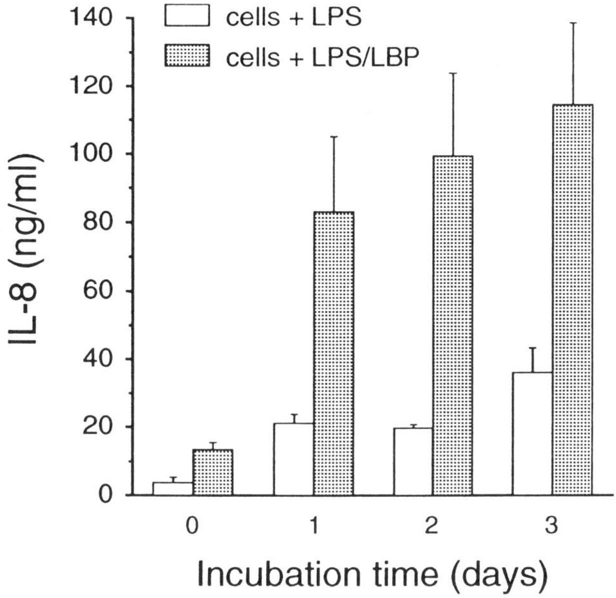

Figure 5 Time-dependent effect of D_3 on the production of IL-8 by THP-1 cells treated with LPS or LPS/LBP complexes. Cells were incubated in complete media with D_3 (1×10^{-7} M) for 0, 24, 48, or 72 h, then washed thoroughly and incubated for an additional 16 h in serum-free media containing LPS (10 ng/ml) in the absence (open bars) or presence (shaded bars) of LBP (100 ng/ml). IL-8 was measured in the supernatants by enzyme immunoassay. The data are the mean ± SE of four separate experiments.

of interleukin-8 (IL-8). This effect can be blocked either by immunodepleting LBP from the serum or by including anti-CD14 monoclonal antibodies in the incubation mixture. These experimental observations indicate that as mononuclear cells mature, they acquire responsiveness to LPS and LBP complexes, and that this is mediated by the CD14 "differentiation antigen" on the cell surface.

Transfection experiments using the 70Z/3 cell line have also provided strong evidence for an important role for CD14 in mediating the effects of

LPS–LBP complexes (29). The 70Z/3 cells are murine B cells that express large amounts of surface IgM in response to stimulation with interferon-γ (IFN-γ), but express IgM only in response to very high concentrations of LPS. The 70Z/3 cells normally lack CD14, but transfection of 70Z/3 cells with a cDNA construct encoding for mature human CD14 resulted in the expression of CD14 on the cell surface (29). The transfectants expressing CD14 responded to much lower concentrations of LPS than did the nontransfected cells (Fig. 6). Treatment with LPS–LBP complexes dramatically enhanced the LPS dose response of the transfected cells. In contrast, transfection of 70Z/3 cells with CD14 did not affect the expression of surface IgM in response to IFN-γ. These experiments indicate that transfection of CD14 can confer LPS responsiveness on cells that normally respond only minimally to LPS. The data also support the important conclusion that the downstream signaling pathways that regulate LPS responses are present in some cells that normally show little response to LPS.

Recently, transgenic mice have been created that overexpress the human CD14 antigen in myeloid cells (30). These mice are more sensitive to the effects of intravenous LPS than mice that lack the transgene, supporting the importance of CD14 in mediating the responses to endotoxin in vivo (30). These experiments also suggest that there is significant interspecies conservation of functional epitopes on the CD14 molecule.

An interesting experiment of nature provides additional support for the importance of CD14 in mediating the responses of monocytes and macrophages to LPS in humans. Patients with paraoxysmal nocturnal hemoglobinuria (PNH) have a chronic hemolytic anemia characterized by episodic hemolysis, recurrent venous thrombosis, and infections that can be life threatening (31–34). These patients lack cell-membrane proteins that are anchored by glycosylinositolphospholipid tails (34). The lack of decay-accelerating factor on erythrocyte membranes explains the unusual sensitivity of erythrocytes to complement-mediated lysis (35). These patients also have absolute or intermediate deficiencies of membrane CD14 (21,36). In the presence of serum, LPS does not bind efficiently to monocytes from patients with PNH (37), consistent with the interpretation that LPS–LBP complexes in serum cannot bind via CD14 to the monocyte surface.

The importance of CD14 in mediating the effects of LPS–LBP complexes on macrophages suggested to us that the response to LPS might be abnormal in such patients. Recently, we had the opportunity to study LPS-induced cytokine production in the peripheral blood of a patient with PNH in Seattle, and we found that the patient has a striking inability to respond to LPS. Peripheral blood was drawn from the patient and a normal volunteer on several occasions when the patient was clinically stable. CD14 was virtually absent from the patient's blood leukocytes by flow cytometry. Mea-

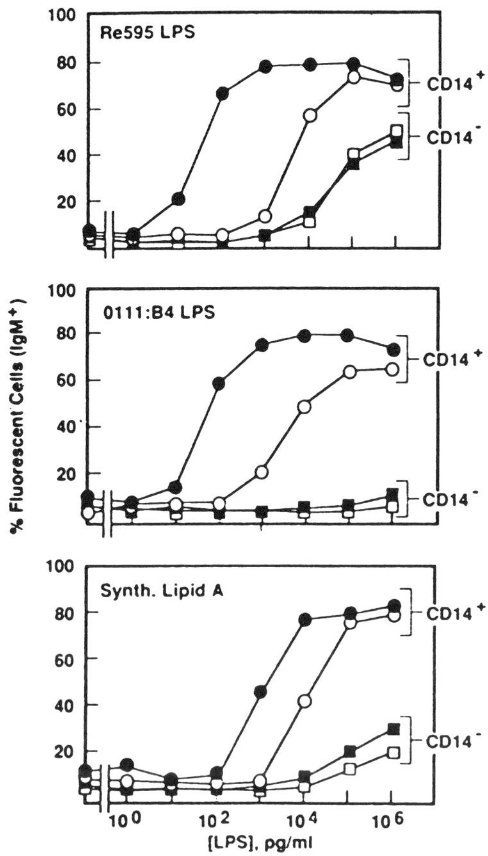

Figure 6 Stimulation of 70Z/3 cells that were transfected with a cDNA for human CD14 with varying doses of two different forms of LPS (*S. minnesota* Re595 LPS, top; *E. coli* O111:B4 LPS, middle), or with lipid A alone (bottom) in the presence or absence of purified rabbit LBP (100 ng/mL). Surface IgM was measured by flow cytometry as an indication of LPS response (vertical axis). The parental strain showed little response to LPS, and at only the highest doses. The transfectants acquired responsiveness to LPS, and the response was substantially augmented by added LBP. (From Ref. 29.)

surement of serum LBP by immunoassay showed that the LBP concentration was normal in the patient's blood (ca. 15 μg/mL). We added 10 ng/mL *S. minnesota* Re595 LPS to aliquots of whole blood, then incubated the blood samples for up to 18 h at 37°C in 5% CO_2-air. At selected times, the plasma from some of the aliquots was harvested and stored for cytokine measurements by immunoassay. Additional aliquots of whole blood were

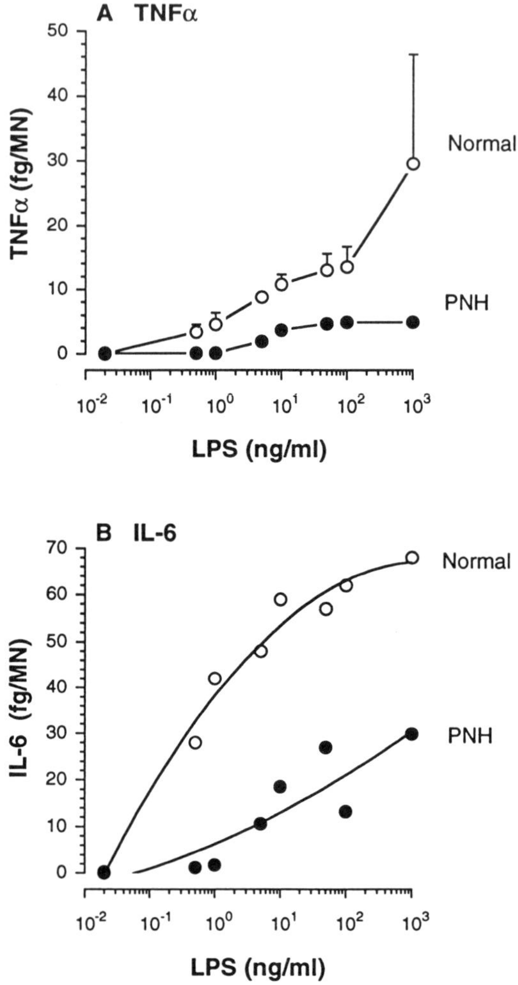

Figure 7 Measurement of TNF-α (A) and IL-6 (B) production in human blood from a normal patient (o--o) and a patient with paroxysmal nocturnal hemoglobinuria (o--o). LPS was added in the indicated concentrations to aliquots of peripheral blood. After 18 h of incubation, the serum was assayed for TNF-α and IL-6 using sensitive and specific immunoassays. The blood of the patient produced lower levels of TNF-α and IL-6 at each concentration of added LPS.

incubated with LPS, and total cellular RNA was harvested from buffy coat leukocytes for the measurement of steady-state mRNA levels for specific cytokines.

Figure 7 shows the production of TNF-α and IL-6 in the blood in response to LPS. The patient with PNH produced substantially less of the proinflammatory cytokines TNF-α and IL-6 in response to LPS. In contrast, cytokine production was normal in the PNH blood in response to PMA, a direct activator of protein kinase C that triggers mononuclear cells independently of surface CD14. A monoclonal antibody to CD14 inhibited TNF-α production in the blood of the normal subject but had no effect in the blood of the patient with PNH (Fig. 8). In fact, TNF-α production in the normal blood treated with the anti-CD14 antibody was similar to that seen in the blood of the patient with PNH. These findings were supported by experiments in which we used the polymerase chain reaction to amplify TNF-α mRNA transcripts in blood leukocytes. Total cellular RNA was recovered and reverse-transcribed to yield cDNA, then cDNA signals for TNF-α and β-actin were amplified using specific primers (38). Although β-actin amplified similarly in the RNA from the normal subject and the patient, amplification of the message for TNF-α was delayed in the sample from the patient with PNH (Fig. 9). This suggests strongly that cytokine gene expression in response to LPS stimulation does not occur normally when CD14 is lacking from the surface of circulating leukocytes. These findings have been conformed in recent studies of two additional patients (39). Thus an experiment of nature confirms experimental studies about the importance of CD14 in cellular activation by LPS.

The mechanism by which CD14 participates in cellular activation by LPS and LPS–LBP complexes is uncertain. Although CD14 facilitates the uptake of radiolabeled LPS by macrophages (40), it turns over slowly on the cell surface (41), and a change in the surface expression of CD14 is not required for LPS-stimulated cytokine production (14). This suggests that internalization of CD14 following ligation with LPS–LBP complexes is not required for cellular activation. LPS–LBP complexes transfer LPS directly to binding sites on CD14 (19). Studies with anti-CD14 monoclonal antibodies with differing specificities indicate that it is possible to block cellular activation via CD14 without blocking the binding of LPS to CD14 (42). Whether CD14 is capable of transmitting direct intracellular signals is uncertain, although other PI-linked proteins have been implicated in direct intracellular signaling. For example, transfection of Thy-1, a PI-linked protein, into lymphocytes caused calcium flux in the transfected cells when Thy-1 antigens were cross-linked using monoclonal antibodies (43). However, we found that activation of alveolar macrophages by the LPS–LBP–CD14 pathway

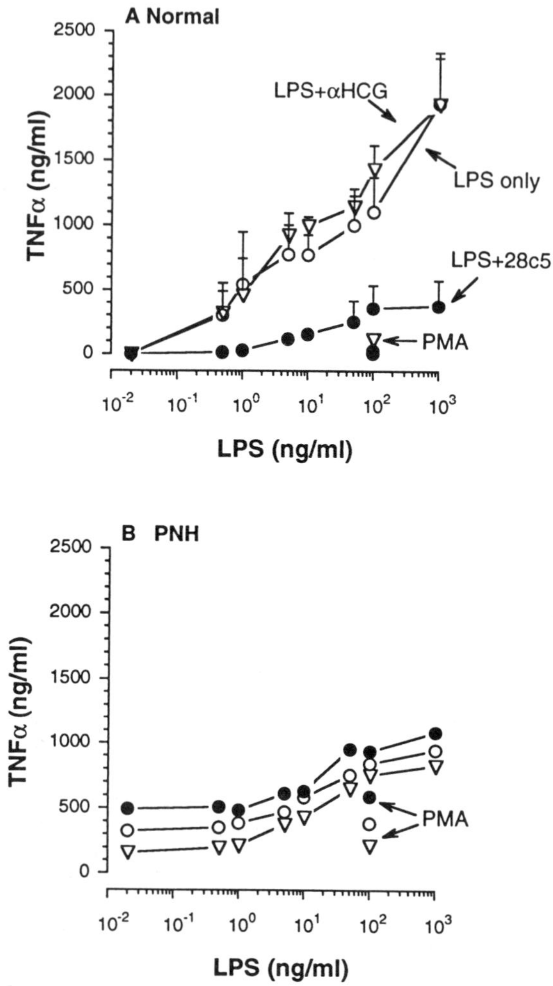

Figure 8 Effect of anti-CD14 antibody on TNF-α production in LPS-stimulated peripheral blood of a normal volunteer and a patient with paroxysmal nocturnal hemoglobinuria. LPS was added in the indicated concentrations to aliquots of peripheral blood in the presence or absence of a murine monoclonal antibody to CD14 (28c5) or a control antibody of the same isotype (antihuman chorionic gonadotropin, αHCG). The αCD14 antibody blocked TNF-α production in the blood of the normal subject but had no effect on TNF-α production in the blood of the patient.

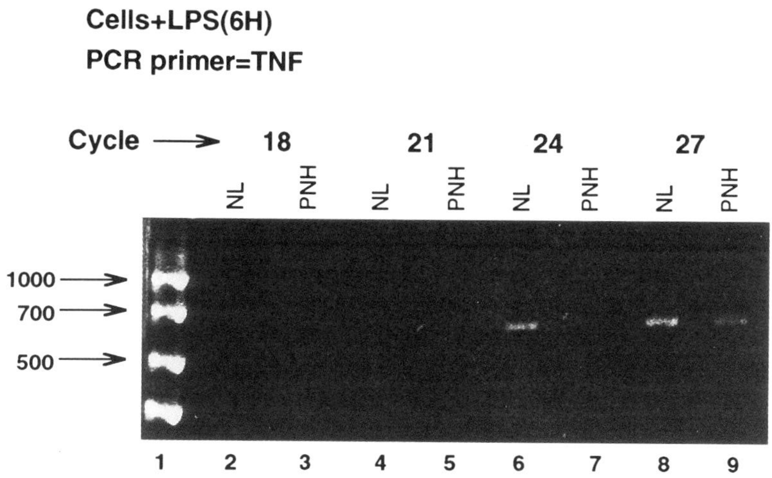

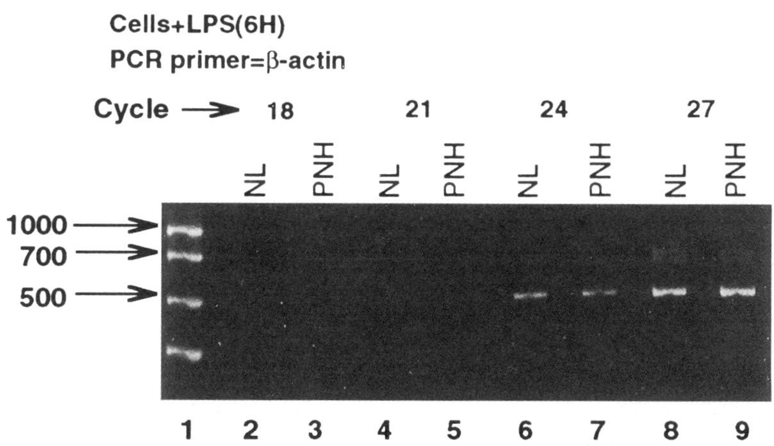

Figure 9 Analysis of mRNA transcripts in peripheral blood leukocytes of a patient with paroxysmal nocturnal hemoglobinuria and a normal volunteer. Peripheral blood was incubated for 4 h with LPS (10 ng/mL), then the total cellular RNA was isolated by CsCl centrifugation and phenol–chloroform extraction. Aliquots of the RNA were analyzed using RT-PCR for the indicated number of cycles, and primers specific for human TNF-α, or β-actin as a control. The β-actin signal amplified at the same rate in the patient with paroxysmal nocturnal hemoglobinuria and the normal volunteer, but amplification of the TNF-α signal was later in the patient than in the normal subject.

is not associated with an initial change in either intracellular Ca^{2+} or pH in alveolar macrophages (14), suggesting that initial changes in intracellular Ca^{2+} and/or pH are not required for LPS to initiate TNF-α production in lung macrophages. Recent evidence suggests that tyrosine kinase pathways are involved in activation of macrophages by LPS, as inhibitors of tyrosine phosphorylation block LPS-induced signaling in macrophages (44). Precipitation of CD14 from cell membranes coprecipitates tyrosine kinase activity, suggesting that CD14 is linked to a tyrosine phosphorylation pathway (45). Experiments with 70Z/3 cells that have been transfected with a construct coding for human CD14 have shown that LPS induces rapid tyrosine phosphorylation of a 38-kDa protein (p38) in these transfectants, as well as in RAW264.7 cells and rabbit peritoneal macrophages (46). The PI anchor is not required for CD14 to participate in cellular activation, as transfection of either transmembrane or GPI anchored forms of CD14 into 70Z/3 cells results in identical patterns of responsiveness to LPS (42).

These observations are consistent with the hypothesis that CD14 does not transmit signals directly into cells in response to ligation with LPS or

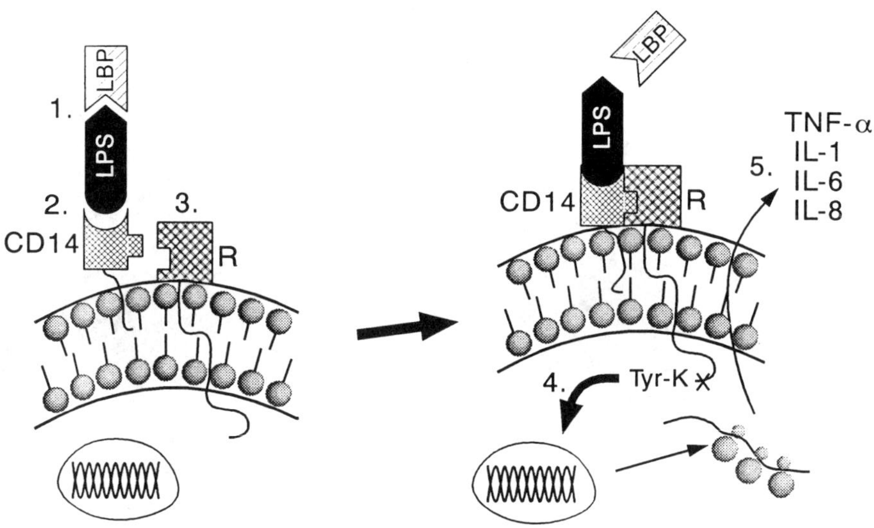

Figure 10 Model for LPS/LBP/CD14 interactions on monocytes and macrophages. (1) LPS in solution interacts with circulating LBP. (2) The LPS–LBP complex interacts with CD14 on the cell surface. (3) This complex may associate with a second, as yet unidentified protein on the cell surface, which results in the transmission of intracellular signals that involve tyrosine phosphorylation reactions (46). (5) This results in the transcription of cytokine genes and the secretion of the mature cytokine proteins.

LPS–LBP complexes, but associates with a second, as yet undefined membrane protein which transmits a signal intracellularly that results in tyrosine phosphorylation. The function of CD14 may be similar to that of the IL-6 receptor, which interacts with another membrane protein, gp130, that transmits an intracellular signal (47). The IL-2 receptor also associates with a second membrane protein to initiate intracellular tyrosine phosphorylation (48). A putative model for interactions between LPS, LBP, and CD14 on monocytes and macrophages is shown in Fig. 10.

VI. Endotoxin-Binding Protein in the Lungs

Recent studies have demonstrated that the LPS–LBP–CD14 pathway is highly relevant for the interaction of LPS with cells in the lungs. Studies with human and rabbit alveolar macrophages have shown that LBP shifts the LPS dose-response curve dramatically to the left, rendering alveolar macrophages sensitive to concentrations of LPS in the nanogram range that do not ordinarily activate these cells (Fig. 11) (14). The shift in the LPS dose-response produced by LBP is associated with a more rapid onset of cytokine production with an earlier peak and an earlier decline in TNF-α production as compared with cells stimulated with LPS alone. This change in the kinetics of TNF-α production is associated with a change in the kinetics and stability of mRNA expression for TNF-α. TNF-α mRNA appears earlier in rabbit alveolar macrophages stimulated with LPS and LBP, peaks earlier, and disappears earlier (14). The change in mRNA kinetics is associated with a prolonged mRNA half-life, indicating increase stability of the mRNA message. This finding has also been reproduced in rabbit peritoneal macrophages (16). The effects of LBP on human alveolar macrophages are completely inhibitable by the anti-CD14 monoclonal antibody 28c5, demonstrating the critical importance of CD14 in mediating the effects of LPS and LBP (14).

As noted earlier, LBP appears to be produced in the liver but not in lungs or alveolar macrophages, despite the remarkable synthetic capacity of stimulated alveolar macrophages. We have identified LBP in bronchoalveolar lavage fluids from the lungs of normal volunteers, patients with acute lung injury, and patients with interstitial lung diseases. Initial immunoprecipitation studies suggested that LBP is present normally in low concentrations in lung lavage fluid, and that its concentration increases in the setting of acute lung injury (14). Subsequent studies using a sensitive immunoassay for LBP have shown that LBP increases substantially in both the serum and the lung lavage fluid of patients with adult respiratory distress syndrome (ARDS) (49). Normally, LBP is detectable at a concentration of approximately 10 ng/mL in bronchoalveolar lavage fluid, whereas in patients with ARDS the concentration of LBP rises more than 10-fold. Patients with ARDS following

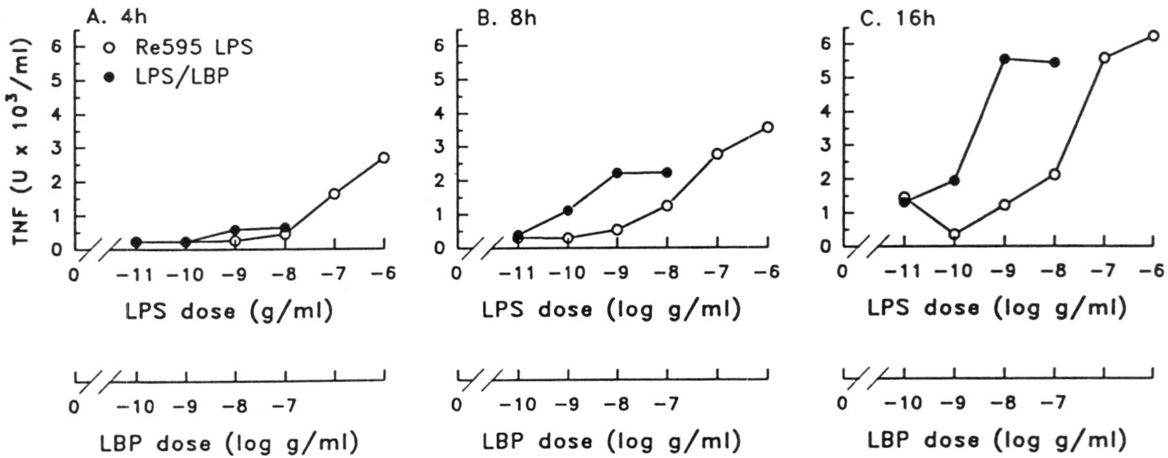

Figure 11 Production of TNF-α by human alveolar macrophages stimulated with LPS and purified rabbit LBP. Human alveolar macrophages were incubated with rough-form *S. minnesota* RE595 LPS for 4, 8, or 16 h in the absence or presence of a 12-fold excess of LBP as indicated. LBP caused TNF-α production to begin sooner and at lower concentrations of LPS than with LPS alone. (From Ref. 14.)

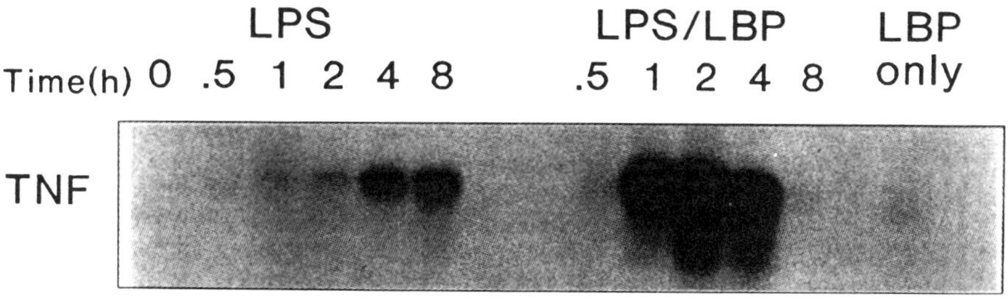

Figure 12 Effect of LPS and LBP on the induction of TNF-α mRNA in rabbit alveolar macrophages. The alveolar macrophages were incubated for up to 8 h with either 12 ng/mL Re595 LPS alone (left), 1 ng/mL Re595 LPS and 12 ng/mL rabbit LBP (middle), or with 12 ng/mL LBP alone for 4 h (right). Total cellular RNA was recovered at the indicated times and Northern analysis was performed using a specific cDNA probe for rabbit TNF-α. In the presence of LBP, the message appeared earlier, reached higher steady-state levels, and disappeared earlier than with LPS alone. (From Ref. 14.)

sepsis syndrome or trauma have similar concentrations of LBP in lung lavage fluid, and the level of LBP that is detectable does not appear to predict outcome. The concentration of LBP in lavage fluid correlates with the total protein concentration in the lavage fluid. Because LBP does not appear to be produced in the lungs, the association between LBP and total protein suggests that LBP moves from the plasma into the airspaces when lung endothelial and epithelial permeability is altered. We have found that the concentration of LBP in lung lavage fluid is typically 100- to 1000-fold higher than the concentration of LPS in the fluid. This provides more than enough LBP to amplify the effect of low concentrations of LPS on cells in the airspaces of the lungs. The concentration of soluble CD14 also increases in the lung fluids of patients with ARDS (49). Compared with normal volunteers, levels of soluble CD14 typically increase more than 10-fold in patients with ARDS. Soluble CD14 correlates strongly with lavage total protein, suggesting that like LBP, soluble CD14 also may move from the plasma compartment into the lungs. However, the presence of CD14 on lung macrophages suggests that shedding of CD14 from alveolar macrophages and monocytes (50) may provide an additional intraalveolar source for soluble CD14. These studies in patients with ARDS demonstrate that all of the components of the LPS–LBP–CD14 system are present in the airspaces of patients with acute lung injury and suggest strongly that this pathway may regulate the responses to LPS in the lungs in these patients. Further studies will help to determine the exact role of this system in the lungs.

VII. Summary

The discovery that endotoxin binds to LBP and that this complex activates macrophages via CD14 has led to a new understanding of the way in which LPS activates mononuclear cells. The discovery of LBP helps to resolve the "endotoxin paradox," as LBP can account for the enhanced responsiveness to LPS in vivo as compared with in vitro. The LPS–LBP–CD14 pathway is important in circulating blood cells, tissue macrophages, and endothelial cell interactions with LPS (39). This important pathway provides the host with a means of recognizing and responding to minute concentrations of LPS that enter tissues. This may be a great advantage in initiating early host defense responses to bacteria in tissues. However, when concentrations of LPS are higher, this pathway may initiate inflammatory responses that have deleterious effects for the host. An improved understanding of the function of this pathway may lead to new ideas for the treatment of inflammatory responses that are initiated by LPS.

Acknowledgments

We thank Steve Mongovin, Venus Wong, John Ruzinski, Nguyet Ky, Lois Kline, Loren Hatlen, Eleanor Wolfson, and Katrin Soldau for their helpful contributions to these studies. Monoclonal antibodies to human CD14 were provided by Didier Leturcq and Ann Moriarty, R.W. Johnson Pharmaceutical Research Institute, San Diego, CA.

References

1. Morrison DC, Ryan JL. Endotoxins and disease mechanisms. Annu Rev Med 1987; 38:417–432.
2. Desch CE, Kovach NL, Present W, Broyles C, Harlan JM. Production of human tumor necrosis factor from whole blood ex vivo. Lymphokine Res 1989; 8:141–146.
3. Parsons PE, Worthen GS, Moore EE, Tate RM, Henson PM. The association of circulating endotoxin with the development of the adult respiratory distress syndrome. Am Rev Respir Dis 1989; 140:294–301.
4. Danner RL, Elin RJ, Hosseini JM, Wesley RA, Parillo J. Endotoxemia in human septic shock. Chest 1991; 99:169–175.
5. Ulevitch RJ, Johnston AR, Weinstein DB. New function for high density lipoproteins. Isolation and characterization of a bacterial lipopolysaccharide–high density lipoprotein complex formed in rabbit plasma. J Clin Invest 1981; 67:827–837.

6. Cavaillon J-M, Fitting C, Haeffner-Cavaillon N, Kirsch SJ, Warren HS. Cytokine response by monocytes and macrophages to free and lipoprotein-bound lipopolysaccharide. Infect Immun 1990; 58:2375–2382.
7. Tobias PS, Ulevitch RJ. Control of lipopolysaccharide–high density lipoprotein binding by acute phase protein(s). J Immunol 1983; 131:1913–1916.
8. Tobias PS, McAdam KPWJ, Soldau K, Ulevitch RJ. Control of lipopolysaccharide–high density lipoprotein interactions by an acute phase reactant in human serum. Infect Immun 1985; 50:73–80.
9. Tobias PS, Soldau K, Ulevitch RJ. Isolation of a lipopolysaccharide-binding acute phase reactant from rabbit serum. J Exp Med 1986; 164:777–793.
10. Schumann RR, Leong SR, Flaggs GW, et al. Structure and function of lipopolysaccharide binding protein. Science 1990; 249:1429–1431.
11. Tobias PS, Mathison JC, Ulevitch RJ. A family of lipopolysaccharide binding proteins involved in responses to gram-negative sepsis. J Biol Chem 1988; 263: 13479–13481.
12. Gray PW, Flaggs G, Leong SR. Cloning of the cDNA of a human neutrophil bactericidal protein. Structural and functional correlations. J Biol Chem 1989; 264:9505–9509.
13. Ramadori G, Meyer zum Buschenfelde K-H, Tobias PS, Mathison JC, Ulevitch RJ. Biosynthesis of lipopolysaccharide-binding protein in rabbit hepatocytes. Pathobiology 1990; 58:89–94.
14. Martin TR, Mathison JS, Tobias PS, et al. Lipolysaccharide binding protein enhances the responsiveness of alveolar macrophages to bacterial lipolysaccharide. Implications for cytokine production in normal and injured lungs. J Clin Invest 1992; 90:2209–2219.
15. Tobias PS, Soldau K, Ulevitch RJ. Identification of a lipid A binding site in the acute phase reactant lipopolysaccharide binding protein. J Biol Chem 1989; 264:10867–10871.
16. Mathison JC, Tobias PS, Wolfson E, Ulevitch RJ. Plasma lipopolysaccharide binding protein. A key component in macrophage recognition of gram-negative LPS. J Immunol 1992; 149:200–206.
17. Wright SD, Ramos RA, Tobias PS, Ulevitch RJ, Mathison JC. CD14, a receptor for complexes of lipopolysaccharide (LPS) and LPS binding protein. Science 1990; 249:1431–1433.
18. Ulevitch RJ. Recognition of bacterial endotoxins by receptor dependent mechanisms. Adv Immunol 1993; 53:267–289.
19. Tobias PS, Soldau K, Kline L, et al. Cross-linking of lipopolysaccharide (LPS) to CD14 on THP-1 cells mediated by LPS-binding protein. J Immunol 1993; 150:3011–3021.
20. Todd RF III, Nadler LM, Schlossman SF. Antigens on human monocytes identified by monoclonal antibodies. J Immunol 1981; 126:1435–1442.
21. Haziot A, Chen S, Ferrero E, Low MG, Silber R, Goyert SM. The monocyte differentiation antigen, CD14, is anchored to the cell membrane by a phosphatidylinositol linkage. J Immunol 1988; 141:547–552.
22. Low MG, Saltiel AR. Structural and functional roles of glycosyl phosphatidylinositol in membranes. Science 1988; 239:268.

23. Weingarten R, Sklar LA, Mathison JC, et al. Interactions of lipolysaccharide with neutrophils in blood via CD14. J Leukocyte Biol 1992; 53:518–524.
24. Labeta MO, Landmann R, Obrecht JP, Obrist R. Human B cells express membrane-bound and soluble forms of the CD14 myeloid antigen. Mol Immunol 1991; 28:115–122.
25. Martin TR, Mongovin SM, Tobias PS, Mathison JC, Moriarty AM, Leturcq DJ, Ulevitch RJ. The CD14 differentiation antigen mediates the development of endotoxin responsiveness during differentiation of mononuclear phagocytes. J Leukocyte Biol 1994 (in press).
26. Tsuchiya S, Yamabe M, Yamaguchi Y, Kobayashi Y, Konno T, Tada K. Establishment and characterization of a human acute monocytic leukemia cell line (THP-1). Int J Cancer 1980; 26:171–176.
27. Auwerx J. The human leukemia cell line, THP-1: a multifacetted model for the study of monocyte–macrophage differentiation. Experientia 1991; 47:22–31.
28. Fleit HB, Kobasiuk CD. The human monocyte-like cell line THP-1 expresses Fc-gamma RI and Fc-gamma RII. J Leukoc Biol 1991; 49:556–565.
29. Lee J-D, Kato K, Tobias PS, Kirkland TN, Ulevitch RJ. Transfection of CD14 into 70Z/3 cells dramatically enhances the sensitivity to complexes of lipopolysaccharide (LPS) and LPS binding protein. J Exp Med 1992; 175:1697–1705.
30. Ferrero E, Jiao D, Tsuberi BZ, et al. Transgenic mice expressing human CD14 are hypersensitive to lipolysaccharide. Proc Natl Acad Sci USA 1993; 90:2380–2384.
31. Rosse WF, Parker CJ. Paroxysmal nocturnal haemoglobinuria. Clin Haematol 1985; 14:105–125.
32. Crosby WH. Paroxysmal nocturnal hemoglobinuria. Relation of clinical manifestations to underlying pathogenetic mechanisms. Blood 1953; 8:769–812.
33. Craddock PR, Fehr J, Jacob HS. Complement-mediated granulocyte dysfunction in paroxysmal nocturnal hemoglobinuria. Blood 1976; 47:931–939.
34. Rosse WF. Phosphatidylinositol-linked proteins and paroxysmal nocturnal hemoglobinuria. Blood 1990; 75:1595–1601.
35. Rosse WF. Paroxysmal nocturnal hemoglobinuria and decay-accelerating factor. Annu Rev Med 1990; 41:431–436.
36. Simmons DL, Tan S, Tenen DG, Nicholson-Weller A, Seed B. Monocyte antigen CD14 is a phospholipid anchored membrane protein. Blood 1989; 73:284–289.
37. Couturier C, Haeffner-Cavaillon N, Caroff M, Kazatchkine MD. Binding sites for endotoxins (lipopolysaccharides) on human monocytes. J Immunol 1991; 147:1899–1904.
38. Brenner CA, Tam AW, Nelson PA, et al. Message amplification phenotyping (MAPPing): a technique to simultaneously measure multiple mRNAs from small numbers of cells. Biotechniques 1989; 7:1096–1103.
39. Pugin J, Ulevitch RJ, Tobias PS. Critical role for myeloid cells and CD14 in endotoxin-induced endothelial cell activation. J Exp Med 1993; 178:2193–2200.
40. Kitchens RL, Ulevitch RJ, Munford RS. Lipolysaccharide (LPS) partial structures inhibit responses to LPS in a human macrophage cell line without inhibiting LPS uptake by a CD14-mediated pathway. J Exp Med 1992; 176:485–494.

41. Bamezai A, Goldmacher V, Reiser H, Rock KL. Internalization of phosphatidylinositol-anchored lymphocyte proteins. I. Documentation and potential significance for T cell stimulation. J Immunol 1989; 143:3107–3116.
42. Lee J-D, Kravchenko V, Kirkland TN, et al. Glycosylphosphatidylinositol-anchored or integral membrane forms of CD14 mediate identical cellular responses to endotoxin. Proc Natl Acad Sci USA 1993; 90:9930–9934.
43. Kroczek RA, Gunter KC, Germain RN, Shevach EM. Thy-1 functions as a signal transduction molecule in T lymphocytes and transfected B lymphocytes. Nature (London) 1986; 322:181–184.
44. Weinstein SL, Gold MR, DeFranco AL. Bacterial lipopolysaccharide stimulates protein tyrosine phosphorylation in macrophages. Proc Natl Acad Sci USA 1991; 88:4188–4152.
45. Stefanova I, Horejsi V, Ansotegui IJ, Knapp W, Stockinger H. GPI-anchored cell-surface molecules complexed to protein tyrosine kinases. Science 1991; 254: 1016–1019.
46. Han J, Lee J-D, Tobias PS, Ulevitch RJ. Endotoxin induces rapid protein tyrosine phosphorylation in 70Z/3 cells expressing CD14. J Biol Chem 1993; 268: 25009–25014.
47. Taga T, Kishimoto T. IL-6 receptor. In: Cochrane CG, Gimbrone MA Jr, eds. Cellular and Molecular Mechanisms of Inflammation. San Diego, CA: Academic Press, 1990:219–243.
48. Mills GB, May C, McGill M. Interleukin 2-induced tyrosine phosphorylation. J Biol Chem 1990; 265:3561–3567.
49. Rubenfeld GD, Steinberg KP, Maunder RJ, et al. Lipolysaccharide binding protein is elevated in bronchoalveolar lavage fluid in patients with adult respiratory distress syndrome. Am Rev Respir Dis 1993; 147:A618.
50. Bazil V, Strominger JL. Shedding as a mechanism of down-modulation of CD14 on stimulated human monocytes. J Immunol 1991; 147:1567–1574.

4

Lipopolysaccharide-Induced Signal Transduction and Gene Transcription

JACEK HAWIGER

Vanderbilt University School of Medicine
Nashville, Tennessee

Lipopolysaccharides (LPSs), unusual components of the outer membrane of the envelope of gram-negative bacteria, have a striking propensity toward membranes of susceptible cells of myelomonocytic, B-cell, endothelial, and megakaryocytic lineages. Although the glycolipid structure of LPS differs ammong gram-negative bacteria, interaction of LPS with susceptible cells is greatly enhanced by a membrane-bound or cell-free pool of CD14. CD14-mediated presentation of LPS to its putative receptor on the membranes of susceptible cells evokes a signal that is transduced by pathway(s) leading to a robust transcriptional activation of cellular and viral genes. They encode tumor necrosis factor α (TNF-α), interleukin-1 (IL-1), and the procoagulant molecule, tissue factor, in monocytes/macrophages. The latter is also expressed in endothelial cells, along with genes encoding cell adhesion molecules, cytokines, and colony-stimulating factors. Expression of human immunodeficiency virus type 1 (HIV-1) in monocytic cells is also stimulated by LPS. So far, two transcriptional factors, NF-κB/Rel and AP-1, have been implicated in LPS-induced expression of these genes. It is plausible that other factors or translation and/or posttranslation mechanisms are also involved in LPS-induced gene expression. The members of the NF-κB/Rel family are regulated at the transcriptional and posttranscriptional level. LPS stimulates synthesis of mRNA for several members of the NF-κB/Rel family and induces phosphorylation of NF-κB inhibitors such as IκBα/MAD3 in the human monocytic cell line THP-1. The identity of kinase(s) involved in the latter process is not established. Among kinases, protein kinase C isozymes and MAP kinases are linked to LPS-induced activation of signal transduction. Although phorbol ester–sensitive protein kinase C isozymes are unlikely signal transducers in the human monocytic cell line THP-1, novel PKC isozymes expressed in this cell line can be involved. In human platelets, protein kinase C isozymes are activated by LPS-derived monophosphoryl lipid A. Delineating signal-transducing pathways in susceptible cells is hampered by the present lack of knowledge of a putative, signal-generating receptor for LPS. Once such a receptor is identified, its link to G proteins, kinases, and other signal transducers will be more easily established. Then connections can be completed between signal(s) generated at the level of the LPS receptor and gene transcription.

I. Introduction

It is generally accepted that LPS-producing gram-negative bacteria are among the most frequent causative organisms of septic shock, which affects approximately 400,000 patients annually in the United States (1). Vascular complications of septic shock, hypotension and disseminated intravascular coagulation (DIC), are related to the rapid activation of blood cells and endothelial cells by LPS; generation of potent cytokines, TNF-α, and IL-1, IL-6 and IL-8; activation of complement cascade and the kinin system; and de novo synthesis of plasma proteins representing an acute-phase response (2). The LPS responsible for the initiation of these changes is a very potent inducer, acting at picogram to nanogram concentrations, detectable in blood by a *Limulus* assay (3). The toxic principle of LPS is lipid A, which is composed of disaccharide with amide- and ester-linked fatty acid residues (4). Lipid A serves as the lipophilic membrane anchor for the lipopolysaccharide moiety of the outer membrane of the envelope of gram-negative bacteria. As such, it differs among gram-negative bacteria in the number and length of its fatty acid residues (4). These differences, as well as the varying complexity of oligosaccharides attached to lipid A, contribute to the differential toxicity of LPS among gram-negative bacteria and to different antigenic reactivity, especially in regard to monoclonal antibodies directed against the lipid A part of LPS. This variability is exemplified by experimental demonstration that LPS from *Neisseria meningitides*, a cause of fulminant purpura associated with DIC and septic shock in susceptible individuals, has a propensity for inducing vascular lesions in small blood vessels of the skin, whereas LPS from *Escherichia coli* usually does not (5). Interestingly, LPS isolated from *Rhodobacter spheroides* is known as nontoxic LPS, with a meager capacity for activating susceptible cells in vitro and causing toxic changes in vivo (6,7). Thus LPS molecules derived from different species of gram-negative bacteria bear different structural features within their lipid A and oligosaccharide portions. Accordingly, LPS can induce either extreme vascular complications or transient episodes of fever and hypotension or be innocuous toward the host. These differences imply that recognition of LPS molecules by putative membrane receptor(s) on susceptible cells and subsequent intracellular signaling pathways are dependent on the structure of LPS. Importantly, such a recognition of LPS by a putative receptor must involve some common structural determinants because the precursors of lipid A, lipid IVa, and lipid X and its analogs, can block interaction of LPS with the membrane of susceptible cells (4). Similarly, nontoxic forms of LPS (e.g., those derived from *R. spheroides*) can interfere with interaction of the toxic form of LPS with susceptible cells (8). Clearly, the recognition specificity and signaling function of a putative LPS receptor on the membrane of susceptible cells hold the key to the most salient features of LPS-induced stimulus–response coupling.

II. LPS Activates Susceptible Cells via a Receptor-Mediated Mechanism: The Role of CD14

Not all cells in the human body are susceptible to LPS. Primarily, monocytes, macrophages, polymorphonuclear leukocytes, B-cells, endothelial cells, and platelets are activated by LPS. Among them, cells of the myelomonocytic lineage express CD14, a 55-kDa glycosylphosphatidylinositol (GPI)-anchored membrane protein that serves as a receptor for LPS in complex with a soluble (serum-derived) LPS-binding protein (LBP) (9,10). However, cells that do not express detectable CD14, such as pre-B-cell lines and endothelial cells, are also susceptible to LPS (11,12). When the murine pre-B-cell line 70Z/3 was transfected with cDNA encoding CD14 to express the GPI-linked membrane form of CD14, LPS sensitivity of the transfected cells increased 10,000-fold (12). We observed that the cell-free pool of CD14 present in human serum can render CD14-negative endothelial cells at least 3000 times more sensitive to LPS (13). Our findings evolved from experiments conducted on primary cultures of human umbilical vein endothelial cells (HUVEC), which do not express detectable CD14 on their surface. Such cells responded to ng/mL concentrations of LPS when human serum was present as the source of CD14. The role of the cell-free soluble CD14 (sCD14) pool was established in our laboratory on the basis of a blocking effect of monoclonal antibodies, lack of an enhancing effect by human serum immunodepleted of sCD14, and finally by setting up a reconstituted serum-free system in which recombinant CD14 was used to render human endothelial cells susceptible to ng/mL concentrations of LPS (13). These experiments are extending observations made in other laboratories on bovine endothelial cells in terms of LPS-induced cytotoxicity and on human endothelial cells in which expression of endothelial leukocyte adhesion molecule (ELAM-1) was measured (14). Importantly, a reconstituted, serum-free system has allowed us to demonstrate that LPB present in serum is not needed for sCD14-mediated activation of human endothelial cells by LPS. We interpret these results as indicating that sCD14 facilitates recognition of LPS by a putative, signal-generating receptor that is more responsive to LPS when the LPS is presented in conjunction with CD14.

The significance of the enhancing role of membrane-bound or cell-free pool of CD14 in rendering susceptible cells more responsive to LPS transcends the problem of vascular lesions mediating septic shock and DIC. Expression of the human immunodeficiency virus type 1 (HIV-1) in monocytic cell lines was stimulated by LPS in conjunction with CD14 (15). In view of a recent report that human liver sinusoidal endothelial cells can harbor HIV in vitro (16), LPS enhanced by sCD14 can play a role in the activation of HIV therein.

The obvious question arises: Where does sCD14 come from to enhance the LPS effect on endothelial cells? It may be shed from monocytic cells via proteolysis and/or phosphatidylinositol-specific phospholipase C (17,18). Interestingly, the level of circulating sCD14 in blood is increased in patients with sepsis (19). Thus the novel mechanism of endothelial activation by LPS in conjunction with sCD14 provides a framework for amplification of LPS-induced vascular injury in septic shock.

III. Signal Transducers Activated by LPS

Although LPS is a pleotropic stimulator of susceptible cells derived from different lineages (monocytes/macrophages, B-cells, endothelial cells, and platelets), signal-transducing pathways are not clearly delineated. The major obstacle is the lack of knowledge about the identity of a putative signal-generating LPS receptor recognizing LPS in conjunction with CD14. CD14 anchored to the membrane of cells of myelomonocytic lineage via GPI lacks the intracellular domain usually involved in signal transduction (20). Interestingly, antibodies against CD14 and other GPI-anchored proteins induce activation of leukocytes (21). Moreover, analysis of immunoprecipitates of CD14 and other GPI-anchored molecules in solubilized membrane preparations demonstrated protein tyrosine kinase activity (22). This implies that CD14 in human peripheral monocytes can be functionally associated with membrane protein(s) possessing intrinsic protein tyrosine kinase (PTK) activity. The identity of such a kinase coprecipitated with CD14 in monocytes has not been established. In human T-cells other GPI-anchored proteins (CD59, CD55, and CD48) coprecipitate with p56^{lck}, a protein tyrosine kinase related to *Src* (22). LPS induces protein tyrosine phosphorylation in the RAW 269.7 murine macrophage cell line and in peritoneal murine macrophages (23).

Herbimycin A, an inhibitor of PTK, also inhibited LPS-stimulated tyrosine phosphorylation and release of arachidonic acid metabolites. Subsequently, two tyrosine phosphoproteins, of 41 and 44 kDa, were identified as isoforms of mitogen-activated protein (MAP) kinase (24). MAP kinase belongs to the group of extracellular signal-regulated kinases (ERKs) involved in cell proliferation, transcription, and secretion (25). MAP kinase phosphorylates and activates the 90-kDa ribosomal S6 protein kinase, the *c-jun* transcription factor, and cytosolic phospholipase A_2 (26–28). The latter is involved in generation of arachidonic acid metabolites, prostaglandins, and leukotrienes, known to be increased in monocytes and macrophages in response to LPS (2). Interestingly, there is no compelling evidence of activation of phospholipase C breaking down phosphatidylinositol in all LPS-responsive cells, including monocytes, macrophages, and platelets (23,29,30).

We have found that human platelets are directly stimulated by endotoxic lipid A of *Salmonella minnesota* Re595 through activation of protein kinase C without the apparent involvement of phospholipase C and cyclooxygenase (30). Also, affinity-purified protein kinase C from human platelets is activated in a cell-free system by endotoxic lipid A (31). Using a chromatographically purified triacyl monophosphoryl lipid A homolog of m/z 1097, we observed significant activation of protein kinase C in human platelets, although serotonin secretion and platelet aggregation were not evoked (32). Moreover, we established that hydroxymyristic acid attached to glucosamine at position 3 is essential for monophosphoryl lipid A activation of platelet protein kinase C (32).

In a murine macrophage cell line, LPS from *Salmonella abortus equi* potentiated the phosphorylation of a major substrate of protein kinase C (PKC), myristoylated alanine-rich kinase substrate (MARKS), in macrophages (33). However, in the human monocytic cell line THP-1, down-regulation of phorbol-sensitive PKC did not prevent LPS-induced activation of NF-κB (34). This finding is consistent with a similar study conducted by Vincenti and colleagues on murine RAW 264 (35). THP-1 cells express at least four isoforms of PKC: α, β, ϵ, and ζ, and phorbol-insensitive isoforms can potentially mediate LPS-induced signal transduction (F. Carlisle and J. Hawiger, unpublished data).

Nitric oxide, a potent activator of soluble guanylate cyclase, is generated in macrophages and bovine endothelial cells in response to LPS (36, 37). Particularly, endothelial cell generation of NO is rapid (within 1 min);

the NO-generating enzyme is NO synthase. In endothelial cells and neurons, a constitutive NO synthase is activated transiently by agonists that elevate intracellular calcium concentrations and promote the binding of calmodulin. In contrast, inducible NO synthase activity in macrophages is progressively increased after stimulation of the cells with cytokines and LPS (38,39). Brain NOS is phosphorylated by cyclic adenosine monophosphate (AMP)-dependent protein kinase (PKA), PKC, and calcium/calmodulin protein kinase II at different phosphorylation sites. PKC drastically reduces NO synthase activity (40). Activation of NO synthase in endothelial cells would require the opening of calcium channels to allow the rise needed for binding to calmodulin. Calmodulin would then activate NO synthase. The NO that is generated will react with iron at the active site of guanylyl cyclase, increasing its catalytic activity and raising the level of cyclic AMP.

This NO synthase pathway in endothelial cells is important for relaxation of vascular smooth muscle cells and decrease in blood pressure. In monocytes/macrophages NO plays an important role as a potent bactericidal molecule targeted toward intracellular pathogens (41). However, cyclic GMP generated by NO-activated guanylate cyclase is not likely to trigger a cascade of phosphorylation reactions via cyclic GMP kinase. So far, kinase inhibitors have been ineffective in blocking LPS-induced activation of NF-κB (34).

A growing number of intracellular signal transducers, such as cellular protooncogenes bearing protein tyrosine kinase activity and hterotrimeric and low-molecular-weight GTP binding proteins (G-proteins), are modified by lipidation and methyl esterification. Such modifications allow these proteins to be redistributed from cytosol to membrane. LPS induces activation of a heterotrimeric G-protein in the murine B lymphoma cell line WEHI-231, the P388D macrophage cell line (42), and in a human promyelocytic cell line, U937 (43). The gamma subunits of the heterotrimeric G-proteins can be methyl-esterified and isoprenylated with a geranylgeranyl group (44). Inhibitors of 3-hydroxy-3-methylglutanyl coenzyme A (HMG-CoA) reductase, such as mevinolin, inhibit protein isoprenylation by depleting cells of mevalonate (45). The mevinolin-treated murine pre-B lymphocyte 70Z/3 cell line was not responsive to LPS but retained responsiveness to phorbol ester in terms of activation of NF-κB (46).

Isoprenylated C-terminal cysteine residues are recognized by protein carboxyl methyltransferases. Therefore, inhibitors of these enzymes [e.g., 5-methyl thioadenosine (MAT)] can interfere with the posttranslational carboxyl methylation of signal transducers (47). MAT inhibited LPS-induced activation of NF-κB DNA binding and κ gene expression in murine pre-B 70Z/3 cells, suggesting the involvement of protein methylation reactions in the LPS signal transduction pathway (46).

LPS induces generation of reactive oxygen intermediates (ROIs) in the murine pre-B 70Z/3 cell line and the EBV-transformed human B-cell line (48). ROIs identified as O_2^- and H_2O_2 were detectable almost immediately after the addition of 100 ng/mL LPS to cell cultures. They increased progressively, reaching 5- to 10-fold higher levels at 60 min (48). The antioxidant, the pyrrolidine derivative of dithiocarbamate (PDTC), attenuated NF-κB activation in the LPS-stimulated murine pre-B-cell line 70Z/3. The role of ROIs and their enzymatic generation in LPS-induced signal transduction need to be elucidated.

IV. LPS-Induced Gene Expression

Interaction of LPS with susceptible cells induces rapid expression of gene products, including cytokines such as TNF-α, IL-1, IL-6, and IL-8; procoagulant molecules, including tissue factor and plasminogen activator inhibitor type 1 (PAI-1); cell adhesion molecules, including ELAM-1 and vascular cell adhesion molecule-1 (VCAM-1); and immunoglobulins in B-cell lines (for a review, see ref. 11 and 49 and the references therein). Importantly, LPS also induces expression of HIV in monocytic cell lines as well as expression of protooncogenes bearing protein tyrosine kinase activity, *hck* and *lyn*, and transcriptional activity, *myc* and *fos* (50–53).

The promoter regions of several LPS-inducible genes contain elements with consensus sequences recognized by NF-κB or AP-1 DNA binding proteins, or both. The human tumor necrosis factor gene promoter contains three NF-κB sites that are not involved in LPS-induced transcription of the TNF-α gene (54). In contrast, the human tissue factor gene promoter contains a 56-bp LPS-responsive element comprised of two segments, one containing distal and proximal AP-1 sites, and a second segment containing the NF-κB site (55). The distal AP-1 site had a role in LPS induction, whereas the proximal site was silent on the basis deletion endpoints and mutagenesis analyses. On the other hand, an NF-κB site was essential for the full transcriptional activity of the LPS-responsive element. Mackman and colleagues speculate that this element is a combinatorial unit that requires the presence of both the NF-κB site constituting "the core" and the distal AP-1 site stabilizing the binding of NF-κB proteins to the core (55).

The AP-1 proteins are encoded by two gene families: *fos* and *jun*. Their dimerization involves a leucine zipper and is presumably initiated by phosphorylation due to activation of kinase(s) by phorbol ester and other stimuli (56–59). Agents that activate adenylate cyclase, thereby raising the intracellular level of cyclic AMP, inhibit the expression of tissue factor induced by LPS in human peripheral blood monocytes (60). This inhibition observed between 1 and 4 h was attributed to the reduction of the rate of tissue factor factor gene transcription.

The characterization of the NF-κB/Rel transcription factor family evolved from the initial discovery of p65 and p50, which bind to DNA as a heterodimeric complex, followed by identification of a growing number of new members, such as *rel* B, *v-rel*, *c-rel* (p85), p49, *lyt* 10, and p50B (61). The aminoterminal "Rel" domain is shared among these proteins as well as with the *Drosophila* protein Dorsal. Different NF-κB/Rel proteins can form distinct complexes with presumably distinct recognition specificity for κB DNA elements in the promoter regions of cellular and viral genes (62–65).

Before NF-κB/Rel proteins can be translocated from cytosol to the nucleus, they are freed from inhibitory proteins termed IκBα/MAD3, IκBβ, pp40, IκBγ, I-*rel*, and the *bcl*-3 gene product (61,66–69). Dissociation of cytoplasmic NF-κB proteins from inhibitory molecules (IκB) was attributed to phosphorylation of IκB induced in a cell-free system by phorbol ester–activated protein kinase C or other kinases (70,71). However, until recently, the evidence that IκB proteins dissociate from NF-κB in vivo during cell activation was lacking. In particular, LPS-induced activation of NF-κB in monocytic and pre-B-cell line 70Z/3 was not characterized in terms of molecular events underlying activation and molecular composition of NF-κB complexes translocated to the nucleus. We demonstrated that LPS induces nuclear expression of c-Rel/p50 heterodimers as well as p50/p65 (NF-κB) κB DNA binding complexes in the human monocytic THP-1 cell line (34). While these complexes appeared in the nucleus, their level in cytosol decreased rapidly. However, 24 h after stimulation with LPS, there was a striking increase in the cytosolic levels of c-Rel, p105, and p50. This rise was preceded by an increase in the amount of mRNA during LPS activation of THP-1 cells. Because the genes encoding c-*rel* and p50/p105 contain κB elements in their promoter regions, our findings suggest the existence of an autoregulatory loop that is set in motion by LPS in monocytic THP-1 cells. However, the subsequent decrease in mRNA encoding c-*rel* and p50/p105, while these proteins remain elevated in the nucleus for at least 24 h, suggest instability of mRNA or inhibition of transcription of these genes. Our data are relevant to the observation that HIV late-terminal repeat-driven constructs are transcriptionally activated by LPS-induced NF-κB/Rel DNA binding (50). Importantly, our demonstration of LPS-induced activation of c-Rel complements a recent finding that c-Rel activates HIV late-terminal repeat-driven constructs (72).

LPS activation of THP-1 cells leads to phosphorylation of inhibitory protein IκBα/MAD3, a rapid increase in IκBα/MAD3 mRNA, and an increase in IκBα/MAD3 protein by 2 h (34). This is the first evidence of IκBα/MAD3 phosphorylation by a biological agonist in an intact cell system. IκBα/

MAD3 bears a consensus PKC phosphorylation site in the C-terminal domain (67). However, phorbol-sensitive protein kinase C isozymes do not appear to be involved in human monocytic THP-1 cells and murine RAW 264 macrophages (34,35). We identified PKC isozymes α, β, ϵ, and ζ in THP-1 cells and established that only some of them are down-regulated by prolonged stimulation with phorbol ester (F. Carlisle and J. Hawiger, manuscript in preparation). The role of the PMA-insensitive protein kinase C isozymes in NF-κB activation is being investigated. Somatic cell genetic approaches were applied to delineate signal-transducing pathways linked to activation of the κ gene in the pre-B cell line 70Z/3 (73). The mutant IB8, with an abnormally slow response to LPS, showed an almost complete loss of activation of NF-κB, although its cytosolic levels were normal.

In primary cultures of HUVEC, LPS-induced activation of NF-κB results in nuclear translocation of at least two distinct DNA-binding complexes. One is composed of the p50/p65 heterodimer and the other is made of a p50 homodimer (8). This pattern of activation is different from that characterized by us in the human monocytic THP-1 cell line, implying distinct activation pathways. Again, phorbol ester–sensitive isoforms of protein kinase C appear not to be involved, although other PKC isozymes can participate in LPS-induced signal transduction.

V. Epilogue

Lipopolysaccharide-induced signal transduction and gene expression play pivotal roles in the molecular mechanism of vascular complications of septic shock and in activation of HIV harbored in human monocytic cells. Advances in understanding LPS-stimulated activation of functionally important genes encoding cytokines, procoagulant molecules, and cell adhesion molecules require precise dissection of signal-transducing pathways leading to activation of transcriptional factors. Although we gained significant insight into this problem through analysis of inhibitory molecule IκBα/MAD3, more work lies ahead to identify kinase(s) involved in phosphorylation of IκBα/MAD3 and other κB inhibitors operational in LPS-responsive cells. Finally, unraveling the identity of a putative, signal-generating LPS receptor will undoubtedly cap efforts made in many laboratories to gain a handle on the fundamental step in LPS-induced vascular injury. It is likely that new and more effective therapeutic measures will evolve from such an analysis of the molecular mechanisms through which cells recognize LPS and respond to its signals.

Acknowledgments

The recent experimental work discussed in this chapter and conducted in my laboratory represents the efforts of my colleagues Margaret Read, Susan

Cordle, Radhika Donald, Ruth Ann Veach, Frederick Carlisle, and Cherlyn Carlisle. It was supported by the NIH NHLBI Merit Award HL-30647 and grants HL-30648 and T32-07186. The assistance of Carol Watler and Jane Kline in the preparation of this manuscript is also acknowledged.

References

1. Centers for Disease Control. Increase in national hospital discharge survey rates for septicemia: United States, 1979–1987. MMWR 1990; 39:31–34.
2. Morrison DC, Ryan JL. Endotoxins and disease mechanisms. Annu Rev Med 1987; 38:417–432.
3. Levin J, Poore TE, Young NS, et al. Gram-negative sepsis: detection of endotoxemia with the *Limulus* test. Ann Intern Med 1972; 76:1–7.
4. Raetz CRH. Biochemistry of endotoxins. Annu Rev Biochem 1990; 59:129–170.
5. Davis CE, Arnold K. Role of meningococcal endotoxin in meningococcal purpura. J Exp Med 1974; 140:159–171.
6. Takayama K, Qureshi N, Beutler B, Kirkland TN. Diphosphoryl lipid A from *Rhodopseudomonas sphaeroides* ATCC 17023 blocks induction of cachectin in macrophages by lipopolysaccharide. Infect Immun 1989; 57:1336–1338.
7. Golenbock DT, Hampton RY, Qureshi N, Takayama K, Raetz CRH. Lipid A-like molecules that antagonize the effects of endotoxins on human monocytes. J Biol Chem 1991; 266:19490–19498.
8. Jones MR, Cordle SR, Hawiger J. Human endothelial cell NF-κB DNA binding complex is activated by lipopolysaccharide (LPS). FASEB J 1992; 6:A1033.
9. Wright SD, Ramos RA, Tobias PS, Ulevitch RJ, Mathison JC. CD14, a receptor for complexes of lipopolysaccharide (LPS) and LPS binding protein. Science 1990; 249:1431–1433.
10. Wright SD, Ramos RA, Hermanowski-Vosatka S, Rockwell P, Detmers PA. Activation of the adhesive capacity of CR3 on neutrophils by endotoxin: dependence on lipopolysaccharide binding protein and CD14. J Exp Med 1991; 173:1281–1286.
11. Pober JS, Cotran RS. Cytokines and endothelial cell biology. Physiol Rev 1990; 70:427–451.
12. Lee JD, Kato K, Tobias PS, Kirkland TN, Ulevitch RJ. Transfection of CD14 into 70Z/3 cells dramatically enhances the sensitivity to complexes of lipopolysaccharide (LPS) and LPS binding protein. J Exp Med 1992; 175:1697–1705.
13. Read MA, Cordle SR, Veach RA, Carlisle CD, Hawiger J. Cell-free pool of CD14 mediates activation of NF-κB by lipopolysaccharide in human endothelial cells. Proc Natl Acad Sci USA 1993; 90:9887–9891.
14. Frey EA, Miller DS, Jahr TG, et al. Soluble CD14 participates in the response of cell to lipopolysaccharide. J Exp Med 1992; 176:1665–1671.
15. Bagasra O, Wright SD, Seshamma T, Oakes JW, Pomerantz RJ. CD14 is involved in control of human immunodeficiency virus type 1 expression in latently infected cells by lipopolysaccharide. Proc Natl Acad Sci USA 1992; 89:6285–6289.

16. Steffan A-M, Lafon M-E, Gendrault J-L, et al. Primary cultures of endothelial cells from the human liver sinusoid are permissive for human immunodeficiency virus type 1. Proc Natl Acad Sci USA 1992; 89:1582–1586.
17. Bazil V, Baudys M, Hilgert I, et al. Structural relationship between the soluble and membrane-bound forms of human monocyte surface glycoprotein CD14. Mol Immunol 1989; 26:657–662.
18. Bazil V, Strominger JL. Shedding as a mechanism of down-modulation of CD14 on stimulated human monocytes. J Immunol 1991; 147:1567–1574.
19. Krüger C, Schütt C, Obertacke U, et al. Serum CD14 levels in polytraumatized and severely burned patients. Clin Exp Immunol 1991; 85:297–301.
20. Simmons DL, Tan S, Tenen DG, Nicholson-Weller A, Seed B. Monocyte antigen CD14 is a phospholipid anchored membrane protein. Blood 1989; 73:284–289.
21. Low MG. The glycosyl-phosphatidylinositol anchor of membrane proteins. Biochim Biophys Acta 1989; 988:427–454.
22. Stefanová I, Horejsí V, Ansotegui IJ, Knapp W, Stockinger H. GPI-anchored cell-surface molecules complexed to protein tyrosine kinases. Science 1991; 254: 1016–1019.
23. Weinstein SL, Gold MR, DeFranco AL. Bacterial lipopolysaccharide stimulates protein tyrosine phosphorylation in macrophages. Proc Natl Acad Sci USA 1991; 88:4148–4152.
24. Weinstein SL, Sanghera JS, Lemke K, DeFranco AL, Pelech SL. Bacterial lipopolysaccharide induces tyrosine phosphorylation and activation of mitogen-activated protein kinases in macrophages. J Biol Chem 1992; 267:14955–14962.
25. Cobb MH, Boulton TG, Robbins DJ. Extracellular signal-regulated kinases: ERKs in progress. Cell Regul 1991; 2:965–978.
26. Sturgill TW, Wu J. Recent progress in characterization of protein kinase cascades for phosphorylation of ribosomal protein S6. Biochim Biophys Acta 1991; 1092:350–357.
27. Pulverer BJ, Kyriakis JM, Avruch J, Nikolakaki E, Woodgett JR. Phosphorylation of *c-jun* mediated by MAP kinases. Nature 1991; 353:670–674.
28. Lin LL, Wartmann M, Lin AY, Knopf JL, Seth A, Davis RJ. $cPLA_2$ is phosphorylated and activated by MAP kinase. Cell 1993; 72:269–278.
29. Wang J, Kester M, Dunn MJ. Involvement of a pertussis toxin-sensitive G-protein-coupled phospholipase A_2 in lipopolysaccharide-stimulated prostaglandin E_2 synthesis in cultured rat mesangial cells. Biochim Biophys Acta 1988; 963:429–435.
30. Grabarek J, Timmons S, Hawiger J. Modulation of human platelet protein kinase C by endotoxic lipid A. J Clin Invest 1988; 82:964–971.
31. Romano M, Hawiger J. Interaction of endotoxic lipid A and lipid X with purified human platelet protein kinase C. J Biol Chem 1990; 265:1765–1770.
32. Grabarek J, Her G-R, Reinhold VN, Hawiger J. Endotoxic lipid A interaction with human platelets. Structure–function analysis of lipid A homologs obtained from *Salmonella minnesota* Re595 lipopolysaccharide. J Biol Chem 1990; 265: 8117–8121.
33. Rosen A, Nairn AC, Greengard P, Cohn ZA, Aderem A. Bacterial lipopolysaccharide regulates the phosphorylation of the 68K protein kinase C substrate in macrophages. J Biol Chem 1989; 264:9118–9121.

34. Cordle SR, Donald R, Read MA, Hawiger J. Lipopolysaccharide induces phosphorylation of MAD3 and activation of c-Rel and related NF-κB proteins in human monocytic THP-1 cells. J Biol Chem 1993; 268:11803–11810.
35. Vincenti MP, Burrell TA, Taffet SM. Regulation of NK-κB activity in murine macrophages: effect of bacterial lipopolysaccharide and phorbol ester. J Cell Physiol 1992; 150:204–213.
36. Iyengar R, Stuehr DJ, Marletta AM. Macrophage synthesis of nitrite, nitrate, and *N*-nitrosamines: precursors and role of the respiratory burst. Proc Natl Acad Sci USA 1987; 84:6369–6373.
37. Salvemini D, Korbut R, Änggård E, Vane J. Immediate release of a nitric oxide-like factor from bovine aortic endothelial cells by *Escherichia coli* lipopolysaccharide. Proc Natl Acad Sci USA 1990; 87:2593–2597.
38. Xie Q-W, Cho HJ, Calaycay J, et al. Cloning and characterization of inducible nitric oxide synthase from mouse macrophages. Science 1992; 256:225–228.
39. Stuehr DJ, Cho HJ, Kwon NS, Weise MF, Nathan CF. Purification and characterization of the cytokine-induced macrophage nitric oxide synthase: an FAD- and FMN-containing flavoprotein. Proc Natl Acad Sci USA 1991; 88:7773–7777.
40. Bredt DS, Ferris CD, Snyder SH. Nitric oxide synthase regulatory sites. Phosphorylation by cyclic AMP-dependent protein kinase, protein kinase C, and calcium/calmodulin protein kinase; identification of flavin and calmodulin binding sites. J Biol Chem 1992; 267:10976–10981.
41. Hibbs JB, Vavrin Z, Taintor RR. L-arginine is required for expression of the activated macrophage effector mechanism causing selective metabolic inhibition in target cells. J Immunol 1987; 138:550–565.
42. Jakway JP, DeFranco AL. Pertussis toxin inhibition of B cell and macrophage responses to bacterial lipopolysaccharide. Science 1986; 234:743–746.
43. Daniel-Issakani S, Spiegel AM, Strulovici B. Lipopolysaccharide response is linked to the GTP binding protein, Gi_2, in the promonocytic cell line U937. J Biol Chem 1989; 264:20240–20247.
44. Yamane HK, Farnsworth CC, Xie H, et al. Brain G protein γ subunits contain an all-*trans*-geranylgeranyl-cysteine methyl ester at their carboxyl termini. Proc Natl Acad Sci USA 1990; 87:5868–5872.
45. Sinensky M, Beck LA, Leonard S, Evans R. Differential inhibitory effects of lovastatin on protein isoprenylation and sterol synthesis. J Biol Chem 1990; 265:19937–19941.
46. Law RE, Stimmel JB, Damore MA, Carter C, Clarke S, Wall R. Lipopolysaccharide-induced NF-κB activation in mouse 70Z/3 pre-B lymphocytes is inhibited by mevinolin and 5'methylthioadenosine: roles of protein isoprenylation and carboxyl methylation reactions. Mol Cell Biol 1992; 12:103–111.
47. Smith DS, King CS, Pearson E, Gittinger CK, Landreth GE. Selective inhibition of nerve growth factor-stimulated protein kinases by K-252a and 5'-*S*-methyladenosine in PC12 cells. J Neurochem 1989; 53:800–806.
48. Schreck R, Meier B, Männel DN, Dröge W, Baeuerle PA. Dithiocarbamates as potent inhibitors of NF-κB activation in intact cells. J Exp Med 1992; 175:1181–1194.

49. Raetz CR, Ulevitch RJ, Wright SD, Sibley CH, Ding A, Mathan CF. Gram-negative endotoxin: an extraordinary lipid with profound effects on eukaryotic signal transduction. FASEB J 1991; 5:2652–2660.
50. Pomerantz RJ, Feinberg MB, Trono D, Baltimore D. Lipopolysaccharide is a potent monocyte/macrophage-specific stimulator of human immunodeficiency virus type 1 expression. J Exp Med 1990; 172:253–261.
51. Boulet I, Ralph S, Stanley E, et al. Lipopolysaccharide- and interferon-gamma-induced expression of *hck* and *lyn* tyrosine kinases in murine bone marrow-derived macrophages. Oncogene 1992; 7:703–710.
52. Introna M, Hamilton TA, Kaufman RE, Adams DO, Bast RC Jr. Treatment of murine peritoneal macrophages with bacterial lipopolysaccharide alters expression of c-Fos and c-Myc oncogenes. J Immunol 1986; 137:2711–2715.
53. Ziegler SF, Wilson CB, Perlmutter RM. Augmented expression of a myeloid-specific protein tyrosine kinase gene (*hck*) after macrophage activation. J Exp Med 1988; 168:1801–1810.
54. Goldfeld AE, Doyle C, Maniatis T. Human tumor necrosis factor alpha gene regulation by virus and lipopolysaccharide. Proc Natl Acad Sci USA 1990; 87: 9769–9773.
55. Mackman N, Brand K, Edgington TS. Lipopolysaccharide-mediated transcriptional activation of the human tissue factor gene in THP-1 monocytic cells requires both activator protein 1 and nuclear factor κB binding sites. J Exp Med 1991; 174:1517–1526.
56. Lee W, Haslinger A, Karin M, Tjian R. Activation of transcription by two factors that bind promoter and enhancer sequences of the human metallothionein gene and SV40. Nature 1987; 325:368–372.
57. Angel P, Imagawa M, Chiu R, et al. Phorbol ester–inducible genes contain a common *cis* element recognized by a TPA-modulated *trans*-acting factor. Cell 1987; 49:729–739.
58. Gentz R, Rauscher FJ III, Abate C, Curran T. Parallel association of Fos and Jun leucine zippers juxtaposes DNA binding domains. Science 1989; 243:1695–1699.
59. O'Shea EK, Rutkowski R, Kim PS. Evidence that the leucine zipper is a coiled coil. Science 1989; 243:538–542.
60. Ollivier V, Houssaye S, Ternisien C, et al. Endotoxin-induced tissue factor messenger RNA in human monocytes is negatively regulated by a cyclic AMP–dependent mechanism. Blood 1993; 81:973–979.
61. Baeuerle PA. The inducible transcription activator NF-κB: regulation by distinct protein subunits. Biochim Biophys Acta 1991; 1072:63–80.
62. Kieran M, Blank V, Logeat F, et al. The DNA binding subunit of NF-κB is identical to factor KBF1 and homologous to the *rel* oncogene product. Cell 1990; 62:1007–1018.
63. Ghosh S, Gifford AM, Riviere LR, Tempst P, Nolan GP, Baltimore D. Cloning of the p50 DNA binding subunit of NF-κB: Homology to *rel* and *dorsal.* Cell 1990; 62:1019–1029.
64. Nolan GP, Ghosh S, Liou H-C, Tempst P, Baltimore D. DNA binding and IκB inhibition of the cloned p65 subunit of NF-κB, a *rel*-related polypeptide. Cell 1991; 64:961–969.

65. Schmid RM, Perkins ND, Duckett CS, Andrews PC, Nabel GJ. Cloning of an NF-κB subunit which stimulates HIV transcription in synergy with p65. Nature 1991; 352:733–736.
66. Davis N, Ghosh S, Simmons DL, et al. Rel-associated pp40: an inhibitor of the Rel family of transcription factors. Science 1991; 253:1268–1271.
67. Haskill S, Beg AA, Tompkins SM, et al. Characterization of an immediate-early gene induced in adherent monocytes that encodes IκB-like activity. Cell 1991; 65:1281–1289.
68. Hatada EN, Nieters A, Wulczyn FG, et al. The ankyrin repeat domains of the NF-κB precursor p105 and the protooncogene *bcl-3* act as specific inhibitors of NF-κB DNA binding. Proc Natl Acad Sci USA 1992; 89:2489–2493.
69. Inoue J-I, Kerr LD, Kakizuka A, Verma IM. IκBγ, a 70 kd protein identical to the C-terminal half of p110 NF-κB: a new member of the IκB family. Cell 1992; 68:1109–1120.
70. Shirakawa F, Mizel SB. In vitro activation and nuclear translocation of NF-κB catalyzed by cyclic AMP-dependent protein kinase and protein kinase C. Mol Cell Biol 1989; 9:2424–2430.
71. Ghosh S, Baltimore D. Activation in vitro of NF-κB by phosphorylation of its inhibitor IκB. Nature 1990; 344:678–682.
72. Muchardt C, Seeler J-S, Nirula A, Shurland D-L, Gaynor RB. Regulation of human immunodeficiency virus enhancer function by PRDII-BF1 and c-*rel* gene products. J Virol 1992; 66:244–250.
73. Rooney JW, Emery DW, Sibley CH. Slow response variant of the B lymphoma 70Z/3 defective in LPS activation of NF-κB. Immunogenetics 1990; 31:65–72.

5

Effects of Endotoxin on Lung Endothelium

MARGARET A. READ

Brigham and Women's Hospital
Boston, Massachusetts

BARBARA O. MEYRICK

Vanderbilt University School of Medicine
Nashville, Tennessee

I. Introduction

One of the leading causes of the adult respiratory distress syndrome (ARDS) is gram-negative endotoxemia. This syndrome is characterized by early microvascular injury, resulting in bilateral pulmonary edema without right heart failure. Increased leakage of radiolabeled tracers from the lung's vasculature (1), increased plasma levels of von Willebrand antigen (2), and plasminogen activator inhibitor (3) have been used as markers of endothelial perturbation in this syndrome. The mechanism for the production of the endotoxin-induced endothelial changes is not fully understood. Suggested mechanisms include direct endothelial perturbation through intracellular generation of free radicals, alterations in levels of cyclic nucleotides, or imbalance of the protease/antiprotease system. In addition, interactions between circulating leukocytes and endothelial cells, as well as activation of alveolar macrophages with release of mediators, such as tumor necrosis factor α, interferon, and interleukin-1β, have been suggested as key events to the onset of the endothelial changes. Whether an endotoxin receptor mediates these effects is not yet certain.

This chapter is a brief account of some of the effects of *Escherichia coli* endotoxin on the pulmonary endothelium. Included are data from whole-animal experiments, mainly sheep, and in vitro data collected in our laboratory using endothelial cells isolated from bovine and ovine main pulmonary artery and from the lung's microvasculature. The clinical importance of this area of research has produced a vast literature, and because of the abundance of data using nonpulmonary endothelium, particularly endothelium from human umbilical veins, much of this chapter focuses on what has been learned of endotoxin-induced endothelial injury from nonpulmonary sources. Where possible, we have made an effort to relate these findings to pulmonary endothelium.

II. Studies in Whole Animals

A. Hemodynamic Studies

The importance of endotoxin to the development of ARDS has been documented in normal adult volunteers where administration of low-dose intravenous endotoxin (ca. 4 ng/kg) results in acute leukopenia, fever, increased lung permeability (4), fibrinolysis (5), and release of inflammatory cytokines (6,7), all features of the syndrome. A number of animals have been used as models of ARDS resulting from endotoxemia (e.g., Refs. 8 to 12). Those studies demonstrated that not all species are equally susceptible to endotoxin-induced pulmonary injury. For example, the rat requires massive dosages of endotoxin (ca. 2 mg/kg intraperitoneally) to elicit a response—early transient increase in lung microvascular permeability (11), marked peripheral blood leukopenia, sequestration of neutrophils in the lung and their later migration into alveoli (13)—while other animals, including sheep, pig, and human, are particularly sensitive (12).

Endotoxemia in sheep has been used extensively as a model of acute lung injury and ARDS, and mimics many of the pulmonary changes seen in patients with ARDS (12). In that model, a single intravenous infusion of endotoxin (ca. 0.5 μg/kg over a 20-min period) causes a two-phase response in the pulmonary vasculature. The early response includes peripheral blood leukopenia, fever, and a transient rise in pulmonary artery pressure. This is followed by a sustained increase in pulmonary vascular permeability, indicating a major breach, and alterations, in the endothelial layer (14,15). Markers of endothelial perturbation in this model include increased levels of various endothelial-derived prostanoids [e.g., prostacyclin (PGI_2) and prostaglandin E_2 (PGE_2)] in lung lymph (12).

B. Structural Studies

Morphological changes are also apparent in the lung of patients with ARDS and in animal models during endotoxemia. Examination of autopsy and lung biopsy tissue from patients with ARDS shows leukocyte accumulation in the lung, diffuse endothelial injury, and pulmonary edema (16). In sheep, the earliest change seen, by 15 min after the start of endotoxin infusion, is marked and diffuse sequestration of both granulocytes and lymphocytes (17,18) in the pulmonary microcirculation (Fig. 1). Many of the sequestered granulocytes show evidence of degranulation (Fig. 2). Whether this accumulation of leukocytes is the result of increased adhesive properties of the leukocyte and/or endothelium (19), or of changes in leukocyte rigidity (20, 21), or both, is currently not certain. In addition, a special population of pulmonary intravascular mononuclear cells becomes more evident (Fig. 1), but whether this is the result of increased cell number of cell hypertrophy is not known (17,22). Further evidence of endothelial alteration is apparent by 30 min when localized areas of interstitial edema are found, although the structure of the adjacent endothelium appears normal. Additionally, from this time, cell profiles suggesting emigration of both granulocytes and lymphocytes across the endothelial layer and into the interstitial space are found. From 1 h, localized endothelial injury is more severe and is apparent as in-

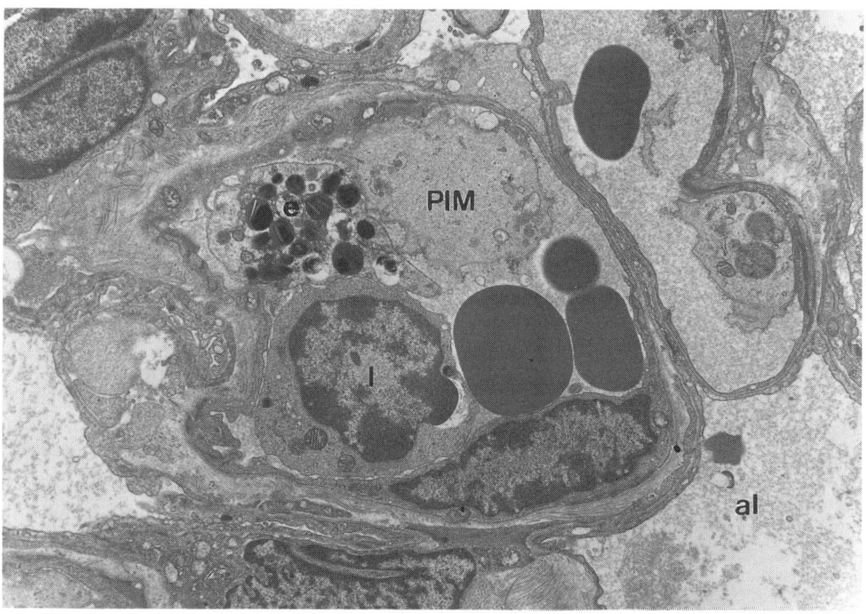

Figure 1 Electron micrograph of sheep lung 30 min following the start of endotoxin infusion. The pulmonary capillary at center contains profiles of an eosinophil (e), lymphocytes (l), and pulmonary intravascular macrophage (PIM). Part of a PIM is also seen in a second capillary. Edema is present in alveoli (al). ×5200.

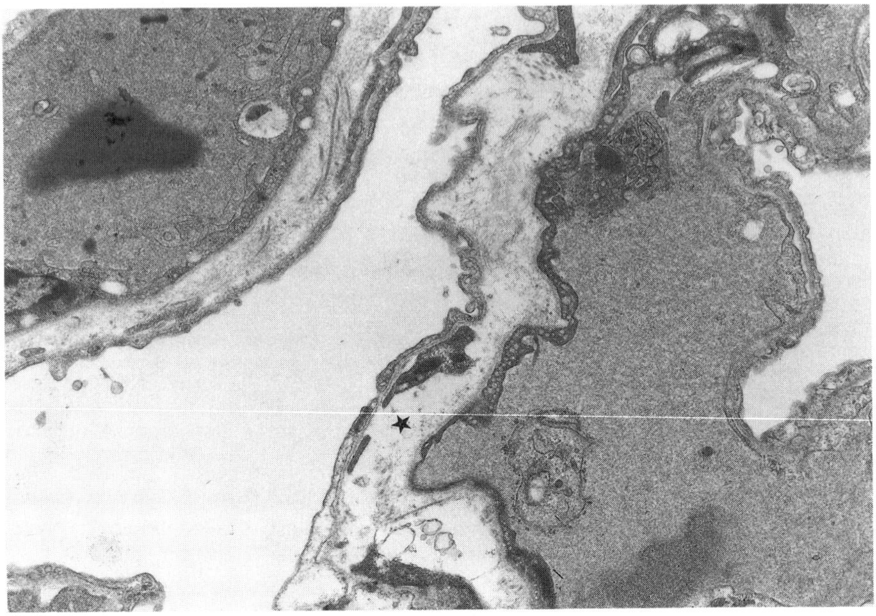

Figure 2 Electron micrograph of sheep lung 2 h following the start of endotoxin infusion. The endothelial cells of the capillaries are more electron dense than normal and contain numerous pinocytic vesicles. Several free neutrophil granules are present in the capillary lumen to the left. The interstitial space (*) is widened by edema. A break is apparent in the alveolar epithelial lining at the arrow. ×10,900.

creased electron density of the cell with increased numbers of pinocytic vesicles (Fig. 2).

Thus clear evidence of endothelial alterations are found during gram-negative endotoxemia. First, the microvascular endothelial integrity is breached, in that interstitial edema and leukocyte emigration is noted. This is followed by increased release of various endothelial-derived prostanoids and structural evidence of endothelial injury. These changes are the forerunners of the increase in pulmonary microvascular permeability.

The differences in responsiveness to endotoxin by various species are not understood. One hypothesis is that sheep, cows, and pigs are particularly sensitive to endotoxin because of their complement of resident pulmonary intravascular macrophages (22). However, even if those cells contribute to the endothelial changes, they would seem not to be the only player, since recent studies also indicate the presence of a mononuclear cell analogous to the pulmonary intravascular macrophage in the "relatively resistant" rat lung following administration of endotoxin (23) as well as in human lung under conditions of systemic injury (24). Additionally, these cells would not explain the findings that exposure of cultured pulmonary artery endothelial cells to endotoxin leads to endothelial injury (25). What is certain from the available data is that an animal's response to endotoxin is complex and almost certainly involves responses to multiple mediators by a range of cell types, some resident within the lung and some from more distant organs. While understanding this mechanism is complicated in the whole animal, in vitro studies with known cell types allow some insights, although those studies have the obvious disadvantage of removing cells from their native environment.

III. Studies with Cultured Endothelium

Just as the response of animals to endotoxin differs in severity, endothelial cells in culture show a range of responses to endotoxin. For example, it has been demonstrated that endothelial cells from bovine aorta, bovine and ovine pulmonary artery (26,27), bovine mesenteric artery and vein (26), human kidney (28), and bovine brain microvasculature (29) are particularly sensitive since these cells undergo cytolysis in the presence of endotoxin. Other endothelial cells, such as those from human umbilical vein, human pulmonary artery, goat aorta, and vena cava (26), have been described as endotoxin insensitive on the basis that they do not lyse. However, we now know that endothelial cells, in general, are responsive to endotoxin, although differences in the severity of each response occurs. The differences seem to be species and organ specific, and more recent studies indicate that endotoxin responsiveness can also vary between different levels of the same organ (27).

A. Response of Lung Endothelium to Endotoxin

Several characteristic events have been reported following exposure of endothelial cells cultured from the main pulmonary artery of sheep and cows to endotoxin (25). First, the cells undergo localized retraction from their neighbors and consequent rounding up (Fig. 3); this is accompanied by increased release of PGI_2, PGE_2 (Fig. 4) (25), and $PGF_{2\alpha}$ (M. Read and B. Meyrick, unpublished observations) and is followed by release of lactate dehydrogenase (LDH) (Fig. 5) and eventual cell death. This cytolytic effect of endotoxin seems to be dependent on new protein synthesis since treatment with cycloheximide markedly enhances this response (26), while prostanoid release is strikingly attenuated (B. Meyrick, unpublished observations). Permeability studies of endothelial monolayers show a reduction in barrier function in response to endotoxin, in that tracer molecules—$[^3H]$water and $[^{125}I]$albumin—

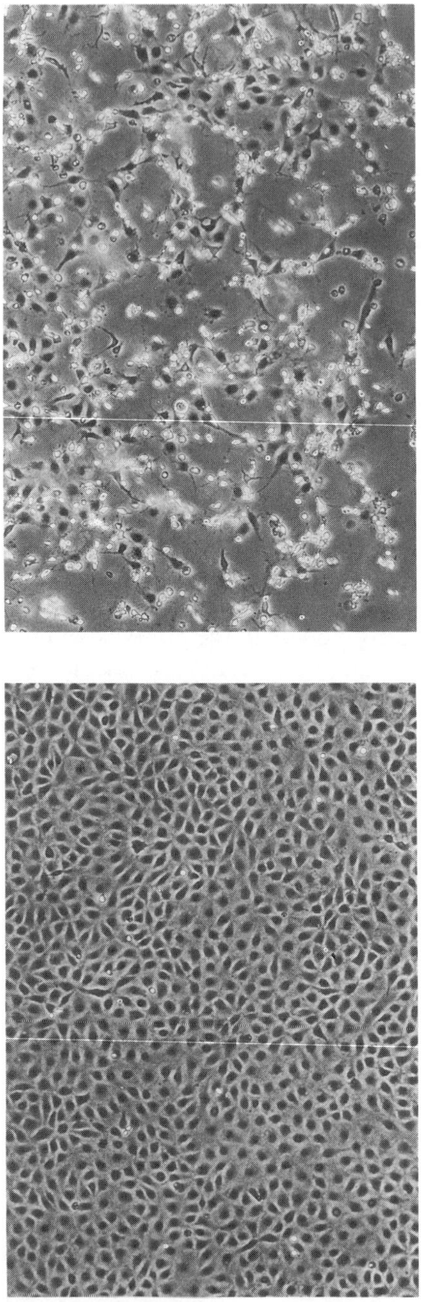

(A)

(B)

Figure 3 Phase-contrast micrographs of cultured bovine pulmonary artery endothelial cells: (A) 24 h of exposure to 0.1 μg/mL endotoxin; (B) control. ×175.

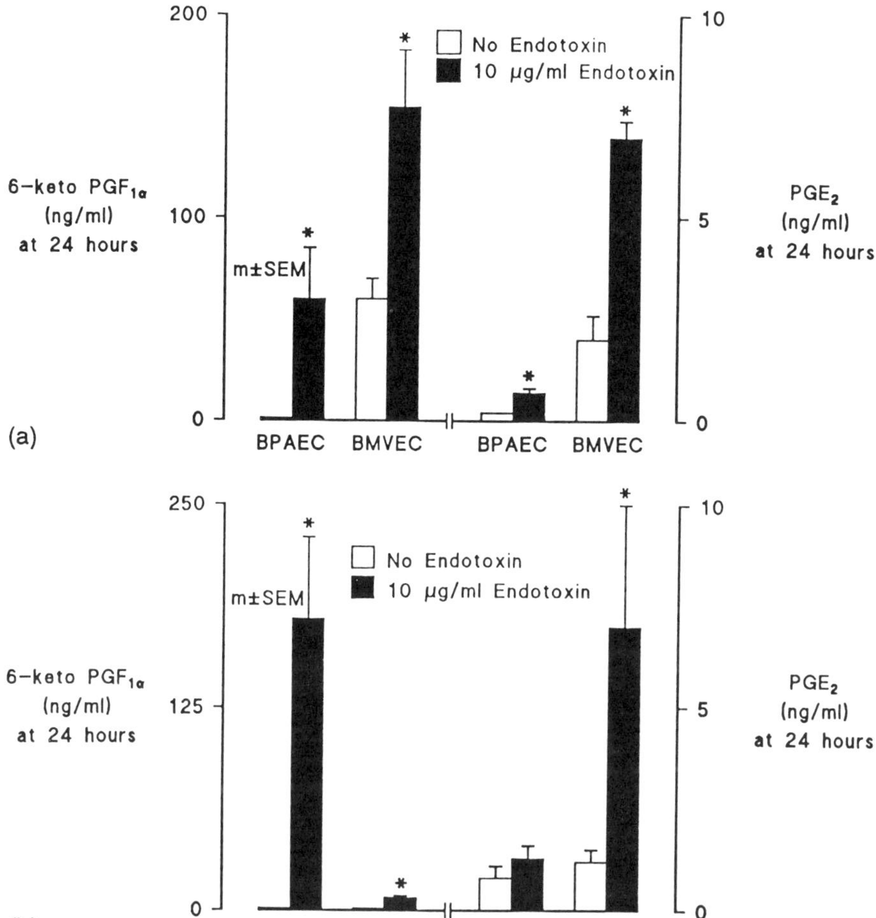

Figure 4 Endotoxin-stimulated 6-keto-$PGF_{1\alpha}$ and PGE_2 release from cultured (a) bovine pulmonary artery (BPAEC) and microvascular endothelial cells (BMVEC) and (b) sheep pulmonary artery (SPAEC) and microvascular endothelial cells (SMVEC). Data shown are for experiments of 24 h duration. Data are shown as mean ± SEM; $n = 6$; $*p < 0.05$. (Data from Ref. 27.)

equilibrate faster across endotoxin-treated endothelial monolayers than do controls (25). We have also shown that endotoxin alters the physical properties (a decrease in cytoplasmic movement in the presence of increased cytoplasmic fluidity) of endothelial cells using a nonoptical magnetometric method, following ingestion of iron oxide particles by the endothelium (30). These endotoxin-induced responses are dose dependent and are a direct effect of endotoxin since they occur in the absence of any other cell type.

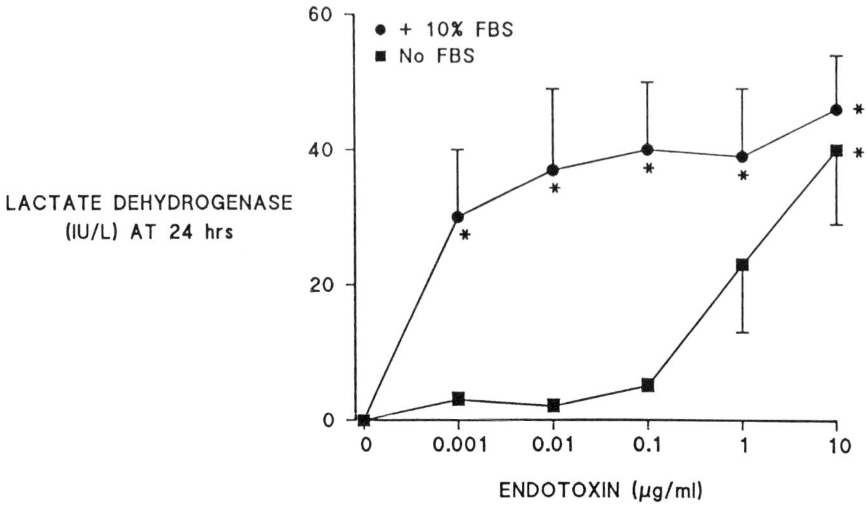

Figure 5 Endotoxin-stimulated lactate dehydrogenase (LDH) from cultured bovine pulmonary artery endothelial cells. In the presence of serum all doses of endotoxin caused increased release of LDH ($n = 15$). In the absence of serum, only the highest dose resulted in toxicity ($n = 10$). Data are shown as mean ± SEM; $*p < 0.05$. (Data from Ref. 25.)

While endotoxin-induced responses occur in the absence of serum, they are enhanced by the presence of serum (Fig. 5). As little as 1% fetal bovine serum is sufficient to potentiate the endotoxin-induced changes (31). In vitro studies with monocytes suggest that certain components of serum may be responsible for this enhancement. In the blood, endotoxin binds to a circulating protein called lipopolysaccharide-binding protein (LBP), which in turn binds to the CD14 receptor on the surface of monocytes (32). Unfortunately, the distribution of CD14 seems to be restricted to monocytic cells and granulocytes, and thus far, cultured endothelial cells have not been shown to express CD14 on their surface. More recent studies with bovine brain endothelial cells exposed to *Haemophilus influenzae* endotoxin in the presence of antibodies to CD14 suggest that a soluble form of CD14 present in serum may contribute to endotoxin-induced endothelial injury (29).

In contrast to cells isolated from the main pulmonary artery, cells from sheep and bovine lung periphery (and even from the same animal as the main pulmonary artery) show a rather different response to endotoxin (27). Peripheral lung microvascular endothelial cells exposed to endotoxin demonstrate increased release of both PGI_2 and PGE_2, with PGE_2 showing the most strikingly elevation (PGI_2 is the major prostanoid released from the main pulmonary artery endothelium) (Fig. 4). By phase-contrast microscopy, the microvascular cells show a transient retraction over the first 3 to 4 h of endotoxin exposure (Fig. 6) but do not undergo endotoxin-stimulated lysis as do the cells from the main pulmonary artery. These studies demonstrate that endothelial cells from different levels in the same organ maintain site-specific phenotypic differences in culture. These studies also may demonstrate that endothelial cells with different "sensitivities" to endotoxin are present within a single organ and, at least in the lung, their response is dependent on location.

Whether the response of the cultured microvascular endothelial cells to endotoxin mimics more closely the lung's vasculature in whole animals than those of the main pulmonary artery is not clear. Most structural studies in experimental animals have examined only the peripheral lung following endotoxin infusion. However, studies of intimal explants of bovine main pulmonary artery demonstrate that endothelial retraction, dilatation of the intercellular junctions, and cell death do occur in large pulmonary arteries in response to endotoxin (25). Additionally, our own studies demonstrate similar dilatations in the interendothelial junctions of muscular pulmonary arteries of sheep following endotoxin administration (Fig. 7). Thus the endothelial cells of both large and small pulmonary vessels do respond to endotoxin, but whether the response is different (e.g., the same adhesion proteins are induced, production of prostanoids is identical at each level) is not well documented. At present, there is no explanation for the differences in susceptibilities of endothelial cells at different sites and different species to endotoxin-induced cytotoxicity.

B. Response of Other Endothelial Cells to Endotoxin

Other endothelial cells from various species and organs that were previously considered unresponsive to endotoxin are now known to respond and undergo a number of functional alterations without evidence of injury. Endothelial cells that are not irretrievably damaged by endotoxin show up-regulation of a number of proteins, including tissue factor (33) and plasminogen activator inhibitor-1 (PAI-1) (34), cytokines [e.g., interleukins-1 (35), -6 (36), and -8 (37)], macrophage colony-stimulating factor (M-CSF) (38), granulocyte CSF (G-CSF) (39), granulocyte-monocyte CSF (GM-CSF) (39), and adhesion molecules, such as intercellular adhesion molecule-1 (ICAM-1), endothelial leukocyte adhesion molecule-1 (ELAM-1), and vascular adhesion molecule-1 (VCAM-1) (40). This altered phenotype, referred to as endothelial activation, may predispose the cells to indirect injury by leukocytes and cell-derived mediators, including cytokines and reactive oxygen species (41). Endotoxin not only induces expression of a number of cytokine genes directly within endothelial cells [e.g., interleukin-1 and interleukin-8 (35,37)] but may also increase the sensitivity of endothelial cells to the action of secon-

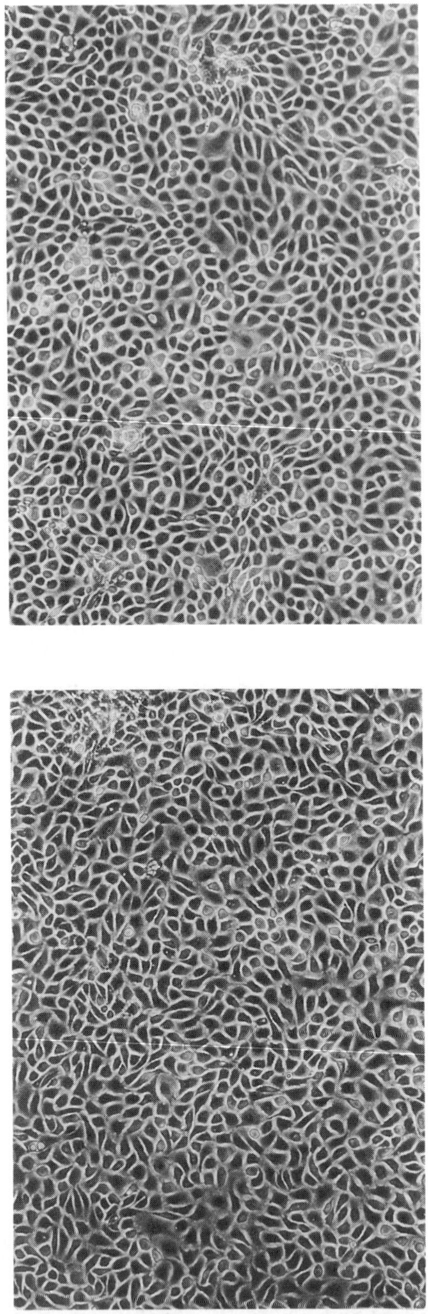

(A)

(B)

Figure 6 Phase-contrast micrographs of bovine microvascular endothelial cells exposed to endotoxin for 24 h: (A) control cells; (B) after 24 h. ×175.

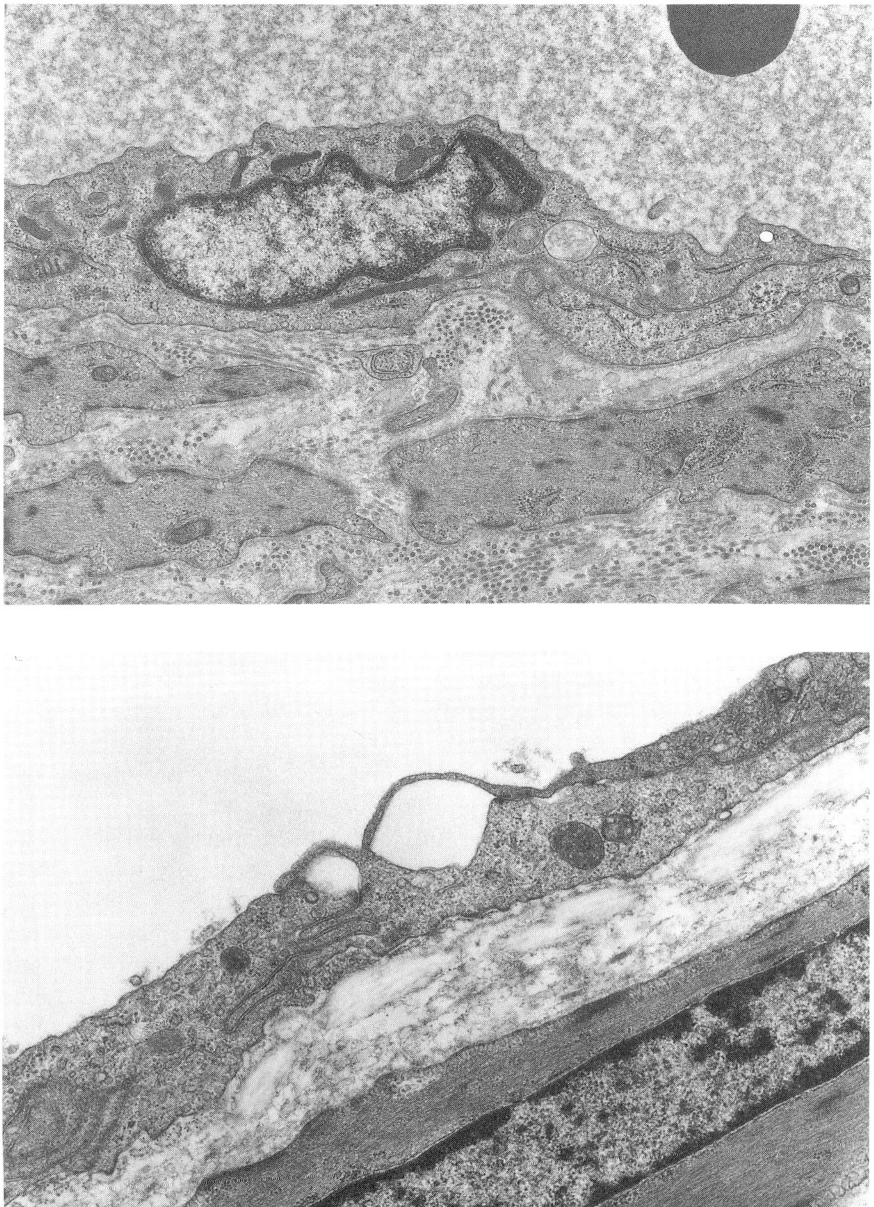

Figure 7 Electron micrographs showing intercellular junctions endothelial in (top) a control sheep muscular pulmonary artery and (bottom) following 60 min after the start of endotoxin infusion. The junction (top) is normal, whereas the junction (bottom) shows dilatations following endotoxin infusion. ×17,000.

dary cytokines and to injury by leukocytes. In many cases, IL-1β and tumor necrosis factor (TNF)-α are amplifiers of the response to endotoxin, and these mediators show a wide spectrum of overlapping biological activities on endothelial cells in culture (41).

A well-documented feature of endothelial activation results in the acquisition of procoagulant properties. Upon exposure to endotoxin, umbilical vein endothelial cells rapidly express tissue factor, a cofactor for the extrinsic coagulation pathway, on their surface (33). Induction of endothelial tissue factor expression is enhanced in the presence of lymphocytes, granulocytes (42), and monocytes (43), suggesting the importance of endothelial–leukocyte interactions in the endothelial response to endotoxin. In addition to tissue factor, endotoxin stimulates endothelial cells from a number of origins to express PAI-1, an inhibitor of fibrinolytic activity (34,44). Studies show that PAI-1 is released from the apical endothelial surface and suggest that localized formation of intravascular thrombi is promoted by decreased fibrinolytic activity at the endothelial surface (45). Thus endothelial cells from a wide range of tissues are responsive to endotoxin. "Responsivity" includes up-regulation of a number of proteins and cytokines by the endothelium, some of which result in enhanced leukocyte–endothelial interactions, some which alter the normal homeostasis of the clotting cascade, and still others (e.g., manganous superoxide dismutase) that may act to protect the cell from subsequent and/or additional damage (see below). Whether endotoxin also results in down-regulation of specific proteins and other mediators is not currently known.

Endotoxin also up-regulates pulmonary endothelial production of a number of lipid mediators, such as PGE_2, PGI_2, and $PGF_{2\alpha}$ (27,31). These mediators are known to affect the tone of the pulmonary vasculature, and alterations in their levels have been linked to hemodynamic changes in the endotoxin response (12). Recent studies also indicate that two other endothelial-derived agents that alter vascular tone are up-regulated by endotoxin: the potent pulmonary vasoconstrictor endothelin-1 (46–48) and the endothelial-derived relaxant factor (EDRF), nitric oxide (NO) (49–51). Whether these agents will be shown to play a role in the regulation of the lung's response to endotoxin, or vice versa, or whether these increases are independent of the response is not presently apparent.

C. Activation of Pulmonary Endothelium by Endotoxin

Although most evidence concerning endothelial activation has been obtained using endothelial cells cultured from human umbilical vein, recent studies in animal models of sepsis and cultured endothelial cells exposed to endotoxin suggest that similar alterations are induced in pulmonary endothelium. A prominent feature of endotoxin and cytokine-activated human umbilical vein endothelium is increased adherence of leukocytes, which is mediated by coordinated up-regulation of the leukocyte adhesion molecules ICAM-1, VCAM-1, and ELAM-1 (40). Our studies using ovine peripheral lung microvascular endothelial cell monolayers show that endotoxin enhances adherence and migration of lymphocytes and suggest that sequestration of lymphocytes in the pulmonary microcirculation following endotoxin infusion is due to increased adhesive properties of the endothelium (52). Whether known adhesion molecules are induced on ovine pulmonary endothelial cells is not yet certain, but experiments examining known lymphocyte adhesion molecules suggest that lymphocytes adhere to pulmonary endothelial cells by mechanisms unique to this cell type. Expression of adhesion molecules has also been examined in the lungs of animals. Studies in baboons made septic by intravenous administration of *E. coli* have revealed increased ELAM-1 expression in numerous vascular beds, with prominent expression in lung vessels, including capillaries, arterioles, arteries, and postcapillary venules (53). These data provide strong evidence that endotoxin both activates pulmonary microvascular endothelial cells and causes up-regulation of adhesion molecules in the lungs during sepsis. These concepts are addressed in greater detail in Chapter 11.

The finding that PAI-1 is up-regulated in endotoxin-stimulated umbilical vein endothelium also is consistent with findings in patients with ARDS and animal models of this syndrome. Abnormally high levels of PAI-1 are observed in the plasma of patients with sepsis and disseminating intravascular coagulopathy, and are thought to be derived mainly from endothelium (3). Recent in vivo studies have revealed that endotoxin treatment of mice results in increased PAI-1 gene expression, with striking early increases observed in the lung (54). Such studies provide further evidence that endotoxin-induced endothelial activation occurs in vivo and may contribute to pulmonary damage in sepsis.

IV. Mechanisms of Endotoxin-Induced Endothelial Injury

Although a wealth of information is available concerning the response of endothelial cells to endotoxin, the mechanism for these changes is not known. Studies to show receptor-mediated recognition are hampered by the lipid nature of the endotoxin molecule. However, it is known that endotoxin initiates many of the changes described above with just 15 min of exposure (26) and that these changes occur even at 4°C (B. Meyrick, unpublished observations). In addition, protein synthesis has been found to be critical to endotoxin-induced responses in endothelial cells (55).

Several candidates for endotoxin receptors have been identified (32,56–58), but none of these exhibit properties of a signal-generating molecule.

Indirect evidence for an endotoxin receptor on endothelial cells comes from studies which show that nontoxic forms of both lipopolysaccharide and its lipid A component block effects of toxic forms on this cell type (59–61). These nontoxic compounds are thought to antagonize the interaction of endotoxin with the cell surface and thereby inhibit the endotoxin-induced endothelial changes. Additionally, as mentioned above, serum factors (which may involve CD14 and LBP) enhance endotoxin-induced endothelial injury (27,29) and expression of cytokines and adhesion molecules (62). Thus evidence for the presence of an endotoxin receptor on endothelial cells is evolving, but the nature of this receptor and its relationship to serum proteins involved in the recognition of endotoxin by pulmonary and other endothelial cells remains a mystery. How endotoxin interacts with the cell surface is addressed in detail in Chapter 2.

A. Regulation of Signal Transduction and Gene Expression

Just as the identity of an endotoxin receptor on endothelium remains unknown, the mechanisms by which endotoxin generates signals that lead to increased gene transcription are poorly understood. Many endotoxin-inducible genes, including tissue factor (63), IL-1 (64), IL-6 (65), IL-8 (66), G-CSF (67), VCAM-1 (68), and ELAM-1 (69), contain elements in their promoter region that are recognized by the nuclear factor-κB (NF-κB) family of transcription factors (70). In particular, induction of the leukocyte adhesion molecules ELAM-1 (71) and VCAM-1 (72) gene transcription in human umbilical vein endothelial cells requires NF-κB sites. Endotoxin and other stimuli, including TNF-α and phorbol esters, trigger nuclear translocation and DNA binding of NF-κB proteins, presumably by stimulating phosphorylation of, and release from, an inhibitory molecule, IκB, which allows the movement of these proteins into the nucleus (70). Protein kinase C (PKC) has been shown to cause phosphorylation of NF-κB in vitro, but the role of PKC in endotoxin-induced signal transduction in endothelial cells is controversial. PKC inhibitors have been shown to block a number of endothelial responses to endotoxin, including expression of ICAM-1 (73) and enhanced granulocyte adherence to umbilical vein endothelial cells (74), and expression of the competence factor gene product, KC, in porcine aortic endothelial cells (75). These studies also showed that activation of PKC by phorbol esters mimicked the effects of endotoxin (73–75). However, endotoxin does not activate PKC in human umbilical vein endothelial cells, as evidenced by a lack of translocation of PKC from the cytosolic to membrane compartment, and NF-κB activation by endotoxin is not affected by PKC inhibition (76). Thus, although studies with PKC inhibitors indicate that PKC may be involved in some endotoxin-mediated endothelial responses, PKC-independent pathways may also be in use. More refined studies are needed to identify which isoenzymes of PKC are present in endothelial cells, and to clarify which are subject to inhibition by the various methods employed so far.

B. Cyclic Nucleotides

Cyclic nucleotides have been suggested to play a role in endotoxin-stimulated signal transduction pathways leading to endothelial injury. We have shown that cotreatment of cultured bovine pulmonary artery endothelial cells with the phosphodiesterase inhibitor 1-methyl-3-isobutylxanthine results in a threefold increase in intracellular cAMP and attenuation of endotoxin-stimulated prostanoid and LDH release (Fig. 8) (77). Addition of dibutyryl cAMP, however, only attenuated LDH release. Addition of either the adenylate cyclase activator forskolin or sodium nitroprusside (an agent known to upregulate cGMP generation) had no effect on the endotoxin response. Thus phosphodiesterase inhibition with subsequent increases in intracellular cAMP seems to protect endothelial cells from endotoxin-induced damage, and phosphodiesterase activity contributes to endotoxin-stimulated prostanoid release. Confirmation of these data comes from sheep studies following endotoxin infusion where increased concentrations of cAMP have been demonstrated in lung lymph (78). Additionally, the phosphodiesterase inhibitor aminophylline and the adenylate cyclase stimulator isoproteranol have been dem-

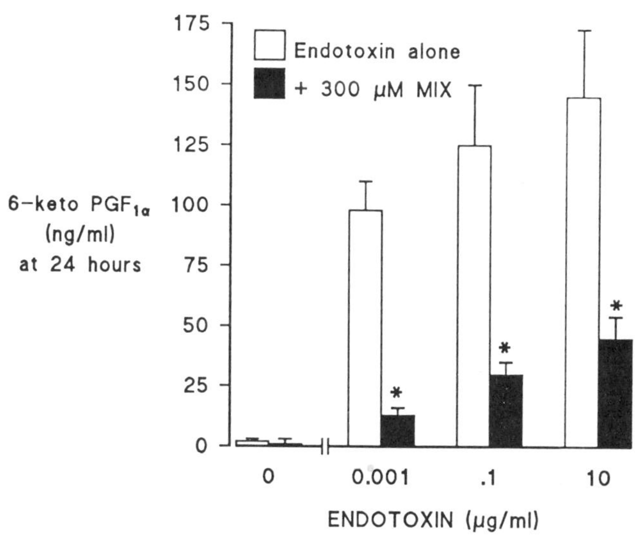

Figure 8 Effect of methyl isobutylxanthine (MIX) on endotoxin-stimulated release of prostacyclin (6-keto-$PGF_{1\alpha}$). Data are shown as mean ± SEM; $n = 7$; $*p < 0.05$. (Data from Ref. 77.)

onstrated to attenuate the lungs' response to endotoxin (79). More recently, direct administration of dibutyryl cAMP has been shown to attenuate the pulmonary vascular responses of endotoxemia (80). In that study cAMP was found to attenuate the endotoxin-stimulated increase in PGI_2 as well as the endotoxin-induced increase in lung lymph flow and protein clearance, again indicating that elevations of cAMP may serve to protect endothelium from injury. Thus studies in animals confirm data obtained from cultured endothelium, although the mechanism of protection by elevated cAMP is unknown.

C. Antiproteases

Elevated protease activity has also been linked to endotoxin-stimulated prostanoid release and endotoxin-induced cytolysis in cultured endothelium (81). A 15-min pretreatment with various antiproteases was shown to attenuate endotoxin-induced endothelial injury. Chloromethyl ketone (CK) derivatives, such as *N*-tosyl-L-lysine-CK (TLCK) (Fig. 9), *N*-tosyl-L-phenylalanine (TPCK), and methoxysuccinyl-Ala-Ala-Pro-Val (SPCK), all attenuated endotoxin-induced LDH and prostanoid release. Other antiproteases, such as *N*-α-p-tosyl-L-arginine-methyl ester hydrochloride (TAME), soybean trypsin inhibitor, and α_1-proteinase inhibitor, had a less striking effect. The protective

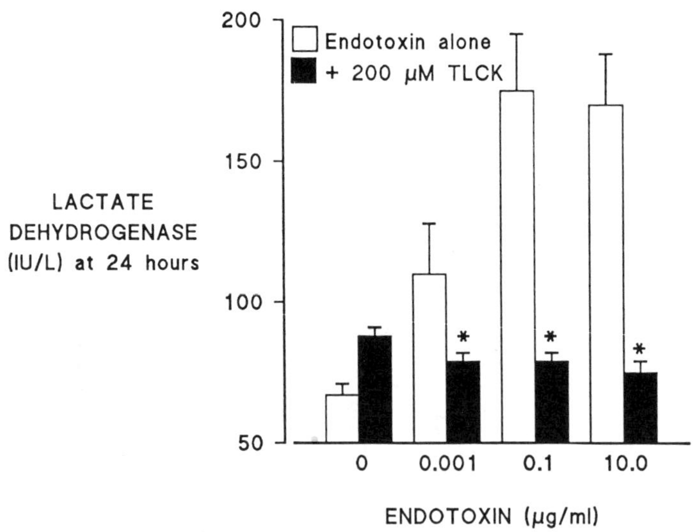

Figure 9 Effect of tosyl-lysyl chloromethyl ketone (TLCK) on endotoxin-stimulated lactate dehydrogenase release. Data are shown as mean ± SEM; $n = 6$; *$p < 0.05$. (Data from Ref. 81.)

effect exhibited by each of these agents was less pronounced in the absence of FBS or in the presence of heat-inactivated FBS, perhaps indicating that the endothelial cells, per se, as well as serum, contributed to the increased protease activity. The appearance of a 70 to 75-kDa serine protease has been reported in sheep lung lymph in several models of acute lung injury, including endotoxemia (82,83). Thus these data confirm the idea that endotoxin-induced endothelial injury may include an imbalance between proteases and normally circulating antiproteases.

D. Oxygen Radicals

The ability of the antiproteases to scavenge superoxide radicals suggests a possible role for toxic oxygen radicals in endotoxin-induced endothelial changes. Cultured endothelial cells from a number of organs are known to express constitutively a number of antioxidant enzymes: for example, manganous superoxide dismutase, copper/zinc superoxide dismutase, glutathione peroxidase, catalase and cytochrome *c* oxidase (84), and xanthine oxidase and xanthine dehydrogenase (85–87). In addition, endothelial cells have measurable basal concentrations of H_2O_2 (87).

Following exposure of bovine pulmonary artery endothelial cells to endotoxin, we have shown a striking increase in intracellular free-radical generation as determined by increased reduction of nitroblue tetrazolium, oxidation of 2′,7′-dichlorofluoroscein, and increased release of lipid peroxidation products (88). Measurements of intracellular free-radical generation by reduction of nitroblue tetrazolium showed an early increase in intracellular free radicals, with peak levels occurring by 15 min of endotoxin exposure (Fig. 10) (K.L. Brigham and B. Meyrick, unpublished observations). We further tested the notion that this elevation in intracellular oxygen radicals may contribute to endotoxin-stimulated changes by pretreating bovine pulmonary artery endothelial cells for 30 min with either of the free-radical scavengers dimethyl sulfoxide or dimethyl thiourea (Fig. 11). Both of these agents attenuated endotoxin-stimulated endothelial injury and prostanoid release (88). Pretreatment with the xanthine oxidase inhibitor allopurinol showed attenuation of only the cytotoxic effects of endotoxin (88). Each of these agents also led to a striking reduction in the endotoxin-stimulated generation of intracellular oxygen radicals and release of lipid peroxidation products (Fig. 12). Thus several lines of evidence suggest that the mechanism of endotoxin-induced endothelial injury involves intracellular generation of toxic oxygen species. However, our studies indicate that not all intracellular oxidant pathways are affected similarly (84). Studies in bovine aortic endothelial cells (87), in bovine pulmonary artery endothelial cells (86), and in rat lung microvascular endothelial cells (89) suggest specifically that endo-

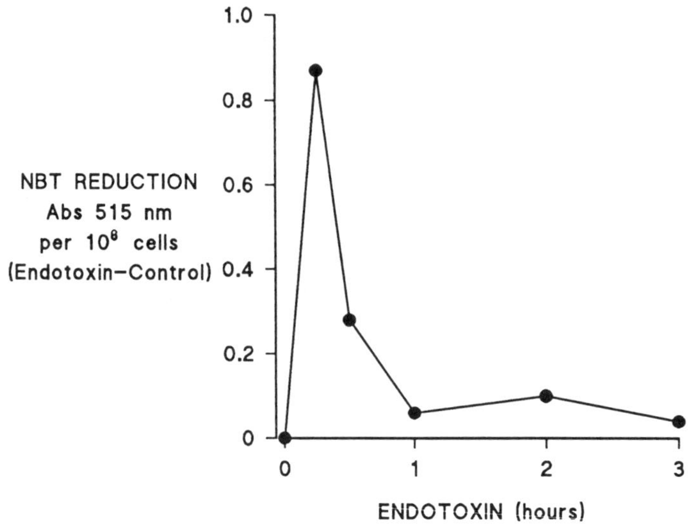

Figure 10 Endotoxin-stimulated (0.1 μg/mL) intracellular generation of free radicals in cultured bovine pulmonary endothelial cells over time. The data are presented as reduction of nitroblue tetrazolium per 10^6 cells.

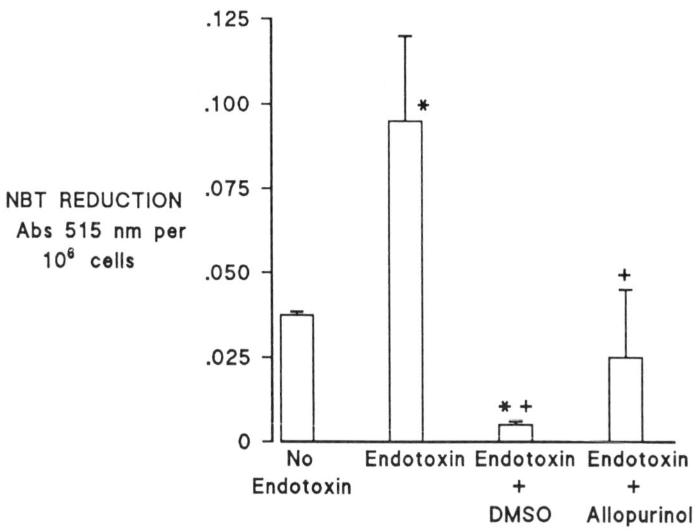

Figure 11 Attenuation of endotoxin-stimulated intracellular free-radical generation in bovine pulmonary artery endothelial cells by the free-radical scavenger dimethyl sulfoxide (DMSO) and the xanthine oxidase inhibitor allopurinol. The endothelial cells were exposed to endotoxin (0.1 μg/mL) for 15 min and the data are presented at reduction of nitroblue tetrazolium per 10^6 cells. Data are presented as mean ± SEM; $n = 6$; *$p < 0.05$ compared to controls; $+ = < 0.05$ compared to endotoxin. (Data from Ref. 88.)

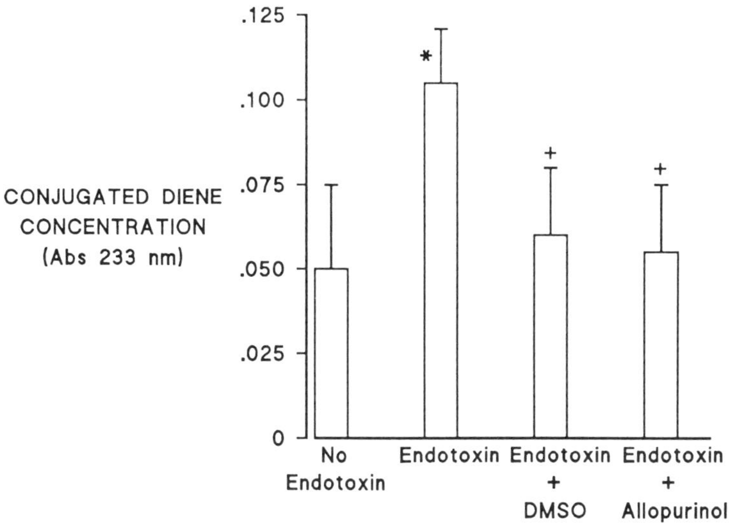

Figure 12 Attenuation of endotoxin-stimulated (0.1 μg/mL) release of conjugated dienes from cultured bovine pulmonary artery endothelial cells by dimethyl sulfoxide (DMSO) and allopurinol. Data are presented as mean ± SEM; $n = 6$; *$p < 0.05$ compared to controls; + $p < 0.05$ compared to endotoxin. (Data from Ref. 88.)

toxin does not affect intracellular levels of H_2O_2 and xanthine dehydrogenase/oxidase activity.

Iron-dependent hydroxyl radical formation following intracellular activation of xanthine oxidase has, however, been suggested to contribute to endotoxin-induced endothelial lysis (86). Studies in which bovine pulmonary artery endothelial cells were preincubated for 48 h with deferoxamine (an iron chelating agent that forms a catalytically inactive complex, thereby inhibiting hydroxyl radical by decreasing the ability of unbound iron to participate in catalysis) showed inhibition of the cytolytic effects of endotoxin. It was shown further that deferoxamine strikingly lowered intracellular concentrations of xanthine oxidase and xanthine dehydrogenase, although endotoxin per se had little effect on these enzymes. These data suggest that iron chelation modifies endotoxin-induced endothelial injury through inhibition of synthesis or bioactivity of constitutive intracellular xanthine oxidase. The data also suggest that maintained constitutive expression of certain oxidant enzymes may be necessary for endotoxin-induced injury to occur.

Although it is possible that those antioxidant enzymes that are not up-regulated may contribute to endotoxin-induced endothelial changes (e.g., deferoxamine study outlined above), the finding that certain endothelial antioxidant enzymes are up-regulated supports the notion that intracellular generation of toxic oxygen radicals contributes to endotoxin-induced endothelial injury. An obvious example of the latter is manganous superoxide dismutase. This enzyme has been shown to have a protective effect in a number of experimental conditions in which oxygen radicals are involved [e.g., ischemia reperfusion injury (90)] and has been suggested to be essential to the resistance of A549 cells to TNF-α (91). Manganous superoxide dismutase has also been shown to protect against endotoxin-induced pulmonary injury in sheep, where it has been demonstrated to attenuate the endothelial injury, seen by a reduction in increased microvascular permeability and release of PGI_2 into lung lymph (92).

Two forms of superoxide dismutase are found in mammalian cells, the copper/zinc form, which is located in the cytosol, and the manganous form, which is found in mitochondria. Intracellular measurements of both of these enzymes in bovine pulmonary artery endothelial cells exposed to endotoxin for varying periods show a gradual increase in manganous superoxide, while that for copper/zinc remained at baseline (Fig. 13) (84). Other mitochondrial antioxidant enzymes failed to show the increase seen for manganous superoxide dismutase (Table 1). More recent studies have shown induction of

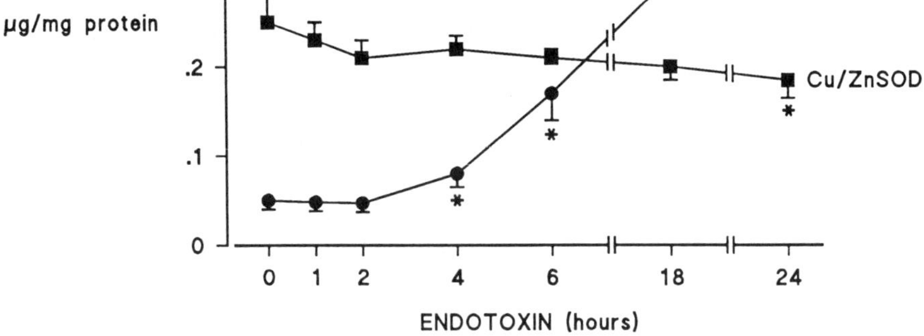

Figure 13 Endotoxin causes a gradual time-dependent increase in the antioxidant enzyme, manganous superoxide dismutase (MnSOD). Levels of copper/zinc superoxide dismutase (CuZnSOD) are relatively unaltered. Data are presented as mean ± SEM; $n = 5$; *$p < 0.05$. (Data from Ref. 84.)

Table 1 Protein and Enzyme Concentrations in Endothelial Cells Exposed to Endotoxin[a]

Hours of endotoxin exposure	Protein	Glutathione peroxidase	Catalase	Cytochrome *c* oxidase	Fumarase
Control					
1	8.42 ± 0.52	34 ± 7	12 ± 1	19 ± 5	123 ± 13
24	8.87 ± 0.54	29 ± 5	12 ± 1	15 ± 1	126 ± 8
Endotoxin					
0.5	9.01 ± 0.76	33 ± 7	12 ± 2	21 ± 2	129 ± 15
1	8.98 ± 0.66	32 ± 5	11 ± 1	19 ± 4	125 ± 11
2	8.93 ± 0.63	32 ± 8	10 ± 1	23 ± 8	119 ± 14
4	9.15 ± 0.49	33 ± 8	11 ± 1	17 ± 5	113 ± 11
6	8.73 ± 0.53	32 ± 6	12 ± 1	18 ± 5	117 ± 13
18	7.99 ± 0.56	25 ± 5	11 ± 1	16 ± 2	118 ± 15
24	7.56 ± 0.60	15 ± 2*	10 ± 2	17 ± 4	104 ± 12

Source: Ref. 84.

[a]Values are mean ± SEM ($n = 5$), data expressed in mg/mL of protein, mU/mg of protein for glutathione-peroxidase and fumarase, and kU/g for catalase and cytochrome *c* oxidase. Student's *t*-test was used for statistical analysis. *$p < 0.05$ endotoxin vs. control at 24 h.

MnSOD mRNA in endothelial cells treated with endotoxin; levels of Cu/ZnSOD were unchanged (93). These data support the notion that endotoxin-stimulated endothelial injury may involve intracellular generation of toxic oxygen radicals other than H_2O_2. The data further demonstrate that following endotoxin exposure, some constitutively expressed endothelial antioxidant enzymes are up-regulated while others are unaltered.

In summary, endotoxin results in pulmonary endothelial changes in animal models and in endothelial cells in culture. Just as animal species vary in their sensitivity to endotoxin, endothelial cells in culture exhibit a wide range of responses to endotoxin. Some endothelial cells are particularly sensitive and undergo lysis in the absence of any other cell type, while others are resistant to endotoxin-mediated cell lysis. In vitro studies suggest further that responsivity of endothelial cells from different organs and even different regions of the same organ may vary. For example, in the lung, endothelial cells from the main pulmonary artery are particularly responsive to endotoxin, whereas those from the lung periphery are more resistant. It is now apparent that endothelial cells, in general, are responsive to endotoxin and show up-regulation of a number of proteins and lipid mediators. Whether these events are mediated by an endotoxin receptor is not certain. In vitro studies suggest several intracellular signaling pathways as pivotal to endotoxin-induced endothelial changes (e.g., intracellular generation of reactive oxygen species, cyclic nucleotides, imbalance of the protease–antiprotease system, protein kinase C, and activation of transcription factors). In animals and endothelial cultures, these effects are complicated by simultaneous generation of a number of cytokines. Up-regulation of these cytokines may potentiate the endotoxin-induced endothelial changes. In animals, these mediators may be released either by cells resident in the lung, including the endothelium, or even by cells from distant organs. Elucidation of these mechanisms is a fertile area of future research.

Acknowledgments

This work was supported by NIH grants HL-34208 and T32-07186.

References

1. Harris Tr, Bernard GR, Brigham KL, et al. Lung microvascular transport properties measured by multiple indicator dilution methods in patients with adult respiratory distress syndrome. A comparison between patients reversing respiratory failure and those failing to reverse. Am Rev Respir Dis 1990; 141:272–280.
2. Rubin DB, Wiener-Kronish JP, Murray JF, et al. Elevated von Willebrand factor antigen is an early plasma predictor of acute lung injury in nonpulmonary sepsis syndrome. J Clin Invest 1990; 86:474–480.
3. Colucci M, Paramo JA, Collen D. Generation in plasma of a fast-acting inhibitor of plasminogen activator in response to endotoxin stimulation. J Clin Invest 1985; 75:818–824.
4. Suffredini AF, Shelhamer JH, Neumann RD, Brenner M, Baltaro RJ, Parrillo JE. Pulmonary and oxygen transport effects of intravenously administered endotoxin in normal humans. Am Rev Respir Dis 1992; 145:1398–1403.
5. Suffredini AF, Fromm RE, Parker MM, et al. The cardiovascular response of normal humans to endotoxin. N Engl J Med 1989; 321:280–287.
6. Martich GD, Danner RL, Ceska M, Suffredini AF. Detection of interleukin-8 and tumor necrosis factor in normal humans following intravenous endotoxin: the effect of anti-inflammatory agents. J Exp Med 1991; 173:1021–1024.
7. Cannon JD, Tompkins RG, Gelfand JA, et al. Circulating interleukin-1 and tumor necrosis factor in septic shock and experimental endotoxin fever. J Infect Dis 1990; 161:79–84.
8. Coalson JJ, Hinshaw LB, Guenter CA. The pulmonary ultrastructure in septic shock. Exp Mol Pathol 1970; 12:84–90.
9. Horn RG, Collins RD. Fragmentation of granulocytes in pulmonary capillaries during development of the generalized Shwartzmann reaction. Lab Invest 1968; 19:451–461.
10. Beutler B, Milsark IW, Cerami A. Passive immunization against cachectin/tumor necrosis factor protects mice from lethal effects of endotoxin. Science 1985; 229:869–870.

11. Matuschak GM, Pinsky MR, Klein EC, Van Thiel DH, Rinaldo JE. Effects of D-galactosamine-induced acute liver injury on mortality and pulmonary responses to *Escherichia coli* lipopolysaccharide. Am Rev Respir Dis 1990; 141:1296–1306.
12. Brigham KL, Meyrick B. Endotoxin and lung injury. State of the art. Am Rev Respir Dis 1986; 133:913–927.
13. Rinaldo JE, Dauber JH, Christman J, Rogers RM. Neutrophil alveolitis following endotoxemia. Enhancement by previous exposure to hyperoxia. Am Rev Respir Dis 1984; 130:1065–1071.
14. Brigham KL, Bowers R, Haynes J. Increased sheep lung vascular permeability caused by *E. coli* endotoxin. Circ Res 1979; 45:292–298.
15. Bradley JD, Roselli RJ, Parker RE, Harris TR. Effects of endotoxemia on the sheep lung microvascular membrane: a two pore theory. J Appl Physiol 1988; 64:2675–2683.
16. Bachofen M, Weibel ER. Structural alterations in lung parenchyma in the adult respiratory distress syndrome. Clin Chest Med 1982; 3:35–56.
17. Meyrick B, Brigham KL. Acute effects of *Escherichia coli* endotoxin on the pulmonary microcirculation of anesthetized sheep. Lab Invest 1983; 48:458–470.
18. Duke SS, Guerry-Force ML, Forbes JT, et al. Acute endotoxin-induced lymphocyte subset sequestration in sheep lungs. Lab Invest 1990; 62:355–362.
19. Pohlman TH, Stanness KA, Beatty PG, Ochs HD, Harlan JM. An endothelial surface factor(s) induced in vitro by lipopolysaccharide, interleukin 1 and tumor necrosis factor-alpha increases neutrophil adherence by a CDw18-dependent mechanism. J Immunol 1986; 136:4548–4553.
20. Doerschuk CM, Allard MF, Hogg JC. Neutrophil kinetics in rabbits during infusion of zymosan-activated plasma. J Appl Physiol 1989; 67:88–95.
21. Worthen GS, Schwab B, Elson EL, Downey GP. Mechanics of stimulated neutrophils: cell stiffening induces retention in capillaries. Science 1989; 245:183–185.
22. Warber AE, Molina RM, Brain JD. Uptake of blood-borne bacteria by pulmonary intravascular macrophages and consequent inflammatory responses in sheep. Am Rev Respir Dis 1987; 136:683–690.
23. Niehaus GD, Karve SV, Mullens DL, Dehring DJ. Microbial products can increase the number of rat vascular phagocytes. FASEB J 1992; 6:A1064.
24. Dehring DJ, Wismar BL. Intravascular macrophages in pulmonary capillaries of humans. Am Rev Respir Dis 1989; 139:1027–1029.
25. Meyrick B, Ryan US, Brigham KL. Direct effects of *E. coli* endotoxin on structure and permeability of pulmonary endothelial monolayers and the endothelial layer of intimal explants. Am J Pathol 1986; 122:140–151.
26. Harlan JM, Harker LA, Reidy MA, Gajusek CM, Schwartz SM, Striker GE. Lipopolysaccharide-mediated bovine endothelial injury in vitro. Lab Invest 1983; 48:269–274.
27. Meyrick B, Hoover R, Jones MR, Berry LC, Brigham KL. In vitro effects of endotoxin on bovine and sheep lung microvascular and pulmonary artery endothelial cells. J Cell Physiol 1989; 138:165–174.
28. Raghu G, Striker LJ, Striker GE. Lipopolysaccharide-mediated injury to cultured human glomerular endothelial cells. Clin Immunol Immunopathol 1986; 38:275–281.

29. Patrick D, Betts J, Frey EA, Prameya R, Dorovini-Zis K, Finlay BB. *Haemophilus influenzae* lipopolysaccharide disrupts confluent monolayers of bovine brain endothelial cells via a serum-dependent cytotoxic pathway. J Infect Dis 1992; 165:865–872.
30. Valberg PA, Meyrick B, Brain JD, Brigham KL. Phagocytic and motile properties of endothelial cells measured magnetometrically: effects of endotoxin. Tissue Cell 1988; 20:345–354.
31. Suttorp N, Galanos C, Neuhof H. Endotoxin alters arachidonate metabolism in pulmonary endothelial cells. Am J Physiol 1987; 253:C384–C390.
32. Tobias PS, Mathison J, Mintz D, et al. Participation of lipopolysaccharide-binding protein in lipopolysaccharide-dependent macrophage activation. Am J Respir Cell Mol Biol 1992; 7:239–245.
33. Colucci J, Balconi G, Lorenzet R, et al. Cultured human endothelial cells generate tissue factor in response to endotoxin. J Clin Invest 1983; 71:1893–1896.
34. Crutchley DJ, Conanan LB. Endotoxin induction of an inhibitor of plasminogen activator in bovine artery endothelial cells. J Biol Chem 1986; 261:154–159.
35. Libby P, Ordovas JM, Auger KR, Robbins AH, Birinyi LK, Dinarello CA. Endotoxin and tumor necrosis factor induce interleukin-1 gene expression in adult human vascular endothelial cells. Am J Pathol 1986; 124:179–185.
36. Jirik FR, Podor TJ, Hirano T, et al. Bacterial lipopolysaccharide and inflammatory mediators augment IL-6 secretion by human endothelial cells. J Immunol 1989; 142:144–147.
37. Sica A, Marsushima K, VanDamme J, et al. IL-1 transcriptionally activates the neutrophil chemotactic factor/IL-8 gene in endothelial cells. Immunology 1990; 69:548–553.
38. Clinton SK, Underwood R, Hayes L, Sherman ML, Kufe DW, Libby P. Macrophage colony-stimulating factor gene expression in vascular cells and in experimental and human atherosclerosis. Am J Pathol 1992; 140:301–306.
39. Quesenberry PJ, Gimbrone MA. Vascular endothelium as a regulator of granulopoiesis: production of colony-stimulating activity by cultured human endothelial cells. Blood 1980; 56:1060–1067.
40. Springer TA. Adhesion receptors of the immune system. Nature 1990; 346: 425–434.
41. Pober JS, Cotran RS. Cytokines and endothelial biology. Physiol Rev 1990; 70:427–451.
42. Lyberg T, Galdal KS, Evensen SA, Prydz H. Cellular cooperation in endothelial cell thromboplastin synthesis. Br J Haematol 1983; 53:85–95.
43. Warren BL, Fitting K, Kunkel SL, Remick DG, Merritt SE, Wiggins RC. Tissue factor expression in endothelial/monocyte cocultures stimulated by lipopolysaccharide and/or aggregated IgG. Mechanisms of cell/cell communication. J Immunol 1991; 146:1437–1445.
44. Loskutoff DJ, Sawdey M, Mimuro J. Type 1 plasminogen activator inhibitor. In: Coller B, ed. Progress in Hemostasis and Thrombosis. Columbia, MD: Bermedica Production, 1989; 9:87–115.
45. Schleef RR, Loskutoff DJ, Podor TJ. Immunoelectron microscopic localization of type 1 plasminogen activator inhibitor on the surface of activated endothelial cells. J Cell Biol 1991; 113:1413–1423.

46. Sugiura M, Inagami T, Kon V. Endotoxin stimulates endothelin release in vivo and in vitro as determined by radioimmunoassay. Biochem Biophys Res Commun 1989; 161:1220–1227.
47. Morel DR, Lacroix JS, Hemsen A, Steinig DA, Pittet J-F, Lundberg JM. Increased plasma and pulmonary lung lymph levels of endothelin during endotoxin shock. Eur J Pharmacol 1989; 167:427–428.
48. Nakamura T, Kikuo K, Sekiguchi Y, et al. Elevation of plasma endothelin concentrations during endotoxin shock in dogs. Eur J Pharmacol 1991; 205:277–282.
49. Kosaka H, Watanabe M, Yoshikara H, Harada N, Shiga T. Detection of nitric oxide production in lipopolysaccharide-treated rats by ESR using carbon monoxide hemoglobin. Biochem Biophys Res Commun 1992; 184:1119–1124.
50. Salvemini D, Korbut R, Anggard E, Vane J. Immediate release of a nitric oxide-like factor from bovine aortic endothelial cells by *E. coli* lipopolysaccharide. Proc Natl Acad Sci USA 1990; 87:2593–2597.
51. Salvemini D, Mollace V, Pistelli A, Anggard E, Vane J. Metabolism of glyceryl trinitrate to nitric oxide by endothelial cells and smooth muscle cells and its induction by *E. coli* lipopolysaccharide. Proc Natl Acad Sci USA 1992; 89:982–986.
52. Jones M, Hoover R, Meyrick B. Endotoxin enhancement of lymphocyte adherence to cultured sheep lung microvascular endothelial cells. Am J Respir Cell Mol Biol 1992; 7:81–89.
53. Redl H, Dinges HP, Buurman WA, et al. Expression of endothelial leukocyte adhesion molecule-1 in septic shock but not traumatic/hypovolemic shock in the baboon. Am J Pathol 1991; 139:461–466.
54. Sawdey MS, Loskutoff DJ. Regulation of murine type 1 plasminogen activator inhibitor gene expression in vivo. Tissue specificity and induction by lipopolysaccharide, tumor necrosis factor-alpha, and transforming growth factor-beta. J Clin Invest 1989; 88:1346–1353.
55. Pohlman TH, Harlan JM. Human endothelial cell response to lipopolysaccharide, interleukin-1, and tumor necrosis factor is regulated by protein synthesis. Cell Immunol 1989; 119:41–52.
56. Kirkland TN, Virca GD, Kuus-Reichel T, et al. Identification of lipopolysaccharide-binding proteins in 70Z/3 cells by photoaffinity cross-linking. J Biol Chem 1990; 265:9520–9525.
57. Lei MG, Stimpson SA, Morrison DC. Specific endotoxic lipopolysaccharide-binding receptors on murine splenocytes. III. Binding specificity and characterization. J Immunol 1991; 147:1925–1932.
58. Hampton RY, Golenbock DT, Penman M, Kreiger M, Raetz CRH. Recognition and plasma clearance of endotoxin by scavenger receptors. Nature 1991; 352: 342–344.
59. Pohlman TH, Munford RS, Harlan JM. Deacylated lipopolysaccharide inhibits neutrophil adherence to endothelium induced by lipopolysaccharide in vitro. J Exp Med 1987; 165:1393–1402.
60. Pohlman TH, Winn RK, Callahan KS, Maier RV, Harlan JM. A glycolipid precursor of bacterial lipopolysaccharide (lipid X) lacks activity against endothelial cells in vitro and is not toxic in vivo. J Surg Res 1988; 45:228–237.

61. Jones MR, Cordle SR, Hawiger J. Human endothelial cell NF-κB DNA-binding complex is activated by lipopolysaccharide. FASEB J 1992; 6:A1033.
62. von Asmuth EJ, Leeuwenberg JF, Ceska M, Buurman WA. LPS and cytokine-induced endothelial cell IL-6 release and ELAM-1 expression; involvement of serum. Eur Cytokine Netw 1991; 2:291–297.
63. Mackman N, Brand K, Edgington TS. Lipopolysaccharide-mediated transcriptional activation of the human tissue factor gene in THP-1 monocytic cells requires both activatior protein 1 and nuclear factor kB binding sites. J Exp Med 1991; 174:1517–1526.
64. Clark BD, Collins KL, Gandy MS, Webb AC, Auron PE. Genomic sequence for human prointerleukin 1 beta: possible evolution from a reverse transcribed prointerleukin 1 alpha gene. Nucleic Acids Res 1986; 20:7877–7914.
65. Libermann TA, Baltimore D. Activation of interleukin-6 gene expression through the NF-κB transcription factor. Mol Cell Biol 1990; 10:2327–2334.
66. Mukaida N, Shiroo M, Matsushima K. Genomic structure of the human monocyte-derived neutrophil chemotactic factor IL-8. J Immunol 1989; 143:1366–1371.
67. Nishizawa M, Nagata S. Regulatory elements responsible for inducible expression of the granulocyte colony-stimulating factor gene in macrophages. Mol Cell Biol 1990; 10:2002–2011.
68. Cybulsky MI, Fries JWU, Williams AJ, et al. Gene structure, chromosomal location, and basis for alternative mRNA splicing of the human VCAM-1 gene. Proc Natl Acad Sci USA 1991; 88:7859–7863.
69. Collins T, Williams A, Johnston GI, et al. Structure and chromosomal location of the gene for endothelial–leukocyte adhesion molecule 1. J Biol Chem 1991; 266:2466–2473.
70. Blank V, Kourilsky P, Israel A. NF-kB and related proteins: Rel/dorsal homologies meet ankyrin-like motifs. Trends Biochem Sci 1992; 242:135–140.
71. Whelan J, Ghersa P, van Huijsduijnen RH, et al. An NFκB-like factor is essential but not sufficient for cytokine induction of endothelial leukocyte adhesion molecule 1 (ELAM-1). Nucleic Acids Res 1991; 19:2645–2653.
72. Neish AS, Williams AJ, Palmer HJ, Whitley MA, Colins T. Functional analysis of the human vascular cell adhesion molecule 1 promoter. J Exp Med 1992; 176:1583–1593.
73. Lane TA, Lamkin GE, Wancewicz EV. Protein kinase C inhibitors block the enhanced expression of intercellular adhesion molecule-1 on endothelial cells activated by interleukin-1, lipopolysaccharide and tumor necrosis factor. Biochem Biophys Res Commun 1990; 172:1273–1281.
74. Magnuson DK, Maier RV, Pohlman TH. Protein kinase C: a potential pathway of endothelial activation by endotoxin, tumor necrosis factor, and interleukin-1. Surgery 1989; 106:216–223.
75. Shen X-Y, Hamilton TA, Dicorleto PE. Lipopolysaccharide-induced expression of the competence gene KC in vascular endothelial cells is mediated through protein kinase C. J Cell Physiol 1989; 140:44–51.
76. Montgomery KF, Osborn L, Hession C, et al. Activation of endothelial–leukocyte adhesion molecule 1 (ELAM-1) gene transcription. Proc Natl Acad Sci USA 1991; 88:6523–6527.

77. Hussein A, Meyrick B, Graber S, Berry LC, Brigham KL. Attenuation of endotoxin-induced cytotoxicity and prostacyclin production in cultured bovine pulmonary artery endothelial cells by phosphodiesterase inhibition. Exp Lung Res 1988; 14:637–654.
78. Snapper JR, Brigham KL, Heflin AC, Bomboy JD, Graber SE. Effects of endotoxemia on cyclic nucleotides in the unanesthetized sheep. J Lab Clin Med 1983; 102:240–249.
79. Foy T, Marion J, Brigham K, Bowers R, Harris T. Isoproteranol and aminophylline reduce capillary filtration during high permeability. J Appl Physiol 1979; 46:146–151.
80. Koyama S. Toyofuku T, Hirai K, Yoshimura K, Sakai A, Kobayashi T. Dibutyryl cyclic AMP attenuates lung responses induced by endotoxin in conscious sheep. Am Rev Respir Dis 1992; 146:32–38.
81. Tumen J, Meyrick B, Berry LC, Brigham KL. Antiproteases protect cultured lung endothelial cells from endotoxin injury. J Appl Physiol 1988; 65:835–843.
82. Orlowski M, Lesser M, Ayala J, Lasdun A, Kam CM, Powers JC. Substrate specificity and inhibitors of of a capillary injury-related protease from sheep lung lymph. Arch Biochem Biophys 1989; 269:125–136.
83. Orlowski M, Lesser M. Lung lymph capillary injury-related protease. Am J Physiol 1989; 257:L202–L208.
84. Shiki Y, Meyrick B, Brigham KL, Burr IM. Endotoxin increases superoxide dismutase in cultured bovine pulmonary artery endothelial cells. Am J Physiol 1987; 252:C436–C440.
85. Phan SH, Gannon DE, Varani J, Ryan US, Ward PA. Xanthine oxidase activity in rat pulmonary artery endothelial cells and its alteration by activated neutrophils. Am J Pathol 1989; 134:1201-1212.
86. Rinaldo JE, Gorry M. Protection by deferoxamine from endothelial injury: a possible link with inhibition of intracellular xanthine oxidase. Am J Respir Cell Mol Biol 1990; 3:525–533.
87. Royall JA, Gwin PD, Parks DA, Freeman BA. Responses of vascular endothelial oxidant metabolism to lipopolysaccharide and tumor necrosis factor-α. Arch Chem Biophys 1992; 294:686–694.
88. Brigham KL, Meyrick B, Berry JC, Repine JE. Antioxidants protect cultured bovine lung endothelial cells from injury by endotoxin. J Appl Physiol 1987; 63:840–850.
89. Dupont GP, Huechsteadt TP, Marshall BC, Ryan US, Michael JR, Hoidal JR. Regulation of xanthine dehydrogenase and xanthine oxidase activity and gene expression in cultured rat pulmonary endothelial cells. J Clin Invest 1992; 89: 197–202.
90. Tamura Y, Chi LG, Driscoll EM Jr, et al. Superoxide dismutase conjugated to polyethylene glycol provides sustained protection against myocardial ischemia/reperfusion injury in canine heart. Circ Res 1988; 63:944–959.
91. Wong GHW, Elwell JH, Oberley JW, Goeddel DV. Manganous superoxide dismutase is essential for cellular resistance to cytotoxicity of tumor necrosis factor. Cell 1989; 58:923–931.

92. Koyama S, Kobayashi T, Kubo K, Sekiguchi M, Ueda G. Recombinant-human superoxide dismutase attenuates endotoxin-induced lung injury in awake sheep. Am Rev Respir Dis 1992; 145:1404–1409.
93. Visner GA, Block ER, Burr IM, Nick HS. Regulation of manganese superoxide dismutase in porcine pulmonary artery endothelial cells. Am J Physiol 1991; 260:L444–L449.

6

Effects of Endotoxin on Human Pulmonary Circulation

JOHN H. NEWMAN

Vanderbilt University School of Medicine
Nashville, Tennessee

I. Introduction

A pulmonary artery catheter is rarely in place when a human is first exposed to endotoxin. Thus much of the available information about the effects of endotoxin on systemic and pulmonary hemodynamics in humans comes from observations made a number of hours or days after the initial insult. Pulmonary artery catheters are routinely placed in patients who suffer shock and diffuse pulmonary injury and are undergoing vigorous resuscitation in intensive care units; thus the hemodynamic information gathered from these patients is undoubtedly influenced by medical therapy as well as contributions of multiorgan damage. Prospective evaluations of the effects of endotoxin on the pulmonary circulation in normal subjects by necessity have used doses of endotoxin that are safe (1–5). Thus these experiments cannot be assumed to reproduce all of the effects of a life-threatening dose of endotoxin. Because of these limitations, more is not known about the effects of endotoxin on the human pulmonary circulation than is known. Humans are exquisitely sensitive to endotoxin. A dose of 2 to 6 ng/kg gives a flulike syndrome with well-defined cardiopulmonary abnormalities (2,5). In sheep, a similar response occurs at approximately 500 ng/kg, whereas the dog and rat require 1000- to 10,000-fold-higher doses to achieve serious injury (6). The purpose of this chapter is to review what is known about the effects of endotoxin on the human pulmonary circulation, the species of greatest clinical interest. The first basic observation is that the pulmonary circulation responds differently to endotoxin than does the systemic circulation. The pulmonary circulation constricts in response to endotoxin, whereas the systemic vascular bed dilates (1,2). This most fundamental observation is still unexplained.

II. Normal Pulmonary Circulation

Analysis of the effects of endotoxin requires some consideration of the normal pulmonary circulation (7). The cardiac output traverses the pulmonary circulation under extremely low driving pressures. The difference between mean pulmonary arterial pressure and left atrial pressure may be only 5 to 10 mmHg. At rest, the mean pulmonary artery pressure is approximately 14 ± 3 STD mmHg. Cardiac output is approximately 5 L/min, yielding a pulmonary vascular resistance of about 2 Woods units, or 150 to 200 dyn/s per cubic centimeter. The stroke work of the right ventricle is approximately 10 times less than that of the left ventricle, whose stroke volume is the same but whose mean driving pressure is far greater.

The right ventricle is capable of pumping blood through the pulmonary circulation at vastly increased flows without the development of failure. During exercise when cardiac output may be 5 to 10-fold greater than at rest, mean pulmonary artery pressure in humans rises to as high as 30 mmHg (8). Because of the ability of the lung vasculature to both recruit and dilate, pulmonary vascular resistance decreases approximately 30 to 40% during high flow. Thus the right ventricle is able to manage large changes in preload. On the other hand, an abrupt increase in afterload to a mean pressure above 40 mmHg may cause the right ventricle to dilate, but it can continue to pump under strain to mean pressures as high as 60 to 70 mmHg in situations such as acute pulmonary embolism (9,10). Because the changes in pulmonary artery pressure that occur during endotoxemia do not ordinarily reach this level of afterload, acute failure of the right ventricle would not appear to be a major contributor to observed hemodynamic complications that result in endotoxic shock.

Despite the fact that the pulmonary circulation is normally a low-resistance conduit, it is capable of increasing vasomotor tone. Table 1 lists the major endogenous vasoactive mediators and their effects on the pulmonary circulation (7). In general, the effects of vasodilators can be demonstrated only when the pulmonary circulation is preconstricted. This is because the pul-

Table 1 Endogenous Mediators of Pulmonary Vascular Tone

Vasoconstriction	Vasodilation
Hypoxia (P_AO_2)	Nitric oxide (NO^-)
Endothelin	Prostacyclin
Angiotensin	PGE_2
Thromboxanes, PGD_2	Adenosine
Leukotrienes	Bradykinin
Norepinephrine	
Platelet activating factor	

Variable Effects	
Histamine:	Constricts resting circulation, dilates preconstricted bed
Acetylcholine:	Constricts via muscarinic receptors, dilates via NO^- release
Epinephrine:	Net effect reflects balance of α and β effects
Serotonin:	Constricts via serotonin receptors, dilates via NO^- release

monary circulation is relaxed at rest and vasodilators have little potential effect. Most of the mediators listed in Table 1 are released in increased amounts during endotoxemia in animals, although none of these has been well established to be important in modulating the human pulmonary response. The most important newly discovered mediators are endothelin and nitric oxide (11–14). Endothelin is a potent vasoconstrictor with long-lasting effects, and it is theoretically an attractive mediator of the prolonged hypertension that occurs in endotoxic lung injury (15). Nitric oxide (EDRF) is an endogenous vasodilator that has potent effects on the pulmonary circulation and may also be partly responsible for the systemic shock that occurs after endotoxemia (13–17).

Hypoxic vasoconstriction is the most important acute and chronic modulator of pulmonary vascular tone (18–21). To cause significant pulmonary hypertension, the effects of hypoxic pulmonary vasoconstriction in general require chronicity such as seen with chronic obstructive lung disease (COPD) or high-altitude dwelling, but hypoxic vasoconstriction also modulates perfusion and improves ventilation–perfusion matching in disease states. Indirect evidence for the importance of hypoxic vasoconstriction in humans with endotoxin-induced lung injury has come from studies using vasodilators in patients, which show that they may worsen hypoxemia by increasing perfusion in poorly ventilated regions of the lung where hypoxic vasoconstriction was active (22–24). In Fig. 1 the effects of local and generalized hypoxic vasoconstriction are depicted. Localized hypoxic vasoconstriction is a useful mechanism for maximizing oxygenation (18). Generalized hypoxic vasoconstriction can raise pulmonary vascular resistance and thus pulmonary artery pressure by causing reduction in the cross-sectional area of the arteriolar bed by diffuse vasoconstriction (21).

The factors that control lung fluid balance are relatively unstressed at rest compared to conditions of high flow, high pressure, or increased permeability (25–27). Because lung capillary pressure is around 8 mmHg at rest and oncotic pressure is approximately 23 mmHg, the hydrostatic-to-oncotic pressure gradient favors intravascular retention of water, solute, and protein. Capillary hypertension occurs in both disease and health, causing the

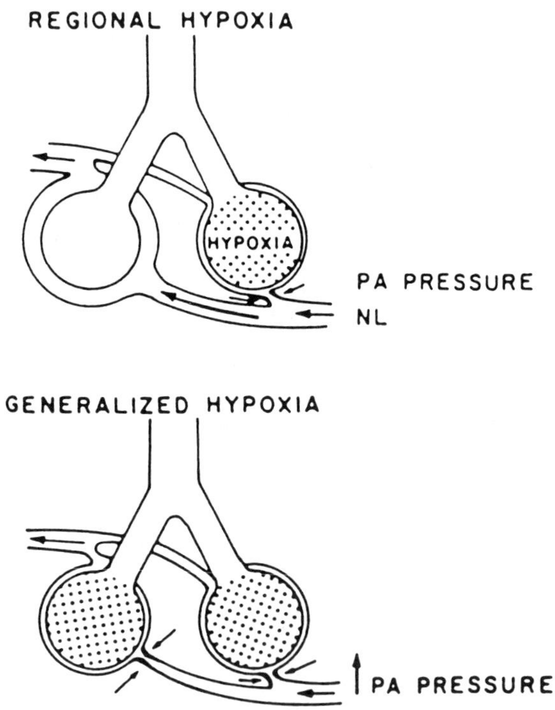

Figure 1 Effects of hypoxic vasoconstriction in the lung. In a situation of focal hypoventilation, hypoxic vasoconstriction acts to distribute blood to regions of higher PO_2, thus maximizing blood oxygenation. Generalized hypoventilation results in diffuse hypoxic vasoconstriction, which can acutely and chemically raise pulmonary arterial pressure. (From Ref. 99.)

hydrostatic pressure to exceed the oncotic back pressure. During exercise in normal people, capillary pressure may rise as high as 30 cmH_2O (8). Endotoxemia may raise capillary pressure to levels that cause edema, left ventricular failure, valvular disease, and pulmonary venous constriction (28). The lung has a great capacity to clear an increase in filtered liquid; estimates extrapolated from animal studies would be up to a 10-fold increase. Although the permeability coefficient of the human lung to plasma proteins is not well established, lungs of most mammals have a reflection coefficient of approximately 0.7 to 0.9 (25,26). This keeps the lung interstitium relatively dry; the lung is an impermeable organ compared to the liver, where the reflection coefficient of proteins may be as low as 0.2 to 0.1, and interstitial liquid protein concentration is similar to that of plasma. Figure 2 depicts the accumulation of lung water as a function of hydrostatic pressure in the normal state, with decreased oncotic pressure and with increased lung vascular permeability. During endotoxin-induced lung vascular injury, all three edemagenic conditions may exist (8).

The overall structure and function of the lung circulation is to spread blood into a very thin layer at a very high surface area to allow rapid excretion of carbon dioxide and uptake of oxygen into red cells. Endothelial cells

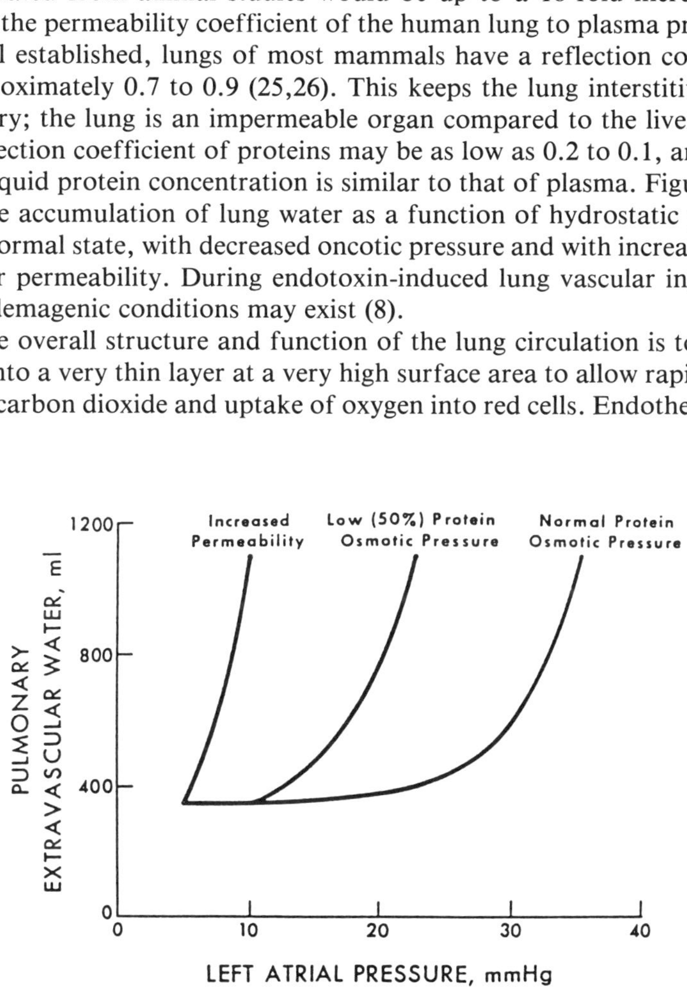

Figure 2 Pulmonary edema (extravascular lung water) as a function of left atrial pressure in the normal state and during either decreased oncotic pressure (low plasma albumin) or increased lung vascular permeability. Hypoalbuminemia and increased lung vascular permeability are features of endotoxemia, increasing the likelihood of pulmonary edema. (From Ref. 98.)

which line the capillaries are pleuri-potential. Their functions are described in other chapters of this book. Probably by virtue of its immense surface area, the lung endothelium appears to be the major source of conversion of angiotensin 1 to angiotensin 2 and inactivation of bradykinin through angiotensin converting enzyme (29). In addition, the lung endothelium is capable of uptake of bioactive amines, such as serotonin and catecholamines and inactivation of prostaglandins (30–32). Probably because the lung has such a large interface with inhaled air and therefore is at a great deal of risk for exposure to infectious agents and chemical irritants, the lung circulation is a major pool for marginated polymorphonuclear leukocytes (33,34) and the lung is abundantly endowed with macrophages (35–37). Some persons appear to possess pulmonary intravascular macrophages, and intravascular macrophages appear to be inducible in sick human lungs, but these cells are found in more abundance in other species (38,39). The inflammatory potential of the lung undoubtedly influences the response to endotoxin. Injury to the vasomotor responses of the lung circulation undoubtedly contributes to the abnormalities of gas exchange that occur in endotoxemia. The alterations in other metabolic functions, such as conversion of angiotensin, production of anti-inflammatory arachidonic acid products, or uptake of amines are of unknown consequence (40).

III. Systemic Response to Endotoxin

The properties and biological effects of bacterial pyrogens in humans have been studied extensively since the mid-nineteenth century and have been fully reviewed at intervals (1–3,41–44). Systematic studies of the effects of endotoxin in humans have been performed over the last 50 years. The need for research was stimulated initially by the problem of febrile reactions in patients receiving intravenous fluids, as this form of therapy began to be delivered in the 1930s (1). More recently, a major stimulus for the study of endotoxemia has been the problem of sepsis, septic shock, and the adult respiratory distress syndrome in patients in surgical and medical intensive care units (8,41–47).

The effects of low doses of intravenously injected endotoxin in humans are highly reproducible. The dose has varied from a 0.45-μg bolus, approximately 6.5 ng/kg, to 4 ng/kg in recent studies (2,5). A latent period occurs in approximately 45 to 90 min, after which patients develop fever and in most cases a shaking chill (1,2). Patients may suffer a degree of lassitude before the chill. Subjects may complain of headache, malaise, and myalgias. The duration of the chill is from 10 to 30 min. Fever reaches its peak between 2 and 4 h after endotoxin infusion, then slowly returns back toward normal over the next 8 h. Some patients have cool cyanotic skin, probably due to cutaneous vasoconstriction. Oxygen consumption (VO_2), minute ventilation, and respiratory rate all increase within 3 h of endotoxin infusion, near the peak of the fever response, and then decrease slowly (Fig. 3) (2). Marked increases in plasma ACTH and epinephrine occur between 2 and 3 h (4), and multiple proinflammatory cytokines and other inflammatory mediators accompany the reaction (4,48–54).

Although changes in cardiac output, heart rate, and mean arterial pressure begin within 1 to 2 h after endotoxin infusion, significant decreases in blood pressure and an increase in cardiac output do not usually occur until 3 to 4 h after exposure to endotoxin (Fig. 4) (2,5). At the low doses given to normal subjects, systemic pressure decreases by approximately 10 to 20%. Larger, and therefore presumably unsafe doses of endotoxin would need to be given to approximate endotoxic shock more closely. Cardiac output increases within 3 to 4 h after low-dose endotoxin infusion in normal subjects, and the increase is in the range of 30 to 50% over baseline. Because of the increase in cardiac output coupled with reduction in systemic blood pressure, reduction in systemic vascular resistance is on the order of 50%. The hemodynamic response to a volume infusion of approximately 6 to 7 L of saline given between 3 and 5 h after 4 ng/kg of endotoxin results in a small increase in mean arterial pressure and a small increase in cardiac index with no discernible effect on systemic vascular resistance or heart rate (5). Thus,

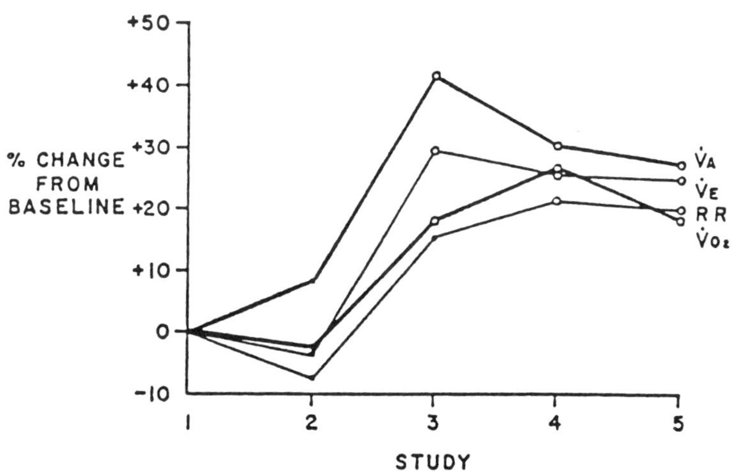

Figure 3 Changes in minute ventilation (Ve), alveolar ventilation (V_A), respiratory rate (RR), and oxygen consumption in 10 normal human subjects after receiving 0.45 ng of endotoxin intravenously. The study period represents baseline and the duration was 6 h. (From Ref. 2.)

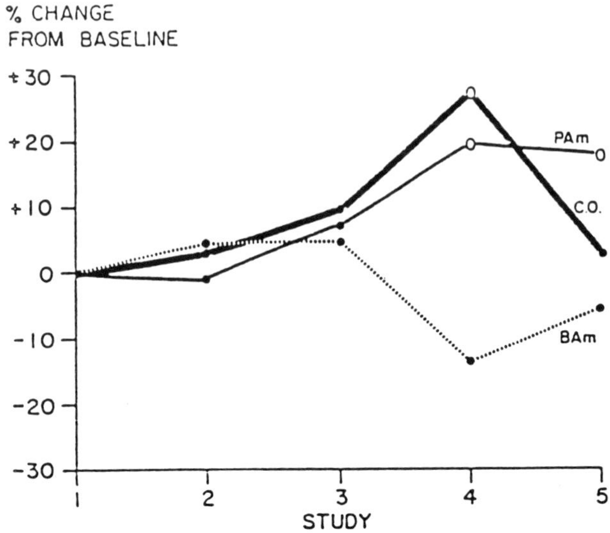

Figure 4 Changes in cardiac output (CO), mean pulmonary artery pressure (PAm), and systemic artery pressure (BAm) in 10 normal subjects after 0.45 ng of endotoxin intravenously. The study period is baseline and the duration was 6 h. (From Ref. 2.)

at least in normal humans, the initial response to endotoxin is a vasodilated, high-cardiac-output state. It is clear that an initial response of this nature in severe sepsis does not necessarily imply a good outcome (41,55).

The cause of the systemic hypotension due to endotoxin is of a great deal of interest because of the organ failure that follows (41–43,47). Cerebral and renal perfusion can quickly be severely compromised by hypotension, resulting in immediate functional abnormalities. Patients who are in septic shock may respond poorly to catecholamine vasoconstrictor agents, including norepinephrine and dopamine (56). Vessels and organs isolated from animals that have been given endotoxin fail to constrict in response to α-agonists, demonstrating abnormalities of α-receptors (57,58). Another potential contributing cause of the observed vasodilation is the release of β-endorphins (59–61). Interleukins and tumor necrosis factor (TNF), which may be released during endotoxemia, are known to cause significant systemic vasodilation and the mechanism may be through the release of nitric oxide (62–64). Experiments in animals and in humans suggest that the systemic hypotension of endotoxemia may in large part be due to the accelerated release and effect of nitric oxide; inhibitors of the production of nitric oxide have been shown to reverse endotoxic shock dramatically in animals (65–68).

Whether or not the reversal of shock will have a major impact on the morbidity and mortality of endotoxemia in human patients remains to be seen. Nonetheless, it seems clear that this mechanism is at least in part responsible for the hypotension seen in the systemic circulation after endotoxemia in humans; there is cause for hope.

A circulating myocardial depression factor released during endotoxemia has been hypothesized for a number of years, in part due to the tachycardia and low blood pressure that occur in response to endotoxin (69,70). It seems clear now that although cardiac output rises during endotoxemia, ventricular function is compromised (5,41,43,47,55,56). The compromise does not appear to be due to ischemia (71), at least in normal hearts, nor due to an abrupt reduction in levels of ionized calcium (5). The cytokine, tumor necrosis factor, appears to cause a response similar to endotoxin and may cause direct myocardial injury (4,72). Blood tumor necrosis factor levels rise in humans following endotoxin infusion (4). During endotoxic injury,

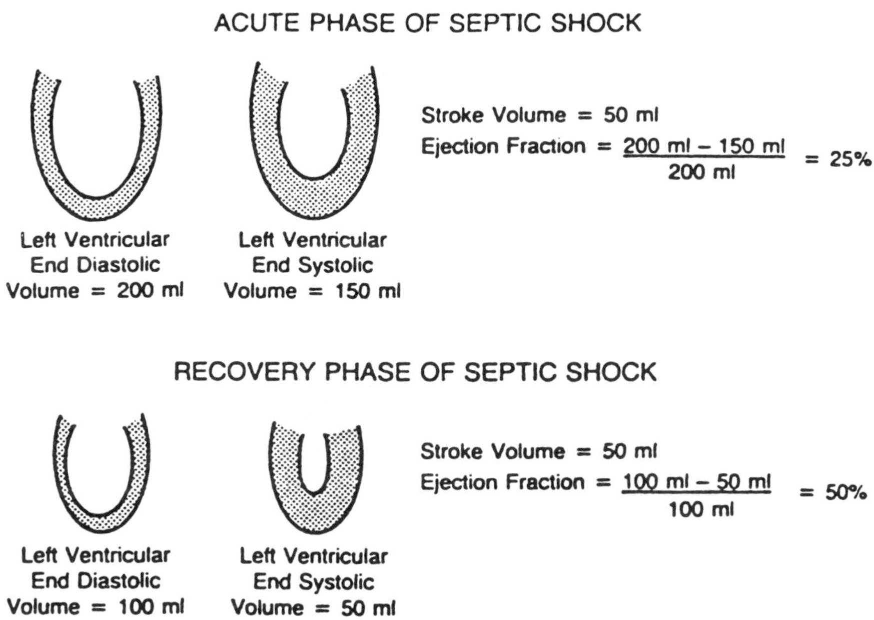

Figure 5 Depiction of the changes in cardiac function during and after recovery from septic shock. Endotoxemia causes an increase in end-diastolic volume and a reduced ejection fraction. Myocardial compliance is increased. The dysfunction occurs with both ventricles. Stroke volume is relatively unchanged. Cardiac output occurs via increased heart rate. (From Ref. 41.)

volume loading with saline reveals that ventricular compliance increases and ejection fraction decreases (5,55,73). The net result is a higher diastolic volume with a lower ejection fraction but a constant ejection stroke volume which, along with increased heart rate, results in increased cardiac output (Fig. 5) (74). By ratioing ejection fraction to the changes in preload as measured by the pulmonary wedge pressure, it is apparent that the contractile response to the volume load is significantly reduced over 3 to 6 h after endotoxin infusion (5). Nonetheless, it seems unlikely that myocardial dysfunction is a major cause of septic shock because the heart is usually capable of increasing the cardiac output (75). Perhaps late in shock, when systemic vascular resistance rises and cardiac output drops, the negative effects of endotoxin on myocardial performance contribute to multiple-organ failure.

IV. Response of the Pulmonary Circulation to Endotoxin

An increase in pulmonary artery pressure and an increase in pulmonary vascular resistance are features of the response to endotoxin in the human pulmonary circulation. At low doses of endotoxin in normal subjects, pulmonary artery pressure rises 3 to 4 h after endotoxin and remains elevated for approximately 8 h (Fig. 4) (2,5). Because cardiac output also rises after endotoxemia, the change in pulmonary vascular resistance is less impressive than the change in pulmonary artery pressure. However, late after endotoxemia when cardiac output has returned toward baseline, pulmonary vascular resistance is clearly elevated (2). This is in clear contrast to systemic vascular resistance, which decreases rapidly after endotoxemia and remains depressed, a consequence of reduction in mean arterial pressure coupled with high cardiac output (2,41–43). At low doses of endotoxin, the pulmonary artery wedge pressure does not change significantly up or down, and changes in the wedge pressure are not responsible for the measured changes in pulmonary vascular resistance. At low doses of endotoxin (4 ng/kg), change in pulmonary artery mean pressure and pulmonary vascular resistance may be on the order of 30 to 80% increased, a modest amount (2,5).

Patients in critical care units with overt sepsis have an even greater degree of pulmonary hypertension and elevated pulmonary vascular resistance, although this is not a universal finding (76). Pulmonary hypertension in septic patients has been well documented in a number of studies, and the average mean pulmonary artery pressure is somewhere around 30 mmHg (22–24,45, 54,77,78). This increase in pulmonary artery pressure is not due to back pressure from a dysfunctional left ventricle because wedge pressure in general is not elevated in sepsis before volume resuscitation. The pulmonary artery diastolic-to-wedge pressure gradient is increased, suggesting upstream causes

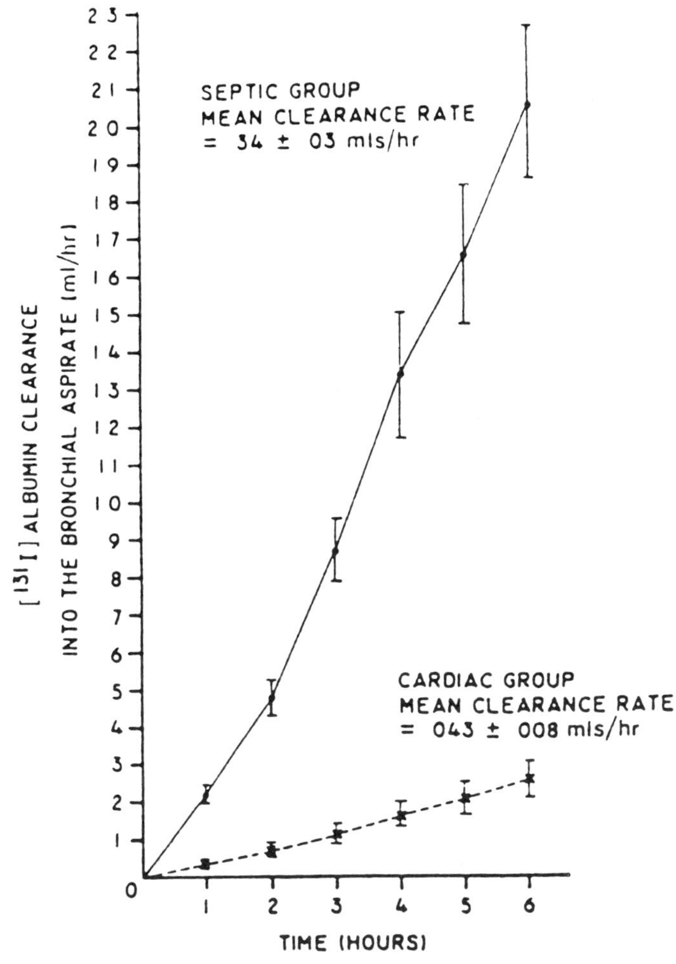

Figure 6 Comparison of the clearance of [^{131}I]albumin into bronchial secretions in patients with septic lung injury and patients with congestive heart failure. Albumin leak into the airways is increased greatly in permeability injury versus hydrostatic stress. (From Ref. 92.)

of vascular obstruction (78). Right ventricular function is also impaired during sepsis (41,47) and afterload is increased, but the usual levels of pulmonary hypertension in septic shock do not appear to be sufficient to grossly impair right ventricular output. In a recent study of 100 patients at our institution (G. Bernard, personal communication), the mean pulmonary artery pressure was greater than 35 mmHg in 18 of the 100 patients, and 8 patients had mean pressures of 40 mmHg or greater. Thus a minority of septic patients may have pulmonary artery pressures which approach a level that may cause dilatation and failure of a normal right ventricle.

The causes of the elevations in pulmonary artery pressure and pulmonary vascular resistance after endotoxemia have not been established (79). Possibilities for these changes include reduction in cross-sectional area due to vascular obstruction by thrombosis (80), plugging by leukocytes and platelets (81), vasoconstriction (22–24), edema (8), and airway changes that may overdistend alveoli and compress the capillary bed (Table 2). In addition, microatelectasis related to diffuse lung injury and surfactant abnormalities (82) may elevate pulmonary vascular resistance by reducing cross-sectional area through collapse and crinkling of the alveolar wall. There is evidence for all of these mechanisms of vascular injury occurring in patients with septic lung injury, although these changes are not unique to septic injury and occur in a variety of situations involving shock and organ failure. After obstruction or obliteration of the vascular bed, vasoconstriction is the next major potential cause of increased pulmonary artery pressure and pulmonary vascular resistance. Nearly every constrictor mediator known has been found to be elevated after endotoxemia in a variety of animals, and it has been tempting to ascribe vasoconstriction to whichever mediator is under general scrutiny at the moment (Table 1) (52,53,56,79). On the other hand, there has been no careful study of normal subjects given small doses of endotoxin to determine which, if any, of the mediators listed in Table 1 are

Table 2 Causes of Pulmonary Hypertension in Septic Lung Injury

Change	Explanation
Increased CO	$P_{PA} = CO \times PVR$
LV failure	$\uparrow P_{LA} \rightarrow \uparrow$ Pcapillary
Pulmonary venoconstriction	$\uparrow P_V \rightarrow \uparrow$ Pcapillary
Pulmonary arteriolar constriction	$\uparrow$ PVR – mediators
Obstruction due to thrombosis, leukocytes, platelets	
Alveolar collapse	Edema, surfactant abnormalities
Capillary compression	Microatelectasis, PEEP
Overhydration	$\uparrow$ Pcapillary

important in modulating pulmonary vascular tone during endotoxemia. In septic humans, marked elevations in thromboxane and prostacyclin occur, but these levels have not been correlated with the level of pulmonary artery pressure, and use of the cyclooxygenase inhibitor ibuprofen in patients with pulmonary hypertension related to sepsis has not been effective at lowering pulmonary arterial pressure (54). Pulmonary artery pressure in septic lung injury does not appear to correlate with outcome (Fig. 7) (79).

That a degree of active pulmonary vasoconstriction occurs in patients with septic lung injury seems clear. These patients respond to vasodilator drugs with a reduction in pulmonary resistance (22–24). Vasodilators may increase cardiac output and, in many cases, reduce oxygenation. This reduction in oxygenation suggests the reversal of hypoxic vasoconstriction in areas of lung that are hypoventilated, leading to an increase in blood flow through poorly ventilated regions. Of a great deal of interest is recent work showing that inhalation of nitric oxide markedly and repeatedly improves oxygenation in patients with acute lung injury (83,84). Inhalation of nitric oxide has the interesting effect of affecting primarily the lung vasculature because of the rapid uptake of nitric oxide into hemoglobin, preventing any systemic vascular reaction (85). Nitric oxide presumably diffuses into the vasculature only through alveoli that are ventilated. By vasodilating the ves-

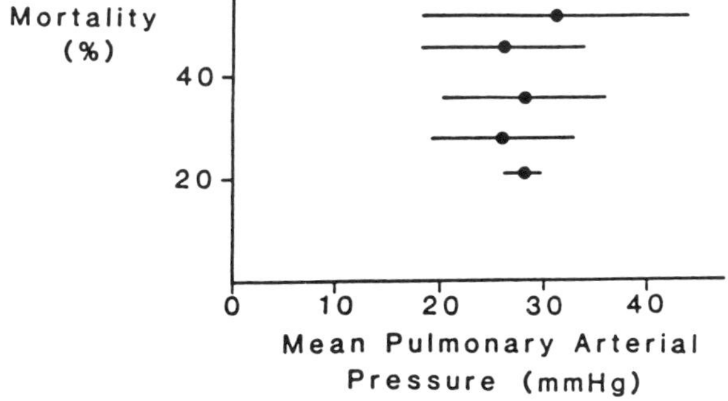

Figure 7 Mean pulmonary arterial pressure versus mortality in patients with acute respiratory failure due to lung injury. (From Ref. 79.)

sels leading to these units, blood is directed away from areas of low V/Q ratios toward better ventilating alveoli. This phenomenon suggests that vasoconstriction exists in the lung and that it exists in zones that may not be hypoxic. Thus although hypoxic vasoconstriction may be functioning in some manner to maximize oxygenation during acute septic lung injury, other causes of vasoconstriction are almost certainly present. The fact that nitric oxide works in this manner suggests that the smooth muscle is able to dilate and raises the question of whether endogenous production of nitric oxide is somehow damaged during endotoxin injury to the lung circulation. This will be an area of great interest in the future. At low doses in normal subjects, hypoxemia is not a major feature of the response to endotoxemia (2).

There are two general mechanisms by which endotoxin could alter lung fluid balance in favor of the development of pulmonary edema: hydrostatic filtration and increased lung vascular permeability (25–28). The first is via hydrostatic changes related to pulmonary venous constriction or left ventricular dysfunction (5,41,55). Pulmonary venous constriction has not been proven to occur as in humans in response to endotoxin, although it is clearly present in some animal species (86). Left ventricular failure or left ventricular and diastolic pressure elevation leading to capillary hypertension is not a problem at low doses of endotoxin in normal subjects and is not usually seen in patients at the initiation of a pulmonary artery catheter after clinical sepsis, but it may be a factor in endstage multiple-organ failure. Finally, overhydration or cardiac dysfunction accelerated by the use of inotropic agents may induce pulmonary edema through hydrostatic mechanisms, independent of the direct effects of endotoxin on the pulmonary circulation.

Increased lung vascular permeability is a well-documented feature of the endotoxin reaction in a variety of species (8,87). Controlled studies using small doses of endotoxin have not been carried out in normal humans to test for its existence. On the other hand, a number of studies in patients with acute lung injury, including the use of indicator dilution by multiple tracers (88), thermal indicator dilution (89,90), positron emission tomography (PET) scanning (91), bronchoalveolar lavage (92,93), and albumin vascular leak (94), lends considerable weight to the likelihood that lung vascular endothelial permeability is increased after endotoxin-induced injury. Increased lung vascular permeability, of course, is not unique to endotoxemia and occurs in any situation in which inflammatory mediators, especially the cytokines interleukin-1 and tumor necrosis factor are released. The effects of endotoxin and the cytokines on endothelial function are reviewed in Chapter 6. Based on some of the inflammatory destruction that occurs with severe acute lung injury related to sepsis, it is likely that outright denudation of endothelium and breaching of the basement membrane occurs, leading to such wholesale destruction of the capillary bed so that assessment of permeability changes may become meaningless (95). On the other hand, small but real increases in the permeability of the endothelium to water, solute, and proteins put an increased burden on the lymphatic clearance system (26,27). Because the epithelial barrier is much tighter than that of the endothelium (25–27), the fact that alveolar edema fluid and cellular debris is found in severe septic lung injury suggests that the strength of the inflammatory response far exceeds the simple alteration of endothelial permeability. Abnormalities and impairment of surfactant are well documented in acute lung injury (82) and undoubtedly contribute to the microatelectasis that is radiographically similar to diffuse pulmonary edema. Surfactant may be efficacious in the treatment of septic lung injury (96). Although the effects of endotoxin on the bronchial circulation in humans have not been studied, bronchial venular leak is a feature of endotoxin exposure in dogs (97) and may contribute to the bronchovascular edema and airway dysfunction that results in gas-exchange abnormalities in humans.

V. Summary

The hemodynamic response to endotoxin is pulmonary vasoconstriction and systemic vasodilation. Alterations in effective circulatory volume and cardiac output influence pressure and resistance. Myocardial dysfunction may not influence hemodynamics unless preexisting cardiac disease results in gross impairment. Most patients with septic lung injury have modest but definable pulmonary hypertension, which appears to be due to vasoconstriction and, in part, elevated cardiac output. In advanced endotoxic lung injury, obstruction and occlusion of the lung vasculature by inflammatory cells, desquamated cells, and thrombosis undoubtedly contributes to increased pulmonary vascular resistance. The mediator(s) of pulmonary vasoconstriction that offer endotoxemia in humans is(are) unknown. Generalized vasodilators may impair oxygenation in septic lung injury by increasing perfusion to poorly ventilated alveoli. Nitric oxide inhalation may improve oxygenation by improving perfusion in ventilated alveoli. Lung vascular permeability is increased in septic lung injury but has not been assessed fully in controlled studies of normal subjects. Further studies in normal subjects and septic patients are needed to address the many features of the response of the human pulmonary circulation to endotoxin.

Acknowledgment

This work is supported by grants No. HL 45107 and HL 41952 from the National Institutes of Health.

References

1. Bennett IL Jr, Beeson PB. The properties and biologic effects of bacterial pyrogens. Medicine 1950; 29:365–400.
2. Moser KM, Perry RB, Luchsinger PC. Cardiopulmonary consequences of pyrogen-induced hyperpyrexia in man. J Clin Invest 1963; 42:626–634.
3. Wolff SM. Biological effects of bacterial endotoxins in man. J Infect Dis 1973; 128:S259–S264.
4. Michie HR, Manogue KR, Spriggs DR, et al. Detection of circulating tumor necrosis factor after endotoxin administration. N Engl J Med 1988; 318:1481–1486.
5. Suffredini AF, Fromm RE, Parker MM, et al. The cardiovascular response of normal humans to the administration of endotoxin. N Engl J Med 1989; 321: 280–287.
6. Newman JH. Sepsis and pulmonary edema. Clin Chest Med 1985; 6:371–391.
7. Fishman AP. Pulmonary circulation. In: Fishman AP, Fisher AB, eds. The Respiratory System Handbook of Physiology. Baltimore: Williams & Wilkins, 1985:93–165.
8. Reeves JT, Dempsey JA, Grover RF. Pulmonary circulation during exericse. In: Weir EK, Reeves JT, eds. Pulmonary Vascular Physiology and Pathophysiology. New York: Marcel Dekker, 1989:107–133.
9. Dalen JE, Banas J Jr, Brooks H, et al. Resolution rate of acute pulmonary embolism in man. N Engl J Med 1969; 280:1194–1198.
10. McGinn S, White PD. Acute cor pulmonale resulting from pulmonary embolism. JAMA 1935; 104:1473–1479.
11. Yanagisawa M, Kurihara H, Kimura S, et al. A novel potent vasoconstrictor peptide produced by vascular endothelial cells. Nature 1988; 332:411–415.
12. Lippton HL, Hauth TA, Summer WR, Hyman AL. Endothelin produces pulmonary vasoconstriction and systemic vasodilation. J Appl Physiol 1989; 66: 1008–1012.
13. Furchgott RF, Zawadzski JV. The obligatory role of endothelial cells in the relaxation of arterial smooth muscle to acetylcholine. Nature 1980; 288:373–376.
14. Ignarro LJ. Biological actions and properties of endothelium-derived nitric oxide formed and released from artery and vein. Circ Res 1989; 65:1–21.
15. Morel DR, Lacroix JS, Hemsen A, Steinig DA, Pittet JF, Lundberg JM. Increased plasma and pulmonary lymph levels of endothelin during endotoxin shock. Eur J Pharmacol 1989; 167:427–428.
16. Beasley D, Cohen RA, Levinsky NA. Endotoxin inhibits contraction of vascular smooth muscle in vitro. Am J Physiol 1990; 258:H1187–H1192.
17. Kilbourn RG, Gross SS, Jubran A, et al. N^G-methyl-L-arginine inhibits tumor necrosis factor-induced hypotension: implications for the involvement of nitric oxide. Proc Natl Acad Sci USA 1990; 8:3639–3732.
18. Euler US, Liljestrand V. Observations on the pulmonary arterial blood pressure in the cat. Acta Physiol Scand 1946; 12:301–320.
19. Voelkel NF. Mechanisms of hypoxic pulmonary vasoconstriction. Am Rev Respir Dis 1986; 133:1186–1195.
20. Burrows B. Arterial oxygenation and pulmonary hemodynamics in patients with chronic airways obstruction. Am Rev Respir Dis 1974; 110:64.

21. Grover RF. Pulmonary circulation in animals and man at altitude. Ann NY Acad Sci 1965; 127:632–639.
22. Weigelt JA, Gewertz BL, Aurbakken CM, Snyder WH. Pharmacologic alterations in pulmonary artery pressure in the adult respiratory distress syndrome. J Surg Res 1982; 32:243–248.
23. Zapol WM, Rie MA, Frikker M, Snider MT, Quinn DA. Pulmonary circulation during adult respiratory distress syndrome. In: Zapol WM, Falke KJ, eds. Acute Respiratory Failure. New York: Marcel Dekker, 1985:241–271.
24. Bihari D, Smithies M, Gimson A, Tinker J. The effects of vasodilation with prostacyclin on oxygen delivery and uptake in critically ill patients. N Engl J Med 1987; 317:397–403.
25. Staub NC. Pulmonary edema. Physiol Rev 1974; 54:678–773.
26. Taylor AE, Parker JC. Pulmonary interstitial spaces and lymphatics. In: Fishman AP, Fisher AB, eds. Handbook of Physiology: The Respiratory System. Baltimore: Williams & Wilkins, 1985:167–230.
27. Staub NC. Pulmonary edema due to increased microvascular permeability to fluid and protein. Circ Res 1978; 43:
28. Newman JH. Sepsis and pulmonary edema. Clin Chest Med 1985; 6:371–391.
29. Ryan JW, Ryan US. Pulmonary endothelial cells. Fed Proc 1977; 36:2683–2691.
30. Gillis CN, Pitt BR. The fate of circulating amines within the pulmonary circulation. Annu Rev Physiol 1982; 44:269–281.
31. Gillis CN. Pulmonary extraction of PGE_1 in the adult respiratory distress syndrome. Am Rev Respir Dis 1988; 137:1–2.
32. Gillis CN, Greene NM, Cronau LH, et al. Pulmonary extraction of 5-hydroxytryptamine and norepinephrine before and after cardiopulmonary bypass in man. Circ Res 1972; 30:666–674.
33. Cartwright GE, Athens JW, Wintrobe MM. The kinetics of granulopoiesis in normal man. Blood 1964; 24:780–803.
34. Hogg JC. The interaction between polymorphonuclear cells and pulmonary endothelium. In: Weir EK, Reeves JT, eds. Pulmonary Vascular Physiology and Photophysiology. New York: Marcel Dekker, 1989:613–642.
35. Fels AOS, Cohn ZA. The alveolar macrophage. J Appl Physiol 1986; 60:353–369.
36. Johnston RB. Monocytes and macrophages. Current concepts: immunology. N Engl J Med 1988; 318:747–752.
37. Lee TH. Interactions between alveolar macrophages, monocytes and granulocytes. Am Rev Respir Dis 1987; 135:S14–S17.
38. Dehring DJ, Wismar BL. Intravascular macrophages in pulmonary capillaries of humans. Am Rev Respir Dis 1989; 139:1027–1029.
39. Staub NC. The Pulmonary Intravascular Macrophage. New York: Futura, 1989.
40. Wiedemann HP, Matthay MA, Gillis CN. Pulmonary endothelial cell injury and altered lung metabolic function. Clin Chest Med 1990; 11:723–736.
41. Parrillo JE, Parker MM, Natanson C, et al. Septic shock in humans. Advances in the understanding of pathogenesis, cardiovascular dysfunction and therapy. Ann Intern Med 1990; 113:227–242.
42. Weil MH, Shubin H, Biddle M. Shock caused by gram-negative microorganisms. Ann Intern Med 1964; 60:384–400.

43. Rackow EC, Astiz ME. Pathophysiology and treatment of septic shock. JAMA 1991; 266:548–558.
44. Mizock B. Septic shock. A metabolic perspective. Arch Intern Med 1984; 144: 579–585.
45. Clowes GHA Jr, Hirsch E, Williams L, et al. Septic lung and shock lung in man. Ann Surg 1981; 181:681–692.
46. Wiedemann HP, Matthay MA, Matthay RA. Adult respiratory distress syndrome. Clin Chest Med 1990; 11:
47. Harris RL, Musher DM, Bloom K, et al. Manifestations of sepsis. Arch Intern Med 1987; 147:1895–1906.
48. Fourrier F, Chopin C, Wallaert B, Mazurier C, Mangalaboyi J, Durocher A. Compared evolution of plasma fibronectin and angiotensin-converting enzyme levels in septic ARDS. Chest 1985; 87:191–195.
49. Clowes GHA Jr, Hirsch E, George BC, et al. Survival from sepsis: the significance of altered protein metabolism regulated by proteolysis inducing factor, the circulating cleavage product of interleukin 1. Ann Surg 1985; 202:446–458.
50. Calandra T, Gerain J, Heumann D, Baumgartner JD, Glauser MP. High circulating levels of interleukin-6 in patients with septic shock: evolution during sepsis, prognostic value, and interplay with other cytokines. Am J Med 1991; 91:23–29.
51. Suffredini AF, Harpel PC, Parrillo JE. Promotion and subsequent inhibition of plasminogen activation after administration of intravenous endotoxin to normal subjects. N Engl J Med 1989; 320:1165–1172.
52. Jacobs ER, Bone RC. Mediators of septic lung injury. Med Clin North Am 1983; 67:701–715.
53. Bone RC. Modulators of coagulation. A critical appraisal of their role in sepsis. Arch Intern Med 1992; 152:1381–1389.
54. Bernard GR, Reines HD, Halushka PV, et al. Prostacyclin and thromboxane A_2 formation is increased in human sepsis syndrome. Am Rev Respir Dis 1991; 144:1095–1101.
55. Parker MM, Suffredini AF, Natanson C, Ognibene FP, Shelhamer JH, Parrillo JE. Responses of left ventricular function in survivors and nonsurvivors of septic shock. J Crit Care 1989; 4:19–25.
56. Chernow B, Zalaga GP. Pharmacologic management of the critically ill patient in the perioperative period: emphasis on the sepsis syndrome. Med Clin North Am 1987; 71:541–549.
57. McMillan M, Chernow B, Roth BL. Alterations in hepatic alpha-1 adrenergic receptors in a rat model of chronic sepsis. Circ Shock 1986; 19:185–191.
58. Ghosh S, Liu M. Changes in α-adrenergic receptors in dog livers during endotoxin shock. J Surg Res 1983; 34:239–245.
59. Holaday JW. Cardiovascular effects of endogenous opiate systems. Ann Rev Pharmacol Toxicol 1983; 23:541–594.
60. Holaday JW, Faden AI. Naloxone reversal of endotoxin hypotension suggests role of endorphins in shock. Nature 1978; 275:450–451.
61. Rock P, Silverman H, Plump D, et al. Efficacy and safety of naloxone in septic shock. Crit Care Med 1985; 13:28–33.

62. Natanson C, Eichenholz PW, Danner RL, et al. Endotoxin and tumor necrosis factor challenges in dogs simulate the cardiovascular profile of human septic shock. J Exp Med 1989; 169:823–832.
63. Hollenberg SM, Cunnion RE, Parrillo JE. The effect of tumor necrosis factor on vascular smooth muscle. Chest 1991; 100:1133–1137.
64. Beasley D, Schwartz JH, Brenner BM. Interleukin 1 induces prolonged L-arginine-dependent cyclic guanosine monophosphate and nitrite production in rat vascular smooth muscle cells. J Clin Invest 1991; 87:602–608.
65. Kilbourn RG, Jubran A, Gross SS, et al. Reversal of endotoxin-mediated shock by N^G methyl-L-arginine, an inhibitor of nitric oxide synthesis. Biochem Biophys Res Commun 1990; 172:1132–1138.
66. Flemming I, Julou-Schaeffer G, Gray GA, Parrat JR, Stoclet JC. Evidence that an L-arginine/nitric oxide dependent elevation of tissue cyclic GMP content is involved in depression of vascular reactivity by endotoxin. Br J Pharmacol 1991; 103:1947–1952.
67. Martin W, Furchgott RF, Villani GM, Jothianandan D. Depression of contractile responses in rat aorta by spontaneously released endothelial-derived relaxing factor. J Pharmacol Exp Ther 1986; 237:529–538.
68. Nava E, Palmer RMJ, Moncada S. Inhibition of nitric oxide synthesis in septic shock: how much is beneficial? Lancet 1991; 338:1555–1557.
69. Udhoji VN, Weil MH. Hemodynamic and metabolic studies on shock associated with bacteremia: observations on 16 patients. Ann Intern Med 1965; 62: 966–978.
70. Lefer AM, Martin J. Origin of myocardial depressant factor in shock. Am J Physiol 1970; 218:1423–1427.
71. Cunnion RE, Schaer GL, Parker MM, Natanson C, Parrillo JE. The coronary circulation in human septic shock. Circulation 1986; 73:637–644.
72. Tracey KJ, Beutler B, Lowry SF, et al. Shock and tissue injury induced by recombinant human cachectin. Science 1986; 234:469–474.
73. Ognibene FP, Parker MM, Natanson C, Shelhamer JH, Parrillo JE. Depressed left ventricular performance. Response to volume infusion in patients with sepsis and septic shock. Chest 1988; 93:903–910.
74. Parker MM, Shelhamer JH, Bacharach SL, et al. Profound but reversible myocardial depression in patients with septic shock. Ann Intern Med 1984; 100:483–490.
75. Hess ML, Hastillo A, Greenfield LJ. Spectrum of cardiovascular function during gram-negative sepsis. Prog Cardiovasc Dis 1981; 23:279–298.
76. Fowler AA, Hamman RF, Zerbe GO, Benson KH, Hyers TM. Adult respiratory distress syndrome: prognosis after onsete. Am Rev Respir Dis 1985; 132: 472–478.
77. Fairman RP, Glauser FL. Pathophysiologic alterations in endotoxemia. Arch Intern Med 1980; 140:788–790.
78. Sibbald WJ, Paterson NAM, Holliday RL, Anderson RA, Lobb TR, Duff JH. Pulmonary hypertension in sepsis. Measurement by the pulmonary arterial diastolic-pulmonary wedge pressure gradient and the influence of passive and active factors. Chest 1978; 73:583–591.

79. Rounds SIS. Pulmonary circulatory control in lung injury. In: Weir EK, Reeves JT, eds. Pulmonary Vascular Physiology and Pathophysiology. New York: Marcel Dekker, 1989:403–433.
80. Greene R, Zapol WM, Snider MT, et al. Early bedside detection of pulmonary vascular occlusion during acute respiratory failure. Am Rev Respir Dis 1981; 124:593–601.
81. Tomashefski JF, Davies P, Boggis C, Greene R, Zapol WM, Reid LM. The pulmonary vascular lesions of the adult respiratory distress syndrome. Am J Pathol 1983; 112:112–126.
82. Ashbaugh DG, Bigelow DB, Petty TL. Acute respiratory distress in adults. Lancet 1967; 2:319–323.
83. Rossaint R, Falke KJ, Keitel M, et al. Successful treatment of severe adult respiratory distress syndrome with inhaled nitric oxide (abstr). Am Rev Respir Dis 1992; 145:A80.
84. Bigatello LM, Hunford WE, Kacmarek RM, Roberts JD, Zapol WM. Inhaled nitric oxide is a selective pulmonary vasodilator in septic patients with severe ARDS (abstr). Am Rev Respir Dis 1992; 145:A185.
85. Frostell C, Fratacci MD, Wain JC, Jones R, Zapol WM. Inhaled nitric oxide, a selective pulmonary vasodilator reversing hypoxic pulmonary vasoconstriction. Circulation 1991; 83:2038–2047.
86. Parker RE, Brigham KL. Effects of endotoxemia on pulmonary vascular resistances in unanesthetized sheep. J Appl Physiol 1987; 63:1058–1062.
87. Brigham KL, Meyrick B. State of the art. Endotoxin and lung injury. Am Rev Respir Dis 1986; 133:913–927.
88. Brigham K, Kariman K, Harris T, et al. Correlation of oxygenation with vascular permeability-surface area but not with lung water in humans with acute respiratory failure and pulmonary edema. J Clin Invest 1983; 72:339.
89. Sibbald WJ, Warshawski FJ, Short AK, Harris J, Lefcoe MS, Holliday RL. Clinical studies of measuring extravascular lung water by the thermal dye technique in critically ill patients. Chest 1983; 83:725–731.
90. Hill SL, Elings VB, Lewis FR. Changes in lung water and capillary permeability following sepsis and fluid overload. J Surg Res 1980; 28:140–150.
91. Kaplan JD, Calandrino FS, Sehnister DP. A positron emission tomographic comparison of pulmonary vascular permeability during ARDS and pneumonia. Am Rev Respir Dis 1991; 143:150–154.
92. Anderson RR, Holliday RL, Driedger AA, Lefcoe M, Reid B, Sibbald WJ. Documentation of pulmonary capillary permeability in the adult respiratory distress syndrome accompanying human sepsis. Am Rev Respir Dis 1979; 119: 869–877.
93. Sibbald WJ, Driedger AA, Moffat JD, Myers ML, Reid BA, Holliday RL. Pulmonary microvascular clearance of radiotracers in human cardiac and noncardiac pulmonary edema. J Appl Physiol Respir Environ Exercise Physiol 1981; 50:1337–1347.
94. Sibbald WJ, Anderson RR, Reid B, Holliday RL, Driedger AA. Alveolocapillary permeability in human septic ARDS. Chest 1981; 79:133–142.

95. Bachofen M, Weibel ER. Structural alterations of lung parenchyma in the adult respiratory distress syndrome. In: Bone R, ed. Clinics in Chest Medicine. New York: W.B. Saunders, 1982:35–56.
96. Lewis JF, Jobe AH. Surfactant and the adult respiratory distress syndrome. Am Rev Respir Dis 1993; 147:218–233.
97. Pietra GG, Szidon JP, Carpenter HA, Fishman AP. Bronchial venular leakage during endotoxin shock. Am J Pathol 1974; 77:387–406.
98. Hopewell PC, Murray JF. Adult respiratory distress syndrome. In: Moser KM, Spragg RG, eds. Respiratory Emergencies. St. Louis, MO: C.V. Mosby, 1982: 106.
99. Newman JH, Ross JC. Chronic cor pulmonale. In: Hurst JW, ed. The Heart. New York: McGraw-Hill, 1989:1220–1229.

7

Effects of Endotoxin on Airway Function

JONATHAN D. PLITMAN and JAMES R. SNAPPER

Vanderbilt University School of Medicine
Nashville, Tennessee

I. Introduction

Airway function can be altered by endotoxins reaching the lungs via inhalation or via the vasculature. In this chapter we first address protective functions of the airspace. A discussion follows on the pathogenetic effects of inhaled endotoxins and their possible role in airways disease. In the remainder of the chapter we address the pathophysiological effects of endotoxemia on airway function and the mechanisms involved in these changes.

II. Limits of Analysis

Measures of disturbed airspace function typically do not yield specific information about the site and mechanism responsible for the changes observed. Altered gas exchange, for example, can result from changes within the airways and alveolar spaces, within the interstitium, or within the vasculature. Ventilation–perfusion abnormalities commonly involve changes in both the air and vascular spaces. Routine lung mechanics measurements do not necessarily distinguish between abnormalities primarily of the larger conducting airways and those of the lungs' periphery; changes in compliance and in resistance to airflow can result from alterations in either or both of these zones (1–3). The potential ambiguity of bronchoalveolar lavage (BAL) data is well recognized (4,5). Concentrations of mediators in BAL fluid, for example, are affected by sampling differences and variable dilutional effects. Increases in the number of inflammatory cells in BAL fluid can result from changes within the conducting airways (bronchitis and bronchiolitis) or from within the alveoli. Assessment of alterations in airspace permeability can reflect changes in the transit of fluid and solutes from the vasculature to the airspace (or vice versa, depending of the methods used) occurring either in the conducting airways (involving the bronchial circulation) (6) or in the alveoli (involving the pulmonary circulation) (7,8). As is true for studies in the pulmonary microvasculature (9), studies of airway fluid and solute exchange are complicated by alterations in the exchange surface area. This summary of difficulties in interpretation is meant only as a reminder of the risks inherent in conclusions that are more precise than the data on which they are based. This is especially important when one studies a generalized lung injury such as is observed in the adult respiratory distress syndrome. Major changes in any one compartment within the lungs (airspaces, interstitium, or vasculature) usually result in significant changes in the others.

The vagaries of interpretation of airway responsiveness deserve special mention. One must distinguish between altered "lung mechanics" and altered "airway responsiveness". When lung mechanics and airway responsiveness are both altered at the same time, there is no reason to assume that the phenomena accountable for the altered mechanics are the same as those accountable for the altered responsiveness. Lung mechanics and airway responsiveness can also vary independently. Asthmatics, for example, can have normal spirometry despite airway hyperresponsiveness on bronchial provocation testing (10). On the other hand, symptoms and abnormal spirometry in asthmatics can probably occur in the absence of laboratory hyperresponsiveness (11,12). We will, however, develop the hypothesis that endotoxin-related alterations in lung mechanics and airway responsiveness are linked, that is, that endotoxins cause airway inflammation with edema,which in turn alters airway responsiveness.

III. Protective Airspace Function

Endotoxins are ubiquitous in the environment. They are present within the cell walls of bacteria or as free lipopolysaccharide in, for example, household dust (13,14), the tap water of most cities (15), and milk produced commercially (15). A significant minority of normal individuals carry gram-negative bacteria in their pharynges (16). Humans have, moreover, markedly altered the quality of the air they breathe, and certain occupations (e.g., those dealing with unprocessed vegetable fibers or animal waste) are particularly associated with increased exposure to endotoxins.

The airspaces escape to a significant degree the adverse local effects of endotoxins. The protective mechanisms are partially understood. Factors such as the upper airways' ability to remove large particulates, the effects of droplet size and velocity on deposition in the airways (17,18), and clearance mechanisms, including cough, mucus secretions, and ciliary action (19–21), are all potentially important. Although the upper airways effectively condition air under normal conditions and remove many particles, the lower airways and alveoli can, over time, receive considerable amounts of endotoxins. Normal lungs tolerate ordinary amounts of inhaled endotoxin well, which stands in notable contrast to the perilous sensitivity of the lung to endotoxins entering via the vascular bed (ref. 22 and below). The alveolar macrophage almost certainly has a regulatory role and is currently the subject of extensive research examining in detail both its ability to attack and remove noxious agents such as bacteria and its possible role in the amplification of lung dysfunction and injury (see ref. 23 and other chapters in this volume). Local immunologic mechanisms such as the release of secretory immunoglobulin A (IgA) are supplemented by systemic production of immunoglobulins (21,24). The interaction of lipid mediators with cytokines and their role in modulating the response to injury are the subject of a number of chapters in this volume. The airspace epithelium also forms a relatively "tight" barrier, blocking the transepithelial movement of endotoxin from the airspace to the vasculature (25). The ability of the lungs to resist the effects of noxious airborne substances is remarkable, especially when viewed from an evolutionary perspective. Many of the changes in the quality of air we breathe and the toxins to which we are exposed are of only recent origin, the products of human labor.

IV. Pathogenetic Responses to Inhaled Endotoxin

Despite these defenses, airborne endotoxins may cause harm under certain circumstances. Sizable inhaled doses (80 to 200 μg) are required to induce moderate (and inconsistent) acute decrements in FEV_1 in human volunteers (26,27). The threshold dose for a spirometric response is probably lower in the presence of preexisting airways disease (i.e., asthma) (28). Transient reductions in diffusion capacity in the absence of spirometric changes have been reported at lower doses (29,30).

Once a threshold dose is attained, pathogenetic mechanisms are initiated. Endotoxins are directly toxic to airway epithelial cells (31), but their principal injurious effects seem to be due to the inflammatory response they trigger. Endotoxins cause airway cells to release soluble factors that are chemotactic to inflammatory cells (32,33). Inhalation of endotoxins can result in rapid accumulation of inflammatory effector cells in the airspaces and the tissues lining them (34–38). [One can speculate that the airspace's relative tolerance to the presence of endotoxin may relate to the relative paucity, under normal conditions, of inflammatory effector cells there, including the near or complete absence of polymorphonuclear leukocytes (39)]. Endotoxins can stimulate the production of an elaborate cascade of mediators, including histamine (40), arachidonic acid metabolites (41,42), platelet activating factor (PAF) (43,44), inflammatory cytokines (45,46), and endothelin (47). The response to inhaled endotoxins is characterized by an increase in airway vascular permeability with exudation of fluid and solutes (48), increased airway responsiveness to nonantigenic stimuli (27,37), and obstruction to airflow (49). Endotoxins can also increase the contractile responsiveness of airway smooth muscle to nonantigenic stimuli (50,51), perhaps by attenuation of the β-adrenergic responsiveness of the muscle (52,53) or by reducing the availability of counterregulatory epithelium-derived relaxing factors (50). Endotoxins impair ciliary action (54) and may alter the amount and composition of airspace lining fluid (55,56).

There is an obvious conceptual connection between these potential effects of airborne endotoxins and the current understanding of the pathogenesis of asthma, which emphasizes the effects of inflammation on the airways (57,58). The local structural effects of endotoxins can both cause an airway response [i.e., precipitate obstruction to airflow (26–28)] and increase airway responsiveness [i.e., heighten the obstructive response to subsequent stimuli (27,28)]. A self-reinforcing asthmalike state could theoretically arise, given serial exposure of sufficient magnitude. This cycle of response and responsiveness is demonstrated, at least symbolically, by the observations that asthmatics are more responsive than normals to inhaled endotoxin (28), and that the intensity of this response is proportional to the subject's airway responsiveness to a dissimilar stimulus such as inhaled histamine (59). It is reasonable to think of endotoxins as one of many pro-inflammatory insults that can adversely affect the airways of a predisposed person. Given the ubiquity of endotoxins in the environment, they would not seem to be a particularly potent pro-inflammatory stimulus. Heavy domestic exposure to airborne endotoxins can, however, worsen asthma by both clinical and spirometric criteria (60).

Most nonasthmatic humans are notably poor responders to moderate exposures to inhaled endotoxins (26,29,61). Substantial chronic exposure, however, may be able to induce an asthmalike state in normals. The study of occupational diseases supplies evidence to support this hypothesis. Byssinosis, a disorder affecting a subset of workers who process raw vegetable fibers such as cotton, flax, and hemp, has several asthmalike features (62). Depending on variable worksite conditions, these workers have been estimated to inhale approximately 5 to 40 μg or more of endotoxins per day (26,29) in concert with dust and various other substances (see below). The disease is characterized by dyspnea, cough, and chest tightness, which typically are most prominent on the first day of work following an absence (the "Monday" phenomenon) (63,64). These symptoms are associated with obstructive spirometric changes during work shifts (65,66) and possibly, over time, reduced FEV_1 and FVC at baseline (67–69). There appears to be a dose-response relationship between disease and the amount of exposure (67,70,71). The disease process may be augmented by the presence of a coirritant such as smoking (68,71,72) or, controversially, atopy (73,74). Byssinotic workers have increased nonspecific airway responsiveness, which tends to be dose related to the degree of occupational exposure (74). Autopsy specimens of byssinotic lungs are notable for changes resembling those of asthmatic bronchitis: increases in mucus glands and goblet cells and smooth muscle hyperplasia (75,76). Emphysema is not common and seems to be explained by age and smoking habit; fibrosis is absent (75,76).

It has long been suggested that byssinosis may be caused by the toxic effects of the numerous gram-negative bacteria that contaminate the raw fibers (77,78). This endotoxin hypothesis has received considerable experimental support. The presence of byssinosis closely correlates with the number of airborne gram-negative bacteria in the workplace (79). Cross-shift (i.e., comparing before and after a work shift) decrements in FEV_1 in cotton mill workers correlate significantly with degree of endotoxin exposure (80). In a group of normal subjects selected for their consistent spirometric response to inhaled cotton dust (roughly one out of three volunteers), decrements in FEV_1 on exposure to cotton dust for several hours correlate strongly with endotoxin exposure, and correlate less well with dust exposure per se (61,81).

Similarly, workers in commercial swine confinement buildings who are exposed to large amounts of dried feces and grain dust report significantly increased frequency of dyspnea, cough, and wheeze (82–84). These workers appear to have baseline reductions in FEV_1 and FVC which are mild but statistically significant in some (83,85,86) but not all (82,87) studies. Cross-shift decrements in expiratory flows or FVC can occur which, again, are mild but statistically significant (84,85). Workers exposed to higher levels of ambient endotoxins seem to have greater cross-shift loss of expiratory flow rate (84). Swine confinement workers also have been observed to have mildly increased airway responsiveness to nonantigenic stimuli (86). Interestingly, the atmosphere inside the confinement buildings seems also to have a deleterious effect on the respiratory health of animals housed therein (88,89).

The primary limitation of these studies (and thus of these diseases as models of endotoxin-induced asthma) is that they represent a respiratory response to a heterogeneous exposure. Both work environments contain pollutants other than endotoxins which could alter airway function [e.g., tannin and methyl piperonylate in the cotton mill (90–92) and ammonia, hydrogen sulfide, antigenic proteins, and a veritable menagerie of microorganisms in the piggery (93,94)]. Endotoxins are somewhat better established as a pathogenetic agent in byssinosis, but even there, they seem to be most toxic in partnership with cigarette smoke. The absence of the type of dramatic airway changes observed with endotoxemia attests once more to the remarkable protective mechanisms in the epithelium-bound space.

V. Airway Responses to Endotoxemia

The adult respiratory distress syndrome (ARDS) is associated with significant abnormalities of lung mechanics ARDS (95,96). Decreased compliance with resultant rapid, shallow breathing and increased deadspace ventilation, not refractory hypoxemia, are often the rate-limiting factors preventing weaning from mechanical ventilation in patients with established ARDS (97). Decreases in compliance are also associated with increased airway pressures in the mechanically ventilated patient and with the development of macroscopic barotrauma (98). ARDS patients have increased airway resistance (99), and increased airway responsiveness to nonantigenic bronchial provocation has been reported in some survivors of ARDS (100). An understanding of mechanisms responsible for the alterations in lung mechanics observed following experimental endotoxemia may suggest rational interventions that could diminish morbidity and mortality in clinical acute lung injury.

One must differentiate between acute and chronic airway changes. Permanent destruction of lung parenchyma can occur in ARDS, and proliferative changes are observed (101). These changes can lead to permanent injury (fibrosis, for example) or can resolve over extended periods of time (102,103). Our knowledge of the mechanisms responsible for the later changes and the potential benefit from interventions designed to prevent these changes is limited. More is known about the acute dysfunctional effects occurring within hours of endotoxemia.

The acute lung injury observed in ARDS is thought of as form of "noncardiogenic" pulmonary edema. Alterations in microvascular permeability and function occur in both the pulmonary and bronchial circulations during acute injury (48,104–106), resulting in the extravasation of plasma liquid and solutes. The resultant edema fluid can be confined to the interstitial space or may collect in the airspace also. The fluid is relatively rich in protein, thus decreasing the osmotic gradient which ordinarily protects against edema formation (107). Under normal conditions, the airway and alveolar epithelia form a considerably tighter barrier than the endothelium to the movement of solutes. Epithelial permeability can markedly increase following endotoxemia. This not only allows the rapid accumulation of airspace edema, but may also diminish a potentially protective increase in interstitial hydrostatic pressure which accompanies the accumulation of interstitial fluid (107). The physical actions of edema fluid may be integral to airspace dysfunction (see below).

The edema fluid observed in ARDS and following endotoxemia contains potentially harmful substances which presumably contribute to the associated lung abnormalities. Many of these factors (including lipid mediators, cytokines, and a variety of inflammatory cells) are the subject of extensive discussions elsewhere in this volume. Where discrete information is available on the role of these agents in endotoxin-induced airway dysfunction, their role will be discussed.

Much of the information on the effects of endotoxin on airway function come from experiments performed on chronically instrumented awake sheep (104). Sheep are remarkably responsive to infused endotoxin—the infusion of relatively high doses of endotoxin (4 μg/kg) can cause severe lung injury, respiratory failure, and death (108). Lower doses of endotoxin (0.5 to 1.0 μg/kg) cause reversible changes in lung mechanics, pulmonary hemodynamics, and lung fluid and solute exchange. It is possible, using the low-dose model, to design intervention studies in which individual sheep serve as their own controls. The vast majority of data on the role of specific mechanisms in endotoxin-induced changes in airway function come from the low-dose ovine model.

Marked early changes in lung mechanics occur within the first hour after endotoxin infusion (109). These changes include a marked decrease in static and dynamic compliance of the lungs and increased resistance to airflow across the lungs. Despite the physiologic evidence for severe bronchoconstriction, thoracic gas volume does not increase. In fact it decreases acutely, by approximately 25%. Concomitant changes in pulmonary hemodynamics occur. Pulmonary artery pressure and pulmonary vascular resistance increase markedly, and there is increased flow of protein-poor lung lymph. This early alteration in lung fluid and solute exchange is felt to result primarily from increased pressure within the pulmonary microvasculature. These changes are followed from 2 to 5 h after endotoxemia by prolonged but less severe abnormalities in lung mechanics. Concurrent with the later changes in lung mechanics is a period of high flow of protein-rich lung lymph, despite relatively normal pulmonary hemodynamics. These later abnormalities in fluid and solute exchange are felt to reflect increased permeability of the pulmonary microvasculature and to be analogous to ARDS in humans (104).

VI. Mediator Effects

Endotoxemia induces a complex cascade of soluble mediator production and release. Endotoxin appears to stimulate a variety of parallel pathways, all of which can result in lung injury. Data support a role not only for eicosanoids and PAF, but also for peptide mediator such as endothelin and the neuropeptides, for the proteinase–antiproteinase system, cytokines, and oxidant injury. Individual chapters in this volume address these mechanisms. Interventions designed to ameliorate the deleterious effects of endotoxemia must take into account the various interactive pathways. Is it more rational to intervene early in the cascade of endotoxin-induced lung injury (e.g., modulation of both PAF and eicosanoid production by inhibiting phospholipase A_2), or should interventions be aimed at later mechanisms (e.g, a thromboxane-receptor antagonist)? Intervention higher in the cascade may blunt a greater number of injurious events but could also interfere with beneficial homeostatic actions by other mediators. Table 1 summarizes the effects of a number of interventions on the sheep's response to endotoxemia. In most cases, the interventions have similar effects on lung mechanics and pulmonary hemodynamics.

Table 1 Effect of Interventions on the Sheep's Response to Endotoxin[a]

Intervention	Early	Late
Cyclooxygenase inhibitors	Yes	No
Thromboxane synthase inhibitors	Yes	No
Thromboxane-receptor antagonist	Yes	No
5-Lipoxygenase inhibitors	Yes	No
PAF-receptor antagonists	Yes	Yes
Hydroxyurea granulocyte depletion	Yes[b]	Yes
N-Acetylcysteine	Yes	Yes
Prostaglandin E_2	Yes	Yes
Verapamil	Yes	Yes
Neutrophil elastase inhibitors	Yes	Yes
Cyclooxygenase inhibitor + corticosteroid	Yes	Yes
Corticosteroids	No	Yes
Antihistamines	No	Yes
Platelet depletion	No	No
Sulfidopeptide leukotriene antagonists	No	No

Source: Adapted from Ref. 125.
[a]See Refs. 109 to 124.
[b]Lung mechanics, not hemodynamics.

A strong case can be made for the role of cyclooxygenase products of arachidonic acid, especially the potent smooth muscle agonist thromboxane, in the early changes induced by endotoxin. Thromboxane-receptor antagonists and thromboxane synthase and cyclooxygenase inhibitor attenuate the early changes in both lung mechanics and pulmonary hemodynamics. A variety of possible mediators of the endotoxin response (cytokines such as tumor necrosis factor; other eicosanoids, such as leukotrienes; and other lipid mediators, such as PAF) all cause secondary production of thromboxane when administered to the sheep. Blocking thromboxane production or blocking the actions of thromboxane attenuates the changes in lung mechanics and pulmonary hemodynamics caused by these exogenous mediators. Interference with the synthesis or action of thromboxane appears, however, not to alter the later changes in lung fluid and solute exchange observed following endotoxemia.

PAF-receptor antagonists, interestingly, attenuate both the early changes in lung mechanics and pulmonary hemodynamics induced by endotoxin and the later increases in lung fluid and solute exchange. PAF-receptor antagonists also decrease the endotoxin-induced production of thromboxane. It is probable, based on these data, that endotoxins cause a cascade of events involving, fairly early, the production of PAF. PAF indirectly causes increased production of cyclooxygenase products, resulting in the observed changes in lung mechanics and pulmonary hemodynamics. These results also suggest that PAF, through a noncyclooxygenase-mediated pathway, contributes to the late endotoxin-induced increases in lung fluid and solute exchange. Preliminary results from the French Sepsis Study Group suggest that a PAF-receptor antagonist may have a beneficial effect in gram-negative sepsis in humans (126).

Although, in general, the early endotoxin-induced changes in lung mechanics are modified by the same interventions that alter the early changes in pulmonary hemodynamics, hydroxyurea-induced granulocyte depletion attenuates the changes in lung mechanics and not the early changes in hemodynamics. In these studies, thromboxane production was attenuated. One possible explanation for the separation between the early endotoxin-induced airway and vascular changes in these experiments is that granulocytes are necessary for the cascade of events leading to the thromboxane production that results in the acute changes in lung mechanics but that sufficient thromboxane is still released, possibly from a different reservoir, such as platelets, to cause maximal early changes in pulmonary hemodynamics. Isolated depletion of platelets, however, does not seem to impair mediator production sufficiently to blunt the response to endotoxemia.

Endotoxemia in sheep also increases airway responsiveness to inhaled nonantigenic bronchoconstrictor agents (127). The mechanism or mechanisms by which endotoxin increases airway responsiveness are only partly understood. In other models, a case has been made for a variety of mediators (e.g., thromboxane, sulfidopeptide leukotrienes, and PAF) in altered airway responsiveness (58). Definitive data do not support a major role for any one mediator in endotoxin-induced increases in airway responsiveness.

VII. Mechanical Effects of Edema

The changes in lung mechanics observed 2 to 5 h following endotoxemia appear to involve both a dynamic and a relatively static component (128). Bronchodilators given late in the response result in a correction toward normal in the observed changes in lung mechanics (129). Histological evidence for airway inflammation and edema also exists (104). Many of the changes observed in lung mechanics and airway responsiveness may result from the structural effects of airway inflammation and edema.

Data exist implying that the structural effects on the airways of endotoxin-induced inflammation (edema and increased airway wall thickness) might contribute to the concomitant alterations of lung mechanics and increases in airway responsiveness. Increased airway responsiveness has been observed in humans with congestive heart failure (130–132) and in animal models of left heart failure (133). When accompanied by pulmonary edema, increased left atrial pressure reduces dynamic compliance and increases airway responsiveness to inhaled histamine in chronically instrumented awake sheep. Acute increases in left atrial pressure without pulmonary edema do neither (134). In left heart failure, as well as in a model of noncardiogenic pulmonary edema using the xenobiotic *Perilla ketone*, the magnitude of the observed increases in airway responsiveness correlate with the severity of the concomitant pulmonary edema (135,136). It is possible that the edema itself, with associated increases in airway wall thickness (136,137), accounts for the observed increases in airway responsiveness. Muscarinic reflexes do not appear to be involved (138).

There are two basic structural mechanisms whereby edema contained *within airway walls* could increase airway responsiveness. The relative importance of these two mechanisms is the subject of debate. In a model elaborated by Hogg, Paré, and their colleagues, edema fluid located on the luminal side of airway smooth muscle could have relatively minor effects on baseline lung mechanics but, when the smooth muscle contracts, cause profound obstruction to airflow. The incompressible edema fluid will cause disproportionate narrowing of the airway lumen (139). The second mechanism, popularized by Permutt, argues that increased airway thickness, irrespective of the location of the edema fluid, will increase airway responsiveness by uncoupling the bronchial smooth muscle from parenchymal and pleural forces which tend to hold the airways open, thus allowing unopposed constriction by airway smooth muscle (140).

Yager and colleagues suggested other possible mechanisms whereby edema fluid contained *with the airway lumen* could have profound effects on airway responsiveness (141). The epithelium of constricted airways, being incompressible, folds to form small interstices (142,143). It has been calculated that as little as 6 mL of fluid could fill the interstices of constricted human airways. Not only can this fluid decrease airway luminal cross-sectional diameter directly (thus increasing resistance to airflow), but the increased surface tension caused by edema fluid could pull adjacent folds closer, thus causing further airway narrowing (141).

In addition to its structural effects, edema fluid could alter the tightly regulated composition of airway surface liquid, resulting in local mediator release and airway narrowing (144-146). It is theoretically possible to manipulate local fluid homeostasis in the airways pharmacologically. Inhaled furosemide, for example, can prevent bronchoconstriction induced by distilled water and both the early and late bronchoconstriction induced by allergen in humans (147,148). The mechanisms of action of inhaled furosemide, however, remain obscure.

It is thus possible that structural changes induced by endotoxin are the cause of the increases in airway responsiveness observed following endotoxemia. Although obviously fraught with hazards, one could extrapolate from the results in sheep to asthma in humans and speculate that, at least in some asthmatics, airway inflammation with increased airway wall thickness causes the increases in airway responsiveness observed. In these patients, increased airway responsiveness would be the result of and a marker of the disease process. Although increased airway responsiveness could exacerbate the disease, it would in these cases be the result, not the cause, of asthma.

VIII. Summary

Endotoxins are ubiquitous in the environment, and the airspace presumably is exposed to them regularly. Inhaled endotoxins may be causal in certain occupational lung disease and may serve as an adjunct pathogen in common asthma. The airspace, however, possesses considerable resistance to the effects of endotoxins, and airborne endotoxin does not regularly cause the type of severe changes in airway physiology noted with endotoxemia. Endotoxin-induced changes in airway function can be a major cause of morbidity and mortality following sepsis in humans. The acute changes in airway function may result from stimulation of thromboxane production. The chronic changes caused by endotoxin probably also involved structural changes, including airway inflammation and edema. Increases in airway wall thickness may contribute to endotoxin-induced increases in airway responsiveness.

Acknowledgment

This work was supported by NIH grants HL 27274, HL 46971, and HL 07123.

References

1. Dodek FM, Paré PD, Minshall DK, et al. Direct injury to the bronchial vasculature in anesthetized sheep. Am Rev Respir Dis 1991; 70:2072–2079.
2. Mitzner W, Blosser S, Yager D, Wagner E. Effect of bronchial smooth muscle contraction on lung compliance. J Appl Physiol 1992; 72:158–167.
3. Snapper JR, Lefferts PL. Effects of aerosol histamine and carbachol on central and peripheral airflow resistance in sheep. J Appl Physiol 1986; 61:760–765.
4. Reynolds HY. Bronchoalveolar lavage. Am Rev Respir Dis 1987; 135:250–263.
5. Schmekel B. A critical view on lavage sampling techniques. Agents Actions Suppl 1990; 30:133–143.
6. McDonald DM. The ultrastructure and permeability of tracheobroncial blood vessels in health and disease. Eur J Respir Dis Suppl 1990; 12:572s–585s.
7. Coggeshall JW, Lefferts PL, Butterfield MJ, et al. Perilla ketone: a model of increased pulmonary microvascular permeability pulmonary edema in sheep. Am Rev Respir Dis 1987; 136:1453–1458.
8. Bradley JD, Roselli RJ, Parker RE, Harris TR. Effects of endotoxemia on the sheep lung microvascular membrane: a two-pore theory. J Appl Physiol 1988; 64:2675–2683.
9. Parker JC, Perry MA, Taylor AE. Permeability of the microvascular barrier. In: Staub NC, Taylor AE, eds. Edema. New York: Raven Press, 1984:143–187.
10. Hargreave FE, Ryan G, Thomson NC, et al. Bronchial responsiveness to histamine or methacholine in asthma: measurement and clinical significance. J Allergy Clin Immunol 1981; 68:347–355.
11. Josephs LK, Gregg I, Mullee MA, Holgate ST. Nonspecific bronchial reactivity and its relationship to the clinical expression of asthma. A longitudinal study. Am Rev Respir Dis 1989; 140:350–357.
12. Pattemore PK, Asher MI, Harrison AC, et al. The interrelationship among bronchial hyperresponsiveness, the diagnosis of asthma, and asthma symptoms. Am Rev Respir Dis 1990; 142:549–554.
13. Peterson RDA, Wicklund PE, Good RA. Endotoxin activity of a house dust extract. J Allergy 1964; 35:134–142.
14. Sirganian RP, Baer H, Hochstein HD, May JC. Allergenic and biologic activity of commercial preparations of house dust extract. J Allergy Clin Immunol 1979; 64:526–533.
15. Di Luzio NR, Friedmann TJ. Bacterial endotoxins in the environment. Nature 1973; 244:49-51.

16. Rosenthal S, Tager IB. Prevalence of gram-negative rods in the normal pharyngeal flora. Ann Intern Med 1975; 83:355–357.
17. Dunnill MS. Pulmonary Pathology, 2nd ed. Edinburgh: Churchill Livingstone, 1987:2.
18. Leopold DA. Pollution: the nose and sinuses. Otolaryngol Head Neck Surg 1992; 106:713–719.
19. Sleigh MA, Blake JR, Liron N. The propulsion of mucus by cilia. Am Rev Respir Dis 1988; 137:726–741.
20. Girod S, Zahm JM, Plotkowski C, et al. Role of the physiochemical properties of mucus in the protection of the respiratory epithelium. Eur Respir J 1992; 5: 477–487.
21. Jacquot J, Gaillard D, Moret S. Functions of proteins and lipids in airway secretions. Eur Respir J 1992; 5:343–358.
22. Wiener-Kronish JP, Albertine KH, Matthay MA. Differential responses of the endothelial and epithelial barriers of the lung in sheep to *Escherichia coli* endotoxin. J Clin Invest 1991; 88:864–875.
23. Sibille Y, Reynolds HY. Macrophages and polymorphonuclear neutrophils in lung defense and injury. Am Rev Respir Dis 1990; 141:471–501.
24. Danielle RP. Immunoglobulin secretion in the airways. Annu Rev Physiol 1990; 52:177–195.
25. Goto H, Rylander R. Kinetics of inhaled lipopolysaccharide in the guinea pig. J Lab Clin Med 1987; 110:287–291.
26. Cavagna G, Foa V, Vigliani EC. Effects in man and rabbits of inhalation of cotton dust or extracts and purified endotoxins. Br J Ind Med 1969; 26:314–321.
27. Rylander R, Bake B, Fischer JJ, Helander IM. Pulmonary function and symptoms after inhalation of endotoxin. Am Rev Respir Dis 1989; 140:981–986.
28. Michel O, Duchateau J, Sergysels R. Effect of inhaled endotoxin on bronchial reactivity in asthmatic and normal subjects. J Appl Physiol 1989; 66:1059–1064.
29. Jamison JP, Lowry RC. Bronchial challenge of normal subjects with the endotoxin of *Enterobacter agglomerans* isolated from cotton dust. Br J Ind Med 1986; 43:327–331.
30. Herbert A, Carvalheiro M, Rubenowitz E, et al. Reduction of alveolar-capillary diffusion after inhalation of endotoxin in normal subjects. Chest 1992; 102:1095–1098.
31. Koyama S, Rennard SI, Claassen L, Robbins RA. Dibutyryl cAMP, prostaglandin E2, and antioxidants protect cultured bovine bronchial epithelial cells from endotoxin. Am J Physiol 1991; 261:L126–L132.
32. Koyama S, Rennard SI, Leikauf GD, et al. Endotoxin stimulates bronchial epithelial cells to release chemotactic factors for neutrophils. A potential mechanism for neutrophil recruitment, cytotoxicity, and inhibition of proliferation in bronchial inflammation. J Immunol 1991; 147:4293–4301.
33. Koyama S, Rennard SI, Leikauf GD, Robbins RA. Bronchial epithelial cells release monocyte chemotactic activity in response to smoke and endotoxin. J Immunol 1991; 147:972–979.

34. Hudson AR, Kilburn KH, Halprin GM. McKenzie WN. Granulocyte recruitment to airways exposed to endotoxin aerosols. Am Rev Respir Dis 1977; 115:89–95.
35. Snella MC, Rylander R. Lung cell reactions after inhalation of bacterial lipopolysaccharides. Eur J Respir Dis 1982; 63:550–557.
36. Venaille T, Snella MC, Holt PG, Rylander R. Cell recruitment into lung wall and airways of conventional and pathogen-gree guinea pigs after inhalation of endotoxin. Am Rev Respir Dis 1989; 139:1356–1360.
37. Pauwels RA, Kips JC, Peleman RA, van-der Straeten ME. The effect of endotoxin inhalation on airway responsiveness and cellular influx in rats. Am Rev Respir Dis 1990; 141:540–545.
38. Sandstrom T, Bjermer L, Rylander R. Lipopolysaccharide (LPS) inhalation in healthy subjects increases neutrophils, lymphocytes and fibronectin levels in bronchoalveolar lavage fluid. Eur Respir J 1992; 5:992–996.
39. The BAL Cooperative Group Steering Committee. Bronchoalveolar lavage constituents in healthy individuals, idiopathic pulmonary fibrosis, and selected comparison groups. Am Rev Respir Dis 1990; 141(5 Pt 2):S169–S202.
40. Norn S, Jensen D, Dahl BT, et al. Endotoxins release histamine by complement activation and potentiate bacteria-induced histamine release. Agents Actions 1986; 18:149–152.
41. Burrell R, Lantz RC, Hinton DE. Mediators of pulmonary injury induced by inhalation of bacterial endotoxin. Am Rev Respir Dis 1988; 137:100–105.
42. Elissalde MH, Beier RC. Stimulation and release of prostaglandins and thromboxane from macrophages by cotton dust associated lipopolysaccharides. Am Ind Hyg Assoc J 1990; 51:652–658.
43. Rylander R, Beijer L. Inhalation of endotoxin stimulates alveolar macrophage production of platelet-activating factor. Am Rev Respir Dis 1987; 135:83–86.
44. Lantz RC, Keller GE, Burrell R. The role of platelet-activating factor in the pulmonary response to inhaled bacterial endotoxin. Am Rev Respir Dis 1991; 144: 167–172.
45. Adams JL, Czuprynski CJ. Bacterial lipopolysaccharide induces release of tumor necrosis factor-alpha from bovine peripheral blood monocytes and alveolar macrophages in vitro. J Leukoc Biol 1990; 48:549–556.
46. Ulich TR, Watson LR, Yin SM, et al. The intratracheal administration of endotoxin and cytokines. I. Characterization of LPS-induced IL-1 and TNF mRNA expression and the LPS-, IL-1-, and TNF-induced inflammatory infiltrate. Am J Pathol 1991; 138:1485–1496.
47. Ninomiya H, Uchida Y, Ishii Y, et al. Endotoxin stimulates endothelin release from cultured epithelial cells of guinea-pig trachea. Eur J Pharmacol 1991; 203: 299–302.
48. Evans TW, Rogers DF, Belvisi MG, et al. Endotoxin-induced plasma exudation in guinea-pig airways in vivo and the effect of neutrophil depletion. Eur Respir J 1990; 3:299–303.
49. Gordon T, Balmes J, Fine J, Sheppard D. Airway oedema and obstruction in guinea pigs exposed to inhaled endotoxin. Br J Ind Med 1991; 48:629–635.
50. Folkerts G, Engels F, Nijkamp FP. Endotoxin-induced hyperreactivity of the guinea-pig isolated trachea coincides with decreased prostaglandin E2 production by the epithelial layer. Br J Pharmacol 1989; 96:388–394.

51. Yamawaki I, Tamaoki J, Kanemura T, et al. Effects of lipopolysaccharide from *Pseudomonas aeruginosa* on airway smooth muscle functions in guinea pigs. Respiration 1990; 57:268–274.
52. Schreurs AJM, Terpstra GK, Raaijmakers JAM, Nijkamp FP. Effect of vaccination with *Haemophilus influenzae* on adrenoceptor function of tracheal and parenchymal strips. J Pharmacol Exp Ther 1980; 215:691–696.
53. van Heuven-Nolsen D, Folkerts G, de Wildt DJ, Nijkamp FP. The influence of *Bordetella pertussis* and its constituents on the beta-adrenergic receptor in the guinea pig respiratory system. Life Sci 1986; 38:677–685.
54. Denny FW. Effect of a toxin produced by *Haemophilus influenzae* on ciliated repiratory epithelium. J Infect Dis 1974; 129:93–100.
55. Tamaoki J, Sakai N, Isono K, et al. Lipopolysaccharide from *Klebsiella pneumoniae* inhibits Na^+ absorption in canine tracheal epithelium. Infect Immun 1991; 59:716–717.
56. Harkema JR, Hotchkiss JA. In vivo effects of endotoxin on intraepithelial mucosubstances in rat pulmonary airways. Quantitative histochemistry. Am J Pathol 1992; 141:307–317.
57. Djukanović R, Roche WR, Wilson JW, et al. Mucosal inflammation in asthma. Am Rev Respir Dis 1990; 142:434–457.
58. Plitman JD, Snapper JR. Nonspecific airway hyperresponsiveness: mechanisms and meaning. In: Simmons DH, Tierney DF, ed. Current Pulmonology. Vol. 13. Chicago: Mosby-Year Book 1992:143–192.
59. Michel O, Ginanni R, Sergysels R. Relation between the bronchial obstructive reponse to inhaled lipopolysaccharide and bronchial responsiveness to histamine. Thorax 1992; 47:288–291.
60. Michel O, Ginanni R, Duchateau J, et al. Domestic endotoxin exposure and clinical severity of asthma. Clin Exp Allergy 1991; 21:441–448.
61. Castellan RM, Olenchock SA, Hankinson JL, et al. Acute bronchoconstriction induced by cotton dust: dose-related responses to endotoxin and other dust factors. Ann Intern Med 1984; 101:157–163.
62. Fishwick D, Pickering CA. Byssinosis: a form of occupational asthma? Thorax 1992; 47:401–403.
63. Schilling RSF. Byssinosis in cotton and other textile workers. Lancet 1956; 2: 261–265.
64. Schilling RSF. Byssinosis in cotton and other textile workers. Lancet 1956; 2: 319–325.
65. McKerrow CB, McDermott M, Gilson JC, Schilling RSF. Respiratory function during the day in cotton workers: a study in byssinosis. Br J Ind Med 1958; 15: 75–83.
66. Berry G, McKerrow CB, Molyneux MKB, et al. A study of the acute and chronic changes in ventilatory capacity of workers in Lancashire cotton mills. Br J Ind Med 1973; 30:25–36.
67. Jones RN, Diem JE, Glindmeyer H, et al. Mill effect and dose-response relationships in byssinosis. Br J Ind Med 1979; 36:305–313.
68. Beck GJ, Maunder LR, Schachter EN. Cotton dust and smoking effects on lung function in cotton textile workers. Am J Epidemiol 1984; 119:33–43.

69. Beck GJ, Schachter EN, Maunder LR. The relationship of respiratory symptoms and lung function loss in cotton textile workers. Am Rev Respir Dis 1984; 130: 6–11.
70. Merchant JA, Lumsden JC, Kilburn KH. Dose response studies in cotton textile workers. J Occup Med 1973; 15:222–230.
71. Berry G, Molyneux MKB, Tombleson JBL. Relationships between dust level and byssinosis and bronchitis in Lancashire cotton mills. Br J Ind Med 1974; 31:18–27.
72. Merchant JA, Lumsden JC, Kilburn KH, et al. An inductrial study of the biologial effects of cotton dust and cigarette smoke exposure. J Occup Med 1973; 15:212–221.
73. Jones RN, Butcher BT, Hammad YY, et al. Interaction of atopy and exposure to cotton dust in the bronchocontrictor response. Br J Ind Med 1980; 37:141–146.
74. Fishwick D, Fletcher AM, Pickering CA, et al. Lung function, bronchial reactivity, atopic status, and dust exposure in Lancashire cotton mill operatives. Am Rev Respir Dis 1992; 145:1103–1108.
75. Pratt PC, Vollmer RT, Miller JA. Epidemiology of pulmonary lesions in nontextile and cotton textile workers: a retrospective autopsy analysis. Arch Environ Health 1980; 35:133–138.
76. Rooke GB. The pathology of byssinosis. Chest 1981; 79(suppl 4):67S–71S.
77. Neal PA, Schneiter R, Caminita BH. Report on acute illness among rural mattress makers using low grade, stained cotton. JAMA 1942; 119:1074–1082.
78. Pernis B, Vigliani EC, Cavagna C, Finulli M. The role of bacterial endotoxins in occupational diseases caused by inhaling of vegetable dusts. Br J Ind Med 1961; 18:120–129.
79. Cinkotai FF, Lockwood MG, Rylander R. Airborne micro-organisms and prevalence of byssinotic symptoms in cotton mills. Am Ind Hyg Assoc J 1977; 38: 554–559.
80. Rylander R, Haglind P, Lundholm M. Endotoxin in cotton dust and respiratory function decrement among cotton workers in an experimental cardroom. Am Rev Respir Dis 1985; 131:209–213.
81. Castellan RM, Olenchock SA, Kinsley KB, Hankinson JL. Inhaled endotoxin and decreased spirometric values. An exposure-response relation for cotton dust. N Engl J Med 1987; 317:605–610.
82. Donham KJ, Zavala DC, Merchant JA. Respiratory symptoms and lung function among workers in swine confinement buildings: a cross-sectional epidemiological study. Arch Environ Health 1984; 39:96–100.
83. Dosman JA, Graham BL, Hall D, et al. Respiratory symptoms and alterations in pulmonary function tests in swine producers in Saskatchewan: results of a survey of farmers. J Occup Med 1988; 30:715–720.
84. Donham K, Haglind P, Peterson Y, et al. Environmental and health studies of farm workers in Swedish swine confinement buildings. Br J Ind Med 1989; 46: 31–37.
85. Haglind P, Rylander R. Occupational exposure and lung function measurements among workers in swine confinement buildings. J Occup Med 1987; 29: 904–907.

86. Zhou C, Hurst TS, Cockcroft DW, Dosman JA. Increased airways responsiveness in swine farmers. Chest 1991; 99:941–944.
87. Schwartz DA, Landas SK, Lassise DL, et al. Airway injury in swine confinement workers. Ann Intern Med 1992; 116:630–635.
88. Donham KJ, Leininger JR. Animal studies of potential chronic lung disease of workers in swine confinement buildings. Am J Vet Res 1984; 45:926–931.
89. Bongers P, Houthuijs D, Remijn B, et al. Lung function and respiratory symptoms in pig farmers. Br J Ind Med 1987; 44:819–823.
90. Ayars GH, Altman LC, O'Neil CE, et al. Cotton dust-mediated lung epithelial injury. J Clin Invest 1986; 78:1579–1588.
91. Lauque DE, Hempel SL, Schroeder MA, et al. Evaluation of the contribution of tannin to the acute pulmonary inflammatory response against inhaled cotton mill dust. Am J Pathol 1988; 133:163–172.
92. Hitchcock M, Piscitelli DM, Bouhuys A. Histamine release from human lung by a component of cotton bracts. Arch Environ Health 1973; 26:177–182.
93. Donham KJ, Popendorf WJ. Ambient levels of selected gases inside swine confinement buildings. Am Ind Hyg Assoc J 1985; 46:658–661.
94. Donham KJ, Scallon LJ, Popendorf W, et al. Characterization of dusts collected from swine confinement buildings. Am Ind Hyg Assoc J 1986; 47:404–410.
95. Ashbaugh DG, Bigelow DB, Petty TL, Levine BE. Acute respiratory distress in adults. Lancet 1967; 2:319–323.
96. Murray JF, Matthay MA, Luce JM, Flick MR. An expanded definition of the adult respiratory distress syndrome. Am Rev Respir Dis 1988; 138:720–723.
97. Yang KL, Tobin MJ. A prospective study of indexes predicting the outcome of trials of weaning from mechanical ventilation. N Engl J Med 1991; 324:1445–1450.
98. Hinson JR, Marini JJ. Principles of mechanical ventilator use in respiratory failure. Annu Rev Med 1992; 43:341–361.
99. Wright PE, Bernard GR. The role of airflow resistance in patients with the adult respiratory distress syndrome. Am Rev Respir Dis 1989; 139:1169–1174.
100. Simpson DL, Goodman M, Spector SL, Petty TL. Long-term follow-up and bronchial reactivity testing in survivors of the adult respiratory distress syndrome. Am Rev Respir Dis 1978; 117:449–454.
101. Hogg JC, Katzenstein A-L. Pulmonary edema and diffuse alveolar injury. In: Thurlbeck WM, ed. Pathology of the Lung. Stuttgart: Thieme, 1988:263–282.
102. Snyder LS, Hertz MI, Harmon KR, Bitterman PB. Failure of lung repair following acute lung injury. Regulation of the fibroproliferative response (Part 1). Chest 1990; 98:733–738.
103. Snyder LS, Hertz MI, Harmon KR, Bitterman PB. Failure of lung repair following acute lung injury. Regulation of the fibroproliferative response (Part 2). Chest 1990; 98:989–993.
104. Brigham KL, Meyrick B. Endotoxin and lung injury. Am Rev Respir Dis 1986; 133:913–927.
105. Lakshminarayan S, Jindal SK, Kirk W, Butler J. Acute increases in anastomotic bronchial systemic to pulmonary blood flow due to generalized lung injury. J Appl Physiol 1987; 62:2358–2361.

106. Traber DL, Lentz CW, Traber LD, Herndon DN. Lymph and blood flow responses in central airways. Am Rev Respir Dis 1992; 146:S15–S18.
107. Snapper JR, Brigham KL. Pulmonary edema: physiology in medicine. Hosp Pract 1993; 21:87–101.
108. Esbenshade AM, Newman JH, Lams PM, et al. Respiratory failure after endotoxin infusion in sheep: lung mechanics and lung fluid balance. J Appl Physiol 1982; 53:967–976.
109. Snapper JR, Hutchison AA, Ogletree ML, Brigham KL. Effects of cyclooxygenase inhibitors on the alterations in lung mechanics caused by endotoxemia in the unanesthetized sheep. J Clin Invest 1983; 72:63–76.
110. Henry CL Jr, Ogletree ML, Brigham KL, Hammon JWJ. Attenuation of the pulmonary vascular response to endotoxin by a thromboxane synthesis inhibitor (UK-38485) in unanesthetized sheep. J Surg Res 1991; 50:77–81.
111. Kuhl PG, Bolds JM, Loyd JE, et al. Thromboxane receptor–mediated bronchial and hemodynamic responses in ovine endotoxemia. Am J Physiol 1988; 254: R310–R319.
112. Coggeshall JW, Christman BW, Lefferts PL, et al. Effect of inhibition of 5-lipoxygenase metabolism of arachidonic acid on response to endotoxemia in sheep. J Appl Physiol 1988; 65:1351–1359.
113. Kuratomi Y, Lefferts PL, Christman BW, et al. Effect of a 5-lipoxygenase inhibitor on endotoxin-induced pulmonary dysfunction in awake sheep. J Appl Physiol 1993; 74:596–605.
114. Christman BW, Lefferts PL, Blair IA, Snapper JR. Effect of platelet-activating factor receptor antagonism on endotoxin-induced lung dysfunction in awake sheep. Am Rev Respir Dis 1990; 142:1272–1278.
115. Hinson JM Jr, Hutchison AA, Ogletree ML, et al. Effect of granulocyte depletion on altered lung mechanics after endotoxemia in sheep. J Appl Physiol 1983; 55:92–99.
116. Bernard G, Lucht WD, Niedermeyer ML, et al. Effect of *N*-acetylcysteine on the pulmonary response to endotoxin in the awake sheep and upon in vitro granulocyte function. J Clin Invest 1984; 73:1772–1784.
117. Brigham KL, Serafin W, Zadoff A, et al. Prostaglandin E_2 attenuation of sheep lung responses to endotoxin. J Appl Physiol 1988; 64:2568–2574.
118. Parker RE, Brigham KL. Verapamil attenuates lung vascular responses to endotoxin in sheep. J Appl Physiol 1988; 65:2138–2143.
119. Gossage JR, Kuratomi Y, Davidson JM, et al. Neutrophil elastase inhibitors, SC-37698 and SC-39026, reduce endotoxin-induced lung dysfunction in awake sheep. Am Rev Respir Dis 1993; 147:1371–1379.
120. Begley CJ, Ogletree ML, Meyrick B, Brigham KL. Modification of pulmonary responses to endotoxemia in awake sheep by steroidal and nonsteroidal anti-inflammatory agents. Am Rev Respir Dis 1984; 130:1140–1146.
121. Brigham KL, Bowers RE, McKeen CR. Methylprednisolone prevention of increased lung vascular permeability following endotoxemia in sheep. J Clin Invest 1981; 67:1103–1110.
122 Brigham KL, Padove SJ, Bryant D, et al. Diphenhydramine reduces endotoxin effects on lung vascular permeability in sheep. J Appl Physiol 1980; 49:516–520.

123. Snapper JR, Hinson JM Jr, Hutchison AA, et al. Effects of platelet depletion on the unanesthetized sheep's response to endotoxemia. J Clin Invest 1984; 74: 1782–1791.
124. Miller RF, Lefferts PL, Snapper JR. Effect of sulfidopeptide leukotriene receptor antagonists on endotoxin-induced pulmonary dysfunction in awake sheep. Am Rev Respir Dis 1992; 146:997–1002.
125. Snapper JR. Airway responses to acute lung injury. In: Brigham KL, Stahlman MT, eds. Respiratory Distress Syndromes: Molecules to Man. Nashville, TN: Vanderbilt University Press, 1990:98–109.
126. Tenaillon A, Dhainut JF, Letulzo B, et al. Efficacy of a P.A.F. antagonist (BN 52021) in reducing mortality of patients with severe gram-negative sepsis. Am Rev Respir Dis 1993; 147:A97.
127. Hutchison AA, Hinson JM Jr, Brigham KL, Snapper JR. Effect of endotoxin on airway responsiveness to aerosol histamine in sheep. J Appl Physiol 1983; 54:1463–1468.
128. Snapper JR. Lung mechanics in pulmonary edema. Clin Chest Med 1985; 6: 393–412.
129. Wright PE, Bernard GR. Mechanisms of late hemodynamic and airway dynamic responses to endotoxin in awake sheep. Am Rev Respir Dis 1989; 140: 672–678.
130. Cabanes LR, Weber SN, Matran R, et al. Bronchial hyperresponsiveness to methacholine in patients with impaired left ventricular function. N Engl Med 1989; 320:1317–1322.
131. Pison C, Malo JL, Rouleau JL, et al. Bronchial hyperresponsiveness to inhaled methacholine in subjects with chronic left heart failure at a time of exacerbation and after increasing diuretic therapy. Chest 1989; 96:230–235.
132. Sasaki F, Ishizaki T, Mifune J, et al. Bronchial hyperresponsiveness in patients with chronic congestive heart failure. 1990; 97:534–538.
133. Kikuchi R, Sekizawa K, Sasaki H, et al. Effects of pulmonary congestion on airway reactivity to histamine aerosol in dogs. J Appl Physiol 1984; 57:1640–1647.
134. Plitman JD, Lefferts PL, Snapper JR. Effect of cardiogenic and non-cardiogenic pulmonary edema on airway responsiveness in awake sheep. Am Rev Respir Dis 1990; 141:A183.
135. Hwang YS, Trochtenberg DS, Plitman JD, et al. Correlation between increased airway responsiveness and severity of pulmonary edema. Am Rev Respir Dis 1992; 145:A610
136. Hwang YS, Lu WS, Lefferts PL, et al. Correlation between increased airway responsiveness and severity of pulmonary edema: a morphometric study. Am Rev Respir Dis 1993; 147:A508.
137. Trochtenberg DS, Hwang YS, Plitman JD, et al. Effect of pulmonary edema on tracheal diameter. Am Rev Respir Dis 1992; 145:A610.
138. Lu WX, Hwang YS, Lefferts PL, et al. Effects of atropine on airway hyperresponsiveness caused by cardiogenic and non-cardiogenic pulmonary edema in awake sheep. Am Rev Respir Dis 1993; 147:A63.
139. Moreno RH, Hogg JC, Paré PD. Mechanics of airway narrowing. Am Rev Respir Dis 1986; 133:1171–1180

140. Drazen JM, Hirschman C, Macklem PT, et al. Mechanisms of bronchoconstriction influenced by inflammation. In: Holgate ST, Howell JBL, Burney PGJ, et al., eds. The Role of Inflammatory Processes in Airway Hyperresponsiveness. Oxford: Blackwell Scientific, 1989:127–133.
141. Yager D, Butler JP, Bastacky J, et al. Amplification of airway constriction due to liquid filling of airway interstices. J Appl Physiol 1989; 66:2873–2883.
142. James AL, Paré PD, Moreno RH, Hogg JC. Quantitative measurement of smooth muscle shortening in isolated pig trachea. J Appl Physiol 1987; 63: 1360–1365.
143. James AL, Hogg JC, Dunn LA, Paré PD. The use of the internal perimeter to compare airway size and to calculate smooth muscle shortening. Am Rev Respir Dis 1988; 138:136–139.
144. Schoeffel RE, Anderson SD, Altounyan REC. Bronchial hyperreactivity in response to inhalation of ultrasonically nebulised solutions of distilled water and saline. Br Med J 1981; 283:1285–1287.
145. Gravelyn TR, Pan PM, Eschenbacher WL. Mediator release in an isolated airway segment in subjects with asthma. Am Rev Respir Dis 1988; 137:641–646.
146. Silber G, Proud D, Warner J, et al. In vivo release of inflammatory mediators by hyperosmolar solutions. Am Rev Respir Dis 1988; 137:606–612.
147. Robuschi M, Gambaro G, Spagnotto S, et al. Inhaled frusemide is highly effective in preventing ultrasonically nebulised water bronchoconstriction. Pulm Pharmacol 1989; 1:187–191.
148. Bianco S, Pieroni MG, Refini RM, et al. Protective effect of inhaled furosemide on allergen-induced early and late asthmatic reactions. N Engl J Med 1989; 321: 1069–1073.

8

Role of Eicosanoids in Endotoxin-Induced Lung Dysfunction

MICHEL MAKHLOUF, MARY DURANDO, JAMES A. COOK, and PERRY V. HALUSHKA

Medical University of South Carolina
Charleston, South Carolina

I. Introduction

Arachidonic acid is a 20-carbon essential fatty acid that is stored in the 2-position of membrane phospholipids and is released from this position by the enzyme phospholipase A_2. Under most circumstances phospholipase A_2 is the rate-limiting step in the synthesis of eicosanoids, the enzymatic products of arachidonic acid. Phospholipase A_2 is activated by a large number of biologically active substances. Included on the list are many of the inflammatory mediators of endotoxic shock and inflammatory stimuli, such as catecholamines, C_{5a}, platelet activating factor (PAF), vasopressin, and angiotensin II. Most of these compounds activate the enzyme through a receptor G-protein-coupled mechanism.

Arachidonic acid is oxygenated to form a myriad of biologically active metabolites. Three major enzymatic pathways are present in cells to catalyze the insertion of molecular oxygen into arachidonic acid. Fatty acid cyclooxygenase catalyzes the insertion of two molecules of O_2 at the C_9 and C_{15} positions of arachidonic acid (1) to form prostaglandin, (PG)G_2. This in turn is metabolized to the unstable endoperoxide PGH_2. PGH_2 is further metabolized by branching enzymes to form prostaglandins and thromboxane, $(TX)A_2$. The ultimate fate of PGH_2 is highly tissue dependent, with different tissues displaying different metabolic profiles. Fatty acid cyclooxygenase is present in two forms: a constituitive enzyme named cyclooxygenase-1 and an inducible enzyme named cyclooxygenase-2 (2). Interleukin-1 (IL-1) and endotoxin are two of the substances known in induce the synthesis of cyclooxygenase-2 (2,3). Thromboxane A_2 is a potent vasoconstrictor, platelet aggregator (4), inhibitor of monocyte chemotaxis (5), stimulator of neutrophil diapedesis (6), and extracellular matrix protein synthesis (7). PGI_2 is a vasodilator and antiaggregatory substance (4). PGE_2 is also a vasodilator and possesses both pro- and anti-inflammatory potential (4). These three fatty acid cyclooxygenase metabolites have been implicated as potential important pathophysiologic mediators of the sequelae of endotoxemia.

5-Lipoxygenase also inserts molecular oxygen into arachidonic acid at the C_5 position to form 5-HPETE. 5-HPETE is metabolized to leukotriene, $(LT)A_4$ which is further metabolized by LTA_4 hydrolase to form LTB_4 or by LTC_4 synthase to form LTC_4. LTC_4 can then be metabolized to LTD_4 and LTE_4. LTC_4 and LTD_4 are also known as SRS-A. LTC_4/D_4 and E_4 are collectively named sulfidopeptide leukotrienes because they contain glutathione, cysteinylglycine or cysteinyl residues covalently linked at the C_6 position. Like the prostaglandins and thromboxanes, the leukotrienes have biological properties that make them excellent candidates as mediators of the pathophysiology of endotoxemia.

5-Lipoxygenase products have been implicated in a variety of inflammatory diseases. They are potent chemotactic/chemokinetic agents for polymorphonuclear leukocytes (PMNs) and eosinophils (8). They are lymphocyte activators and modulate their response immunologically (8–10), contract airway smooth muscle (11) and vascular smooth muscle (8), and cause postcapillary venule dilation, airway hyperreactivity (11), increased vascular permeability (12–14), and increased mucus secretion (11). They also bind to distinct receptors on PMN cell membranes and stimulate PMN adherence and generate superoxides and the release of lysosomal enzymes. The exact role played by 5-lipoxygenase products in the development of the pulmonary injury associated with endotoxic and septic shock is still uncertain, but numerous studies have investigated the effects of specific leukotriene receptor antagonists or lipoxygenase inhibitors in experimentally induced lung injury.

In addition to the 5-lipoxygenase pathway, arachidonic acid is acted on by one of several other types of lipoxygenases (15). 8-, 12-, or 15-Lipoxygenase insert an O_2 into the 8, 12, or 15 C to produce 8-, 12-, or 15-hydroperoxy, 8,11,14-eicosatetraenoic acid (HPETE). These may be further metabolized to active metabolites.

Although more is currently known about the 5-lipoxygenase products, and they appear to play a prominent role in inflammatory diseases, the 12- and 15-lipoxygenase systems do have specific cellular effects. 12-Lipoxygenase, primarily in platelets and leukocytes, forms 12-HPETE from arachidonic acid. This is reduced to 12-HETE and can cause secretion of PMN specific granules, enhance IgE-mediated mast cell release, enhance chemotaxis, and enhance tumor cell adhesion. However, they do not appear to be very active biologically; high concentrations are generally required to produce pharmacological effects (15).

15-Lipoxygenase is found primarily in leukocytes and reticuloendothelial cells, and converts arachidonic acid to 15-HPETE. 15-HPETE can be reduced to 15-HETE, and either of these compounds can yield lipoxins A and B. Lipoxin A may play a role in patients with certain pulmonary diseases (16). 15-HPETE and/or 15-HETE can modify cellular immune functions, modulate other oxygenation enzymes, are mitogenic, inhibit natural killer cell activity, and activate neutrophils and protein kinase C (15).

A third metabolic pathway that is gaining increased interest and importance is the cytochrome P_{450} oxidation of arachidonic acid (17). This pathway also leads to a myriad of oxygenated fatty acids that have diverse biological activities. It is unknown at this time whether or not these compounds have any potential role in mediating the actions of endotoxin.

II. Synthesis of the Cyclooxygenase Products in Endotoxemia and Sepsis

Endotoxin administration to a variety of animal species causes the release of a wide array of inflammatory mediators. Included among them are the fatty acid cyclooxygenase products of arachidonic acid. Nearly all of the cyclooxygenase metabolite levels increase in response to endotoxin; however, TXA_2 and prostacyclin (PGI_2) metabolites are the most frequently measured. Their levels are determined by their stable but inactive metabolites, TXB_2 and 6-keto-prostaglandin $F_{1\alpha}$ (6-keto-$PGF_{1\alpha}$), respectively. The pattern of their release in response to endotoxin depends on the method of endotoxin administration.

When endotoxin is injected intravenously, TXB_2 levels rise immediately, peak within 2 to 35 min, and decline rapidly. In contrast, 6-keto-$PFG_{1\alpha}$ levels rise more slowly, peak within 75 to 120 min, and show a slower decline. This pattern of release in response to endotoxin has been well documented in a number of species (18–21).

TXB_2 and 6-keto-$PGF_{1\alpha}$ have also been measured after repeated intravenous infusions of endotoxin. TXB_2 levels peak within 1 h after each infusion and subsequently return to baseline. However, with repeated infusions, the TXB_2 peak becomes smaller (22) and the TXB_2 response to LPS may be lost (23). 6-Keto-$PGF_{1\alpha}$ levels show a much less consistent response. Godsoe et al. (23) found no changes in 6-keto -$PGF_{1\alpha}$ levels in sheep after any of the multiple infusions of *E. coli* endotoxin. Demling et al. (22) showed an increase in 6-keto-$PGF_{1\alpha}$ levels after the first infusion with only baseline levels following the repeated infusions, while Klosterhalfen et al. (24) noted a higher peak after the second infusion of endotoxin than after the first. These differences in PGI_2 repsonse to endotoxin may be due to species differences (25), the physiologic state of the animal, and the rate and dose of endotoxin infusion.

III. Cyclooxygenase Products and Lung Injury

The role of the various cyclooxygenase products in endotoxin-induced lung injury is not completely understood. Species differences, differences in experimental procedures, and the involvement of a multitude of mediators [tumor necrosis factor (TNF), interleukins, platelet activating factor] with their multiple and varied effects have made unraveling the role of each cyclooxygenase metabolite rather difficult. However, certain effects of the prostaglandins have been well established, and manipulating these effects may prove helpful in the management of gram-negative sepsis and its sequelae.

Following endotoxin infusion, there is marked pulmonary hypertension, increase vascular resistance, and edema associated with the early increase in TXB_2 levels. The early pulmonary edema is of low protein content, reflecting changes in microvascular pressure but not in permeability (26). Later, the lymph flow becomes richer in protein, indicating increased endothelial permeability (23). Increased airway resistance and reduced lung compliance are also seen (27) associated with immediate and sustained drops in pH and PO_2. Endotoxin also induces acute thrombocytopenia by causing platelet sequestration in the pulmonary vasculature. These effects are caused, at least in part, by TXA_2. TXA_2 in the lung produces arteriolar vasoconstriction (28) and venoconstriction (29), which could increase the pulmonary artery and capillary pressures; although its relative effects on each pulmonary vascular segment may be species related. The vascular effects of TXA_2 can be reversed by the vasodilator PGI_2 (28). Thromboxane synthase inhibitors and thromboxane receptor antagonists ameliorate the pulmonary function abnormalities by blocking or attenuating pulmonary hypertension, early lymph flow (29,30), improving pulmonary airflow and compliance (27), and reducing lymph protein clearance (30). Although increased vascular permeability and protein lymph are believed to be due to endothelial injury by neutrophils and not by TXA_2 (31,32), thromboxane receptor inhibition may have an indirect effect on pulmonary capillary permeability by preventing neutrophil adhesion. The shunting of PGH_2 toward PGI_2 during thromboxane synthase inhibition (26) would also contribute to reducing pulmonary hypertension. However, the effects of PGI_2 are not always beneficial. In newborn lambs, PGI_2 can potentiate TXA_2 mimetic-induced pulmonary edema (33) and in adult rabbit lungs PGI_2 causes cyclooxygenase-dependent vasoconstriction (34). PGI_2, via its vasodilating effect, may cause systemic hypotension that could enhance the lethal sequelae of endotoxin.

Leukocytes play an important role in endotoxin-induced lung injury. Inducing neutropenia in sheep by treatment with hydroxyurea reduced lung lymph protein clearance and ameliorated altered pulmonary mechanics without affecting the pulmonary artery pressure (31,32). Depletion of neutrophils with nitrogen mustard, also in sheep, attenuated the pulmonary hypertensive response to endotoxin, with a concomitant decline in TXB_2 levels (35). Neutrophils sequestered in the rat lung by endotoxin produce large amounts of TXB_2 and induce intense vasoconstriction when stimulated with the chemotactic tripeptide FMLP (36). These observations suggest that vascular neutrophils may not be necessary for pulmonary dysfunction, but when activated by endotoxin and other products, they induce severe lung injury.

Unlike TXA_2 and PGI_2, prostaglandin E_2 may have an anti-inflammatory effect. PGE_2 prevents the release of LTB_4 and superoxide ion from neutrophils (37) and IL-2 from lymphocytes (38). PGE_2 can reduce the levels of TNF-α produced by rat macrophages (discussed later). Administration of PGE_2 to sheep significantly reduced the endotoxin-induced pulmonary hypertension, hypoxemia, lymph flow, and lymph protein clearance and the elevated levels of TXB_2 and 6-keto-$PGF_{1\alpha}$ (39). In a limited clinical study, Holcroft et al. (40) demonstrated a marked increase in survival of ARDS patients treated with PGE_1.

The use of nonsteroidal anti-inflammatory drugs (NSAIDs) (cyclooxygenase inhibitors) and thromboxane synthase inhibitors has beneficial effects in experimental endotoxemia and sepsis. In addition to improving pulmonary function, NSAID generally improve survival and survival time, reduce cardiac dysfunction, and improve indices of tissue ischemia in endotoxemia (41–43). Thromboxane synthase inhibitors and receptor antagonists also attenuate many of the shock sequelae induced by endotoxin (26,27,30,43,44). Thromboxane synthase inhibitors do not improve survival in septic neonatal rats (45), in adult rats (46,47) in endotoxic baboons (48), or in rabbits with hyperdynamic endotoxemia (49). The lack of protection afforded by these agents may be a consequence of different profiles of eicosanoid synthesis, species differences, or effects of mediators other than thromboxane.

IV. Lipoxygenase Products and Lung Injury

Measurements of elevated levels of leukotriene metabolites in bronchoalveolar lavage fluid and in urine from patients with adult respiratory distress syndrome (ARDS) and/or multiple-organ-system failure has supported their role in the pathophysiology of pulmonary compromise. The urinary excretion of LTE_4 was found to be increased in patients with multiple trauma without ARDS when compared with healthy controls and was markedly greater in trauma patients who developed ARDS (49). Another study measured the urinary excretion of LTE_4 and compared ARDS patients with febrile postoperative subjects, healthy volunteers, and mild asthmatics. They found significantly higher levels of urinary LTE_4 in ARDS patients than in other groups. There was also a trend toward increased LTE_4 excretion in the more severe ARDS cases (50).

Elevated levels of the sulfidopeptide leukotrienes have been detected in bronchoalveolar lavage fluid, pulmonary edema fluid, and blood from patients with ARDS (51–53). Seegar et al. found LTB_4 and two of its metabolites (20-COOH-LTB_4 and 20-OH-LTB_4) in bronchoalveolar lavage fluid from 15 of 17 samples in patients with ARDS. In contrast, they were unable to detect leukotrienes in healthy volunteers or in patients with cardiogenic pulmonary edema (54). Another study measured plasma levels of LTB_4 prospectively as a prognostic indicator for the development of ARDS in acutely ill patients. Patients who developed ARDS had significantly enhanced LTB_4 levels at admission when compared with patients at risk who did not develop ARDS (55). A study that used essential fatty acid–deficient rats (EFAD) in an oleic acid–induced model of pulmonary injury found that depletion of arachidonate substrate resulted in decreased lung permeability index, and no hypoxemia, in comparison with control rats (56). Rats fed a control diet had increased levels of LTB_4, LTC_4, and LTD_4 in bronchoalveolar lavage fluid after oleic acid compared with saline controls. Interestingly, the EFAD rats had lower levels of LTB_4 than those of the oleic acid control rats, whereas their levels of LTC_4/LTD_4 were not decreased.

In sheep and rat models of ARDS and endotoxic shock, 5-lipoxygenase inhibitors have exerted protective effects. They have improved mortality, reduced systemic hypotension, decreased pulmonary vascular permeability, decreased pulmonary hypertension, decreased bronchoconstriction, and increased arterioalveolar oxygen differences (57–61). Using specific receptor antagonists, investigators have been able to demonstrate roles for both LTB_4 and sulfidopeptide leukotrienes. An LTB_4-receptor antagonist attenuated systemic hypotension, hemoconcentration, leukopenia, and thrombocytopenia (62) in endotoxemia. LTD_4/E_4 receptor antagonists attenuated microvascular permeability, systemic hypotension, bronchoconstriction, and pulmonary hypertension (63–66).

Studies that have demonstrated increased generation of leukotrienes in ARDS patients, and experimental models that have shown amelioration of some of the clinical signs associated with pulmonary damage by specific receptor antagonists or lipoxygenase inhibitors, strongly implicate a role for leukotrienes in the pathogenesis of ARDS and multiple-organ-system failure. Further studies, such as large multicenter clinical trials, are needed to assess the efficacy of some of the treatments, and much work is still needed to elucidate the exact role of the leukotrienes in the pulmonary failure associated with septic/endotoxic shock.

V. Interaction with Platelet Activating Factor

Platelet activating factor (PAF) is a phospholipid with a wide variety of proinflammatory actions. Its infusion into animals causes physiologic changes similar to those seen with endotoxin (67). Its levels are increased in endotoxin-treated rats (68,69) and administration of PAF-receptor antagonists prolong the survival from shock (70). It appears that eicosanoids may mediate many of the effects of PAF. PAF has been shown to activate phospholipase A_2 or phospholipase C (71), and act as second messenger in the release of eicosanoids and superoxide anions from leukocytes and endothelial cells (72). In cultured equine peritoneal macrophages, it induces increased TXA_2 synthesis (73). In vivo administration of the PAF-receptor antagonist SRI 63-675 to pigs prior to and during endotoxin administration ameliorated the abnormal pulmonary function parameters and reduced the plasma levels of TXB_2 and 6-keto-$PGF_{1\alpha}$ (74) compared to endotoxin alone. Administration of the thromboxane synthase inhibitor OKY046 to sheep reduced the pulmonary hypertension induced by the first injection of PAF. However, with subsequent PAF infusion, OKY046 was of no effect (67), suggesting that PAF may have direct and indirect effects via mediators other than TXA_2. Jancar et al. (75) demonstrated a contractile response to PAF on guinea pig lung parenchyma strips that was independent of TXA_2. In the guinea pig, PAF receptor antagonists attenuated endotoxin-induced platelet sequestration in the lung, whereas the cyclooxygenase inhibitor indomethacin did not (76).

Not only is PAF synthesis induced by endotoxin, it synergizes with the latter in producing lung injury. Infusion of either of the above at low doses in rabbit lungs produced only minimal changes in lung hemodynamics. Exposure of the lungs to endotoxin 2 h prior to PAF caused marked pulmonary changes when PAF was administered, including hypertension, edema, increased protein lymph clearance and elevated TXB_2, and 6-keto-$PGF_{1\alpha}$ levels. These changes were inhibited by thromboxane synthase inhibitors and receptor antagonists (77). Similar results are obtained in the rat, where PAF primes the rats for endotoxin-induced TNF-α and other hemodynamic changes (78). It is noteworthy to mention that in the guinea pig lung, TNF induces neutrophil-dependent increased microvascular pressure and permeability mediated by TXA_2 and PAF (79). These observations suggest that in shock, endotoxin induces inflammatory cells to release PAF and TNF, which would potentiate the effects of endotoxin and each other.

VI. Tumor Necrosis Factor

Tumor necrosis factor (TNF, cachectin) is a cytokine produced in great abundance by macrophages upon stimulation with bacterial endotoxin (80). It is now recognized as one of the principal endogenous mediators of endotoxic shock (80,81). Injection of purified human recombinant TNF in rats or mice produces a constellation of lethal shock sequelae, including altered carbohydrate homeostasis, diarrhea, metabolic acidosis, and multifocal microthrombi or ischemic necrosis of the bowel, lungs, and liver (82). Increased eicosanoid production is one element of toxicity in animals shocked by TNF administration. Tumor necrosis factor–induced shock sequelae are attenuated and survival markedly enhanced in animals pretreated with the NSAIDs ibuprofen, indomethacin, or aspirin. The kinetics of TNF release in humans administered endotoxin is similar to that seen in laboratory animals (83). As with animal studies, pretreatment of humans with ibuprofen ameliorates many of the symptoms of endotoxemia, including fever, tachycardia, and release of stress hormones, but does not alter plasma TNF levels.

Several studies have demonstrated direct effects of TNF on tissue cyclooxygenase activity. Tumor necrosis factor is a potent stimulant of vascular endothelial cell PGI_2 synthesis, which may play a role in the depressed blood pressure associated with TNF-induced shock (84). This cytokine also stimulates collagenase and PGE_2 production from macrophages and interacts in a synergistic fashion with interleukin-1 (IL-1) to augment fibroblast PGE_2 production (85,86). The pyrogenic effect of TNF has been attributed both to direct TNF stimulation of hypothalamic PGE_2 synthesis as well as to TNF-stimulated endogenous interleukin-1 release (83).

The effect of TNF on the lipoxygenase pathway is less well defined. TNF does not cause direct release of LTB_4 or LTC_4 from human neutrophils or eosinophils. It does, however, prime these cells for enhanced production of LTB_4 or LTC_4 (tenfold and fourfold, respectively) following stimulation with calcium ionophore A23187. Administration of TNF in rats also enhances endogenous leukotriene production as determined by increased biliary *N*-acetyl-LTE_4 levels (87).

The mechanism whereby TNF induces eicosanoid synthesis is uncertain. In human synovial cells, TNF decreases tritium-labeled arachidonic acid content in the phospholipids PC, PS, and PI, indicating that it directly or indirectly activates phospholipase A_2 (88). In this study neither dibutyryl cyclic adenosine monophosphate (cAMP) nor PGE_2-depressed TNF induced tritium-labeled arachidonic acid release, suggesting that phospholipase activation was cAMP independent (88). The production of PGE_2, however, may have important modulating effects on TNF synthesis by macrophages. Low doses of PGE_2 (0.1 to 10 ng/mL) stimulated TNF-α production by rat peritoneal macrophages (89). This effect required de novo protein synthesis and a rise in cellular cyclic guanosine monophosphate (cGMP). The PGE_2 effect, however, was biphasic because higher doses of PGE_2 ($>$10 ng/mL) or agents that induce cAMP depressed TNF-α production by macrophages. These observations suggest that TNF-α synthesis in macrophages may be up-regulated by cGMP and down-regulated by cAMP in the classical yin–yang fashion.

VII. Oxygen Radicals

There is an increasing body of evidence for involvement of oxygen radicals in circulatory shock sequelae and in reperfusion injury (90). These unstable oxygen metabolities include the univalent radical hydroxyl, superoxide anion, and hydrogen peroxide. Oxygen free radical–induced tissue injury is frequently associated with increased eicosanoid production. In vitro neutrophil-derived H_2O_2 induced release of PGI_2 from cultured endothelial cells (91). In the ex vivo perfused rabbit lung, H_2O_2 induces increased perfusate levels of TXB_2 (92). Pulmonary hypertension and hydrostatic edema in this form of lung injury are significantly reduced by inhibiting thromboxane formation (93). In unanesthetized sheep, catalase, a specific scavenger for H_2O_2, prevents TXB_2 formation and reduces endotoxin-induced lung injury (94). The lipoxygenase products are also implicated as secondary mediators of free radicals. Increased levels of LTB_4, LTC_4, and LTD_4 are detectable following oxidant-induced injury in the perfused rabbit lung (93). Evidence for participation of leukotrienes in tissue injury comes from experiments in isolated rat lung in which the leukotriene receptor antagonist FPL55712, the lipoxygenase inhibitor U60257, and the dual inhibitor BW755C attenuated lung edema induced by H_2O_2 (95).

The relationships between eicosanoid formation and oxygen-free-radical formation remain to be clearly defined. Ward et al. (96) studied functional responses (eicosanoid production, enzyme secretion, and generation of O_2) in rat neutrophils to several agonists, such as phorbol myristate acetate (FMLP), calcium ionophore, and IgG immune complexes. Enzyme secretion was enhanced by NSAIDs (indomethacin, ibuprofen, and piroxicam) and reduced by the lipoxygenase inhibitors (NDGA and nafazatrom). Superoxide anion formation, however, was reduced by of inhibitors both pathways.

Oxygen-free-radical formation may be linked to the major enzymes of the eicosanoid cascade. Either purified cyclooxygenase or lipoxygenase enzyme acts on arachidonic acid or linoleic acid substrate in the presence of reduced nicotinamide adenine dinucleotide (NADH) or reduced nicotinamide adenine dinucleotide phosphate (NADPH) to generate superoxide anion (97). Indomethacin inhibits superoxide anion formation from cyclooxygenase but does not inhibit lipoxygenase. Thus, either of these enzymes in the presence of a suitable reducing substrate can generate the superoxide radical. Both arachidonic acid and PGG_2 generate oxygen-free-radical formation in sarcoplasmic reticulum of muscle under acidic conditions, an effect also inhibited by indomethacin (98). In neutrophils, arachidonic acid stimulates oxygen free radicals through activation of NADPH oxidase, apparently through a phospholipase A_2–dependent mechanism (99).

Certain arachidonic acid end products may modulate oxygen-free-radical formation. Prostaglandin E_2 and PGE_1 and PGD_2 depress agonist-induced generation of superoxide anions by neutrophils. The thromboxane synthetase inhibitor U63,577A reduces oxygen-free-radical formation in neutrophils and platelets, possibly by increased shunting to prostaglandins (100). In contrast, certain lipoxygenase products may stimulate oxygen-free-radical formation. Leukotriene B_4 induces a rapid but transient neutrophil chemiluminescence, superoxide anion formation, and aggregatory response (101). Pertussis toxin, an inhibitor of guanine nucleotide binding proteins, blocked the LTB_4-induced superoxide anion activity and PI cycle activity, suggesting that LTB_4 acts through a guanine nucleotide binding protein coupled to phospholipase C (102). Leukotriene D_4 has been shown to generate superoxide anions in the guinea pig which induces bronchial smooth muscle hyperreactivity, an effect reduced by the free-radical scavenger superoxide dismutase (103).

VIII. Summary

There is a large body of experimental evidence implicating a pathophysiologic role for cyclooxygenase and lipoxygenase metabolites of arachidonic acid in the pathogenesis of endotoxin-induced lung injury. Clearly, further clinical studies with inhibitors of the synthesis or action of some of the eicosanoids is indicated to further elucidate their role in ARDS.

References

1. Samuelsson B, Goldyne M, Granstrom E. Prostaglandins and thromboxanes. Annu Rev Biochem 1978; 47:997–1029.
2. Xie WL, Chipman JG, Robertson DL, Erikson RL, Simmons DL. Expression of a mitogen-responsive gene encoding prostaglandin synthase is regulated by mRNA splicing. Proc Natl Acad Sci USA 1991; 88(7):2692–2696.
3. Meade EA, Smith WL, DeWitt DL. Pharmacological profiles of the PGH synthase-1 and PGH synthase-2. In: Book of Abstracts. 8th International Conference on Prostaglandins and Related Compounds, 1992:79.
4. Halushka PV, Mais DE, Mayeux PR, Morinelli TA. Thromboxane, prostaglandin and leukotriene receptors. Annu Rev Pharmacol Toxicol 1989; 29(1):213–239.
5. Campbell PB, Tolson TA. Modulation of human monocyte leukotactic responsiveness by thromboxane A2 and 12-hydroxyheptadecatrienoic acid (12-HHT). J Leukoc Biol 1988; 43(2):117–124.
6. Goldman G, Welbourne R, Klausner JM, Valeri CR, Shepro D, Hechtman HB. Thromboxane mediates diapedesis after ischemia by activation of neutrophil adhesion receptors interacting with basally expressed intercellular adhesion molecule-1. Circ Res 1991; 68(4):1013–1019.
7. Bruggeman LA, Horigan EA, Horikoshi S, Ray PE, Klotman PE. Thromboxane stimulates synthesis of extracellular matrix proteins in vitro. Am J Physiol 1991; 261 (3 Pt 2) pf 488–494.
8. Garcia JG, Noonan TC, Jubiz W, Malik AB. Leukotrienes and the pulmonary microcirculation. Am Rev Respir Dis 1987; 136(1):161–169.
9. Rola PM, Borgeat P, Sirois P. Leukotriene B4 induces human suppressor lymphocytes. Biochem Biophys Res Commun 1982; 108(4):1531–1537.
10. Rola PM, Gagnon L, Sirois P. Leukotriene B4 augments human natural cytotoxic cell activity. Biochem Biophys Res Commun 1983; 113(2):531–537.
11. Drazen JM, Austen KF. Leukotrienes and airway responses. Am Rev Respir Dis 1987; 136(4):985–998.
12. Brigham KL. Metabolites of arachidonic acid in experimental lung vascular injury. Fed Proc 1985; 44 (1 Pt 1) p 43–45.
13. Joris I, Majno G, Corey EJ, Lewis RA. The mechanism of vascular leakage induced by leukotriene E4. Endothelial contraction. Am J Pathol 1987; 126(1); 19–24.
14. Hua XY, Dahlen SE, Lundberg JM, Hammarstrom S, Hedqvist P. Leukotrienes C4, D4 and E4 cause widespread and extensive plasma extravasation in the guinea pig. Naunyn Schmiedebergs Arch Pharmacol 1985; 330(2):136–141.
15. Holtzman MJ. Arachidonic acid metabolism. Implications of biological chemistry for lung function and disease. Am Rev Respir Dis 1991; 143(1):188–203.
16. Lee TH, Crea AE, Gant V, et al. Identification of lipoxin A4 and its relationship to the sulfidopeptide leukotrienes C4, D4, and E4 in the bronchoalveolar lavage fluids obtained from patients with selected pulmonary diseases. Am Rev Respir Dis 1990; 141(6):1453–1458.

17. McGiff JC. Cytochrome P-450 metabolism of arachidonic acid. Annu Rev Pharmacol Toxicol 1991; 31(1):339–369.
18. Watkins WD, Huttemeier PC, Kong D, Peterson MB. Thromboxane and pulmonary hypertension following *E coli* endotoxin infusion in sheep: effect of an imidazole derivative. Prostaglandins 1982; 23(3):273–285.
19. Bottoms GD, Templeton CB, Fessler JF, et al. Thromboxane, prostaglandin I2 (epoprostenol), and the hemodynamic changes in equine endotoxin shock. Am J Vet Res 1982; 43(6):999–1002.
20. Harris RH, Zmudka M, Maddox Y, Ramwell PW, Fletcher JR. Relationship of TXB2 and 6-keto PGF1 alpha to the hemodynamic changes during baboon endotoxic shock. Adv Prostaglandin Thromboxane Res 1980; 7(1):843–849.
21. Hales CA, Sonne L, Peterson M, Kong D, Miller M, Watkins WD. Role of thromboxane and prostacyclin in pulmonary vasomotor changes after endotoxin in dogs. J Clin Invest 1981; 68(2):497–505.
22. Demling RH, Lalonde CC, Jin LJ, Albes J, Fiori N. The pulmonary and systemic response to recurrent endotoxemia in the adult sheep. Surgery 1986; 100(5): 876–883.
23. Godsoe A, Kimura R, Herndon D, et al. Cardiopulmonary changes with intermittent endotoxin administration in sheep. Circ Shock 1988; 25(2):61–74.
24. Klosterhalfen B, Hortmann JK, Vogel P, et al. Hemodynamic variables and plasma levels of PGI2, TXA2 and IL-6 in a porcine model of recurrent endotoxemia. Circ Shock 1991; 35(4):237–244.
25. Yellin SA, Nguyen D, Quinn JV, Burchard KW, Crowley JP, Slotman GJ. Prostacyclin and thromboxane A2 in septic shock: species differences. Circ Shock 1986; 20(4):291–297.
26. Winn R, Harlan J, Nadir B, Harker L, Hildebrandt J. Thromboxane A2 mediates lung vasoconstriction but not permeability after endotoxin. J Clin Invest 1983; 72(3):911–918.
27. Schutzer KM, Lindholm E, Haglund U, Falk A. Cardiopulmonary function as related to thromboxane A2 synthesis in experimental septic shock. Circ Shock 1988; 26(1):27–40.
28. Barnard JW, Ward RA, Adkins WK, Taylor AE. Characterization of thromboxane and prostacyclin effects on pulmonary vascular resistance. J. Appl Physiol 1992; 72(5):1845–1853.
29. Yoshimura K, Tod ML, Pier KG, Rubin LJ. Role of venoconstriction in thromboxane-induced pulmonary hypertension and edema in lambs. J Appl Physiol 1989; 66(2):929–935.
30. Kuhl PG, Bolds JM, Loyd JE, Snapper JR, FitzGerald GA. Thromboxane receptor-mediated bronchial and hemodynamic responses in ovine endotoxemia. Am J Physiol 1988; 2(1):2.
31. Heflin AC Jr, Brigham KL. Prevention by granulocyte depletion of increased vascular permeability of sheep lung following endotoxemia. J Clin Invest 1981; 68(5):1253–1260.
32. Hinson JM Jr, Hutchison AA, Ogletree ML, Brigham KL, Snapper JR. Effect of granulocyte depletion on altered lung mechanics after endotoxemia in sheep. J Appl Physiol 1983; 55 (1 Pt 1) p 92–99.

33. Yoshimura K, Tod ML, Pier KG, Rubin LJ. Effects of a thromboxane A2 analogue and prostacyclin on lung fluid balance in newborn lambs. Circ Res 1989; 65(5):1409–1416.
34. Kaapa P, Raj JU, Ibe BO, Anderson J. Vasoconstrictor response to prostacyclin in rabbit pulmonary circulation. Respir Physiol 1991; 85(2):193–204.
35. Huttemeier PC, Watkins WD, Peterson MB, Zapol WM. Acute pulmonary hypertension and lung thromboxane release after endotoxin infusion in normal and leukopenic sheep. Circ Res 1982; 50(5):688–694.
36. Voelkel NF, Czartolomna J, Simpson J, Murphy RC. FMLP causes eicosanoid-dependent vasoconstriction and edema in lungs from endotoxin-primed rats. Am Rev Respir Dis 1992; 145(3):701–711.
37. Ham EA, Soderman DD, Zanetti ME, Dougherty HW, McCauley E, Kuehl FJ. Inhibition by prostaglandins of leukotriene B4 release from activated neutrophils. Proc Natl Acad Sci USA 1983; 80(14):4349–4353..
38. Minakuchi R, Wacholtz MC, Davis LS, Lipsky PE. Delineation of the mechanism of inhibition of human T cell activation by PGE2. J Immunol 1990; 145(8): 2616–2625.
39. Brigham KL, Serafin W, Zadoff A, Blair I, Meyrick B, Oates JA. Prostaglandin E2 attenuation of sheep lung responses to endotoxin. J Appl Physiol 1988; 64(6): 2568–2574.
40. Holcroft JW, Vassar MJ, Weber CJ. Prostaglandin E1 and survival in patients with the adult respiratory distress syndrome. A prospective trial. Ann Surg 1986; 203(4):371–378.
41. Ogletree ML, Brigham KL. Effects of cyclooxygenase inhibitors on pulmonary vascular responses to endotoxin in unanesthetized sheep. Prostaglandins Leukot Med 1982; 8(5):489–502.
42. Halushka PV, Wise WC, Cook JA. Protective effects of aspirin in endotoxic shock. J Pharmacol Exp Ther 1981; 218(2):464–469.
43. Wise WC, Halushka PV, Knapp RG, Cook JA. Ibuprofen, methylprednisolone, and gentamicin as conjoint therapy in septic shock. Circ Shock 1985; 17(1):59–71.
44. Halushka PV, Cook JA, Wise WC. Beneficial effects of UK 37248, a thromboxane synthetase inhibitor, in experimental endotoxic shock in the rat. Br J Clin Pharmacol 1983; 15(1):1335–1395.
45. Short BL, Miller MK, Stroud CY, Fletcher JR. Comparison of thromboxane inhibitors to cyclooxygenase inhibitors on survival in a newborn rat model for group B streptococcal sepsis. Adv Prostaglandin Thromboxane Leukot Res 1983; 12(1):113–116.
46. Butler RJ, Wise WC, Halushka PV, Cook JA. Gentamicin and indomethacin in the treatment of septic shock: effects on prostacyclin and thromboxane A2 production. J Pharmacol Exp Ther 1983; 225(1):94–101.
47. Cook JA, Wise WC, Butler RR, Reines HD, Rambo W, Halushka PV. The potential role of thromboxane and prostacyclin in endotoxic and septic shock. Am J Emerg Med 1984; 2(1):28–37.
48. Casey LC, Fletcher JR, Zumdka MI, Ramwell PW. Prevention of endotoxin-induced pulmonary hypertension in primates by the use of a selective thromboxane synthetase inhibitor, OKY 1581. J Pharmacol Exp Ther 1982; 222:441–447.

49. Fink MP, Morrissey PE, Stein KL, Clement RE, Fiallo V, Gardiner WM. Systemic and regional hemodynamic effects of cyclo-oxygenase and thromboxane synthetase inhibition in normal and hyperdynamic endotoxemic rabbits. Circ Shock 1988; 26(1):41–57.
50. Bernard GR, Korley V, Chee P, Swindell B, Ford HA, Tagari P. Persistent generation of perptido leukotrienes in patients with the adult respiratory distress syndrome. Am Rev Respir Dis 1991; 144(2):263–267.
51. Loyd JE, Tagari P, Chapel H, Hussein A, Boulin DJ, Lee GdJ. Slow reacting substances are present in bronchoalveolar lavage fluid of patients with adult respiratory distress syndrome. Am Rev Respir. Dis 1984; 29:A118.
52. Matthay MA, Eschenbacher WL, Goetzl EJ. Elevated concentrations of leukotriene D4 in pulmonary edema fluid of patients with the adult respiratory distress syndrome. J Clin Immunol 1984; 4(6):479–483.
53. Antonelli M, Bufi M, De BR, et al. Detection of leukotrienes B4, C4 and of their isomers in arterial, mixed venous blood and bronchoalveolar lavage fluid from ARDS patients. Intensive Care Med 1989; 15(5):296–301.
54. Seeger W, Grimminger F, Barden M, et al. Omega-oxidized leukotriene B4 detected in the broncho-alveolar lavage fluid of patients with non-cardiogenic pulmonary edema, but not in those with cardiogenic edema. Intensive Care Med 1991; 17(1):1–6.
55. Davis JM, Meyer JD, Barie PS, et al. Elevated production of neutrophil leukotriene B4 precedes pulmonary failure in critically ill surgical patients. Surg Gynecol Obstet 1990; 170(6):495–500.
56. Ball HA, Cook JA, Spicer KM, Wise WC, Halushka PV. Essential fatty acid-deficient rats are resistant to oleic acid-induced pulmonary injury. J Appl Physiol 1989; 67(2):811–816.
57. Matera G, Cook JA, Hennigar RA, et al. Beneficial effects of a 5-lipoxygenase inhibitor in endotoxic shock in the rat. J Pharmacol Exp Ther 1988; 247(1):363–371.
58. Coggeshall JW, Christman BW, Lefferts PL, et al. Effect of inhibition of 5-lipoxygenase metabolism of arachidonic acid on response to endotoxemia in sheep. J Appl Physiol 1988; 65(3):1351–1359.
59. Turner CR, Quinlan MF, Schwartz LW, Wheeldon EB. Therapeutic intervention in a rat model of ARDS. I. Dual inhibition of arachidonic acid metabolism. Circ Shock 1990; 32(3):231–242.
60. Suematsu M, Miura S, Suzuki M, et al. 5-lipoxygenase inhibitor (AA-861) attenuates neutrophil-mediated oxidative stress on the venular endothelium in endotoxemia. J Clin Lab Immunol 1988; 25(1):41–45.
61. Rogers F, Dunn R, Nolan P, Phunangsab A, Barrett J. Diethylcarbamazine, a leukotriene inhibitor, improves survival of endotoxemia in the rat. Circ Shock 1985; 16(1):52.
62. Li EJ, Cook JA, Wise WC, Jackson WT, Halushka PV. Effect of LTB4 receptor antagonists in endotoxic shock in the rat. Circ Shock 1991; 34(4):385–392.
63. Cook JA, Li EJ, Spicer KM, Wise WC, Halushka PV. Effect of leukotriene receptor antagonists on vascular permeability during endotoxic shock. Circ Shock 1990; 32(3):209–218.

64. Zellner JL, Cook JA, Reines HD, Smith EF 3d, Kessler LD, Halushka PV. Effect of a LTD4 receptor antagonist in porcine septic shock. Eicosanoids 1991; 4(3):169–175.
65. Ahmed T, Wasserman MA, Muccitelli R, Tucker S, Gazeroglu H, Marchette B. Endotoxin-induced changes in pulmonary hemodynamics and respiratory mechanics. Role of lipoxygenase and cylooxygenase products. Am Rev Respir Dis 1986; 134(6):1149–1157.
66. Krausz MM, Dahan JB, Gross D. Effect of the leukotriene receptor antagonist LY-171883 on endotoxemia in awake sheep. Circ Shock 1988; 26(4):431–441.
67. Toyofuku T, Kobayashi T, Koyama S, Kusama S. Pulmonary vascular response to platelet-activating factor in conscious sheep. Am J Physiol 1988; 3(1):2.
68. Chang SW, Feddersen CO, Henson PM, Voelkel NF. Platelet-activating factor mediates hemodynamic changes and lung injury in endotoxin-treated rats. J Clin Invest 1987; 79(5):1498–1509.
69. Inarrea P, Gomez CJ, Pascual J, Ponte MC, Hernando L, Sanchez CM. Synthesis of PAF-acether and blood volume changes in gram-negative sepsis. Immunopharmacology 1985; 9(1):45–52.
70. Herbert JM, Lespy L, Maffrand JP. Protective effect of SR 27417, a novel PAF antagonist, on lethal anaphylactic and endotoxin-induced shock in mice. Eur J Pharmacol 1991; 205(3):271–276.
71. Kawaguchi H, Yasuda H. Effect of platelet-activating factor on arachidonic acid metabolism in renal epithelial cells. Biochim Biophys Acta 1986; 875(3): 525–534.
72. Stewart AG, Dubbin PN, Harris T, Dusting GJ. Platelet-activating factor may act as a second messenger in the release of icosanoids and superoxide anions from leukocytes and endothelial cells. Proc Natl Acad Sci USA 1990; 87(8): 3215–3219.
73. Morris DD, Moore JN. Equine peritoneal macrophage production of thromboxane and prostacyclin in response to platelet activating factor and its receptor antagonist SRI 63-441. Circ Shock 1989; 28(2):149–158.
74. Olson NC, Joyce PB, Fleisher LN. Role of platelet-activating factor and eicosanoids during endotoxin-induced lung injury in pigs. Am J Physiol 1990; 6(1):2.
75. Jancar S, Theriault P, Provencal B, Cloutier S, Sirois P. Mechanism of action of platelet-activating factor on guinea-pig lung parenchyma strips. Can J Physiol Pharmacol 1988; 66(9):1187–1191.
76. Beijer L, Botting J, Crook P, Oyekan AO, Page CP, Rylander R. The involvement of platelet activating factor in endotoxin-induced pulmonary platelet recruitment in the guinea-pig. Br J Pharmacol 1987; 92(4):803–808.
77. Salzer WL, McCall CE. Primed stimulation of isolated perfused rabbit lung by endotoxin and platelet activating factor induces enhanced production of thromboxane and lung injury. J Clin Invest 1990; 85(4):1135–1143.
78. Rabinovici R, Esser KM, Lysko PG, et al. Priming by platelet-activating factor of endotoxin-induced lung injury and cardiovascular shock. Circ Res 1991; 69(1):12–25.
79. Hocking DC, Phillips PG, Ferro TJ, Johnson A. Mechanisms of pulmonary edema induced by tumor necrosis factor-alpha. Circ Res 1990; 67(1):68–77.

80. Beutler B, Cerami A. Cachectin: more than a tumor necrosis factor. N Engl J Med 1987; 316(7):379–385.
81. Mathison JC, Wolfson E, Ulevitch RJ. Participation of tumor necrosis factor in the mediation of gram negative bacterial lipopolysaccharide-induced injury in rabbits. J Clin Invest 1988; 81(6):1925–1937.
82. Talmadge JE, Bowersox O, Tribble H, Lee SH, Shepard HM, Liggitt D. Toxicity of tumor necrosis factor is synergistic with gamma-interferon and can be reduced with cyclooxygenase inhibitors. Am J Pathol 1987; 128(3):410–425.
83. Dinarello CA, Cannon JG, Wolff SM, et al. Tumor necrosis factor (cachectin) is an endogenous pyrogen and induces production of interleukin 1. J Exp Med 1986; 163(6):1433–1450.
84. Kawakami M, Ishibashi S, Ogawa H, Murase T, Takaku F, Shibata S. Cachectin/TNF as well as interleukin-1 induces prostacyclin synthesis in cultured vascular endothelial cells. Biochem Biophys Res Commun 1986; 141(2):482–487.
85. Eliza JA, Gustilo K, Baeder W, Freundlich B. Synergistic stimulation of fibroblast prostaglandin production by recombinant interleukin 1 and tumor necrosis factor. J Immunol 1987; 138(11):3812–3816.
86. Dayer JM, Beutler B, Cerami A. Cachectin/tumor necrosis factor stimulates collagenase and prostaglandin E2 production by human synovial cells and dermal fibroblasts. J Exp Med 1985; 162(6):2163–2168.
87. Roubin R, Elsas PP, Fiers W, Dessein AJ. Recombinant human tumour necrosis factor (rTNF)2 enhances leukotriene biosynthesis in neutrophils and eosinophils stimulated with the CA^{2+} ionophore A23187. Clin Exp Immunol 1987; 70(2): 484–490.
88. Godfrey RW, Johnson WJ, Hoffstein ST. Recombinant tumor necrosis factor and interleukin-1 both stimulate human synovial cell arachidonic acid release and phospholipid metabolism. Biochem Biophys Res Commun 1987; 142(1); 235–241.
89. Renz H, Gong JH, Schmidt A, Nain M, Gemsa D. Release of tumor necrosis factor-alpha from macrophages. Enhancement and suppression are dose-dependently regulated by prostaglandin E2 and cyclic nucleotides. J Immunol 1988; 141(7):2388–2393.
90. Hammond B, Kontos HA, Hess ML. Oxygen radicals in the adult respiratory distress syndrome, in myocardial ischemia and reperfusion inury, and in cerebral vascular damage. Can J Physiol Pharmacol 1985; 63(3):173–187.
91. Harlan JM, Callahan KS. Role of hydrogen peroxide in the neutrophil-mediated release of prostacyclin from cultured endothelial cells. J Clin Invest 1984; 74(2): 442–448.
92. Tate RM, Morris HG, Schroeder WR, Repine JE. Oxygen metabolites stimulate thromboxane production and vasoconstriction in isolated saline-perfused rabbit lungs. J Clin Invest 1984; 74(2):608–613.
93. Gurner G, Sciuto AM, Freeland H. leukotriene inhibitors attenuate rat lung injury induced by hydrogen peroxide. Circulation 1985; 72:106.
94. Seekamp A, Lalonde C, Zhu DG, Demling R. Catalase prevents prostanoid release and lung lipid peroxidation after endotoxemia in sheep. J Appl Physiol 1988; 65(3):1210–1216.

95. Burghuber OC, Strife RJ, Zirrolli J, et al. Leukotriene inhibitors attenuate rat lung injury induced by hydrogen peroxide. Am Rev Respir Dis 1985; 131(5): 778–785.
96. Ward PA, Sulavik MC, Johnson KJ. Activated rat neutrophils. Correlation of arachidonate products with enzyme secretion but not with(0)2-generation. Am J Pathol 1985; 120(1):112–120.
97. Kukreja RC, Kontos HA, Hess ML, Ellis EF. PGH synthase and lipoxygenase generate superoxide in the presence of NADH or NADPH. Circ Res 1986; 59(6): 612–619.
98. Okabe E, Kato Y, Kohno H, Hess ML, Ito H. Inhibition by free radical scavengers and by cyclooxygenase inhibitors of the effect of acidosis on calcium transport by masseter muscle sarcoplasmic reticulum. Biochem Pharmacol 1985; 34(7):961–968.
99. Maridonneau PI, Tauber AI. Activation of NADPH-oxidase by arachidonic acid involves phospholipase A2 in intact human neutrophils but not in the cell free system. Biochem Biophys Res Commun 1986; 138(3):1099–1105.
100. Mehta JL, Lawson D, Mehta P. Modulation of human neutrophil superoxide production by selective thromboxane synthetase inhibitor U63,557A. Life Sci 1988; 43(11):923–928.
101. Palmblad J, Gyllenhammar H, Ringertz B, Nilsson E, Cottell B. Leukotriene B4 triggers highly characteristic and specific functional responses in neutrophils: studies of stimulus specific mechanisms. Biochim Biophys Acta 1988; 971(1): 92–102.
102. Holian A. Leukotriene B4 stimulation of phosphatidylinositol turnover in macrophages and inhibition by pertussis toxin. *Febs* Lett 1986; 201(1):15–19.
103. Weiss EB, Bellino JR. Leukotriene-associated toxic oxygen metabolites induce airway hyperreactivity. Chest 1986; 89(5):709–716.

9

Baboon Models of Three Variants of the Hemostatic Response to Inflammatory Stimuli

FLETCHER B. TAYLOR, JR.

Oklahoma Medical Research Foundation
Oklahoma City, Oklahoma

I. Introduction

Our studies began with the study of the mechanism of the disseminated intravascular coagulant response to *Escherichia coli* in the baboon. It became clear that this coagulant response of the hemostatic system was closely linked to the inflammatory response to *E. coli*. In these previous studies the baboon response to *E. coli* was divided into four stages: inflammatory, coagulant, cell injury, and cell degeneration, the mechanisms of which are described by Taylor et al. (1,2). In the course of these studies it became evident that the hemostatic system could respond to inflammatory stimuli in other ways besides through the disseminated intravascular coagulant response. As a result we have developed and studied the following baboon models of the hemostatic response to inflammatory stimuli: deep vein thrombosis, disseminated intravascular coagulation, and microvascular thrombosis. In this chapter we place all three of these variants of the hemostatic response to inflammation in the framework of a single concept that is carried out through this chapter. The four topics covered in this chapter include the concept of the four functional domains of the hemostatic system, the models of deep vein thrombosis, (DVT), disseminated intravascular coagulation, and microvascular thrombosis.

II. Four Functional Domains of the Hemostatic System

Figure 1 is a conceptual diagram in which thrombin is viewed as playing a central role in both clot formation and clot removal. In this scheme, thrombin effects are controlled by a hierarchy of affinities for its receptors and substrates. These range from a K_d value of 10^{-7} for the thrombomodulin receptor on endothelial cells (3) to 10^{-6} and 10^{-5} for receptors on platelets (4) and for fibrinogen (5), respectively. Although this scheme is oversimplified and the K_d values listed are largely symbolic, representing complex regulatory factors, it does provide a basis for assessing the hierarchy of both the anticoagulant and coagulant functions of thrombin in perspective. Thus, wherever the endothelium is functioning normally, the high affinity of thrombin for thrombomodulin (TM) predominates. Under these circumstances the thrombin–thrombomodulin complex activates protein C. The activated protein C (APC) with protein S as a cofactor attenuates the generation of additional thrombin by inactivating factors Va and VIIIa, thus acting as a powerful anticoagulant. This assures that under normal circumstances activation

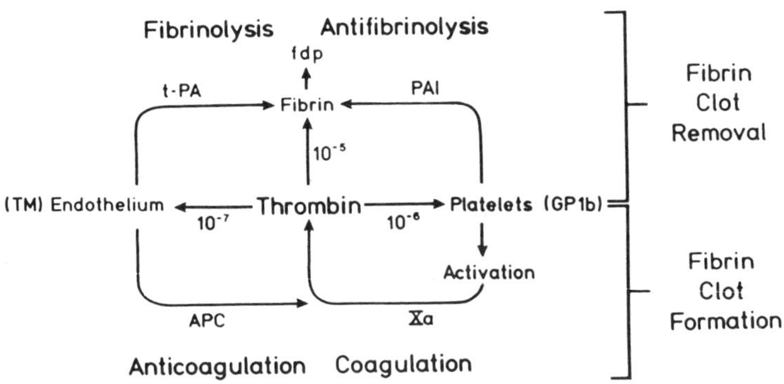

Figure 1 Four-domain concept of the unperturbed hemostatic system: The hemostatic system can be viewed as consisting of four functional domains (coagulation, anticoagulation, fibrinolysis, antifibrinolysis), all of which are regulated by thrombin. The coagulant and anticoagulant domains regulate clot formation and the fibrinolytic and antifibrinolytic domains regulate clot removal. The mechanistic aspects of this figure are reviewed in the accompanying text.

of protein C is the first response to the generation of any thrombin and thus sustains the fluidity of blood *and* the concentration of circulating platelets (6). The latter point is supported by the observation that inhibition of the endogenous protein C system with antibodies to protein C or protein S or with C4b-binding protein results in a 50% decline in platelet numbers within 24 h (7). In circumstances, however, where the endothelium is absent or the endothelium is made dysfunctional by inflammatory agents (8,9), the procoagulant steps in the hierarchy can come into play. In these circumstances, generation of thrombin is unimpeded by the protein C system. The thrombin–glycoprotein interactions (i.e., GPIa, GPV) initiate platelet activation (10,11), generation of factor Xa (Xa), the assembly of the prothrombinase complex, and hence more thrombin (12). Under these circumstances, sufficient thrombin eventually is generated to hydrolyze fibrinogen and induce fibrin formation. Thus the two functional domains controlling clot formation under normal conditions balance each other to maintain the fluidity of blood while providing a coagulant system ready to clot the blood at a local site of endothelial injury or dysfunction at a moment's notice.

The same balance of functions also pertains to fibrinolytic and antifibrinolytic control of fibrin or clot removal. The thrombin-induced release of tissue plasminogen activator (t-PA) from endothelial cells (13) and of plasminogen activator inhibitor (PAI) from platelets (14,15) and endothelium can be viewed as subject to conditions similar to those of the generation of activated protein C and factor Xa.

III. Deep Vein Thrombosis

A. General Principles of DVT Model

Using the four-domain hemostatic construct, the conditions favoring formation of a physiologic clot are compared with those favoring formation of a pathologic thrombus (Fig. 2). Figure 2 (top) shows the effect of the removal of the endothelium overlying collagen. In the absence of endothelium such as would be the case at the site of an injury, neither the anticoagulant (APC) nor fibrinolytic (t-PA) domains of the hemostatic system can function. Following adherence and aggregation of platelets to collagen, procoagulant (factor Xa) and antifibrinolytic (PAI from platelets) factors function unopopposed to seal off the injury by forming a physiologic clot.

Figure 2 (bottom) shows what we believe happens in the model of deep vein thrombosis in which systemic inflammatory factors rather than the local removal of endothelium renders that endothelium dysfunctional. Virchow's triad attributes deep vein thrombosis to changes in (1) the vein wall, (2) the coagulability of blood, and (3) flow (16). In this baboon model of venous

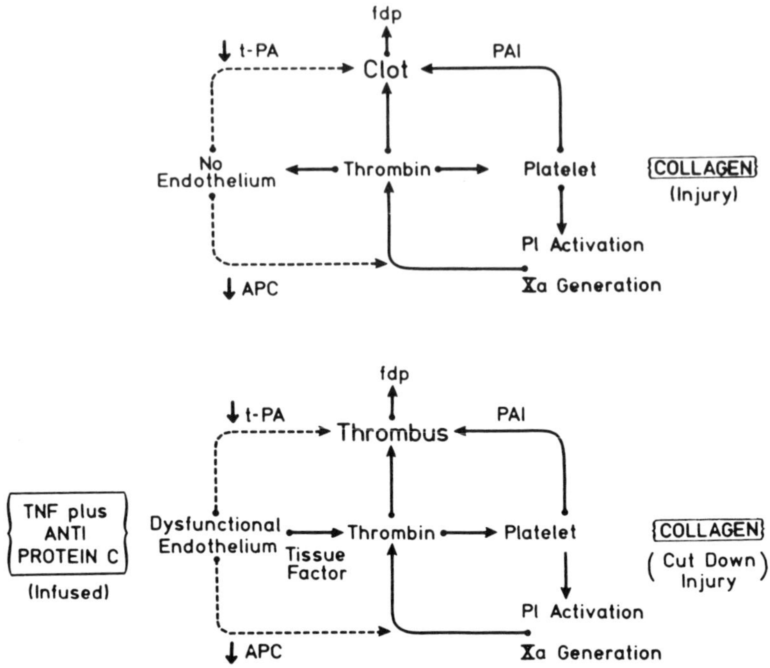

Figure 2 Physiologic clot response to injury (top) compared to pathologic thrombotic (DVT) response to injury and inflammation (bottom). Using the four-domain hemostatic construct the conditions favoring formation of a physiologic clot are compared to those favoring formation of a pathologic thrombus (deep vein thrombosis). We hypothesize that inflammation (TNF, anti-PC) of the normal endothelium at the margins of injury or cutdown makes the endothelium dysfunctional and that this determines whether a physiologic clot will advance beyond these margins and grow into a pathologic thrombus (deep vein thrombosis). The mechanistic aspects of this figure are reviewed in the text.

thrombosis we hypothesize the following: (1) infusion of tumor necrosis factor (TNF) changes the vein wall by rendering the venous endothelium dysfunctional [this includes an increased expression of tissue factor and release of plasminogen activator inhibitor (PAI) together with a reduced release of t-PA] (17); (2) infusion of anti-protein C antibody (anti-protein C) changes the coagulability of blood by inhibiting the generation of APC; and (3) ligation of the superficial femoral vein following catheterization changes the flow of blood, thus completing the Virchow triad. Under these circumstances inflammation of the venous endothelium leads to an increase in coagulant activity and a reduction of both its anticoagulant and fibrinolytic activity, which permits the physiologic clot formed at the site of cutdown or injury (COLLAGEN) to grow into a pathologic thrombus. This sequence of events mimics the deep vein thrombosis seen in clinical practice.

B. Specific Observations of the DVT Model

What are the specifics of the model, and what is the evidence supporting the concept shown in Fig. 2? Table 1 and Fig. 3 summarize the experimental data gathered from this model thus far. Table 1 shows that both a change in the vein wall and in the coagulability of blood by TNF *and* anti-protein C antibody (anti PC), respectively, are necessary before a change of flow by catheterization and ligation of superficial femoral vein can produce a thrombus (18). Reanastemosis after ligation of this superficial femoral vein prevents

Table 1 Deep Vein Thrombotic Model: Establishing Conditions and Controlling Variables in the Baboon Model of Deep Vein Thrombosis

Agent and study conditions	Thrombin–antithrombin (TAT) complexes	Fibrinogen consumption	Thrombus
Catheter/ligation superficial femoral vein only	–	–	–
Catheter/ligation + TNF (100 μg/kg)	+	–	–
Catheter/ligation + anti-PC (2 mg/kg)	+	–	–
Catheter/ligation + anti-PC (10 mg/kg)	+	–	+
Catheter/ligation + TNF (100 μg/kg) + anti-PC (2 mg/kg)	+	–	+
Catheter/reanstamosis + TNF, + anti-PC	+	–	–

Note: In addition to cannulation and ligation, tumor necrosis factor (TNF) and anti-protein C antibody (anti-protein C) are necessary for development of deep vein thrombosis. Details of procedures are as reported by Taylor et al. (18). Briefly, the femoral artery and vein are cannulated under anesthesia. The TNF and anti-protein C are infused as a bolus through a catheter in the cephalic vein at T - 10 and T - 5 min, respectively. The animals are observed (vital sign, blood CBC, chemistries, ultrasound, and [^{125}I]fibrinogen scan) for 6 h and again at 24 or 48 h. The animals are sacrificed, autopsy is performed, and tissue samples from the area of thrombosis are harvested for light- and electron-microscopic examination and immunostaining. The thrombus is then dissected free, dried, and weighed.

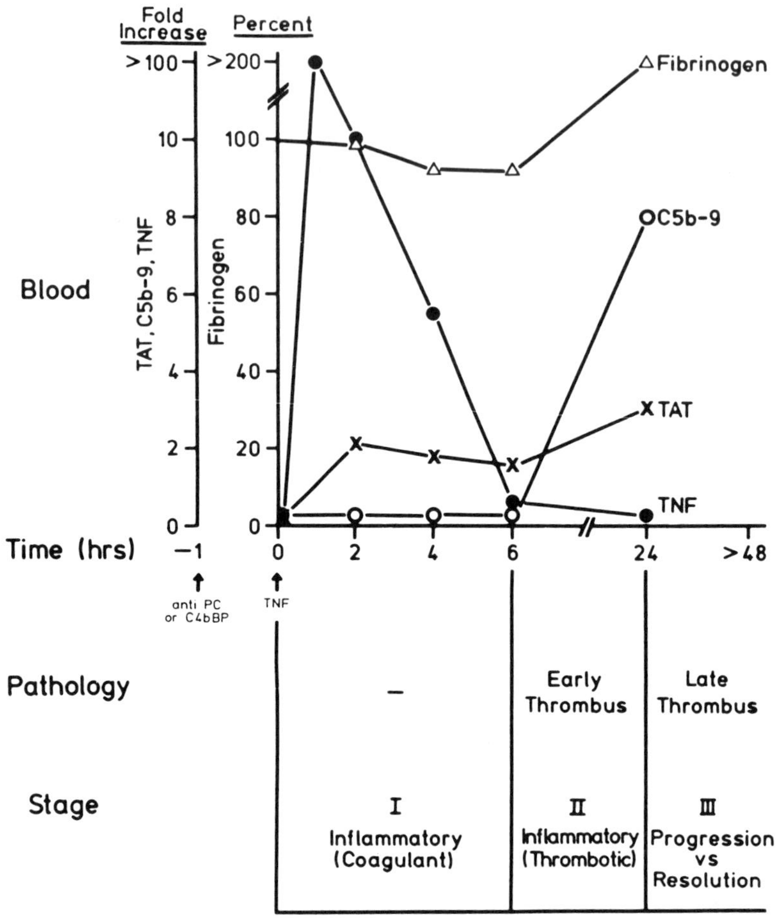

Figure 3 Deep vein thrombotic model: description of sequence of events and stages (1 to 3) in the development of deep vein thrombosis (DVT) following infusion of tumor necrosis factor (TNF) and antiprotein C antibody (anti-PC) or C4b-binding protein (C4bBP). This figure shows that other inflammatory mediators (C5b-9) as well as TNF are associated with the development of DVT. It also shows that increasing amounts of thrombin–antithrombin (TAT) continue to be generated during the course of development of the venous thrombus.

development of deep vein thrombosis, presumably by reestablishing flow through the superficial femoral vein as well as that already existing from the deep femoral system (i.e., flow is maintained through the deep femoral system in all cases). Note that high concentrations of anti-protein C antibody obviate the need for tumor necrosis factor. Finally, note that there is generation of thrombin–antithrombin (TAT) complexes in all cases except following catheterization/ligation alone and that in no case is there consumption of fibrinogen.

Figure 3 summarizes the progression or natural history of the baboon model of deep vein thrombosis (18). There are three stages. The first stage (0 to 6 h) is characterized by the expected rise and peaking of TNF and a rise of TAT complexes. No thrombus appears as determined by doppler scan. There is, however, evidence by [^{125}I]fibrinogen scan of early fibrin deposition during this period. No disruption of endothelium, however, is demonstrable by microscopy along the path of catheter insertion up to and including the vena cava (19). The second stage (6 to 24 h) is characterized by a rise and peaking of C5b-9 attack complex of complement. Thrombus formation in stage II by doppler scan is clearly evident. The thrombus contains patches of leukocytes (19). The adjacent vein wall also shows early infiltration of leukocytes. We believe that this represents a second phase of inflammatory activity involving both the thrombus and vein wall in response to the infused tumor necrosis factor. The third stage (24 to >48 h) is characterized by a peaking of thrombus growth followed by either resolution with recanalization, much as is seen in wound healing or by uncontrolled inflammatory activity involving both thrombus and the vein wall. In the latter case the inflammation can lead to development of phlegmasia cerula dolerans. The timing, progression, and pathologic findings of the deep vein thrombosis in this model closely mimic that seen in clinical practice. Furthermore, this model differs from other models in that inflammatory agents (instead of clotting factors such as factor Xa combined with stasis) are used to initiate the thrombotic response. We believe, therefore, that this illustrates the role that inflammation can play and more closely approximates at least some forms of deep vein thrombosis seen clinically.

IV. Disseminated Intravascular Coagulation Induced by a Noninflammatory Stimulus (Factor XaPC/PS)

A. General Principles of the XaPC/PS DIC Model

Using the four-domain hemostatic construct, the effects of infusing factor Xa phosphatidylcholine/phosphatidylserine (XaPC/PS) infused alone are compared with those of factor XaPC/PS infused with anti-protein C (Fig. 4). Figure 4 (top) suggests that the coagulant effects of factor Xa PC/PS infused alone are balanced by an equally intense generation of the anticoagulant enzyme, APC. Platelet sequestration, and the consumption of fibrinogen to form fibrin, are therefore attenuated. The host's regulatory factors all are operative, resulting in a compensated response to the stress of the infused factor XaPC/PS. This includes the generation of some fibrin and fibrin

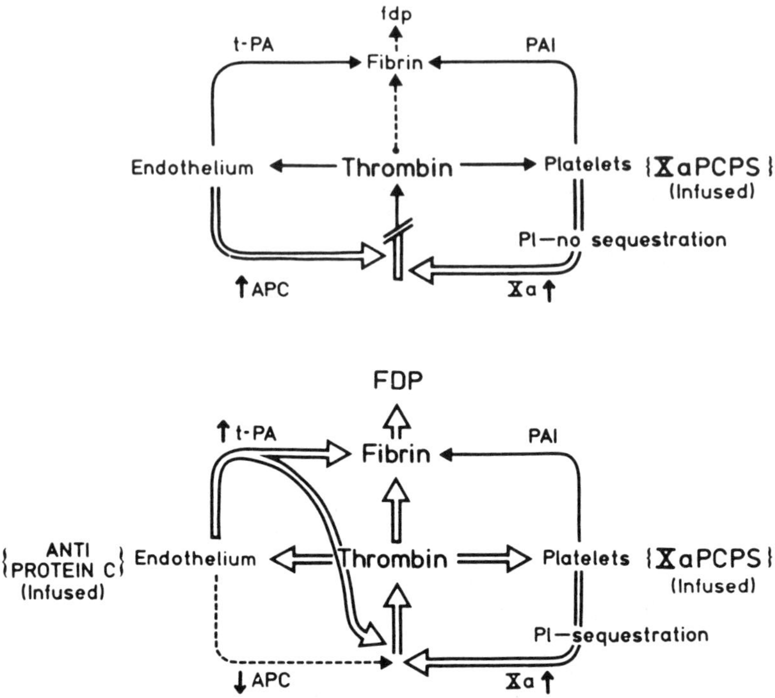

Figure 4 Disseminated intravascular coagulant (DIC) response to infusion of factor XaPC/PS (top) and to factor XaPC/PS plus anti-protein C antibody (bottom). Using the four-domain hemostatic construct, the response to factor XaPC/PS is compared to the response to factor XaPC/PS and anti-protein C antibody. We hypothesize that the DIC response to factor XaPC/PS under normal circumstances is prevented by the endogenous activated protein C (APC). If, however, this protein C anticoagulant system is inhibited with anti-protein C antibody (anti-protein C), we hypothesize that an intense activation of the fibrinolytic system (t-PA) from the large amounts of thrombin produced substitutes for the protein C system and prevents a catastrophic systemic thrombosis by hydrolyzing the fibrin thrombi in the microvasculature as they are formed. The mechanistic aspects of this figure are reviewed in the text.

degradation products (FDPs), and a balanced output of both t-PA and PAI, as well as an increased generation of APC in response to infused factor XaPC/PS.

Figure 4 (bottom) illustrates what we believe happens when the protein C anticoagulant system is shut down with anti-protein C. In the absence of activation of protein C, a superabundance of thrombin is generated which leads to platelet activation and sequestration, and to the consumption of fibrinogen with generation of fibrin degradation products (FDP) (20). In contrast to the compensated response to factor XaPC/PS (Fig. 4, top), this is an uncompensated full-blown coagulopathic [disseminated intravascular coagulation (DIC)] response (Fig. 4, bottom). The surprising result, however, is that the occlusive effects of this coagulopathy are rapidly reversed by an equally intense release of t-PA, activation of the fibrinolytic system, and cleavage not only of fibrin but also of clotting factors. Thus the events shown in Fig. 4 (bottom) suggest that the fibrinolytic system, though far less specific, acts as a backup to the protein C anticoagulant system—at least in so far as reversing the occlusive thrombotic phenomena which otherwise would occur. The sequence of events mimics the events following the bite of Russell's viper. It is a pure uncompensated malfunction of the coagulant system, with no inflammatory component.

B. Specific Observations of XaPC/PS DIC Model

What are the specifics of this model, and what is the evidence supporting the concept shown in Fig. 4? Table 2 and Fig. 5 summarize the experimental data gathered from this model thus far. Table 2 shows that on the one hand, infusion of factor XaPC/PS is followed by generation of TAT complexes but only a limited decrease (consumption) of fibrinogen and no decrease (sequestration) of platelets or increase of t-PA or plasmin–antiplasmin (PAP) complexes in plasma. On the other hand, this infusion of factor XaPC/PS

Table 2 Disseminated Intravascular Coagulation (DIC) Model: Description of the Baboon Response to Factor XaPC/PS and to Factor XaPC/PS Plus Anti-Protein C Antibody (Anti-protein C)

Agent	TAT	Fibrinogen	Platelets	APC/ INH	t-PA (PAP)	FVa (cleaved)	FV (cleaved)
XaPCPS (compensated response)	+	↓	−	+ +	−	+	−
XaPCPS + anti-protein C (uncompensated response)	+ +	↓↓	↓↓	−	+ +	+ +	+ +

Note: The response to infusion of factor XaPC/PS alone is restricted to generation of thrombin–antithrombin (TAT) complexes, activated protein C/inhibitor complexes and cleavage of factor Va, whereas the response to infusion of factor XaPC/PS plus anti-protein C antibody is dominated by a (1) disseminated intravascular coagulation (increased TAT and consumption of fibrinogen), (2) an intense activation of the fibrinolytic system with increased concentrations of tissue plasminogen activator (t-PA) and plasmin–antiplasmin (PAP) complexes, and (3) cleavage of *both* factors Va and V.

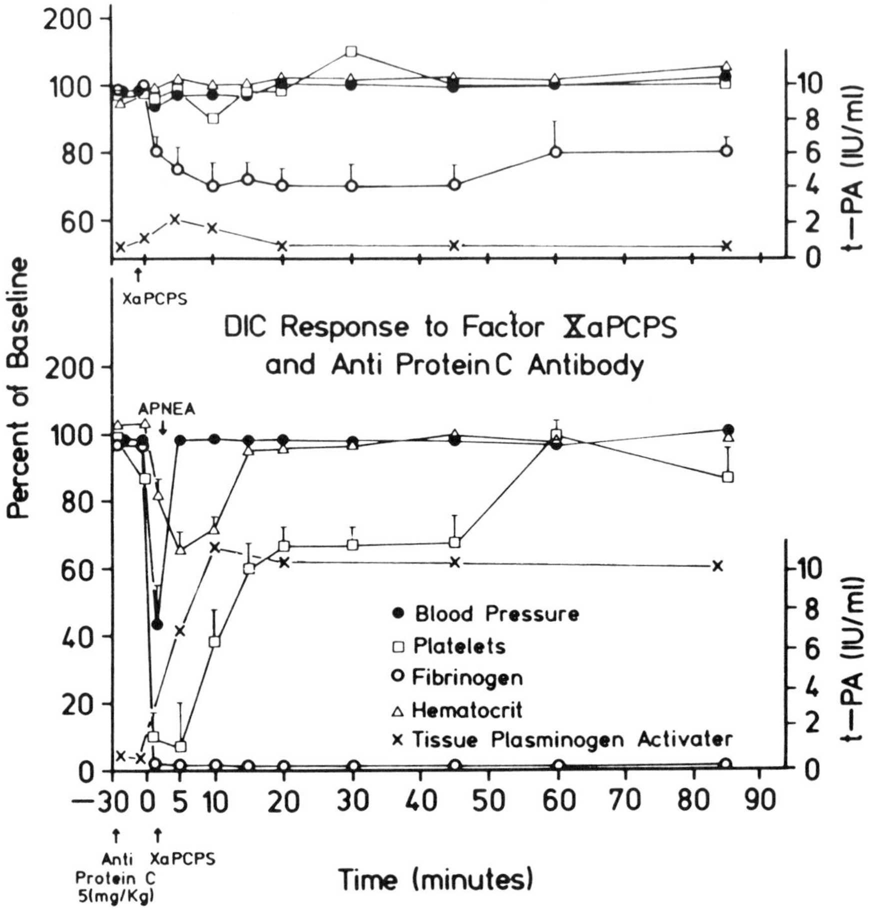

Figure 5 Disseminated intravascular coagulation (DIC) model: Comparison and sequence of events following infusion of factor XaPC/PS alone (top) and factor XaPC/PS plus anti-protein C antibody (anti-protein C) (bottom). This figure shows the effect on the response to factor XaPC/PS of inhibiting the endogenous protein C anticoagulant system with anti-protein C antibody. There is a profound transitory coagulant event probably concentrated in the pulmonary vascular bed, which, however, is rapidly reversed. We believe that this reversal is secondary to the release of tissue plasminogen activator and intense fibrinolytic activity from the vascular endothelium induced by the explosive generation of thrombin.

is followed by the appearance of activated protein C/inhibitor complexes together with evidence that factor Va but not factor V is cleaved and inactivated by activated protein C (20). These results suggest that the protein C anticoagulant system is activated and engaged and that it has attenuated what otherwise would be an uncompensated coagulopathic response. Infusion of factor XaPC/PS with anti-protein C, however, is followed by the appearance of larger amounts of TAT complexes, unrestricted consumption of fibrinogen and sequestration of platelets, and larger amounts of t-PA and PAP complexes (20). Under these circumstances no activated protein C/ inhibitor complexes appear, and there is cleavage and inactivation of *both* factor Va and factor V by plasmin. This suggests that in the absence of the endogenous protein C system the coagulant/fibrinolytic response to factor XaPC/PS is uncompensated.

Figure 5 summarizes and compares the sequence of events following the infusion of factor XaPC/PS and of factor XaPC/PS plus anti-protein C. Figure 5 (bottom) emphasizes the profound but transitory effect on platelet count, hematocrit, fibrinogen, blood pressure, and tissue plasminogen activator produced by infusion of factor XaPC/PS plus anti-protein C antibody. The sudden apnea that occurs between 2 and 3 min and coincides with a fall in platelet count, hematocrit, and blood pressure suggests that these changes could be due to transient sequestration of platelets, red blood corpuscles, and deposition of fibrin in the pulmonary vascular bed. The reversal of these sudden changes coincides with the rapid release of a large amount of tissue plasminogen activator. Based on this and on the information presented in Table 2, we believe that the fibrinolytic system is rapidly activated following the explosive generation of large amounts of thrombin (21), and that it is this fibrinolytic response that reverses the thrombotic events in the pulmonary vascular bed and rescues the animal from a catastrophic thrombotic demise.

V. Disseminated Intravascular Coagulation Induced by an Inflammatory Stimulus (LD_{100} *E. coli*)

A. General Principles of the *E. coli* DIC Model

This is the second model of disseminated intravascular coagulation. It differs from the factor XaPC/PS model of disseminated intravascular coagulation in that inflammatory factors, not coagulant factors, initiate and amplify the consumptive coagulopathy. Figure 6 is a conceptual illustration in which the infusion of *E. coli* activates formed elements of the blood (leukocytes, platelets, macrophages, monocytes) directly and also indirectly through activation of plasma components (i.e., complement). In this scheme the vascular endothelium bears the brunt of the aberrant inflammatory response of these plasma components and formed elements. Rather than succumbing to this attack, however, the endothelium and macrophages/mono-

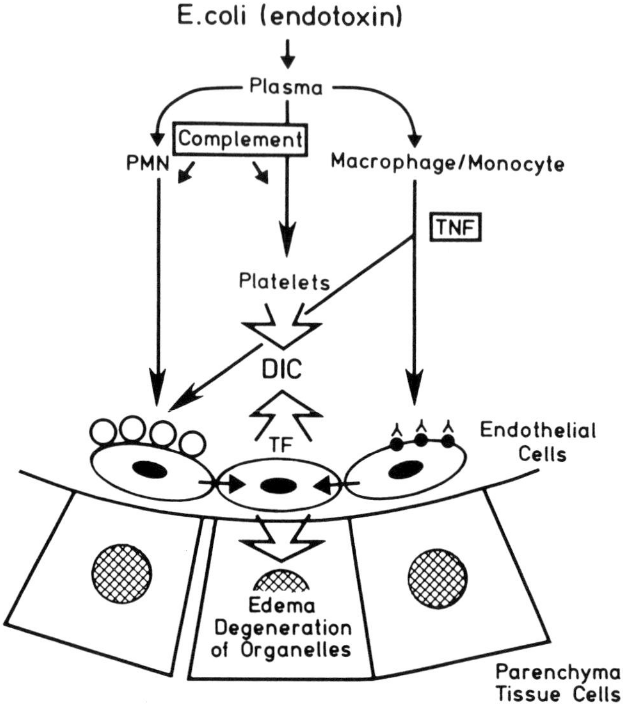

Figure 6 Disseminated intravascular coagulant (DIC) model of baboon response to LD_{100} *E. coli*: *E. coli* activates complement components, neutrophils, and macrophage/monocytes that release mediators. These mediators perturb the vascular endothelium to initiate and amplify activation of coagulant factors (tissue factor, TF) to produce a massive systemic disseminated intravascular coagulation (DIC) response. Finally, in response to these explosive events occurring within the vasculature, the adjacent parenchymal tissue accumulates fluid (edema) and ultimately undergoes degeneration.

cyte mount a massive coagulant/fibrinolytic counterattack (1,2). It is a more complex system in which we believe that (1) the coagulant activity is initiated not by infusion of exogenous factor Xa, but by endogenous factor Xa produced following the expression of tissue factor (TF) by endothelium and by macrophage/monocytes induced by inflammatory agents such as TNF; (2) the anticoagulant protein C activity is reduced not by anti-protein C antibody but by inflammatory agents such as TNF; and (3) the coagulant activity once initiated is amplified not by infused phospholipid vesicles but by vesiculation of platelets and cell membranes induced by inflammatory agents such as the C5b-9 attack complex of complement (Fig. 6).

Using the four-domain hemostatic construct, the effects of infusing factor XaPC/PS and anti-protein C antibody (Fig. 7, top) are compared with

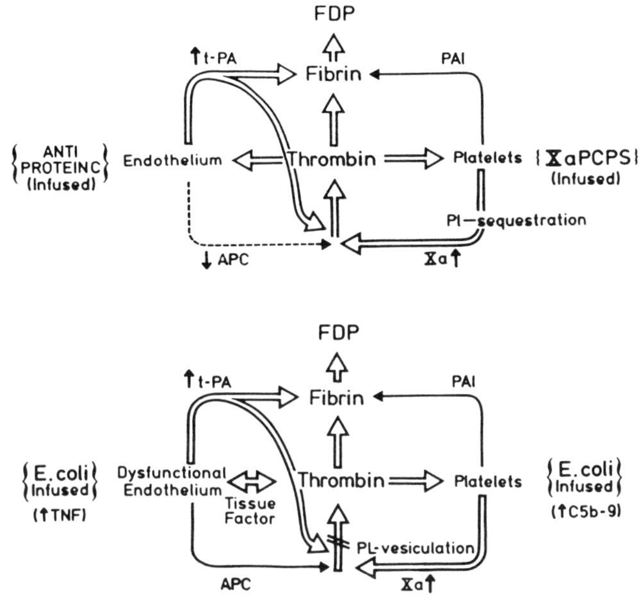

Figure 7 Disseminated intravascular coagulant (DIC) response to infusion of factor XaPC/PS plus anti-protein C antibody (top) and to LD_{100} *E. coli* (bottom): Using the four-domain hemostatic construct, the response to factor XaPC/PS and anti-protein C antibody is compared to the response to LD_{100} *E. coli*. We hypothesize that the inflammatory effects of LD_{100} *E. coli* mimics the consumptive DIC response to factor XaPC/PS. We believe that there is an intense up-regulation of coagulant (tissue factor) (Xa) activity and down-regulation of anticoagulant (APC) activity followed by an explosive generation of thrombin and fibrinolytic (t-PA) activity such as is seen following infusion of factor XaPC/PS. The only differences are that (1) the host in responding to LD_{100} *E. coli* uses endogenous factors to reproduce the effects of factor XaPC/PS and anti-protein C antibody, and (2) the inflammatory response to LD_{100} *E. coli*, in addition to triggering the DIC response, triggers a succession of self-propagating responses (i.e., free-radical injury) that are irreversible and ultimately kill the host. The mechanistic aspects of this figure are discussed in the text.

those of LD_{100} *E. coli* (Fig. 7, bottom). The diagram illustrating the effects of LD_{100} *E. coli* is similar to that illustrating the effects of factor XaPC/PS except that TNF and C5b-9 substitute for anti-protein C antibody and phospholipid vesicles. With respect to LD_{100} *E. coli*, we hypothesize that the hypercoagulable state is induced by cytokines (TNF) (i.e., down-regulation of thrombomodulin and up-regulation of tissue factor expression by endothelium and macrophage/monocytes) (22), while it is amplified by C5b-9 complement attack complex [i.e., assembly of prothrombinase complex (23) and vesiculation of platelet and other cell membranes (24)].

The initiation part of this hypothesis is illustrated on the left of Fig. 7 (bottom). We hypothesize that TNF and other inflammatory mediators produce a dysfunctional endothelium. This includes the expression of tissue factor and down-regulation of thrombomodulin and reduced generation of APC. Together, these effects constitute a hypercoagulable state. The amplification part of this hypothesis is illustrated on the right of Fig. 7 (bottom). We hypothesize that complement C5b-9 attack complex induces vesiculation of dysfunctional platelets, endothelium, and other cells in contact with blood. This results in a shower of procoagulant vesicles being released into the bloodstream, vastly increasing the surface area exposed to blood. This leads to the relatively unopposed generation of thrombin and an explosive amplification in which coagulant factors are consumed and in which anticoagulant factors (protein C and S) are both overridden and consumed. The counterresponse part of this hypothesis also is illustrated in Fig. 7 (bottom). We hypothesize that as a result of the explosive generation of thrombin, the endothelium releases a large amount of t-PA just prior to release or accumulation of PAI. This is sufficient to produce a burst of fibrinolytic activity that clears the vasculature of fibrin thrombi and inactivates factors V and VIII much as APC would have done were it not overridden and consumed. This part of the hypothesis is identical to the diagram of Fig. 7 (top), in which infusion of factor XaPC/PS plus anti-protein C antibody is followed by an intense fibrinolytic response (20). In both cases, fibrin thrombi are hypothesized to be cleared. The inflammatory factors, however, go on to kill the host in this model of *E. coli*-disseminated intravascular coagulation. It should be noted that this fibrinolytic activity is reflected only briefly by an increased rate of in vitro clot lysis. The level of fibrin degradation products and PAP complexes, however, continue to rise, suggesting that fibrinolytic activity continues but is restricted to the microenvironment of the surface of the vascular wall (2). This is an oversimplification of a complex process, to emphasize that an inflammatory stimulus such as *E. coli* activates several inflammatory effector systems at once and that we believe that they act in a coordinated manner to override the anticoagulant factors and to initiate and then amplify the procoagulant response and the fibrinolytic counterresponse. This sequence of events mimics the explosive consumption coagulopathy that can follow septicemia in obstetrical practice (i.e., attempted abortion or retained fetus).

B. Specific Observations of the *E. coli* DIC Model

What are the specifics of this model, and what is the evidence supporting the concepts shown in Fig. 7? Tables 3 to 6 and Fig. 8 summarize the experimental data gathered from the model thus far. Figure 8 illustrates the four-stage response to *E. coli*: stage I (inflammatory stage), stage II (coagulopathic

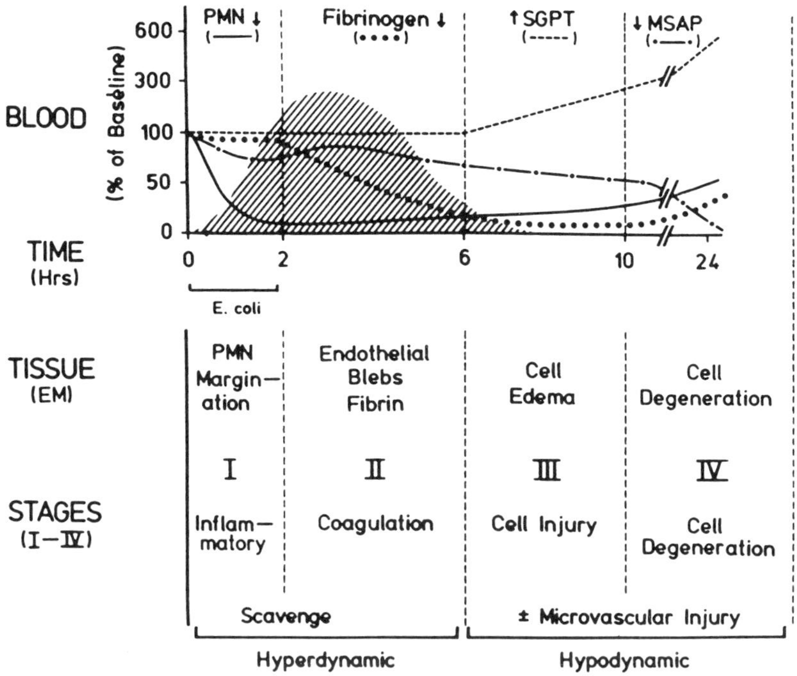

Figure 8 Disseminated intravascular coagulation (DIC) model: description of the sequence of events and stages (I to IV) following infusion of LD_{100} *E. coli*: The sequence of events in this model can be divided into four stages: inflammatory stage 1 (0 to 2 h), coagulopathic (DIC) stage 2 (2 to 6 h), and cell injury and cell degeneration, stages 3 and 4 (6 to 10 and 10+ h, respectively). The hatched area depicts the concentration of *E. coli* organisms in the blood. They reach a maximum concentration of approximately 1×10^7 CFU/mL blood at 2 h, after which they are cleared by 8 h. Stage I is dominated by the fall in leukocyte count and the appearance of tumor necrosis factor, elastase, and tissue plasminogen activator in plasma, all of which reach peak concentrations by $T+2$ h. Stage II is dominated by the fall in fibrinogen concentration and rise in fibrin degradation products. Stages III and IV are dominated by a rise in concentration of markers of cell injury (SGPT) and a steadily falling platelet count and blood pressure (MSAP). Animals die between 16 and 30 h.

stage), stage III (cell injury), stage IV (cell degeneration) (1,2). The legend to Fig. 8 describes these stages in more detail, and the legend to Table 3 gives the details of the experimental model.

Table 3 lists inflammatory agents which we believe may be among the agents that play a role in initiating and amplifying the disseminated intra-

Table 3 Disseminated Intravascular Coagulant (DIC) Model of Baboon Response to LD_{100} *E. coli*: Intervention Studies with Anti-TNF Antibody and Anti-Complement C4bBP/Protein S

Agent	Stage I	Stage II	Stages III–IV	Survival (7 days)
Anti-TNF (initiate)	PMN ↓ No TNF C5b-9 ↑	+	−	+
Anti-complement (amplify)	PMN TNF ↑ No C5b-9	−	−	+

Note: While anti-tumor necrosis factor antibody (anti-TNF) protects and attenuates the stage III and IV responses to LD_{100} *E. coli*, it does not attenuate the stage I fall in neutrophil (PMN) count, the appearance of C5b-9 attack complex, or the stage II disseminated intravascular coagulant (DIC) response. Anti-complement C4bBP/protein S, on the other hand, not only protects and attenuates stages III and IV but also the stage I fall in leukocyte count and the stage II disseminated intravascular coagulant response. The details of the experimental procedure are reported by Taylor et al. (31). Briefly, both the femoral artery and vein are cannulated under anesthesia. The reagents under study are infused as a bolus at $T - 30$ to $T - 5$ min followed by the 2-h infusion of LD_{100} *E. coli* (4×10^{10} CFU/kg). The animals are observed hourly (vital signs, blood, CBC, and chemistries) for 10 h and if they survive, daily thereafter for 7 days.

vascular coagulopathic response to LD_{100} *E. coli*. Infusion of anti-tumor necrosis factor (anti-TNF) with LD_{100} *E. coli* blocked the appearance of TNF but not of C5b-9 complexes in plasma, nor the fall in leukocyte count during the stage I inflammatory response (25). Although it attenuated the stage III and IV cell injury responses to *E. coli* and protected the animal, however, anti-tumor necrosis factor failed to inhibit the stage II–disseminated intravascular coagulant response (25). In contrast, infusion of anticomplement reagents not only gave protection and attenuated the appearance of complement activation products and the fall in leukocyte count of stage I, but also attenuated the stage II–disseminated intravascular coagulant response (F. B. Taylor and C. T. Esmon, unpublished observations). Our interpretation, although oversimplified, is that the cytokines and complement components act together to induce and then amplify the disseminated intravascular coagulant (DIC response) (Fig. 6). Tumor necrosis factor is known to induce expression of tissue factor by endothelial cells in vitro (8,9), and LD_{100} *E. coli* has been shown in this baboon model to induce expression of tissue factor by the capillary endothelium of the spleen in vivo (26). We believe that complement amplifies this procoagulant effect by C5b-9 induction of vasiculation of cell membranes (platelets, endothelium, macrophages, hepatocytes) bearing prothrombinase assemblies generated previously by the TNF-induced expression of tissue factor. This belief is based on the fact that is has been demonstrated in vitro that (1) platelets or endothelial cells exposed to C5b-9 complex can express factor Va, which leads to the assembly of prothrombinase complex and generation of thrombin (23), and (2) C5b-9 induces release of nonadherent vesicles from platelets that are rich in C5b-9 and prothrombinase (24). Thus this concept holds that disseminated intravascular coagulation is a two-step phenomenon in which procoagulant activity is expressed (i.e., tumor necrosis factor induction of tissue factor expression) and then is amplified (i.e., exposure of blood to increased surface area by complement-induced vesicles) in order for the coagulopathic (DIC) response to occur. This argument, however, is limited by two facts: (1) circulating vesicles have not been demonstrated following *E. coli* infusion, and (2) anti-tumor necrosis factor (Table 3) does not inhibit the disseminated intravascular coagulant response (25). Both these limitations, however, are addressed in part by the fact that we have found that infusion of TNF alone induces only the appearance of factor Xa activation peptides and thrombin–antithrombin complexes. This suggests that TNF alone can induce the generation of some thrombin through tissue factor and factor Xa. If, however, phospholipid (PCPS) vesicles plus TNF plus anti-protein C are infused, a disseminated intravascular coagulant response like that seen following infusion of factor XaPC/PS follows (20,27). The overlapping functions of the inflammatory cytokine systems and efficiency of the amplification systems may explain why anti-TNF alone did not inhibit the coagulopathic response to *E. coli*.

Table 4 lists the coagulant agents that we believe mediate the disseminated intravascular coagulant response to LD_{100} *E. coli*. Infusion of anti-factor

Table 4 Disseminated Intravascular Coagulant (DIC) Model of Baboon Response to LD_{100} *E. coli*: Intervention Studies with Anti-factor XII, Anti-receptor (GPIIb-IIIa, P-selectin, CD18, and ELAM), Anti-tissue Factor Antibodies, and with Tissue Factor Pathway Inhibitor (TFPI)

Agent	Stage I	Stage II	Stages III–IV	Survival (7 days)
Anti-factor XII	+	+	+	–
Anti-PL, PMN, EC receptor	+	+	+	–
Anti-tissue factor or tissue pathway inhibitor	±	–	–	+

Note: While anti-tissue factor antibody and tissue factor pathway inhibitor protect and attenuate stage II as well as stages III and IV, antibodies against factor XII and antibodies against platelet, neutrophil, and endothelial cell receptors do not protect and do not attenuate any of stages I to IV.

XII or anti-receptor antibodies (i.e., anti-GPIIb-IIIa) antibodies failed to attenuate this response (28), whereas anti-tissue factor antibody not only attenuated this response but also protected the animal (29). None of these antibodies attenuated the inflammatory stage. The fact that similar results were obtained using tissue factor pathway inhibitor (TFPI) and the fact that tissue factor is expressed in splenic capillaries following *E. coli* infusion suggests that tissue factor is a major mediator of the coagulopathic response to LD_{100} *E. coli*. The failure of anti-GPIIb-IIIa antibody to inhibit platelet consumption following LD_{100} *E. coli* infusion suggests that platelet aggregation per se does not contribute to the coagulopathic response in this model. This view is supported by the facts that (1) thrombi were seen adherent to an intact vascular endothelium in the absence of platelets either at the base of or incorporated in the clot (30), and (2) that in contrast to the factor XaPC/PS model, the platelet count in the LD_{100} *E. coli* model does not follow the consumption of fibrinogen (1,2). This observation is consistent with the in vitro data that despite the fact that C5b-9 induces vesiculation of platelet membranes rich in prothrombinase, these platelets do not express GPIIb-IIIa activity and therefore are nonadherent (24). This interpretation might explain what otherwise seems a paradoxical situation in which there is a massive consumption of fibrinogen probably amplified by vesicles in the presence of a slow decline in platelet count relative to the rate of fibrinogen consumption. Another seeming paradox should be noted in that while the coagulopathic response is intense and fibrin is laid down on vascular endothelium as determined by immunostaining (26), there are relatively few occlusive thrombi seen (3). This is probably because of the intense fibrinolytic counter response.

Table 5 lists the anticoagulant factors which we consider regulate both the coagulant and inflammatory coagulant response to LD_{100} *E. coli*. Both protein C and protein S are consumed in responses to an infusion of LD_{100} *E. coli* (31). The consumption of protein C and the appearance of activated protein C/inhibitor complexes is inhibited by coinfusion of anti-protein C antibody. This suggests that the protein C system is actively participating in regulating the response to LD_{100} *E. coli*. Furthermore, although it does not attenuate the stage I inflammatory response, co-infusion of activated protein C does inhibit the stage II-disseminated intravascular coagulant response *and* protects animals receiving either LD_{100} *E. coli* or sublethal *E. coli* plus anti-protein C antibody (this antibody inhibits activation of protein C but not of activated protein C) (31).

Protein S and C4b-binding protein have also been studied extensively. Coinfusion of anti-protein S antibody or C4b-binding protein (which binds and neutralizes the protein S cofactor necessary for optimal activated protein C function) with sublethal *E. coli* reproduces the lethal response to LD_{100} *E. coli*, including the appearance of TNF in plasma and the disseminated

Table 5 Disseminated Intravascular Coagulation (DIC) Model of Baboon Response to *E. coli*: Intervention Studies with Activated Protein C (APC); Reconstitution Studies with Anti-protein C or Anti-protein S Coinfused with Sublethal *E. coli*

Agent	Stage I	Stage II	Stages III–IV	Survival (7 days)
APC	+	−	−	+
Anti-protein C, or anti-protein S alone	−	± (↓ platelets)	−	+
Sublethal *E. coli* alone	↓ PMN No TNF No C5b-9	−	−	+
Anti-protein C, or anti-protein S + sublethal *E. coli*	↓ PMN ↑ TNF ↑ C5b-9	+	+	−

Note: Activated protein C coinfused with lethal *E. coli* can protect and attenuate stages II to IV. This table also shows that neutralization of endogenous protein C or protein S with their respective antibodies will render an infusion of a sublethal concentration of *E. coli* lethal with full expression of stages I to IV.

intravasacular coagulant response (32). In contrast, coinfusion of a nonneutralizing antibody to protein S or of C4b-binding protein that has been saturated with exogenous protein S with sublethal *E. coli* does not reproduce this lethal response (33). This strongly suggests that the neutralization of protein S by these reagents amplifies the inflammatory and coagulant response to sublethal *E. coli.* We concluded that protein C and protein S play an important role in down-regulating both the inflammatory and coagulopathic stages of the responses to LD_{100} *E. coli* and that high concentrations of C4b-binding protein (an acute phase protein) could prevent this down-regulation by binding and neutralizing protein S.

Protein S may, however, have another function in addition to that of cofactor to the activated protein C anticoagulant. C4b binding protein may bind protein S for reasons other than just to neutralize it. This may be especially important since C4b-binding protein also is an inhibitor of C4b (34), which is necessary for assembly of the C3 convertase and for full expression of complement activity. Is it possible that protein S also facilitates this function by acting as a cofactor for C4b-binding protein inhibition of this assembly of the complement C3 convertase. Studies by Dahlback et al. demonstrated that protein S with its gla domain could serve as an anchor for C4b-binding protein to phospholipid membranes (35). These findings raised the possibility that such a C4bBP/PS complex bound to the negatively charged membranes of perturbed target cells would be more efficient than C4b-binding protein in fluid phase in protecting these membranes from injury by "incoming"

C4b molecules. This seemed especially likely because normally about 40% of the circulating C4b-binding protein is bound to protein S (36). Therefore, we coinfused C4b-binding protein saturated with protein S with LD_{100} *E. coli* based on the following hypothesis: For protein S to protect against disseminated intravascular coagulation and lethal effects of LD_{100} *E. coli*, two conditions must be met. The total protein S concentration must be high, relative to C4b-binding protein such that there is (1) sufficient free protein S to facilitate the anticoagulant action of activated protein C, and (2) sufficient protein S to complex to C4b-binding protein to facilitate the anticomplement action of C4b-binding protein. Thus as proposed by Dahlback et al., protein S may have two beneficial functions (34), one of which involves its free form and one of which involves its complexed form (Fig. 9). This suggests why it circulates in plasma in both the free and bound forms. The paradox, however, is that if there is insufficient total protein S, and if the levels of C4b-binding protein rise to bind and neutralize the limited amount of free protein S available, the beneficial effects of the anticomplement function of the complexed protein S are outweighed by loss of the anticoagulant function of the free protein S. Furthermore, even the anticomplement function favored by the increased C4bBP/protein S complex is insufficient to meet the challenge of a LD_{100} *E. coli* infusion if total protein S levels are relatively low. Thus it would be important to supplement endogenous C4bBP/protein S complex with free protein S and C4bBP/protein S.

Given the above rationale, we coinfused C4b binding protein with an excess of protein S (20 mg/kg C4bBP plus 3.5 mg/kg of protein S equilibrated at 37°C for 2 h before infusion) with LD_{100} *E. coli.* The inflammatory and coagulant responses were attenuated, and the cell injury and cell

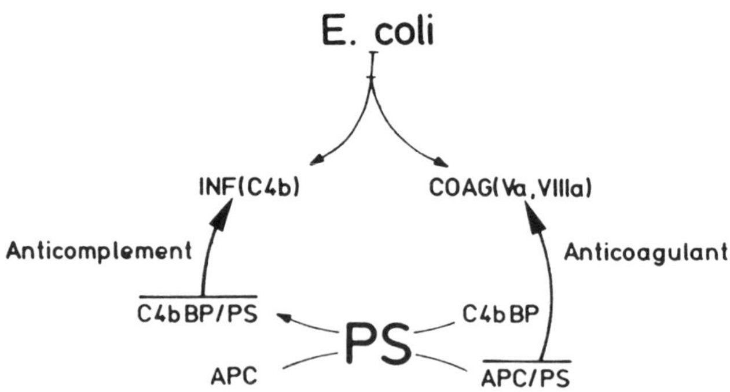

Figure 9 Dual functions of protein S: protein S serves as a cofactor for both C4bBP anti-complement and activated protein C anticoagulant regulation of the effects of LD_{100} *E. coli.*

degeneration stages III and IV were completely inhibited (Fig. 10, top). Three of three animals so treated were permanent survivors (Fig. 10a), whereas none of the control animals survived (Fig. 10b). Coinfusion, however, of protein S alone at concentrations calculated to be in excess did not inhibit the response to lethal *E. coli*. These studies suggest that the C4bBP/protein S complex is the active principle and that it is acting by inhibiting the incorporation of C4b into the C4bC3b convertase. In short, the C4bBP/PS complex may inhibit C4b of the classical complement pathway. This possibility is supported by the fact that the concentration of cleaved C4b (C4b/c) and

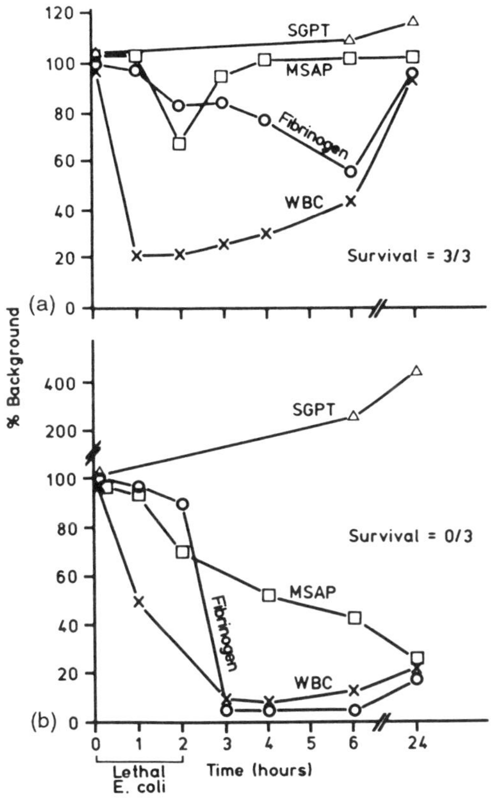

Figure 10 Response of baboon to LD_{100} *E. coli* (b) and to LD_{100} *E. coli* plus C4b binding protein/protein S complex (C4bBP/protein S) (a). This figure shows that C4bBP/protein S (20/3.5 mg/kg) protects, and attenuates not only the consumption of fibrinogen and rise in markers of cell injury (SGPT) but also the fall in leukocyte (WBC) count and blood pressure (MSAP).

cleaved C3b (C3b/c) are reduced by 80% in the treated animals compared to their untreated controls. These observations in turn raise the following questions which have not been addressed in this model: (1) to what extent do anti-endotoxin antibodies form immune complexes with whole *E. coli* organisms or with endotoxins derived from these organisms, and (2) to what extent do these immune complexes and complement participate in the injurious effects of LD_{100} *E. coli* infusion vis-à-vis those effects that are produced by endotoxin alone? This issue will need to be addressed initially by careful assay of circulating anti-endotoxin antibodies and endotoxin such as has been done in vervet monkey by Wessels et al. (37). Additional studies on the role of complement using other agents (i.e., C1-esterase inhibitors) also need to be studied.

Table 6 lists those agents that based on the foregoing observations might be expected to protect against LD_{100} *E. coli* but do not. Factor Xa, the active site of which is inhibited by a chloromethyl ketone (DEGR-Xa), might be expected to protect against LD_{100} *E. coli* because like activated protein C, anti-tissue factor, and tissue factor pathway inhibitor (TFPI), it inhibits the disseminated intravascular coagulant response. The fact is, however, that while it can completely inhibit this response, it does *not* protect against the lethal stage III and IV effects (38). Thus the disseminated intravascular coagulant response is not a part of the lethal chain of events in this specific model of LD_{100} *E. coli* sepsis. These results therefore also raise the question of what properties of these other agents besides their anticoagulant properties

Table 6 Disseminated Intravascular Coagulation (DIC) Model of Baboon Response to *E. coli*: Summary of Intervention Studies with Anti-receptor Antibodies, Including Anti-GPIIa-IIIa, P-selectin, CD18, and ELAM Antibodies and with Factor Xa Inactivated with Chloromethyl Ketone (DEGR-Xa)

Agent	Stage I	Stage II	Stages III–IV	Survival
Anti-PL, PMN, EC receptor	+	+	+	–
DEGR-Xa	+	–	+	–

Note: Blocking receptor function with large concentrations (10 mg/kg) of antibodies (Fab and IgG) to the receptors above does not protect or attenuate stages I to IV. In vitro assays demonstrated that those antibodies raised against human receptors cross reacted with their baboon counterparts. Active-site-inhibited factor Xa (DEGR Xa), while completely inhibiting the stage II consumptive coagulopathic (DIC) response, did not protect or attenuate stages I, III, and IV. The fact that DEGR Xa inhibits stage II but does not protect, whereas activated protein C, anti-tissue factor antibody, and tissue factor pathway inhibitor (which also inhibits the DIC response), do protect suggests that this protective effect is due to some function other than their anticoagulant function.

act to protect as well as anticoagulate? Table 6 also lists antibodies against cell receptors, including ELAM, CD18, GPIIb-IIIa, and P-selectin. These agents again might be expected to protect against LD_{100} *E. coli* because the decrease in leukocyte and platelet counts, and leukocyte adhesion appears to be a prominent feature of the stage I inflammatory response to LD_{100} *E. coli*. The fact is, however, that while attenuating the decrease in leukocyte or platelet counts, antibodies against these receptors do not protect against the lethal stage III and IV effects (F. B. Taylor, Jr., B. Coller, et al., unpublished observations). Thus the activation and recruitment of leukocytes and platelets do not appear to be a part of the lethal chain of events in this specific coagulopathic (DIC) model of *E. coli* sepsis. This is quite different from the microvascular thrombotic model of *E. coli* sepsis reviewed in the next section. Modification of leukocyte, platelet, and coagulant factor responses in that model does protect, much as modulation of these factors protects in models of ischemia reperfusion injury.

VI. Microvascular Thrombosis Induced by Inflammatory Stimuli (LD_{100} *E. coli* Plus Anti-TNF Antibody or Sublethal *E. coli* Plus an Inhibitor of the Protein C Anticoagulant System)

A. General Principles of *E. coli* MVT Model

We discuss next a model in which the host is partially protected by intervention measures (i.e., anti-tumor necrosis factor or anti-tissue factor antibody) and survives the disseminated intravascular coagulant response to LD_{100} *E. coli* only to succumb to a microvascular thrombotic (MVT) response. Figure 11 is a conceptual illustration similar to Fig. 6 in which the vascular endothelium again bears the brunt of an aberrant inflammatory response of activated plasma components and formed elements. The initial DIC response and the inflammatory/coagulant factors that drive it are similar to those described in Section V except that the effects of tumor necrosis factor (TNF) are attenuated. Rather than succumbing within 30 h to the LD_{100} *E. coli*, however, the host survives over a period of days only to suffer the effects of an inflammatory/coagulopathic process that culminates in a generalized microvascular thrombotic response. The concept of the MVT response differs from that of the preceding consumptive coagulopathic (DIC) response in that (1) not only is there expression of TF and down-regulation of thrombomodulin (TM) but also depletion and neutralization of fibrinolytic and anticoagulant factors by rising concentrations of PAI and acute-phase proteins, including C4b-binding protein (C4bBP), and (2) as a consequence, an environment is created in which platelets and fibrin aggregate and poly-

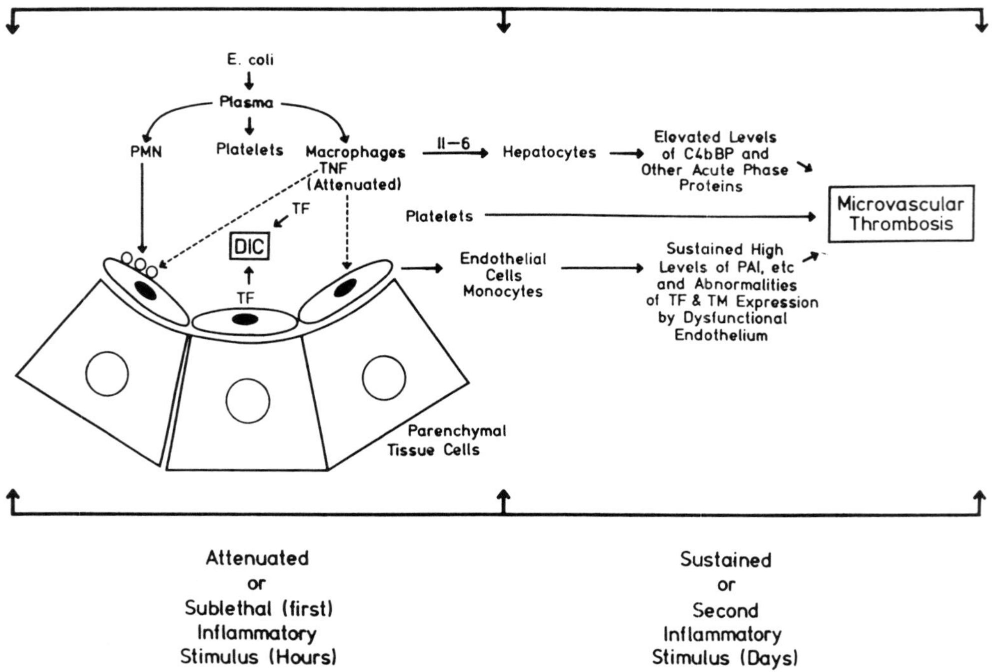

Figure 11 Microvascular thrombotic (MVT) model of the baboon response to suboptimal concentrations of anti-TNF antibody and LD_{100} *E. coli*: diagram of a possible mechanism of the delayed response to *E. coli*. This is a two-step response in which the host is primed and then challenged by continuation of the inflammatory response. Priming involves an inflammatory stage I and coagulopathic stage II response which in this case is not sufficient to produce lethal shock or thrombosis. The host survives long enough to experience a rise of acute-phase proteins, including C4bBP and plasminogen activator inhibitor (PAI). This rise, combined with continued inflammation or a second challenge stimulus, is sufficient to produce a widespread microvascular thrombosis such as that seen in hemolytic uremic syndrome (stage V).

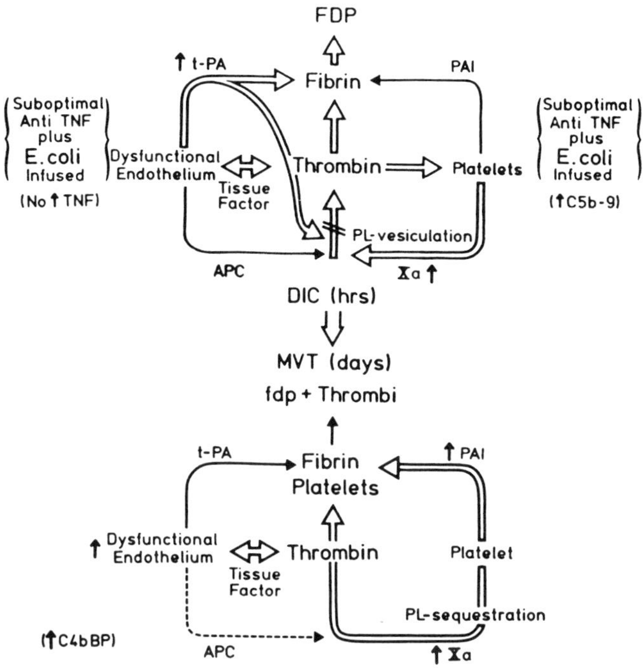

Figure 12 Progression of a disseminated intravascular coagulant (DIC) response into a microvascular thrombotic response following infusion of LD_{100} *E. coli* plus suboptimal concentrations of anti-TNF antibody. Using the four-domain hemostatic construct, the progression of the DIC response to LD_{100} *E. coli* plus suboptimal concentrations of anti-TNF antibody (anti-TNF) into a MVT response is followed. We hypothesize that the combination of LD_{100} *E. coli* plus suboptimal concentrations of anti-tumor necrosis factor (TNF) antibody protects the host from the immediate (10 to 30 h) lethal effects of the inflammatory response (i.e., ↑TNF) such that the stage II coagulopathic DIC response can progress into a coagulopathic MVT response. We believe that following vesiculation of platelets and DIC there is a growing and sustained sequestration of platelets on a dysfunctional endothelium and release of high local concentrations of plasminogen activator inhibitor (PAI). This event, combined with rising levels of C4b-binding protein (↑C4bBP), further neutralize already depleted fibrinolytic and anticoagulant systems to such an extent that there is no clearance of fibrin thrombi. Thus, in contrast to the DIC response to inflammation, the MVT response includes a massive deposition of fibrin, profound thrombocytopenia, anemia, schistocytosis, and renal failure.

merize on the surface of an increasingly dysfunctional capillary endothelium to produce a massive microvascular thrombosis, thrombocytopenia, anemia, and renal failure (39).

This concept using the four-domain hemostatic construct is illustrated in Fig. 12, in which the microvascular thrombotic (MVT) response (bottom) is shown following the disseminated intravascular coagulant response (top) to LD_{100} *E. coli*. In the case of the MVT response, we hypothesize that there is an almost complete neutralization and/or depletion of anticoagulant and fibrinolytic activity due to (1) rising levels of C4b-binding protein, which bind the protein S cofactor for APC, and (2) to a rising and sustained level of PAI, which inhibits the t-PA activ;ation of plasminogen. This depletion and neutralization of anticoagulant activity leaves platelets completely unprotected from thrombin-induced aggregation and adherence to an endothelium that is becoming increasingly dysfunctional to the point of irreversible injury. The massive recruitment of platelets into the microvascular thrombotic process, in turn, is accompanied by a release of platelet PAI (14,15) locally in relatively high concentrations such that the fibrinolytic activity that cleaved the fibrin deposited in the initial DIC response is no longer present. Thus the MVT is pictured in this construct as being amplified by platelet *sequestration* (adherence and aggregation) and secretion, whereas the disseminated intravascular response is pictured as being amplified by *vesiculation* of cell membranes, including those of platelets. The paradox is that platelet consumption runs behind fibrinogen consumption in the DIC response, while platelet consumsption runs ahead of fibrinogen consumption in this MVT response. This sequence of events mimics the hemolytic uremic syndrome and thrombotic thrombocytopenic purpura seen in clinical practice (40). Figure 13 illustrates this concept in another way, as a continuum in which a sequence of events leads from DIC to MVT.

B. Specific Observations of the *E. coli* MVT Model

What are the specifics of this model, and what is the evidence supporting the concepts shown in Figs. 12 and 13? Figure 14 shows the five-stage (DIC/MVT) response to LD_{100} *E. coli* in animals that have been coinfused with a suboptimal concentration of anti-TNF antibody at T + 30 minutes. Stages I to IV are identical to those described previously except that the appearance of TNF in plasma, the intensity of the inflammatory stage, and the fall in mean systemic arterial pressure (MSAP) is attenuated. This is not sufficient, however, to reduce the intensity of the initial disseminated intravascular coagulant response, including the generation of thrombin–antithrombin and plasmin–antiplasmin complexes and the consumption of fibrinogen, protein S, and protein C. The host therefore survives beyond the first 24 h

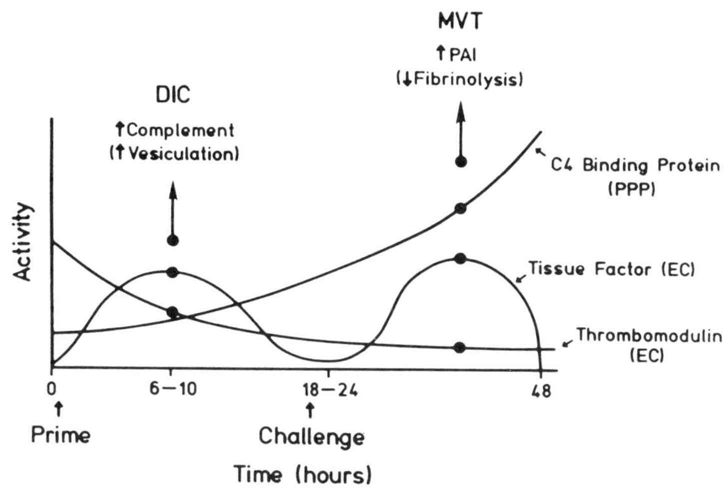

Figure 13 Concept of microvascular thrombotic phenomena: The primary stimulus (TNF, endotoxin) up-regulates the expression of the procoagulant receptor, tissue factor, and down-regulates the anticoagulant receptor (thrombomodulin) by the vascular endothelium. The increase in tissue factor, however, is temporary and peaks at 6 to 10 h, while the decrease in thrombomodulin expression may be sustained. In addition, this prime stimulus induces a rise in the concentrations of C4b-binding protein (C4bBP) and plasminogen activator inhibitor (PAI), which continue to rise over the 48-h period. C4bBP inhibits the protein S cofactor necessary for the anticoagulant activity of activated protein C (protein C is activated to activated protein C by the thrombin–thrombomodulin complex). PAI inhibits tissue plasminogen activator (t-PA) activation of the fibrinolytic (plasminogen) system. Thus, at the time of the challenge stimulus, conditions are set for an extensive microvascular thrombosis. That is, tissue factor expression is elevated for the second time under conditions in which thrombomodulin expression is still depressed and in which protein S and t-PA activity are inhibited by high levels of C4bBP and PAI, respectively. The challenge stimulus could range from translocated *E. coli* organisms or secondary bacterial infection following a viremia.

long enough to enter stage V. It is at this point that the C4b-binding protein (C4bBP) and plasminogen activator inhibitor levels begin to rise and the level of free protein S is seen to decrease even further. At the same time the platelet count is driven to an extreme low, while the fibrinogen concentration returns toward normal! It is also at this point that schistocytes begin to appear in the circulation and the hematocrit also decreases to an extreme low. The arterial pressure (MSAP), however, remains in the normal range while the blood urea nitrogen (BUN) and creatinine rise steadily, approaching 200 to 300 and 6 to 8,

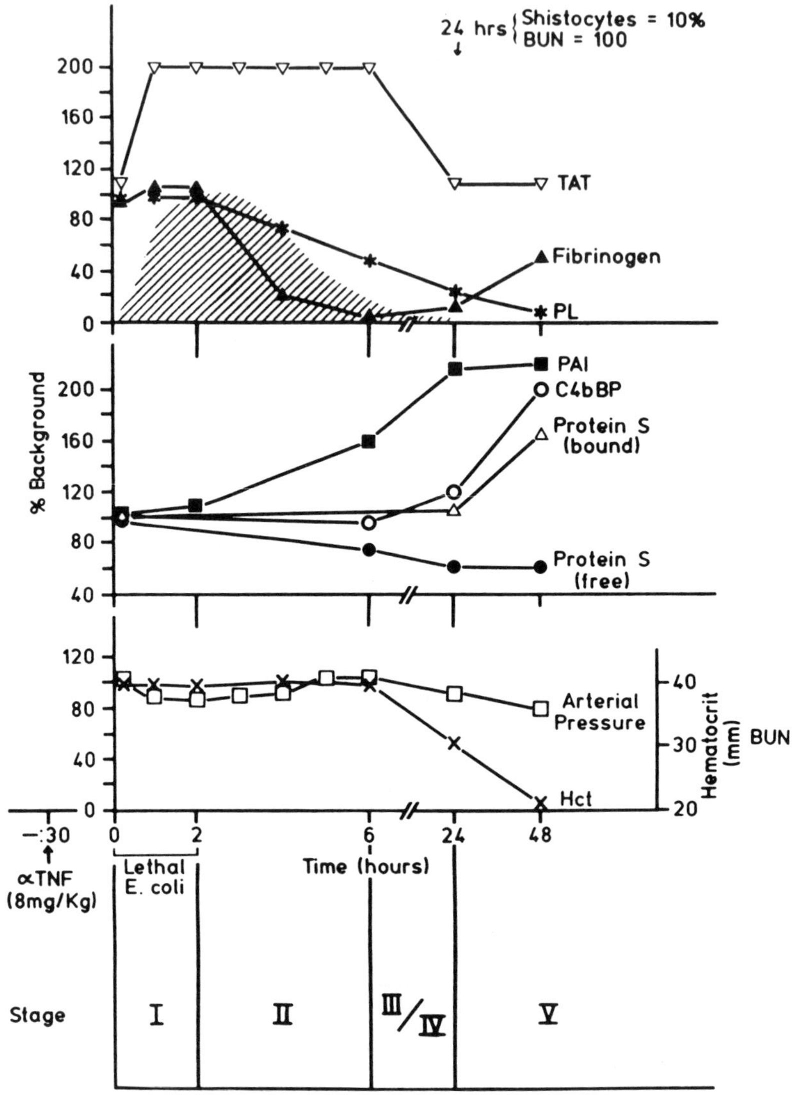

Figure 14 Microvascular thrombotic (MVT) model: description of the sequence of events and stages I to V following infusion of suboptimal anti-TNF antibody and LD_{100} *E. coli*. This figure shows that the first four stages of this model of the microvascular thrombotic response to suboptimal anti-tumor necrosis factor and LD_{100} *E. coli* are similar to those seen in the model of the disseminated intravascular coagulant (DIC) response to LD_{100} *E. coli*. The host, however, survives long enough for (1) concentration of the C4b binding protein (C4aBP) to rise and bind free proteins S, and (2) the concentration of plasminogen activator inhibitor (PAI) to rise and bind tissue plasminogen activator. We hypothesize that after 2 or 3 days this completely shuts down the already depleted anticoagulant and fibrinolytic systems. We hypothesize that this, in turn, leads to an unopposed activation, aggregation and sequestration of platelets and deposition of fibrin, widespread microvascular thrombosis, thrombocytopenia, anemia, and renal failure.

respectively, by 48 h. The renal tissues show the picture of cortical thrombosis, infarction, hemorrhage, and tubular necrosis associated with microvascular thrombosis (39). Animals receiving LD_{100} *E. coli* alone (no anti-TNF) survive only the first four stages (16 to 30 h). They do not exhibit the profound changes in platelet count, hematocrit, schistocytes, or BUN, nor do they exhibit the glomerular thrombotic renal pathology associated with microvascular thrombosis. The legend to Fig. 14 describes these stages in more detail.

Tables 7 to 9 summarize the experimental data gathered from this model so far. Table 7 lists the agents that can change the status of the host such that the microvascular thrombosis shown in Fig. 13 is reproduced. They all are inhibitors of either protein C or protein S of the anticoagulant system. Most important of these is C4b-binding protein (C4bBP). Doubling the concentration of C4bBP can convert a harmless sympathomimetic, mild inflammatory response to sublethal *E. coli* into a more severe inflammatory, coagulopathic (DIC) response followed by a lethal microvascular thrombotic response (32).

Table 7 illustrates not only the possible importance of acute-phase proteins such as C4b-binding protein in modulating the response to *E. coli* but also the role of the protein C system and a regulation of this response. Thus

Table 7 Microvascular Thrombotic (MVT) Model of the Baboon Response to (1) Suboptimal Concentrations of Anti-tumor Necrosis Factor Antibody and LD_{100} *E. coli* and (2) Reconstitution Studies with Anti-protein C, Anti-protein S Antibodies, or C4a Binding Protein (C4bBP) Coinfused with Sublethal *E. coli*[a]

Agent	Stage I	Stage II (DIC)	Stages III–IV	Stage IV (MVT)	Survival (7 days)
Anti-protein C, anti-protein S, or C4bBP alone	–	± (platelet ↓)	–	–	+
Sublethal *E. coli* alone	PMN ↓ No TNF No C5b-9	–	–	–	+
Anti-protein C, anti-protein S, or C4bBP + sublethal *E. coli*	PMN ↓ TNF ↑ C5b-9 ↑	+	±	+	3–5 days

Note: The disseminated intravascular coagulant (DIC) response to suboptimal concentrations of anti-tumor necrosis factor, which then progresses over 3 to 4 days into a microvascular thrombotic response can be duplicated by coinfusing inhibitors of the protein C, protein S anticoagulant systems with sublethal *E. coli*. C4b-binding protein (C4bBP), an inhibitor of protein S, is of special interest because it is an endogenous acute-phase protein, increasing concentrations of which could favor a microvascular thrombotic response to an inflammatory stimulus (i.e., sublethal *E. coli*).

[a]The results are reversed with coinfusion of activated protein C, or protein S.

antibodies to protein S or protein C or C4b-binding protein convert the response to sublethal *E. coli* into a lethal one. This effect is obviated by supplementation with either protein S (33) or activated protein C (31). Finally, these results show the other face of C4b-binding protein. It shows that the protective anticomplementary action of C4bBP/protein S complex described in Section V.B can be lost to the lethal anticoagulant action of C4b-binding protein if its concentration rises to a point where it binds all the free protein S necessary for function of the protein C anticoagulant system.

Table 8 lists the agents that inhibit the microvascular thrombotic response to infusates of sublethal *E. coli* and C4b-binding protein. Antibodies to platelet receptors GPIIb-IIIa and P-selectin, as well as to ELAM and CD18, are effective inhibitors of the microvascular thrombotic response (F. B. Taylor, Jr., B. Coller, et al., unpublished observations). This is in stark contrast to the disseminated intravascular coagulant response to LD_{100} *E. coli* described in Section V.B, where these agents have no inhibitory effect at all. It should be emphasized that like the response to anti-TNF plus LD_{100} *E. coli*, the response to C4b-binding protein plus sublethal *E. coli* includes an initial disseminated intravascular coagulant response followed by a microvascular thrombotic response. In neither case do the antiplatelet agents inhibit the disseminated intravascular coagulant response. They only inhibit the microvascular thrombotic response that follows. These results again support the concept shown in Figs. 11 to 13 that platelets (and leukocytes) are important links in the chain of events in the microvascular thrombotic response, whereas they are not in the disseminated intravascular coagulant response.

Finally, Table 9 shows that infusion of interleukin-6 (IL-6) at T-24 h is followed by a rise in C4bBP and reduction in free protein S such that infusion of sublethal *E. coli* at T-0 is followed by a microvascular thrombotic

Table 8 Microvascular Thrombotic (MVT) Model of the Baboon Response to Coinfusions of C4bBP with Sublethal *E. coli*; Intervention Study with Antireceptor Antibodies, Including AntiGPIIa-IVa, P-Selectin, CD18, and ELAM Antibodies

Agent	Stage I	Stage II (DIC)	Stages III–IV	Stage IV (MVT)	Survival (7 days)
Anti-PL, PMN, EC receptor	PMN ↓ TNF ↑	+	–	–	+

Note: These results are just the opposite of those shown in Table 6. All these antireceptor antibodies protect and attenuate stage V (MVT) response as well as stages III and IV. They do not attenuate the coagulopathic (DIC) stage II response. This result coincides with those shown in Table 6 and strongly suggests that in contrast to the MVT response, the DIC response does not require the formed elements (PMN, platelets) for full expression.

Table 9 Microvascular Thrombotic (MVT) Model of the Baboon Response to Infusion of IL-6 at $T - 24$ Hours Followed at $T - 0$ by an Infusion of LD_{100} *E. coli*

Agent	Stage I	Stage II (DIC)	Stages III–IV	Stage IV (MVT)	Survival (7 days)
IL-6 ($T - 24$ h) + LD_{100} *E. coli* ($T - 0$)	PMN ↓ TNF ↑ C5b-9 ↑	+	±	+	3–5 days

Note: Pretreatment with IL-6 will prime the animal so that its response to LD_{100} *E. coli* is like that seen following either suboptimal concentrations of anti-tumor necrosis antibody plus LD_{100} *E. coli* or C4b-binding protein plus sublethal *E. coli*. Since acute-phase proteins, including fibrinogen and endogenous C4b-binding proteins, rise during this 24 h, it is postulated that the IL-6 induction of this phenomenon is responsible for the microvascular thrombotic response to LD_{100} *E. coli*.

response identical to that following the coinfusion of C4bBP and sublethal *E. coli*. This supports the concept that IL-6 could be a link in the chain of events in which the host, should it survive an initial inflammatory challenge and experience a rise in IL-6 acute-phase proteins (C4bBP), could exhibit a microvascular thrombotic response to a second or continued inflammatory challenge (Figs. 11 to 13). These studies therefore support the concept shown in Figs. 11 to 13 that rising levels of acute-phase proteins, including C4bBP, can lead to microvascular thrombosis provided that the host survives the initial effects of *E. coli*. This would be the case if the C4b-binding protein concentration was already elevated due to a prior single sublethal challenge or due to a preexisting or chronic sustained inflammatory state (parasites, etc.).

VII. Summary; Applicability of Baboon Models of the Hemostatic Response to Inflammatory Stimuli to Study of Pulmonary Diseases

In this review we describe and rationalize three primate models of disorders of hemostasis following inflammatory stimuli. Dividing the hemostatic system into four conceptual domains all of which are regulated by thrombin permits (1) examination of the effect of inflammation on the hemostatic system as an integrated unit (Figs. 1, 2, 4, 7, and 12), and (2) study of the various ways in which disorders of the hemostatic systems induced by inflammation can present (DVT, DIC, MVT).

Deep vein thrombosis (DVT) can lead to pulmonary thromboembolism. Approximately 30% of those animals in which deep vein thrombosis has been induced also have pulmonary emboli by the second or third days. Microvascular thrombosis (MVT) such as seen in hemolytic uremic syndrome (40) can occur in the pulmonary vasculature in inflammatory disorders such as

Goodpasture's syndrome (41). Finally, the adult respiratory distress syndrome (ARDS) associated with gram-negative sepsis can include not only capillary leak and inflammation of the pulmonary interstitium but also a hemostatic response characterized by deposition of fibrin in pulmonary alveolae (hyaline membranes). The mechanism of this phenomenon as seen in the baboon includes (1) expression of tissue factor not only by the vascular endothelium but also by the alveolar epithelium (26), (2) down-regulation of thrombomodulin, and finally, (3) transudation of plasma into the alveolae followed by interaction of the plasma clotting factors with tissue factor and surfactant (42). Current studies suggest that the ARDS can be reproduced in the baboon by priming with nebulized gamma interferon followed by intravenous infusion of sublethal *E. coli.* Observations such as those just listed suggest that the baboon models described in this review can be manipulated at will to amplify the involvement of specific target organs, and that this may include models of pulmonary thromboembolism and of microvascular and alveolar fibrin deposition.

References

1. Taylor FB Jr, Esmon CT, Hinshaw LB. Summary of staging mechanisms and intervention studies in the baboon model of *E. coli* sepsis. J Trauma 1991; 30 (suppl):S–197.
2. Taylor FB Jr. Studies of the natural history and mechanism of the primate (baboon) response to *Escherichia coli.* In: Ryan JL, Morrison DC, eds. Bacterial Endotoxic Lipopolysaccharides. Vol. II. Immunopharmacology and Pathophysiology, Boca Raton, FL: CRC Press, 1991:239.
3. Esmon CT. The roles of protein C and thrombomodulin in the regulation of blood coagulation. J Biol Chem 1989; 264:4743.
4. White GC, Workman EF, Lundblad RL. Platelet thrombin interactions. The platelet is a substrate for thrombin. In: Lundblad RI, Fenton JW, Mann KG, eds. The Chemistry and Biology of Thrombins, Ann Arbor, MI: Ann Arbor Press, 1977:479–498.
5. Parkinson JF, Bang NV, Garcia JGN. Recombinant human thrombomodulin attenuates the inflammatory response of thrombin-stimulated human endothelial cells. Thromb Haemost 1991; 65(6):825.
6. Esmon CT. Protein C. In: Spaet TH, ed. Progress in Hemostasis and Thrombosis. New York: Grune & Stratton, 1984:25.
7. Dahlback B, Esmon CT, Chang ACK, Lockhart MS, Taylor FB Jr. Role of protein S in regulating the coagulant response to *E. coli.* Manuscript in preparation.
8. Pohlman TH, Stannes KA, Beatley PG, Ochs DD, Harlan JM. An endothelial cell surface factor(s) induced in vitro by lipopolysaccharide, interleukin 1, tumor necrosis factor-α increases neutrophil adherence by a CD2-18-dependent mechanism. J Immunol 1986; 136:4548.

9. Nawroth PP, Stern DM. Modulation of endothelial cell hemostatic properties by tumor necrosis factor. J Exp Med 1986; 163:740.
10. Okumura T, Hasitz M, Jamieson GA. Platelet glycalicin interactions with thrombins and its role as thrombin receptor of the platelet surface. J Biol Chem 1978; 253:3435–3443.
11. Mosher DF, Valheri A, Choate JJ, Ghamberg CG. Action of thrombin on surface glycoproteins of human platelets. Blood 1979; 53:437–445.
12. Davis EW, Fujikawa K, Kisiel W. The coagulation cascade. Initiations, maintenance, and regulation. Biochemistry 1991; 30:10366.
13. Loskutoff DJ, Edgington TS. Synthesis of a fibrinolytic activator and inhibitor of endothelial cells. Proc Natl Acad Sci USA 1977; 74:3903. Sprengos ED, Kluft C. Plasminogen activator inhibitors. Blood 1987; 69:381–387.
14. Erickson LA, Ginsberg MG, Loskutoff DJ. Detection and partial characterization of an inhibitor of plasminogen activator in human platelets. J Clin Invest 1984; 74:1465.
15. Booth NA, Anderson JA, Bennett B. Platelet release proteins which inhibits plasminogen activators. J Clin Pathol 1985; 38:825.
16. Virchow R. Cellular pathology as based upon physiological and pathological histology. 2nd ed. of the original, by Frank Chance, with notes and numerous amendations, principally from manuscript notes of the author. London: Churchill, 1860; XXVII:554.
17. Hanss M, Collen D. Secretion of tissue-type plasminogen activator and plasminogen activator inhibitor by cultured human endothelial cells: modulation by thrombin, endotoxin, and histamine. J Lab Clin Med 1987; 109:97.
18. Taylor FB Jr, Esmon CT, Chang A. Primate model of deep vein thrombosis induced by inflammatory mediators. Circulation 1990; 82(abstr 3060):III-770.
19. Wakefield TW, Wrobleski SK, Sarpa MS, Taylor FB Jr, Esmon CT, Greenfield LJ. Deep venous thrombosis in the baboon: an experimental model. J Vasc Surg 1991; 14:587.
20. Taylor FB, Hoogendorn H, Chang ACK, et al. Anticoagulant and fibrinolytic activities are promoted, not retarded, in vivo after thrombin generation in the presence of a monoclonal antibody that inhibits activation of protein C. Blood 1991; 79:1720–1728.
21. Taylor FB, Chang A, Hoogendorn H, Maraganore JM. Evidence that thrombin mediates secondary fibrinolysis in disseminated intravascular coagulation. Thromb Haemost 1991; 65:1238.
22. Moore KL, Esmon CT, Esmon NL. Tumor necrosis factor leads to internalization and degradation of thrombin from the surface of bovine aortic endothelial cells in culture. Blood 1989; 73:159.
23. Wiedmer T, Esmon CT, Sims PJ. Complement proteins C5b-9 stimulate procoagulant activity through platelet prothrombinase. Blood 1986; 68:875.
24. Sims PJ, Faioni EM, Wiedmer T, Shattil S. Complement proteins C5b-9 cause release of membrane vesicles from the platelet surface that are enriched in the membrane receptor for coagulation factor Va and express prothrombinase activity. J Biol Chem 1988; 263:18205.

25. Hinshaw LB, Tekamp-Olsen P, Chang ACK, et al. Survival of primates in LD_{100} septic shock following therapy and antibody to tumor necrosis factor (TNF). Circ Shock 1990; 30:279.
26. Drake TA, Cheng J, Change A, Taylor FB Jr. Expression of tissue factor, thrombomodulin and E-selectin in baboons with lethal *E. coli* sepsis. Am J Pathol. 1983; 142:1458.
27. Taylor FB Jr, He S, Chang ACK, Peer G, Esmon CT. Infusion of phospholipid vesicles converts the local thrombotic response to TNF and anti-protein C into a systemic consumptive response. Blood. Submitted.
28. Pixley RA, de la Cadena R, Page JD, et al. The contact system contributes to hypotension but not disseminated intravascular coagulation in lethal bacteremia: in vivo use of monoclonal anti-factor XII antibody to block contact activation in baboons. J Clin Invest. In Press.
29. Taylor FB Jr, Chang A, Ruf W, et al. Lethal *E. coli* septic shock is prevented by blocking tissue factor with monoclonal antibody. Circ Shock 1991; 33: 127.
30. Voss BL, DeBault LE, Blick KE, et al. Sequential alterations in septic shock in the primate. Circ Shock 1991; 33:142.
31. Taylor FB Jr, Chang A, Esmon CT, D'Angelo A, Vigano-D'Angelo S, Blick KE. Protein C prevents the coagulopathic and lethal effects of *Escherichia coli* infusion in the baboon. J Clin Invest 1987; 79:918.
32. Taylor F, Chang ACK, Ferrell G, et al. C4b-binding protein exacerbates the host response to *E. coli*. Blood 1991; 78:357.
33. Taylor FB Jr, Dahlback B, Esmon CT, Chang ACK, Lockhart MS, Hatanaka K. The role of free protein S and C4bBP protein in regulation to the coagulant response to *E. coli*. Blood, submitted.
34. Dahlback B. Interaction between vitamin K-dependent pretein S and the complement protein, C4b-binding protein. A link between coagulation and the complement system. Semin Thromb Hemost 1984; 10:139.
35. Schwalbe RA, Dahlback B, Nelsestuen GL. Independent association of serum amyloid P component, protein S, and complement C4b with complement C4b-binding protein and subsequent association of the complex with membranes. J Biol Chem 1990; 265:21749.
36. Comp PC, Esmon CT. Recurrent venous thromboembolism in patients with a partial deficiency of protein S. N Engl J Med 1984; 311:1525.
37. Wessels BC, Wells MT, Gaffin SL, Brock-Utne JG, Gathiram P, Hinshaw LB. Plasma endotoxin concentration in healthy primates and during *E. coli*–induced shock. Crit Care Med 1988; 16:601.
38. Taylor FB Jr, Chang AK, Mather T, et al. DEGR-factor Xa blocks disseminated intravascular coagulation initiated by *E. coli* without preventing shock or organ damage. Blood 1991; 78:364.
39. Taylor FB, Kosanke S, Emerson T, et al. Retrospective description and experimental reconstitution of three different responses of the baboon to lethal *E. coli*. Circulatory Shock, in press.

40. Kawaan HC. Introduction: thrombotic microangiopathy. Semin Hematol 1987; 24:69.
41. Taylor FB Jr, Abrams ME. The effect of surface active lipoprotein on clotting, fibrinolysis, and of fibrinogen on the surface tension of surface active lipoprotein. Am J Med 1966; 40:340.

10

Oxidant–Antioxidant Balance in Endotoxin-Induced Oxidative Injury and Tolerance to Oxidative Injury

JOHN E. REPINE

Webb-Waring Institute for Biomedical Research
and University of Colorado Health Sciences Center
Denver, Colorado

POLLY E. PARSONS

University of Colorado Health Sciences Center
Denver, Colorado

I. Introduction

A delicate balance exists between oxidants and antioxidants with respect to susceptibility and resistance to disease. In this chapter we address some aspects of this intriguing balance which affect the pathogenesis of acute lung injury (adult respiratory distress syndrome, ARDS). Initially, we review briefly the nature of oxidants and antioxidants in the lung. We then describe some alterations in the oxidant–antioxidant balance that affect not only lung injury but also tolerance mechanisms that render the lung resistant to oxidants. We conclude by summarizing the findings regarding known alterations in oxidants and antioxidants in patients with ARDS—a condition where endotoxin may contribute in a substantial way.

II. Background: The Nature of Oxidants and Antioxidants in the Lung

Oxygen radicals are normally made in only small amounts in the lung (1). Superoxide anion (O_2^-) leaks from the mitochondrial electron transport system during the routine metabolism of oxygen (respiration). O_2^- may also be generated by NADPH oxidase activity in the relatively quiescent and small numbers of macrophages and neutrophils that reside in the lung under normal conditions for the purpose of providing host defense against inhaled microorganisms and particulates. Production of O_2 radicals under these circumstances is probably noninjurious because only low levels are made and a relative wealth of counteracting antioxidant enzymes is available. Interestingly, aging, which increases tissue iron levels and decreases cellular glutathione (GSH) levels, as well as inhalation of oxidants in the form of cigarette smoke (even second-hand) or polluted air, may augment lung oxidative stress even under these "normal" disease-free conditions (2). Of note, tissue iron levels not only increase with age, but are different between men and women and between premenopausal and postmenopausal women. The latter may render men more susceptible to oxidant injury than women and postmenopausal women more susceptible than premenopausal women.

The principal O_2 radical species in biological systems are diagrammed in Fig. 1. Superoxide anion (O_2^-) is a relatively nontoxic oxidant that traverses cells via anion channels (3). Accordingly, O_2^- is reasonably confined and relatively specific in its influence on biological processes. O_2^- reduces ferric iron (Fe^{3+}) to ferrous iron (Fe^2) and mobilizes Fe^{3+} from storage in ferritin. However, O_2^- usually dismutates rapidly, both enzymatically and nonenzymatically, to hydrogen peroxide (H_2O_2) and O_2. H_2O_2 is slightly more hazardous than O_2^- to tissues and, in contrast to O_2^-, can diffuse directly and

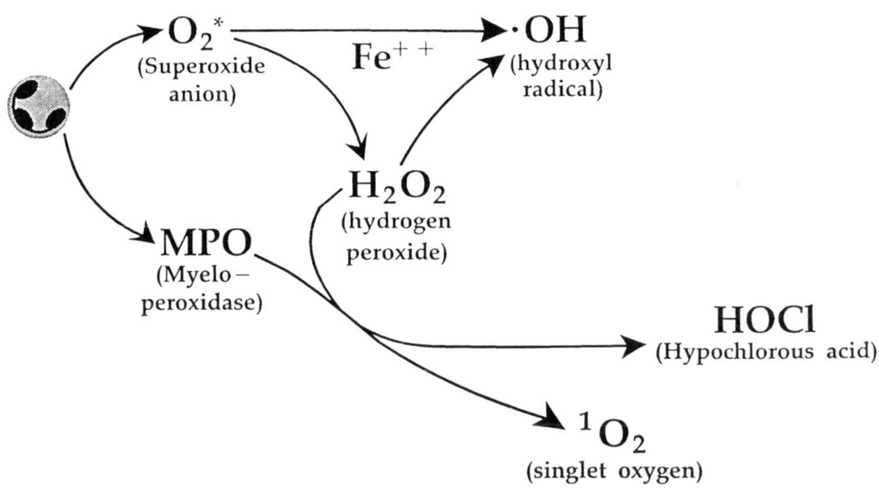

Figure 1 Oxygen radicals produced by neutrophils.

readily across cell membranes (4). More important, the toxicity of H_2O_2 is amplified when it is transformed into more toxic species, such as the hydroxyl radical (·OH), by a process that involves H_2O_2 accepting an electron from metals, such as Fe^{2+} (by the Haber–Weiss reaction or Fenton chemistry). ·OH is one of the most potent oxidants known and the most deleterious radical in biological situations (5). Furthermore, ·OH is indiscriminate and attacks any molecule. Because of its exceptional reactivity, ·OH damages whatever is at the site of its formation. A fourth oxygen free-radical species is hypochlorous acid (HOCl), which is an especially potent reactant formed from H_2O_2 and a halide, usually chloride, by a mechanism involving myeloperoxidase (MPO) from neutrophil granules. Another toxic oxygen species is singlet oxygen ($^1O_2$), an additional product of the MPO system. However, it is unclear if $^1O_2$ is produced by phagocytes, and the potential consequences of $^1O_2$ in biological systems is presently based on only limited indirect evidence.

Multiple mechanisms exist for both enzymatic (Fig. 2) and nonenzymatic dismutation of $O_2^{\cdot-}$ (3). Enzymatic dismutation of $O_2^{\cdot-}$ to H_2O_2 occurs via three specialized species of superoxide dismutase (SOD); copper–zinc SOD, the primary species in the cytoplasm; manganese SOD, the primary species in the mitochondria; and extracellular SOD, the major form of SOD in extracellular fluids (6–8). Nonenzymatic reduction of $O_2^{\cdot-}$ to H_2O_2 occurs primarily via spontaneous dismutation.

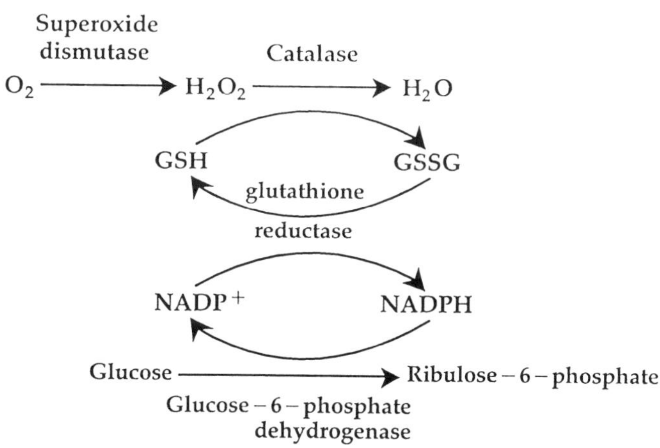

Figure 2 Enzymatic detoxification of oxygen radicals by superoxide dismutase (SOD), catalase, and the glutathione redox system. (Modified from Ref. 2.)

In addition to SOD, all cells contain two additional major antioxidant systems to counteract O_2 radical toxicity; and not surprisingly, both limit the detrimental effects of H_2O_2 (9–12). Indeed, inactivation of H_2O_2 constitutes a strategic point for preventing more severe tissue injury caused by ·OH or HOCl. First, all cells have a redox system that maintains high concentrations of reduced glutathione (GSH), which can detoxify H_2O_2 and lipid hydroperoxides. Glutathione synthetase, peroxidase, and reductase enzymes keep GSH levels high and GSSG levels low. This redox system is ubiquitous in the cytosol. In the face of oxidative stress, GSH is converted to GSSG, which is then recycled to GSH. As shown in Fig. 2, this reaction requires NADPH, which is supplied by the hexose monophosphate shunt via glucose-6-phosphate dehydrogenase. Under extreme oxidative stress, GSSG accumulates (1). Because high concentrations of GSSG can be cytotoxic, mechanisms are available for removing GSSG from the cell (1,8).

Catalase is a second enzymatic system that reduces the toxicity of H_2O_2. By reducing H_2O_2 to H_2O, catalase prevents formation of ·OH and HOCl. Catalase is located primarily in cellular peroxisomes, which are also a site of H_2O_2 generation. In contrast to the GSH system, which is a relatively more effective scavenger at lower concentrations of H_2O_2, catalase is a more effective scavenger against higher H_2O_2 concentrations (7). The catalase content of tissues is variable, but overall, the lung has one of the lowest catalase concentrations (12). Small quantities of catalase reside in plasma and plasma catalase levels vary in certain disease states, one of which is ARDS (13).

The aforementioned enzymatic systems not only provide the principal antioxidant defense mechanism but are also interdependent and affected by oxygen radicals [e.g., $O_2^{\bar{\cdot}}$ can inactivate catalase (14) and peroxidase (15) while H_2O_2 can inactivate SOD (1)]. Conversely, O_2 radicals may alter antioxidant activity by lysing cells, such as red blood cells, which are rich in antioxidants (16). Erythrocyte cell lysis is a two-edged sword, however, as it causes release of Fe^{2+} as well as catalase and SOD. The net effects of alterations in the balance between oxidants and antioxidants is not easy to deduce in complex biological systems (17).

Other compounds have nonenzymatic antioxidant capacity. The potency of nonenzymatic molecules may be relatively limited, however, since they are usually consumed molecule for molecule by oxidants and, as a consequence, are apt to decrease when confronted with a severe or ongoing oxidative stress. Vitamin E, a lipid-soluble molecule that prevents lipid peroxidation, is concentrated in the cell membrane. The importance of vitamin E is emphasized by the enhanced sensitivity of vitamin E–deficient animals to oxidants. Vitamin C is a water-soluble antioxidant that is generally available throughout the body. Although vitamin C can reduce oxidative stress, it can also function as a pro-oxidant by reducing Fe^{3+} to Fe^{2+}, contributing to ·OH formation (18). Uric acid, another water-soluble antioxidant, can scavenge peroxyl radicals formed during lipid peroxidation (19,20). Uric acid is made by xanthine dehydrogenase (XD)/xanthine oxidase (XO)—an endothelial cell-based enzyme system where the XO form generates $O_2^{\bar{\cdot}}$ and H_2O_2 during uric acid production. Vitamin A (β-carotene) quenches singlet oxygen ($^1O_2$) (21,22). Bilirubin and methionine have antioxidant effects. Taurine, a widely distributed B-amino acid, can detoxify HOCl (23). Ceruloplasmin, an acute-phase protein, also acts as an antioxidant in the extracellular space. Ceruloplasmin binds and prevents reduction of Fe^{3+}, which limits Fenton reactions involving H_2O_2 (24). Albumin is the best known of a group of high-molecular-weight compounds which serve as "sacrificial" antioxidants (25). Finally, glucose may scavenge ·OH and reduce its indiscriminate toxicity.

Oxidants are increasingly appreciated for their effects on pulmonary function (Fig. 3). Key lung molecules are highly susceptible to oxidants. O_2 radicals can damage vital cellular DNA (DNA strand breaks), lipids (lipid peroxidation) and protein molecules (e.g., enzyme inactivation). Proteins that have encountered oxidants are especially susceptible to degradation by proteases (26).

Oxygen radicals also have a number of specific effects that might affect the lung in secondary ways (Fig. 3). These responses include α_1-antitrypsin (α_1Pi) inactivation, which enhances the toxicity of neutrophil elastase (27), generation of neutrophil chemotaxins from serum (28), and formation of active molecules from arachidonic acid (29). For example, the latter interaction may involve generation of vasoconstrictive molecules, such as throm-

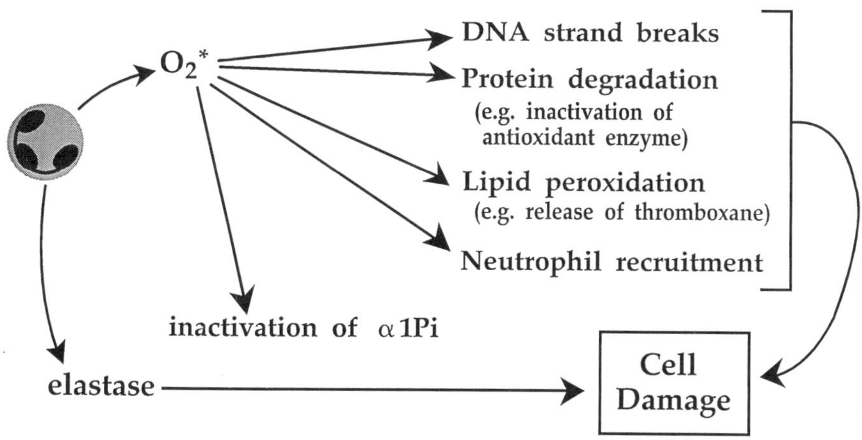

Figure 3 Some consequences of oxygen radicals in biological systems.

boxane (which is increased in patients with ARDS), stimulation of $PGF_{2\alpha}$-dependent reactions, which can initiate mucus secretion (30), and generation and inactivation of leukotriene molecules, which can affect neutrophil recruitment and activation (31). Finally, O_2^- may react with nitric oxide (NO), thereby limit vasodilating effects mediated by NO. This reaction can generate peroxynitrate (OONO), a highly reactive molecule whose fate is presently unknown. Peroxynitrate may be intrinsically toxic or lead to ·OH formation (32).

III. Endotoxin-Mediated Acute Lung Injury Depending on Oxidants and Antioxidants

Direct and indirect effects of endotoxin can rapidly alter the precarious balance between oxidants and antioxidants. Within the circulation, a key effect of endotoxin is priming of neutrophils. As described in detail in Chapter 13, neutrophil responses, including oxygen radical production, secretion of granule contents, and adherence, can be enhanced by prior exposure to endotoxin (33), cytokines (34,35), platelet activating factor (36), colony-stimulating factors (37), or other agents. Although this priming mechanism has not been clearly delineated, neutrophil responsiveness to secondary stimuli, including chemotactic fragments and formyl methionine leucine phenylalanine (FMLP), is similarly enhanced following priming (38). The potential importance of endotoxin priming has been observed in intact animal models (39,40). First, in one animal model, neither administration of low doses of endotoxin nor complement fragments (Cf) alone caused acute lung injury, while LPS and Cf together caused lung leak (39). Second, neutrophils pretreated with endotoxin, washed, and then infused into rabbits are retained more frequently within the lung than are untreated control neutrophils—evidence of neutrophil priming by endotoxin (39).

Although difficult to demonstrate convincingly, endotoxin priming of neutrophils may occur in humans who develop ARDS. Briefly, endotoxemia has been associated with the development of multiple-organ failure in patients with sepsis (41,42) and with the development of ARDS (43). However, whether or not circulating neutrophils from patients at risk for and with ARDS have been primed for oxidant production remains controversial. Some studies report that release of O_2^-, chemotaxis, and chemiluminescence are increased in neutrophils from affected patients (44), while other studies indicate that unstimulated neutrophils neither produce increased amounts of O_2^- nor are primed for FMLP stimulated O_2^- production (45). Explanations abound for the apparent discrepancies between various clinical studies in ARDS regarding neutrophil priming and oxygen radical generation. The possibilities include differences in the patients studied, the timing of neutrophil procurement, and/or the methods used to define increased oxidant production. It is also conceivable that neutrophils which are primed and/or activated within the circulation are then rapidly sequestered within the lung and therefore are no longer available in the circulation. If true, neutrophils being recovered from the peripheral blood are a subpopulation that may not reflect the activity of neutrophils in the lung.

Endotoxin has other major effects in the circulation that may enhance production of oxidants by neutrophils. These events involve stimulation of monocytes, activation of complement, and injury to endothelial cells. Analogous to the priming of neutrophils, endotoxin primes monocytes for release of oxidants (46). Endotoxin also stimulates monocytes to release cytokines, including tumor necrosis factor (TNF) and interleukin-1 (IL-1), which prime neutrophils for enhanced oxidant production in vitro (47). TNF and IL-1 also stimulate alveolar macrophages (AM) to produce oxidants as well as other chemotaxins and injurious mediators. It is not known whether TNF and IL-1 generated within the circulation reaches the alveolar space. However, when infused intravenously, IL-1 does not cause lung injury, whereas intratracheal administration of IL-1 produces an oxidant-dependent, neutrophil-mediated lung injury in rats (48) and isolated rat lungs (49). In patients with ARDS, IL-1 levels are increased in bronchoalveolar lavages (50), and alveolar macrophages release more IL-1, both at baseline and following endotoxin stimulation, than do macrophages from normal subjects (51). Moreover, infusion of IL-1 into normal volunteers mimics the hemodynamic effects of endotoxin infusion (52). Nevertheless, the results are variable when IL-1 levels are measured in the serum of patients at risk for and with ARDS (50,53,54), and a clear association between IL-1 levels (when measurable) and the severity of lung injury is not manifest. Finally, neutrophils from patients with sepsis have increased expression of IL-1 receptors, and there appears to be a relationship between IL-1 receptor expression and ARDS (55).

Activation of the complement cascade by endotoxin can initiate and propagate acute lung injury (as discussed in detail in Chapter 10). Early investigations suggested that ARDS involved complement-mediated neutrophil-dependent lung injury. Although subsequent studies have diminished this possibility, complement fragments circulate in virtually all patients at risk for the development of ARDS (56,57). Complement fragments can stimulate release of $O_2^{\cdot}$ from primed neutrophils, stimulate monocytes and macrophages, and may promote conversion of XD to XO (58). XO may be an important source of oxidants in acute lung injury. Moreover, XO is heavily concentrated within endothelial cells and is released by direct stimulation and injury (59). Endotoxin itself causes oxidant-dependent endothelial injury (59,60) and stimulated neutrophils may convert XD to XO in endothelial cells in vitro (61). XO has been implicated in numerous insults, including intestinal ischemia–reperfusion, shock, hyperoxia, and burn, which may contribute to acute lung injury in animals (62–64).

Direct exposure of endothelial monolayers to endotoxin causes LDH release and morphologic changes which are both inhibited by allopurinol (suggesting a role for XO) and by DMSO [suggesting involvement of oxidants (59)]. Endothelial injury induced by activated neutrophils is also associated with conversion of XD to XO, but conversion is not inhibited by SOD or catalase and it cannot be reproduced by simple addition of stimulated neutrophil supernatants or H_2O_2 to endothelial cells (60). However, the effect can be reversed by addition of elastase inhibitors and reproduced by addition of elastase (61).

Studies in intact animals provide compelling evidence that links endotoxin and oxidants. Early studies using a sheep model revealed that endotoxin-induced lung injury was, at least in part, neutrophil dependent (65–67), and it was hypothesized that oxygen radicals released from activated neutrophils contributed to damage. This premise was supported when infusion of either *n*-acetylcysteine (68) or catalase (69) attenuated the pathophysiological changes associated with endotoxin administration. Moreover, infusion of catalase appeared to decrease lung lipid peroxidation following endotoxin administration (69), and treatment with dimethylthiourea prevented endotoxin-induced lung injury (70). There is some evidence that the contribution of oxidants to endotoxin-induced lung injury may be, in part, species dependent. In a porcine model of endotoxin-induced acute lung injury, infusion of dimethylthiourea (71), but not polyethylene glycol–conjugated SOD and catalase (72), reduced the pulmonary effects of endotoxin.

IV. Endotoxin-Mediated Tolerance Dependent on Oxidants and Antioxidants

In the preceding section we reviewed the ability of endotoxin to initiate a variety of responses that may contribute to the development of acute lung injury. In this section we deal with a less well appreciated but perhaps equally important possibility, which is that endotoxin can initiate responses that render tissues resistant to oxidative stress. In this section we address evidence that supports this proposition, focusing primarily on pulmonary oxygen toxicity—a condition caused by O_2 radicals. Moreover, since many of the effects of endotoxin may be mediated by cytokines, specifically, TNF and IL-1, we will also examine the effects of these cytokines on the induction of tolerance. Finally, we relate a few studies from in vitro and nonpulmonary systems which further indicate mechanisms that may account for tolerance.

Frank and his colleagues observed initially that endotoxin pretreatment could protect the lung against hyperoxia (73). They found that all rats pretreated with relatively small amounts of endotoxin survived following exposure to hyperoxia, while all untreated control rats died. Subsequently, these investigators reported that pretreatment with even a single 500 μg/kg dose of endotoxin given as late as 36 h after the onset of exposure to 100% O_2 significantly increased survival rates and decreased lung injury in rats. In parallel, lungs from endotoxin-pretreated rats developed increased SOD, catalase, and glutathione peroxidase activity. A crucial experiment showed that simultaneous administration of diethyldithiocarbamate (a copper chelator and an inhibitor of Cu-Zn SOD activity) prevented endotoxin-mediated protection against O_2 toxicity, and accordingly, it was concluded that tolerance was dependent on SOD activity. This observation involving inhibition of tolerance by DDC remains the principal mechanistic link between endotoxin-mediated tolerance to oxygen toxicity and increased lung antioxidant enzyme activity.

The next relevant study of pulmonary tolerance was the finding that pretreatment with TNF and IL-1 decreased lung-oxidized glutathione (GSSG) levels, lung injury, and mortality in rats subjected continuously to hyperoxia (74). Nearly all TNF/IL-1 pretreated rats lived >144 h in hyperoxia, while unpretreated control rats only survived >64 h. This observation was important for the following reasons: First, since endotoxin is a potent stimulant of TNF and IL-1, and many of the effects of endotoxin are thought to be mediated by cytokines, it suggested that tolerance caused by endotoxin may be due to TNF and IL-1. Second, the work established that oxidative injury, as specifically reflected by increased lung GSSG levels, was decreased in lungs of TNF/IL-1-pretreated rats compared to untreated rats. This experiment solidified the belief that tolerance was associated with decreased oxidative stress. Although this work did not reveal increased lung antioxidant enzyme activities during the initial 72-h period of exposure to hyperoxia, lung antioxidant enzymes increased subsequently in TNF- and IL-1-pretreated rats, but for the most part, only after exposure to hyperoxia for 72 h, a time when most control rats had already succumbed to the detrimental effects of hyperoxia (75). The latter finding further supports the point that increased lung antioxidant activity, in this case, increased G6PD, SOD, catalase, glutathione reductase, and glutathione peroxidase activity, may be involved in tolerance. It is noteworthy, however, that antioxidant enzyme activities were determined in whole-lung homogenates. Unfortunately, this approach is neither optimal for detecting small but biologically significant increases in enzymes (i.e., increases that do not appreciably increase total lung antioxidant enzyme activity), nor effective at discerning small increases in antioxidant enzyme activity that might be critically located (e.g., in crucial endothelial or epithelial locations that are prime sites for oxygen radical attack).

Pretreatment with either TNF or IL-1 also renders rats resistant to oxygen toxicity (76,77). Moreover, increased survival in O_2 occurred in rats given IL-1 intratracheally. Furthermore, protection was dependent on the pretreatment dose of IL-1, reflecting a precise dose-response relationship between the amounts of IL-1 given and the degree of protection. In addition, IL-1 pretreatment increased lung MnSOD, Cu-Zn SOD, catalase, and glutathione peroxidase activity, and the levels of mRNA for MnSOD (but not Cu-Zn SOD) increased in rats given IL-1 intratracheally. These were excellent studies, but again, enzyme activity analysis was conducted on whole-lung homogenates. As the authors suggest, this approach is not optimal. In situ hybridization is needed to localize and fully determine the significance of increased lung antioxidant enzyme activity.

Tolerance to hyperoxia usually is associated with and, probably requires, an increase in lung antioxidant enzyme activity. However, antioxidant enzyme induction may involve synergistic and complex interactions, as demonstrated by a recent study of the effects of endotoxin and O_2 (78). Briefly, it was reported that pretreatment with endotoxin alone increased mRNAs for MnSOD and glutathione peroxidase without increasing their stability, while pretreatment with O_2 alone elevated mRNA activity for MnSOD and increased mRNA stability. Finally, pretreatment with O_2 and endotoxin together increased both the concentration and stability of MnSOD, catalase, and glutathione mRNAs. The implication is that various antioxidants are regulated differently at the genetic level. For example, mice pretreated with IL-1 had increased survival when given endotoxin, but still developed increased TNF levels. This finding suggests a mechanism for endotoxin tolerance that does not require TNF, even though TNF can cause tolerance. Finally, IL-6 enhances TNF- and IL-1-induced increases in MnSOD mRNA and O_2 tolerance (79). Undoubtedly, an intricate system is responsible for the induction of tolerance.

A myriad of studies involving cultured cells, including endothelial cells, supports a relationship between endotoxin, cytokines, and increases in antioxidant enzyme activities in vitro (80–84). For example, TNF induces MnSOD activity and resistance to TNF and cycloheximide in cultured human embryonic kidney cells. IL-1, TNF, and LPS in additive ways increased MnSOD, but not Cu-Zu SOD, enzyme activity in both rat pulmonary artery and microvascular endothelial cells. In the latter work, MnSOD in RNA induction was dependent on a trascriptional event and did not require de novo protein synthesis.

Studies of endotoxin-, TNF-, and IL-1-induced tolerance to pulmonary oxygen toxicity are closely paralleled by studies showing that pretreatment with endotoxin, TNF, or IL-1 renders isolated rat hearts resistant to ischemia–reperfusion (85–88). This generalizes the phenomenon to another organ and to another insult, although ischemia–reperfusion undeniably involves generation of O_2 radicals. Again, a selectivity in responses to specific cytokines was apparent. For example, pretreatment with endotoxin, TNF, or IL-1 each caused tolerance to ischemia–reperfusion, but endotoxin pretreatment increased myocardial catalase activity, while IL-1 or TNF pretreatment increased G6PD or catalase activity, respectively.

A possible mechanism for induction of tolerance was elucidated in the IL-1 studies of ischemia–reperfusion (87). It appeared that a smaller neutrophil-dependent antecedent oxidative stress was involved in antioxidant induction and the resistance to a larger subsequent oxidative stress inflicted by ischemia–reperfusion (Fig. 4). In addition, because rats needed to be pretreated 36 h prior to the insult to develop resistance, a mechanism involving enzyme induction was implicated (85). Finally, there is another protective event called preconditioning. In this regimen, brief episodes of coronary artery occlusion (ischemia) with intermittent reperfusion reduce the infarct size produced by subsequent episodes of sustained ischemia. The relationship of preconditioning to tolerance, if any, is unknown.

In summary, much needs to be learned about self-protecting antioxidant systems as well as other mechanisms for decreasing oxidant injury. These intrinsic elaborate defenses hold substantial insights into how to re-

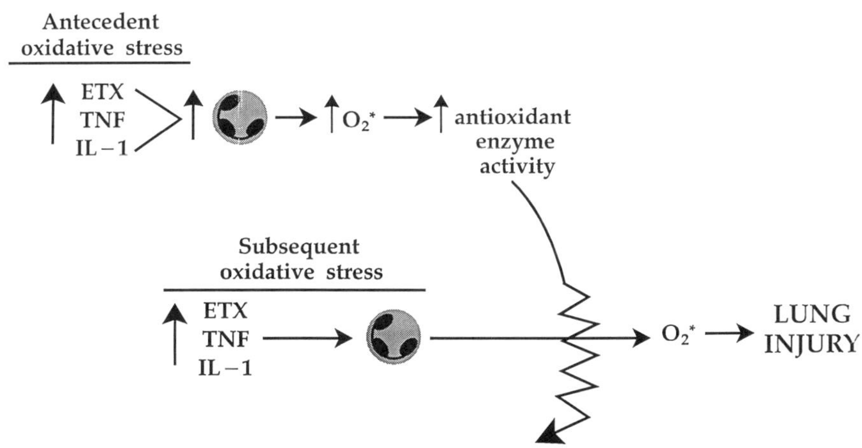

Figure 4 Proposed mechanism for the induction of tolerance involving a smaller, antecedent neutrophil-dependent oxidative stress which increases lung antioxidant enzyme activity and confers tolerance to a subsequent larger oxidative insult mediated by ischemia–reperfusion. (Modified from Ref. 85.)

duce human disease. Undoubtedly, in the healthy state, nature has already optimized these systems, positioned them properly, and limited their toxicity.

V. Oxidant–Antioxidant Alterations in ARDS Patients

Alterations in oxidant–antioxidant balance continue to be recognized in patients with acute lung injury (ARDS) (89). It is increasingly clear that oxidant activity is increased and that antioxidant enzyme activities are perturbed and, most likely, not optimal to combat the enormous oxidative challenge at hand in these critically ill patients. Patients with ARDS have increased exhaled H_2O_2 concentrations (90,91), increased serum xanthine oxidase (XO) concentrations (92), and increased plasma lipid peroxidation products (93). Inactivated antiproteases have been measured in both the blood and bronchoalveolar lavages of ARDS patients, and inactivation is probably a consequence of oxidation of a key methionine site in the molecule (94,95). Parenthetically, recognition of this abnormality prompted development of genetically engineered molecules, in which susceptible methionine sites have been replaced with valine, which resists oxidation.

Although little is known about crucial intracellular antioxidant activities, antioxidant capacity is also apparently disturbed in patients with ARDS. Blood levels of catalase (96), manganese SOD (96), and ceruloplasmin (97) are increased, while GSH (98) and vitamin E levels (99) are decreased in ARDS patients, probably as a consequence of oxidant stress. Ceruloplasmin (97) and myeloperoxidase (100) levels are also increased in bronchoalveolar lavages of patients with ARDS, but their significance is unclear.

These findings can be linked together to form a complex pathophysiologic scheme for endotoxin-mediated ARDS (Fig. 5). Briefly, endotoxin stimulates release of numerous inflammatory stimuli, including complement fragments, interleukin-1, tumor necrosis factor, platelet activating factor, interleukin-6, and interleukin-8, and other factors that are increased in the lung and/or circulation of patients with and/or at risk for ARDS. When primed and/or stimulated by these mediators, neutrophils, which are increased in lungs of patients with ARDS, release $O_2^{\bar{\cdot}}$ and H_2O_2, which along with degranulation products (MPO) damage the lung. Notably, MPO is released as a result of many of these agents stimulating neutrophil degranulation and may serve as a marker of neutrophil recruitment and/or activation. Another potential source of oxidants is the endothelial cell. Endotoxin itself, or endotoxin-primed or endotoxin-stimulated neutrophils, may stimulate conversion of XD to XO in endothelial cells. Injured endothelial cells then leak XO. Subsequently, XO released may adhere to endothelial cells, and as a result, contribute to further injury or neutrophil recruitment at tissue sites distant from the site of XO release, such as the lung. Oxygen radicals from

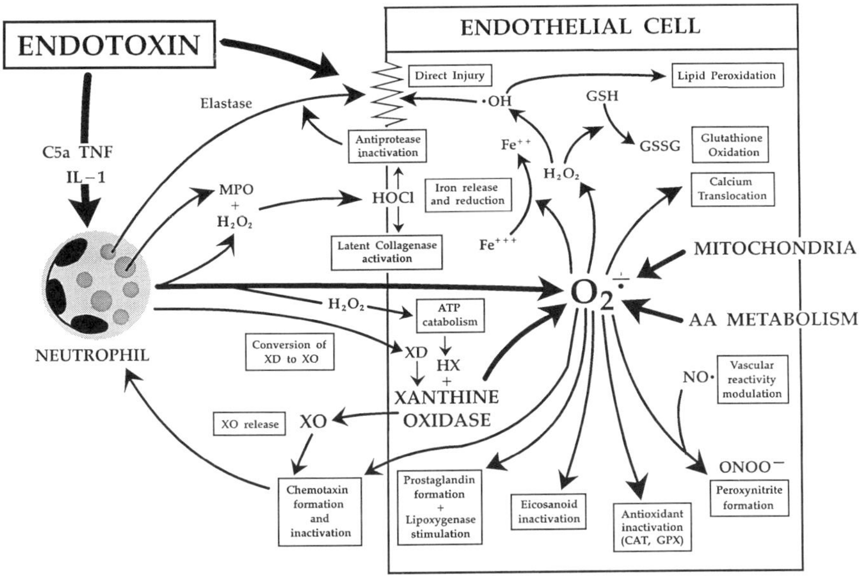

Figure 5 Schematic depiction of the effects of endotoxin on neutrophils and endothelial cells focusing on alternatives in oxidant–antioxidant balance. (Modified from Ref. 89.)

neutrophils and XO then inactivate antiproteases, initiate lipid peroxidation, and damage the endothelial cell in ways that perpetuate additional inflammation and lung injury. Ironically, if shock or multiple-organ failure ensues, the gut may become ischemic and additional endotoxin may be released into the circulation.

VI. Future Directions and Speculation

Many challenges remain in this rapidly evolving area of research. First, techniques are desperately needed for measuring oxidants in biological systems. Regrettably, because of their high reactivity, O_2 radicals cannot be assessed directly, and as a consequence, determination of the postulated relationship between O_2 radicals and acute lung injury has been hindered. For example, it has not been possible to define the relationship between the amount and location of O_2 radical production and various related aspects of damage. Furthermore, not all O_2 radicals have the same biochemical "footprints." Thus, methods for measuring and pinpointing precisely the nature and location of each specific O_2 radical species are needed to solidify the premise that O_2 radicals are responsible for oxidative tissue injury.

Second, more understanding is required to discern which molecules are particularly sensitive to O_2 radicals and which molecules are the most meaningful with respect to clinically significant biological function. Achieving this objective will involve a quest for more knowledge regarding the specific location and function of antioxidants. One direct approach that addresses this concern will involve assessment of the effects of antioxidant interventions in ARDS patients. Clinical trials are being proposed to treat ARDS patients with agents that increase cellular GSH levels (procysteine, *N*-acetylcysteine), agents that augment antioxidant enzyme systems (e.g., SOD, SOD mimics, and genetically designed SOD variants), agents that chelate iron and other molecules which enhance O_2 radical toxicity (e.g., desferioxamine), agents that decrease lipid peroxidation (vitamin E and lazaroids), agents that inhibit XO (allopurinol), and/or agents that reduce neutrophil recruitment and activation (IL-1ra, soluble TNF antibodies, CT1501R, antiadhesion molecule antibodies, etc.) and release of O_2 radicals (NADPH oxidase inhibitors). Careful assessment of the influence of these agents alone, and in aggregate, will improve our understanding of the role of oxidants and antioxidants in the development of acute lung injury.

A third goal will be to determine the contribution of genetic responses that affect susceptibility and resistance to oxidative injury. A number of possibilities may affect oxidant production and antioxidant defenses. It will be necessary to evaluate both systems to determine the net effects of alterations in either or both components. It will also be necessary to assess increases in antioxidant defenses in the face of oxidative stress. Why do only a few individuals with ostensibly the same risk factors and inciting insults develop ARDS while most people do not develop the syndrome? Similarly, why do only certain smokers develop vascular complications or malignancies, whereas the majority of smokers do not? Are these protected individuals inherently resistant and/or do they evoke resistance (adaptation) in ways that are not present in other people? Why, in most cases, does it take so many years to develop damage? Does this depend on the accumulation of iron or some other factor? If we can identify people who are intrinsically resistant to oxidative stress, we may be able to determine why they are resistant and use this knowledge to augment appropriate responses in susceptible persons. This approach may involve genetic therapy: for example, delivery of antioxidant-producing genes via viral vectors.

Finally, because of the importance of a proper oxidant–antioxidant balance for physiologic processes, we must continue to identify and maintain the beneficial effects of O_2 radicals. Presently, this is usually focused on preserving the integrity of host defense, specifically maintaining the production of O_2 radicals by neutrophils for bactericidal action, a necessity well supported by the increased and life-threatening infections of patients with

Figure 6 "Yin-yang" effects of endotoxin in biological systems.

chronic granulomatous disease (CGD) whose neutrophils fail to make O_2 radicals because of a defective NADPH oxidase enzyme complex. However, other beneficial effects of O_2 radicals are being discovered. O_2 radicals may modulate neurotransmission, neutrophil recruitment, and vascular reactivity. Therefore, another goal must be to localize the effects of antioxidant therapy so as to limit undesirable side effects of exogenous manipulation of oxidant–antioxidant balance.

VII. Conclusions

Considerable evidence suggests that endotoxin contributes to the development of acute lung injury in both human subjects and relevant models of acute lung injury. In this chapter we have summarized current knowledge which supports the concept that the multiple effects of endotoxin in this complex setting involve alterations in oxidant–antioxidant balance (Fig. 6). We also suggest that endotoxin can stimulate tolerance mechanisms that reduce susceptibility to oxidant injury. Thus, the beneficial effects of endotoxin need to be considered when interventions directed at neutralizing its negative influences are being developed and tested in human subjects.

References

1. Heffner JE, Repine JE. Pulmonary strategies of antioxidant defense. Am Rev Respir Dis 1989; 140:531–534.

2. Cook JD, Finch CA, Smith NJ. Evaluation of the iron status of a population. Blood 1976; 48:449–455.
3. Fridovich I. Biological effects of the superoxide radical. Arch Biochem Biophys 1986; 247:1–28.
4. Chance B, Sies H, Boveris E. Hydroperoxide metabolism in mammalian organs. Physiol Rev 1979; 59:527–583.
5. Schraufstatter IU, Cochrane CG. Oxidants: types, sources and mechanism of injury in lung injury. In Crystal RG, West JB, eds. The Lung. New York: Raven Press, 1992:43–50.
6. Fridovich I. Superoxide radical: an endogenous toxicant. Annu Rev Pharmacol Toxicol 1983; 23:239–257.
7. White CW, Repine JE. Pulmonary antioxidant defense mechanisms. Exp Lung Res 1985; 8:81–96.
8. Marklund SL. Extracellular superoxide dismutase in human tissues and human cell lines. J Clin Invest 1984; 74:1398–1403.
9. Sies H. Antioxidant activity in cells and organs. Am Rev Respir Dis 1987; 136: 478–480.
10. Ross D, Norbeck K. Moldeus P. The generation and subsequent fate of glutathionyl radicals in biological systems. J Biol Chem 1985; 260:5028–5032.
11. Maddepati KA, Gasparski C, Marnett LJ. Characterization of the hydroperoxide-reducing activity of human plasma. Arch Biochem Biophys 1987; 254:9–17.
12. Hartz JW, Funakoshi S, Deutsch HF. The levels of superoxide dismutase and catalase in human tissues determined immunochemically. Clin Chem Acta 1973; 46:125–131.
13. Leff JA, Parsons PE, Day CE, et al. Increased hydrogen peroxide scavenging and catalase activity in serum from septic patients who subsequently develop ARDS. Am Rev Respir Dis 1992; 146:985–989.
14. Kono Y, Fridovich I. Superoxide radical inhibits catalase. J Biol Chem 1982; 247:5751–5754.
15. Blum J, Fridovich I. Inactivation of glutathione peroxidase by superoxide radical. Arch Biochem Biophys 1985; 240:500–508.
16. Toth KM, Clifford DP, White CW, Repine JE. Intact human erythrocytes prevent hydrogen peroxide mediated damage to isolated perfused rat lungs and cultured bovine pulmonary artery endothelial cells. J Clin Invest 1984; 74:292–295.
17. Toth KM, Berger EM, Beehler CJ, Repine JE. Erythrocytes from cigarette smokers contain more glutathione and catalase and protect endothelial cells from hydrogenperoxide better than erythrocytes from nonsmokers. Am Rev Respir Dis 1986; 134:281–184.
18. Halliwell B, Gutteridge JMC. Oxygen toxicity, oxygen radicals, transition metals and disease. Biochem J 1984; 219:1–14.
19. Heys AD, Dormandy TL. Lipid peroxidation in iron over-loaded spleens. Clin Sci 1981; 60:295–301.
20. Ames BN, Cathcart A, Schwiers E, Hochstein P. Uric acid provides an antioxidant defense in humans against oxidant- and radical-caused aging and cancer: a hypothesis. Proc Natl Acad Sci USA 1981; 78:6858–6862.

21. Krinsky NI. Carotenoid chromophobe length and protection against photosensitization. Photochem Photobiol 1974; 19:217–222.
22. Burton GW, Ingold KU. Beta-carotene: an unusual type of lipid antioxidant. Science 1984; 224:569–573.
23. Emudianughe TS, Caldwell J, Smith RL. The utilization of exogenous taurine for the conjugation of xenobiotic acids in the ferret. Xenobiotica 1983; 13:133–138.
24. Gutteridge JMC, Stocks J. Ceruloplasmin: physiological and pathological perspectives. CRC Crit Rev Lab Sci 1981; 14:257–329.
25. Brown JM, Beehler CJ, Berger FM, et al. Albumin decreases hydrogen peroxide and reperfusion injury in isolated rat hearts. Inflammation 1989; 13:583–589.
26. Davies KJA, Goldberg AL. Proteins damaged by oxygen radicals are rapidly degraded in extracts of red blood cells. J Biol Chem 1987; 262:8227–8234.
27. Baird BR, Cheronis JC, Sandhaus RA, Berger EM, White CW, Repine JE. O_2 metabolites and neutrophil elastase synergistically cause edematous injury in isolated rat lungs. J Appl Physiol 1986; 61:2224–2229.
28. Petrone WF, English DK, Wong K, McCord JM. Free radicals and inflammation: superoxide-dependent activation of a neutrophil chemotactic factor in plasma. Proc Natl Acad Sci USA 1980; 77:1159–1163.
29. Tate RM, Morris HG, Schroeder WR, Repine JE. Oxygen metabolites stimulate thromboxane generation and cause vasoconstriction in isolated saline-perfused rabbit lungs. J Clin Invest 1984; 74:608–613.
30. Adler KB, Holden-Stauffer WJ, Repine JE. Oxygen metabolites stimulate release of high molecular weight glycoconjugates by cell and organ cultures of rodent respiratory epithelium via an arachidonic acid-dependent mechanism. J Clin Invest 1990; 85:75–85.
31. Martin TR, Pistorese BP, Chi EY, Maunder FJ. Effects of leukotriene B4 in the human lung: recruitment of neutrophils into the alveolar space without a change in protein permeability. J Clin Invest 1989; 84:1609–1619.
32. Beckman JS, Crow JP. Pathological implications of nitric oxide, superoxide and peroxynitrite formation. Biochem Soc Trans 1993; 21:330–334.
33. Haslett C, Guthrie LA, Kopaniak MM, Johnston RB, Henson PM. Modulation of multiple neutrophil functions by preparative methods and trace amounts of bacterial lipopolysaccharide. Am J Pathol 1985; 119:101–110.
34. Kharazmi A, Nielsen H, Bendtzen K. Recombinant interleukin-1 alpha and beta prime human monocyte superoxide production but have no effect on chemotaxis and oxidative burst response of neutrophils. Immunobiology 1988; 177: 32–39.
35. Berkow RL, Wang D, Larrick JW, Dodson RW, Howard TH. Enhancement of neutrophil superoxide production by preincubation with recombinant human tumor necrosis factor. J Immunol 1987; 139:3783–3791.
36. Shalit M, Dabiri GA, Southwich FS. Platelet activating factor both stimulates and primes human polymorphonuclear leukocytes for actin filament assembly. Blood 1987; 70:1921–1927.
37. DiPersio JF, Billing P, Williams R, Gasson JC. Human granulocyte-macrophage colony stimulating factor and other cytokines prime human neutrophils

for enhanced arachidonic acid release and leukotriene B4 synthesis. J Immunol 1988; 140:4315–4322.

38. Parsons PE, Worthen GS, Henson PM. Injury from inflammatory cells. In: Crystal RG, West JB, eds. The Lung. New York: Raven Press, 1992.
39. Worthen GS, Haslett C, Rees AJ, Gumbay RS, Henson JE, Henson PM. Neutrophil pulmonary vascular injury: synergistic effects of trace amounts of lipopolysaccharide and neutrophil stimuli on vascular permeability and neutrophil sequestration in the lung. Am Rev Respir Dis 1987; 136:19–28.
40. Anderson BO, Poggetti RS, Shanley PF, et al. Primed neutrophils injure rat lung through a platelet-activating factor-dependent mechanism. J Surg Res 1991; 50:510–514.
41. Brandtzaeg P, Keirulf P, Gaustad P, et al. Plasma endotoxin as a predictor of multiple organ failure and death in meningococcal disease. J Infect Dis 1989; 159:195–204.
42. Danner RL, Elin RJ, Hosseini JM, Wesley RA, Reilly JM, Parillo JE. Endotoxemia in human septic shock. Chest 1991; 99:169–175.
43. Parsons PE, Worthen GS, Moore EE, Tate RM, Henson PM. The association of circulating endotoxin with the development of the adult respiratory distress syndrome. Am Rev Respir Dis 1989; 140:294–301.
44. Zimmerman GA, Renzetti AD, Hill HR. Functional and metabolic activity of granulocytes from patients with ARDS. Am Rev Respir Dis 1983; 127:290–300.
45. Parsons PE, Worthen GS, Moore FA, Moore EE, Henson PM. Activity levels of circulating neutrophils in acute lung injury. Am Rev Respir Dis 1990; 141: A55a.
46. Pabst MJ, Hedegaard HB, Johnston RB. Cultured human homocytes require exposure to bacterial products to maintain an optimal oxygen radical response. J Immunol 1982; 128:123–128.
47. Newton RC. Human monocyte production of interleukin-1: parameters of the induction of interleukin-1 secretion by lipopolysaccharides. J Leukoc Biol 1986; 39:299–234.
48. Leff JA, Baer JW, Bodman ME, et al. Interleukin 1 α induced lung neutrophil accumulation and oxygen metabolite mediated lung leak in rats. J Appl Physiol. Accepted for publication.
49. Guidot DM, Stevens EE, Repine MJ, Lucca-Broco AE, Repine JE. Intratracheal but not intravascular interleukin-1 causes acute edematous injury in isolated neutrophil-perfused rat lungs through an oxygen radical mediated mechanism. J Lab Clin Med. Accepted for publication.
50. Siler TM, Swierkosz JE, Hyers TM, Fowler AA, Webster RO. Immunoreactive interleukin-1 in bronchoalveolar lavage fluid of high risk patients and patients with ARDS. Exp Lung Res 1989; 15:881–894.
51. Jacobs RF, Tabor DR, Burks AW, Campbell GD. Elevated interleukin-1 release by human alveolar macrophages during the adult respiratory distress syndrome. Am Rev Respir Dis 1989; 140:1686–1692.
52. Fisher E, Marano MA, Barber A, et al. A comparison between the effects of interleukin-1 alpha administration and sublethal endotoxemia in primates. Am J Physiol 1991; 261:442–452.

53. Girardin E, Graw GE, Dayer JM, Roux-Lombard P, Group TJ, Lambert PH. Tumor necrosis factor and interleukin-1 in the serum of children with severe infectious purpura. N Engl J Med 1988; 319:397–400.
54. Waage A, Halstensen A, Espevik T. Association between tumor necrosis factor in serum and fatal outcome in patients with meningococcal disease. Lancet 1987; 1:355–357.
55. Fasano MB, Cousart S, Neal S, McCall CE. Increased expression of the interleukin-1 receptor on blood neutrophils of humans with the sepsis syndrome. J Clin Invest 1991; 88:1452–1459.
56. Parsons PE, Gicias PC. The terminal complement complex (C5b-9) is not specifically associated with the development of the adult respiratory distress syndrome. Am Rev Respir Dis 1985; 132:490–493.
57. Parsons PE, Worthen GS, Moore EE, Tate RM, Henson PM. The association of circulating endotoxin with the development of the adult respiratory distress syndrome. Am Rev Respir Dis 1989; 140:294–301.
58. Warren JS, Johnson, Ward PA. Immunoglobulin- and complement-mediated immune injury in lung injury. In: Crystal RG, West JB, eds. The Lung. New York: Raven Press, 1992:179–186.
59. Brigham KL, Mayrick B, Berry LC, Repine JE. Antioxidants protect cultured bovine lung endothelial cells from injury by endotoxin. J Appl Physiol 1987; 63:840–850.
60. Phan SH, Gannon DE, Varani J, Ryan US, Ward PA. Xanthine oxidase activity in rat pulmonary artery endothelial cells and its alteration by activated neutrophils. Am J Pathol 1989; 134:1201–1211.
61. Phan SH, Gannon DE, Wards PA, Karmiol S. Mechanism of neutrophil-induced XD to XO conversion in endothelial cells: evidence of a role for elastase. Am J Respir Cell Mol Biol 1992; 6:270–278.
62. Till GO, Guilds LS, Mahrougui M, Firedl HP, Trentz O, Ward PA. Role of xanthine oxidase in thermal injury of skin. Am J Pathol 1989; 165:195–202.
63. Rodell TC, Cheronis JC, Ohnemus CL, Piermattei DJ, Repine JE. Xanthine oxidase mediates elastase induced injury to isolated lungs and endothelium. J Appl Physiol 1987; 63:2159–2163.
64. Terada LS, Dormish JJ, Shanley PF, Leff JA, Anderson BO, Repine JA. Circulating xanthine oxidase increases following intestinal ischemia-reperfusion and mediates lung neutrophil sequestration. Am J Physiol Lung Cell Mol Physiol 1992; 263:394–401.
65. Heflin AC, Brigham KL. Prevention by granulocyte depletion of increased vascular permeability of sheep lung following endotoxemia. J Clin Invest 1981; 68: 1253–1260.
66. Hinson JM, Brigham KL, Hutchinson AA, Snapper JR. Granulocytes participate in the early changes in lung mechanics caused by endotoxemia. Am Rev Respir Dis 1982; 125:275–284.
67. Lloyd JE, Newman JG, English D, Ogletree ML, Meyrick BO, Brigham KL. Lung vascular effects of phorbol myristate acetate in awake sheep. J Appl Physiol 1983; 54:267–276.

68. Bernard GR, Lucht WD, Niedermeyer ME, Snapper JR. Ogletree ML, Brigham RL. Effect of *N*-acetylcysteine on the pulmonary response to endotoxin in the awake sheep and upon in vitro granulocyte function. J Clin Invest 1984; 73: 1772–1784.
69. Demling MD, Lalonde C, Seekamp A, Fiore N. Endotoxin causes hydrogen peroxide-induced lung lipid peroxidation and prostanoid production. Arch Surg 1988; 123:1337–1341.
70. Wong C, Fox R, Demling RH. Effect of hydroxyl radical scavenging on endotoxin-induced lung injury. Surgery 1985; 97:300–307.
71. Olson NC, Anderson DL, Grizzle MK. Dimethylthiourea attenuates endotoxin-induced acute respiratory failure in pigs. J Appl Physiol 1987; 63:2426–2432.
72. Olson NC, Grizzle MK, Anderson DL. Effect of polyethylene glycol–superoxide dismutase and catalase on endotoxemia in pigs. J Appl Physiol 1987; 63:1526–1532.
73. Frank L, Summerville J, Massaro D. Protection from oxygen toxicity with endotoxin. Role of the endogenous antioxidant enzymes of the lung. J Clin Invest 1980; 65:1104–1110.
74. White CW, Ghezzi P, Dinarello CA, Caldwell SA, McMurtry IJ, Repine JE. Recombinant tumor necrosis factor/cachectin and interleukin-1 pretreatment decreases lung oxidized glutathione accumulation, lung injury and mortality in rats exposed to hyperoxia. J Clin Invest 1987; 79:1868–1873.
75. White CW, Ghezzi P, McMahon S, Dinarello CA, Repine JE. Cytokines increase rat lung antioxidant enzymes during exposure to hyperoxia. J Appl Physiol 1989; 66:2003–1007.
76. Tsan MF, White JE. Kinetics of pulmonary superoxide dismutase in interleukin-1-induced oxygen-tolerant rats. Am J Physiol (Lung Cell Mol Physiol) 1992; 263: 342–347.
77. Tsan MF, Lee CY, White JE. Interleukin-1 protects rats against oxygen toxicity. J Appl Physiol 1991; 71:688–697.
78. Clerch LB, Massaro D. Tolerance of rats to hyperoxia: lung antioxidant enzyme gene expression. J Clin Invest 1993; 91:499–508.
79. Tsan MF, White JE, DelVecchio PJ, Shaffer JB. IL-6 enhances TNF-α- and IL-1-induced increase of Mn superoxide dismutase mRNA and O_2 tolerance. Am J Physiol (Lung Cell Mol Physiol) 1992; 263:22–26.
80. Gerritsen ME, Bloom CM. Endothelial cell gene expression in response to injury. FASEB J 1993; 7:523–532.
81. Wong GH, Elwell JH, Oberley LW, Goeddel DV. Manganous superoxide dismutase is essential for cellular resistance to cytotoxicity of tumor necrosis factor. Cell 1989; 58:923–931.
82. Visner GA, Chesrown SE, Monnier J, Ryan US, Nick HS. Regulation of manganese superoxide dismutase. IL-1 and TNF induction in pulmonary artery and microvascular endothelial cells. Biochem Biophys Res Commun 1992; 188: 453–462.
83. Lu D, Maulik N, Moraru II, Kreutzer DL, Das DK. Molecular adaptation of vascular endothelial cells to oxidative stress. Am J Physiol 1993; 264:715–722.
84. Visner GA, Block ER, Burr IM, Nick HS. Regulation of manganese superoxide dismutase in porcine pulmonary artery endothelial cells. Am J Physiol (Lung Cell Mol Physiol) 1991; 260:444–449.

85. Repine JE. Oxidant-antioxidant balance: some observations from studies of ischemia-reperfusion in isolated perfused rat hearts. Am J Med 1991; 91 (suppl 3C):45S–53S.
86. Brown JM, Grosso MA, Terada LS, et al. Endotoxin pretreatment increases endogenous myocardial catalase activity and decreases ischemia-reperfusion injury of isolated rat hearts. Proc Natl Acad Sci USA 1989; 86:2516–2520.
87. Brown JM, White CW, Terada LS, et al. Interleukin-1 pretreatment decreases ischemia-reperfusion injury. Proc Natl Acad Sci USA 1990; 87:5026–5030.
88. Brown JM, Anderson BO, Repine JE, et al. Neutrophils contribute to TNF induced myocardial tolerance to ischemia. J Mol Cell Cardiol 1992; 24:485–495.
89. Repine JE. Scientific perspectives on adult respiratory distress syndrome. Lancet 1992; 339:466–469.
90. Baldwin SA, Simon RH, Grum CM, et al. Oxidant activity in expired breath of patients with adult respiratory distress syndrome. Lancet 1986; 1:11–14.
91. Sznajder JI, Fraiman A, Hall JB, et al. Increased hydrogen peroxide in the expired breath of patients with acute hypoxemic respiratory failure. Chest 1989; 96:606–612.
92. Grum CM, Ragsdale RA, Ketai LH, Simon RH. Plasma xanthine oxidase activity in patients with ARDS. J Crit Care 1987; 2:22–26.
93. Deby C. Differences in tocopherol–lipid ratios in ARDS and non-ARDS patients. Intensive Care Med 1989; 15:877–893.
94. Lee CT, Fein AM, Lippman M, Holtzman H, Kimbel P, Weinbaum G. Elastolytic activity in pulmonary lavage fluid from patients with adult respiratory distress syndrome. N Engl J Med 1981; 304:192–196.
95. Cochrane CG, Spragg RG, Revak SD, Cohen AB, McGuire WW. The presence of neutrophil elastase and evidence of oxidation activity in bronchoalveolar lavage fluid of patients with adult respiratory distress syndrome. Am Rev Respir Dis 1983; 127:25–27.
96. Leff JA, Parsons PE, Day CE, et al. Serum antioxidants as predictors of adult respiratory distress syndrome. Lancet 1993; 341:777–780.
97. Krsek-Staples JA, Kew RR, Webster RO. Ceruloplasmin and transferrin levels are altered in serum and bronchoalveolar lavage fluid of patients with the adult respiratory distress syndrome. Am Rev Respir Dis 1992; 145:1009–1015.
98. Bernard GR. N-acetylcysteine in experimental and clinical acute lung injury. Am J Med 1991; 91:54S–59S.
99. Richard C, Lemonnier F, Thibault M, Auzepy P. Vitamin E deficiency and lipoperoxidation during adult respiratory distress syndrome. Crit Care Med 1990; 18:4–9.
100. Idell S, Cohen AB. Bronchoalveolar lavage in patients with the adult respiratory distress syndrome. Clin Chest Med 1985; 6:459–471.

11

Endotoxin-Induced Interactions of Inflammatory Cells with the Lungs

SAM R. SHARAR, ROBERT K. WINN, and JOHN M. HARLAN

University of Washington
Seattle, Washington

I. Introduction

The association between endotoxin and acute pulmonary injury has been documented repeatedly during the last decade through observations of human disease, investigations of in vivo animal models, and experiments with in vitro cell culture systems. Overwhelming clinical evidence suggests an association between gram-negative bacteremia or sepsis syndrome and the subsequent development of adult respiratory distress syndrome (ARDS), with an 18 to 38% incidence of ARDS in this setting (1). Although laboratory studies have documented a direct effect of endotoxin in producing both systemic (2–4) and pulmonary endothelial cell injury (5), the predominant emphasis of these studies has focused on the role of inflammatory cells, particularly neutrophils, as mediators of the injury process. The evidence implicating neutrophils as mediators of endotoxin-induced lung injury includes observations that neutrophil numbers are increased in lung lavage from animals and humans with sepsis- or endotoxin-induced ARDS (6–9), that blood neutrophil function is abnormal in patients with ARDS (10–13), that pulmonary capillary neutrophil sequestration occurs in animals and humans with ARDS (14–17), and that either neutrophil depletion or inhibition of toxic neutrophil products ameliorates to varying degrees endotoxin-induced acute lung injury in animal models (18–23).

Despite this considerable body of data, there still is some reluctance to proclaim the neutrophil as the causative agent in endotoxin-induced lung injury (24). The intrapulmonic accumulation of neutrophils in the setting of ARDS does not necessarily prove that they are responsible for cellular lung injury (25,26). Also, recent evidence suggests that neutrophil function is deficient in the setting of ARDS (27), and that this loss of beneficial microbicidal activity may be a complicating factor in the evolution of ARDS whether or not neutrophils initiate the process.

Whatever specific role neutrophils turn out to play in the evolution of endotoxin-induced acute lung injury, it is clear that inflammatory cells are intimately involved with the injury and repair processes. In this chapter we focus on neutrophils, monocytes, and alveolar macrophages, beginning with a discussion of the physiologic trafficking of neutrophils and monocytes and the margination/sequestration in the lung that occurs under normal conditions. We will then review the normal sequence of neutrophil and monocyte emigration, discussing the "marginal pool" of circulating phagocytes in terms of leukocyte–endothelial adherence, and outlining the signals and events associated with transendothelial emigration. In the second section we review the consequences of neutrophil and mononuclear phagocyte interactions in the lung, first considering the beneficial effects of neutrophils and alveolar macrophages in the context of bacterial host defense. Finally, we examine the deleterious effects that endotoxin-induced phagocyte interactions have on both the pulmonary endothelium and epithelium, reviewing data from observations of animal and human settings.

II. Mechanisms of Phagocyte Recruitment

In the normal human being, neutrophils comprise approximately 60% of the circulating, intravascular leukocyte pool. In contrast, only 2% of leukocytes lavaged from the lungs of normal individuals are neutrophils (28). Conversely, monocytes, precursors to tissue macrophages, comprise only about 5% of circulating leukocytes, yet 93% of normal extravascular lung leukocytes are alveolar macrophages. In animals stimulated with intravenous endotoxin, the extravascular distribution of neutrophils and monocytes in the lung reverses rapidly, due to the emigration of large numbers of neutrophils into the alveolar space (29). For these phagocytes to maintain this imbalanced distribution under normal conditions, yet respond with rapid emigration of neutrophils in acute inflammatory conditions, the traffic of neutrophils and monocytes from the vasculature to the alveolar space must be highly regulated. Emigration is initiated with neutrophil and monocyte adherence to endothelium in the pulmonary microcirculation, followed by diapedesis between endothelial cells, migration through matrix and interstitium, and finally, transepithelial migration. Since the physiologic margination of leukocytes within the microcirculation is probably a mechanistically different phenomenon from leukocyte adherence to endothelium in preparation for emigration, we address physiologic margination and adherencc-dependent margination separately.

A. Physiologic Trafficking of Neutrophils and Monocytes

Neutrophils

Neutrophils originate from the bone marrow stem cell pool as myeloblasts, and progress over an 8- to 12-day period through both the mitotic and maturation/storage pools in the marrow. The marrow storage pool of neutrophils is large, containing approximately 10 times as many cells as in the blood. Once released into the circulation, the neutrophil half-life is only 4 to 10 h (30). Neutrophils probably do not emigrate to tissue unless stimulated to do so. Once in tissue they have a very short life span, measured in minutes to hours. Not all blood neutrophils circulate freely. Approximately 50% of neutrophils in peripheral blood are thought to be loosely associated with the vessel wall throughout the microcirculation and are referred to collectively as the marginated pool. The number of circulating neutrophils is determined by the rate of production in the marrow, the rate of release from the marrow, the partitioning of cells in the vasculature between circulating and marginated pools, and the rate of cell emigration/removal from the vasculature. Increased release from the large marrow storage pool and demargination are two mechanisms that provide for rapid increases in circulating neutrophil number.

Neutrophils are stimulated to leave the circulation in response to a variety of inflammatory stimuli. In the context of host defense against bacterial organisms, the neutrophil first loosely adheres to activated endothelium at or adjacent to the site of bacterial invasion. Following activation by chemoattractants or endothelial surface molecules (e.g., platelet activating factor), the neutrophil adheres tightly, migrates along the endothelial surface, and diapedeses between endothelial cells. It then migrates through the extravascular matrix in response to chemoattractants derived from bacteria (e.g., formylated peptides) or generated by the host (e.g., complement fragments, lipoxygenase products, or chemokines). Phagocytosis and vacuolization of pathogens is mediated by opsonins (e.g., antibody, complement, and fibronectin) and surface receptors (e.g., Fc, complement, integrins). Intracellular killing of the pathogen results from discharge of cytoplasmic granules containing a variety of proteolytic enzymes and oxidants into the phagocytic vacuole. The killed organism is digested by hydrolytic enzymes prior to extracellular release of remaining phagocytized products and excess enzymes. This sequence of events is illustrated in Fig. 1.

Monocytes

Monocytes are derived from the same myeloid stem cell as neutrophils, but enter the mitotic and maturation pools in the bone marrow as promonocytes. Monocytes do not have a marrow reserve pool analogous to neutrophils. Mature monocytes released into the circulation have a longer life span than that of neutrophils, with a mean circulating half-life of up to 3 days (31). Unlike neutrophils, monocytes that emigrate into tissue retain their capacity to divide, have a life span measured in days, are incapable of reentering the circulation, and differentiate into tissue macrophages. Once in

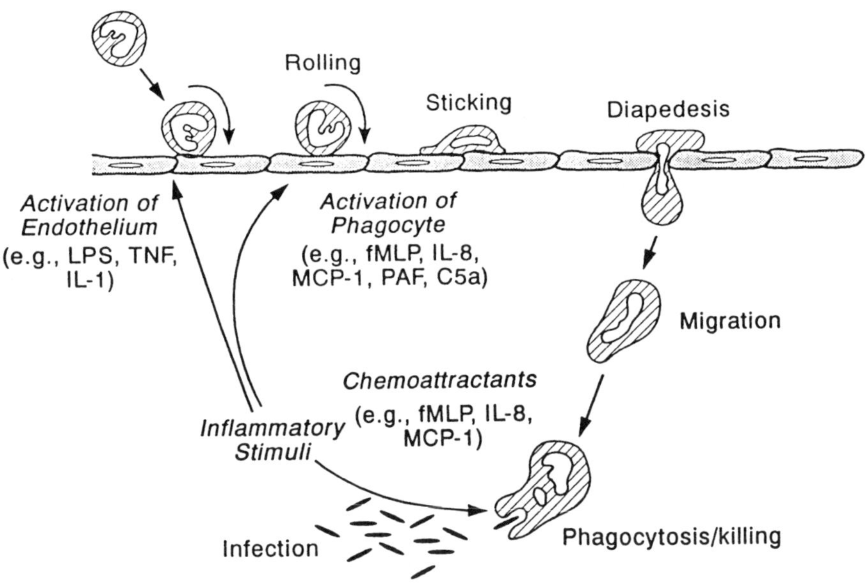

Figure 1 Inflammatory stimuli produce endothelial cell and/or circulating phagocyte activation, resulting in a sequence of activities by the phagocyte. In the setting of bacterial tissue infection, for example, neutrophils contact and adhere loosely to endothelium ("rolling") along vessel walls through selectin-mediated adherence, then attach firmly ("sticking") by integrin-mediated adherence. Following transendothelial migration, the neutrophil phagocytoses and digests the invading organisms, often releasing a variety of additional inflammatory stimuli. See the text for further details.

tissue, macrophages increase their metabolism and size and become more efficient phagocytes. However, the mononuclear phagocytes are generally less efficient than neutrophils at this task. Macrophages are found primarily in the lung, liver, spleen, and peritoneum, and possess a number of functions important to host defense, including bacteriocidal activity and neutrophil recruitment. Macrophages are potent effector cells, producing a variety of cytokines [e.g., interleukin-1 (IL-1), IL-6, tumor necrosis factor (TNF), IL-8], enzymes (e.g., lysozyme, collagenase), eicosanoids (e.g., thromboxane A_2, leukotrienes), and other proteins (e.g., α_1-proteinase inhibitor, fibronectin) (32).

Monocytes are less deformable than neutrophils and are more likely to be retained during passage through polycarbonate filters in vitro (33). Animal studies have also shown that circulating monocytes are less able to pass through the pulmonary microcirculation and have greater retention than do neutrophils (34). Although the circulating pool of monocytes is smaller than that of neutrophils, these data suggest that proportionally more monocytes than neutrophils marginate in the lung. This large marginated pool of monocytes may contribute to the rapid increase in alveolar macrophages seen in response to intravenous endotoxin (35). Alveolar macrophage numbers in the lung may also increase by cell division and differentiation, independent of increased monocyte emigration. This intrapulmonary proliferation is probably minimal in the normal state, but becomes significant during periods of monocytopenia (36) or acute inflammation (37).

B. Neutrophil and Monocyte Sequestration in the Lung

As described above, a marginated pool of noncirculating neutrophils and monocytes exists under normal conditions. The lung is the major site of the marginated neutrophil pool (38,39), containing an estimated 0.6 to 3 times the total number of circulating neutrophils (40,41). The most intimate contact between neutrophils and the endothelial surface occurs in capillary segments. This contact has been shown by in vivo microscopy to be both reversible and dependent on shear forces between the neutrophil and endothelium that vary in intensity with stages of the cardiac cycle (38). Due to this retardation of neutrophil flow in the lung, it has been shown in sheep that the greatest number of neutrophils in the lung localize to alveolar microvessels, with the next greatest excess of cells in the arteries (40). In the human lung, these segments have a diameter of 4 to 7 μm (42), compared with neutrophil diameters of 5 to 8 μm (43). Because of this discrepancy, neutrophils are not only in close contact with the lumen, but most deform to pass somewhat impeded through these segments. Although of comparable size, erythrocytes are more deformable than phagocytes, resulting in a more rapid transit time for erythrocytes through the pulmonary microcirculation (44).

The multisegmented nature of the pulmonary microvasculature allows the more deformable erythrocyte to stream past the less deformable neutrophil that is obstructing a capillary temporarily. This results in a relative concentration of neutrophils in the pulmonary microcirculation, without invoking any adhesive interactions between the neutrophil and the endothelial surface.

Retention of neutrophils in the pulmonary microvasculature has been shown to increase when blood flow is reduced, indicating that pulmonary flow has a marked effect on the uptake and release of phagocytes from the marginated pool (45). This mechanism potentially explains the observations that increased shear forces that accompany increases in cardiac output and blood flow (e.g., exercise, epinephrine infusion) result in concurrent neutrophil demargination in the lung and increased neutrophil numbers in circulating blood (46,47). This mechanism may also contribute to increased leukocyte sequestration in the lung during intravenous endotoxin stimulation when pulmonary vasoconstriction due to inflammatory mediators (e.g., thromboxane A_2) leads to maldistribution of pulmonary blood flow (21).

Changes in the mechanical properties of neutrophils induced by endotoxin and other stimuli may be important in neutrophil sequestration in the lung in acute inflammation, independent of changes in adhesive interactions between the neutrophil and the endothelial cell. Endotoxin stimulation decreases neutrophil deformability and increases neutrophil retention in vitro through the reorganization of f-actin (48). This process appears to be directly induced by endotoxin since antibody blockade of CD14 (the neutrophil receptor for the endotoxin–lipopolysaccharide binding protein complex; see Chapter 4) inhibits the changes in deformability and retention. Pretreatment of neutrophils with cytochalasin D, an agent that disrupts cellular actin organization, also abolishes these changes in deformability and the retention observed following endotoxin (48) or f-met-leu-phe (fMLP) (49) stimulation. In vivo studies of neutrophil activation have demonstrated a rapid, profound decrease in the circulating neutrophil count within 5 min of infusing these stimulants, with a return to normal after several hours (49,50). The immediate neutropenia is not prevented by pretreatment with a monoclonal antibody that inhibits CD11/CD18-mediated neutrophil adherence to endothelium (51,52), suggesting that pulmonary sequestration of neutrophils can result from changes in neutrophil deformability alone (see below). However, it appears that this endotoxin-induced, CD18-independent immediate neutropenia is followed by changes in neutrophil adherence that are CD18-dependent, resulting in prolonged neutrophil sequestration in the pulmonary vasculature (52).

C. Role of Cell Adhesion in Phagocyte Emigration

Adhesion Molecules

In vivo observations by a number of investigators have established that leukocyte adherence to endothelium is a hallmark of the acute inflammatory response. For many decades it was a matter of controversy whether changes in the endothelium (Cohnheim) or the leukocyte (Metchnikoff) initiated this adherence, but it now appears that alterations in both cellular surfaces are responsible (53). Leukocyte–endothelial adherence in the inflammatory response is mediated by three distinct families of surface molecules: β_1 and β_2 integrins (54–56), selectins (57), and the immunoglobulin superfamily (58). Table 1 shows the major leukocyte–endothelial adhesion molecules.

Neutrophil Adhesion Molecules

Of the three β_2 integrins (CD11a/CD18, CD11b/CD18, and CD11c/CD18), CD11b/CD18 appears most important in mediating firm neutrophil adherence, and is constitutively expressed at low levels on the cell surface under normal conditions. The endotoxin-induced macrophage product TNF-α is one of the most potent of a variety of cytokines that can increase CD18-dependent adherence, both through a conformational change in the glycoprotein itself (59,60) and by translocation of additional stored glycoprotein from secondary granules to the cell surface (61,62). The qualitative change in CD11/CD18-mediated adherence is reversible, peaking at 15 min after stimulation and returning to baseline within 30 min in vitro (63).

Table 1 Leukocyte–Endothelial Cell Adhesion Molecules

Adhesion molecules	Leukocyte	Endothelial cell
Integrin–immunoglobulin[a]	CD11a/CD18 (LFA-1)	ICAM-1 (CD54), ICAM-2 (CD102)
	CD11b/CD18 (Mac-1, Mo1, CR3)	ICAM-1 (CD54), ? other
	CD11c/CD18 (p150,95)	?
	VLA-4 (CD49d/CD29)	VCAM-1 (CD106)
Selectin–carbohydrate[b]	L-selectin (CD62L)	GlyCAM-1[c]
	Sialylated Lewis^x (CD15s), ? other	E-selectin (CD62E)
	Sialylated Lewis^x (CD15s), ? other	P-selectin (CD62P)

[a]Leukocyte integrin receptors bind to ligands on the endothelial cell that are members of the immunoglobulin supergene family.

[b]Selectin receptors expressed on both cell types recognize specific carbohydrate counterstructures.

[c]GlyCAM-1 is a ligand on murine high endothelial venules; the ligand for L-selectin on systemic endothelium has not been identified.

Initially, L-selectin (CD62L) was identified as a homing receptor directing lymphocyte traffic to peripheral lymph nodes (64). L-selectin interaction with its counterstructure is mediated by carbohydrate or lectin recognition, and can be partially blocked by a variety of polysaccharides. L-selectin is broadly distributed, being found on neutrophils, monocytes, lymphocytes, and other myeloid cells. It is constitutively expressed on the normal neutrophil surface, but is rapidly shed upon stimulation, including endotoxin stimulation in vitro (65).

Sialyl Lewis^X (SLe^X, CD15s) is a surface carbohydrate found on neutrophils that serves as a ligand for the endothelial-expressed E-selectin (CD62E) (66,67) and P-selectin (CD62P) (68). Other carbohydrate structures may also serve as ligands for E- and P-selectins, but they have not been as well characterized. Recently, Etzioni et al. (69) described a new leukocyte adhesion deficiency (LAD) syndrome designated LAD type II. The two affected patients have a general defect in fucose metabolism and do not fucosylate proteins or lipids, resulting in the absence of SLe^X on neutrophils. Their cells fail to bind to E- or P-selectin, demonstrating that fucose is a critical determinant on physiologic ligands to these selectins.

Monocyte Adhesion Molecules

As descendents from the same myeloid stem cell precursor as neutrophils, monocytes express similar surface adhesion molecules, including all three heterodimers of the CD11/CD18 complex (70). Monocytes up- or down-regulate their surface expression with stimulation similar to that of neutrophils. SLe^X is also found on mature monocytes (71), and probably participates in E-selectin-dependent binding of monocytes to endothelial cells (72).

Monocytes differ importantly from neutrophils in that they express the β_1 integrin receptor VLA-4 (CD49d/CD29). This integrin receptor is also found on human T and B lymphocytes, NK cells, basophils, and eosinophils (58,73–75). The endothelial ligand for VLA-4-mediated leukocyte adherence is vascular cell adhesion molecule-1 (VCAM-1, CD106), a member of the immunoglobulin superfamily (58,76–79). VLA-4 binding to VCAM-1 may represent an alternative, CD18-independent pathway for monocyte emigration, as evidenced by the observations that some mononuclear phagocytes were noted at sites of bacterial infection individuals congenitally deficient in CD11/CD18 (80), and that monocytes accumulated in inflamed peritoneum in rabbits treated with CD18-blocking monoclonal antibody (81).

Endothelial Adhesion Molecules

Endothelial surface molecules that bind leukocytes include members of (1) the immunoglobulin superfamily [ICAM-1 (intercellular adhesion molecule-1, CD54), ICAM-2 (CD102), VCAM-1] recognized by leukocyte integrin receptors; (2) the selectin family (E-selectin and P-selectin), recognizing specific carbohydrate counterstructures on leukocytes [e.g., SLe^X (66,67), L-selectin (82), or CD66 (83)]; and (3) a carbohydrate-presenting murine glycoprotein (Gly-CAM-1) on high endothelial venules in peripheral lymph nodes (84), recognized by L-selectin.

ICAM-1 is a primary ligand for the β_2 integrins, particularly CD11a/CD18 and CD11b/CD18. ICAM-1 is expressed at low levels on normal human endothelial cells in vitro and in vivo (85), as well as on mononuclear leukocytes. Surface expression of ICAM-1 is up-regulated in vitro following exposure to endotoxin, TNF, IL-1, or interferon-γ (86–89). This increase in surface expression is protein synthesis-dependent, requires 6 to 8 h for peak expression, and is maintained for at least 48 h (85). Recently, thrombin has been shown to increase ICAM-1 surface expression in a protein synthesis–independent manner (90). ICAM-2 exhibits noninducible, high levels of expression on resting endothelium and may be a ligand on unactivated endothelium for lymphocytes expressing CD11a/CD18 (58). The constitutive expression of ICAM-2 contrasts with the low but highly inducible expression of ICAM-1, and suggests a more important role of ICAM-1 in inflammatory and immune responses.

VCAM-1 is the primary ligand for leukocyte VLA-4, is expressed minimally on resting endothelial cells in vitro, and is up-regulated in a protein synthesis–dependent manner following activation with endotoxin, TNF, IL-1, or IL-4 (78,91–93). Maximal surface expression of VCAM-1 is observed 4 to 6 h after stimulation and expression is maintained at lower levels with continued incubation.

Phagocyte binding to E- and P-selectin involves the interaction of the terminal lectin domain of these endothelial glycoproteins with specific carbohydrates on the leukocyte. E-selectin is not present on resting endothelial cells but is transiently expressed in response to endotoxin, TNF, or IL-1 (94, 95). Stimulated expression is protein synthesis–dependent and maximal 2 to 6 h after stimulation. Expression declines somewhat by 24 h, even in the presence of continued stimulation. E-selectin binding of neutrophils (94) and monocytes (72) is CD18-independent.

P-selectin is not expressed on resting endothelium, but rather, it is stored in the cytoplasmic secretory granules (Weibel–Palade bodies) (96). P-selectin is redistributed to the surface membrane within seconds to minutes following activation with thrombin, histamine, or phorbol esters, and mediates binding of neutrophils and monocytes (97–99). While endothelial cells normally reinternalize P-selectin within minutes of activation (100), free radicals (e.g., those released by activated, bound neutrophils) prolong the expression of this selectin and may prolong the duration of P-selectin-mediated adhesion (101). Recently, in the mouse P-selectin has also been shown to be induced by endotoxin (102).

The endothelial ligand for leukocyte L-selectin on systemic vasculature has yet to be identified. However, using L-selectin IgG chimera (103) and cDNA cloning (84), a 50-kDa glycoprotein likely to be the ligand on high venular endothelium in lymph nodes has been identified and designated GlyCAM-1 (glycosylation-dependent cell adhesion molecule-1). It is a mucin-like molecule that acts to present carbohydrate ligands to the lectin domain of murine L-selectin.

Phagocyte Adherence to Endothelium

As discussed previously, a large marginated pool of neutrophils exists in the lungs under normal conditions, and stimulation by intravenous endotoxin results in further neutrophil sequestration concomitant with a circulating neutropenia. The initial sequestration of phagocytes is probably due to reduced neutrophil deformability and changes in pulmonary blood flow that cause neutrophil accumulation in the lung independent of changes in adherence. However, changes in neutrophil adherence are then required both to sustain phagocyte sequestration and to initiate their transendothelial migration.

Intravital microscopy in systemic vascular beds has demonstrated that neutrophils "roll" along the vessel wall in postcapillary venules with minimal inflammation (104,105), a process that is markedly increased in the initial stages of an acute response (106,107). Atherton and Born (108) proposed that rolling was the resultant of shear forces produced by flow and low-affinity adhesive interactions between leukocytes and the vessel wall. With sufficient inflammation, the rolling cells often "stick," change shape (becoming ruffled, flattened, and bipolar), and then emigrate into surrounding tissue.

Neutrophil rolling and neutrophil emigration subsequently have been shown to be distinct phenomena, mediated by different sets of adhesion molecules. At physiological flow rates, low-affinity adherence manifested by neutrophil rolling (a local "margination" within the vessel itself) appears to be a selectin-dependent process, whereas high-affinity, firm adherence or sticking, and subsequent transendothelial migration is a CD11/CD18–ICAM-1-dependent process (109–115). Using either in vivo conditions of microvascular flow or in vitro endothelial cell monolayers within a flow chamber, investigators have observed evidence for these two distinct adhesion phenomena. Using intravital microscopy in the rabbit, Arfors et al. (109) first demonstrated that a blocking CD18 monoclonal antibody (60.3) completely inhibited neutrophil sticking and transmigration in response to fMLP, C5a, or leukotriene B_4 (LTB_4) but had no effect on neutrophil rolling. Perry and Granger (110) observed similar results to graded intravascular flow in cat mesenteric venules in the presence or absence of another blocking CD18 monoclonal antibody (IB_4). Further evidence that firm neutrophil adherence and emigration were mediated by interaction between CD18 and ICAM-1 was produced with the demonstration that an anti-ICAM-1 monoclonal antibody (R6.5) inhibited transmigration of neutrophils across activated endothelial monolayers under conditions of physiologic flow rates (111). Importantly, CD18-dependent neutrophil adherence was found to be quite sensitive to shear forces. Stimulated adherence was not observed at shear rates found in postcapillary venules, suggesting that other adherence mechanisms must slow the neutrophil before CD11/CD18-ICAM-1 interactions could occur.

As blockade of CD18 or ICAM-1 inhibited neutrophil sticking and transmigration but did not affect leukocyte rolling, attention was directed to selectin receptors as mediators of the low-affinity, shear-resistant adherence. Because it is constitutively expressed and rapidly mobilized, P-selectin may be in a position to "catch" previously unstimulated neutrophils very early after endothelial activation. Using in vitro bilayers containing purified P-selectin, Lawrence and Springer (112) demonstrated that unstimulated neutrophils roll on P-selectin. Since endotoxin is a stimulus for P-selectin expression in the mouse, it is possible that P-selectin is involved in endotoxin-induced phagocyte emigration. Several investigators have shown that rolling of activated neutrophils is markedly reduced by L-selectin blockade both in vitro (96,113) and in vivo (114,115), thereby confirming the role of L-selectin in the rolling process. Intravital microscopy or monolayer/flow chamber experiments examining the role of E-selectin in activated neutrophil rolling have yet to be reported, although work in animal models showing that monoclonal antibodies to E-selectin inhibit neutrophil emigration suggests that E-selectin also mediates initial adhesive interactions (116,117).

Not only have rolling and firm adherence of neutrophils been shown to depend on different adherence pathways, but it appears that selectin-mediated rolling is a prerequisite to integrin-mediated sticking (i.e., low-affinity selectin–carbohydrate interactions initiate slowing of neutrophils on the endothelial cell surface and allow high-affinity integrin–immunoglobulin ligand firm adhesion) (118). Using neutrophil-perfused bilayers containing various combinations of purified P-selectin and ICAM-1, Lawrence and Springer found that under conditions of flow, activated neutrophils will not firmly adhere to ICAM-1 unless preceded by P-selectin-mediated rolling (112). Similar results were observed with intravital microscopy by von Andrian et al. (115). Furthermore, monoclonal antibody blockade of L-selectin did not impair firm adherence to and transmigration across cytokine-stimulated endothelial monolayers under static conditions (119,120), suggesting that the principal role for L-selectin on neutrophils is to mediate initial, low-affinity adherence under conditions of flow. The topography of the leukocyte membrane also suggests this hypothesis, with L-selectin being concentrated on microvillus projections of the unstimulated neutrophil surface (82), whereas CD18 appears on flat areas of the membrane as well as the surface projections (121).

Transendothelial Migration

Transendothelial migration of neutrophils follows firm CD11/CD18-mediated adherence and results from ameboid movement of phagocytes between endothelial cells. The motile process involves the formation and retraction of various cell surface protrusions that are coordinated with the formation and severing of adhesive attachments of the neutrophil membrane to other cells and to extracellular matrix. The changes in neutrophil shape and deformability induced by endotoxin are mediated by changes in f-actin polymerization (48,49). Directional movement of phagocytes along and between endothelial cells probably involves mutual capping of adhesion–receptor ligand pairs on the two cell surfaces (122), with CD11a/CD18 and ICAM-1 already shown to be of primary importance in lymphocyte migration (123). Cell migration is directed by concentration gradients of chemoattractants released by endothelial cells, epithelial cells, or other phagocytes (macrophages, in particular). Specific to endotoxin, products of gram-negative bacteria (e.g., fMLP) and other inflammatory products (C5a and C5a desArg) are potent neutrophil chemoattractants. In addition, endotoxin exposure induces the release of multiple "second messenger" chemoattractants, including IL-8 (124) and platelet activating factor (125) from pulmonary endothelial cells, LTB_4 from pulmonary epithelial cells (126,127), and a variety of substances (including LTB_4 and IL-8) from macrophages (128–133). Interestingly, several of these chemoattractants also induce neutrophil activation, suggesting an intimate relationship between cell activation and cell motility (134).

Adhesion molecules play an important role in endotoxin-induced transendothelial leukocyte migration. As reviewed by Smith (135), chemotactically stimulated neutrophils can use both CD11a/CD18 and CD11b/CD18 for locomotion. CD11b/CD18 appears to provide the neutrophil with greater flexibility, allowing adhesion to and migration across artificial surfaces and cells. It binds to an array of different proteins and thereby allows neutrophils to adhere to a wide variety of surfaces. Although not yet proven, the intracellular storage pool of CD11b/CD18 appears to provide for replacement of molecules lost during locomotion. CD18-mediated adherence is also predominant in monocyte migration, as monoclonal antibody blockade of CD18, but not E-selectin or VLA-4, greatly reduces transendothelial monocyte migration in vitro (136). Finally, adhesion molecules can also mediate migration through intercellular matrix, as evidenced by VLA-6 (a β_1 integrin)-mediated neutrophil adherence to laminin (137).

Phagocyte Emigration in the Pulmonary Circulation

The current model of leukocyte–endothelial interaction in inflammation—selectin-dependent rolling followed by integrin-mediated adherence and transendothelial migration—was developed from observations in the systemic microcirculation where emigration occurs in postcapillary venules. In the lung significant emigration also occurs in the capillaries (38). Since neutrophils are in intimate contact with endothelium during their transit of capillaries, it could be questioned whether there is a requirement for selectins to initiate adhesive interactions before integrin receptors can be engaged. However, the significant inhibition of neutrophil emigration into the lung observed with monoclonal antibodies to E- (117) and P-selectin (138) indicates that selectin interactions are still quite important in the lung.

Although the CD11/CD18 adhesion receptors are very important in the generalized scheme of transendothelial phagocyte migration described previously, a number of investigations have shown that phagocyte emigration in the pulmonary circulation, including endotoxin-induced emigration, may occur by CD18-independent mechanisms. Doershuck et al. reported that the CD18 monoclonal antibody (60.3) caused a 98% inhibition in neutrophil emigration to endotoxin-soaked sponges in the systemic circulation (abdominal wall) of the rabbit, but only a 72% inhibition of emigration to intrapulmonary endotoxin (139). This difference was most striking with gram-positive organisms, suggesting that pulmonary neutrophil–endothelial adherence is both CD18-independent and stimulus-specific. The latter is also exemplified by the observation that IgG- but not IgA immune complex–induced lung injury is CD18-dependent (140).

The existence of a CD18-independent pathway of neutrophil emigration in the lung has recently been confirmed in humans. Hawkins et al. (141) reported that neutrophils were observed in alveoli in postmortem lung specimens of a severe CD11/CD18-deficient patient, although neutrophils were not detected at extravascular sites of infection in the systemic circulation. The molecular basis of the CD18-independent adherence pathway in the lungs is incompletely understood, although proadhesive factors released by tissue macrophages may be involved (142).

III. Consequences of Phagocyte Interactions in the Lung

Neutrophils and monocytes provide for a rapid, relatively nonspecific mechanism of defense against pathogenic organisms. They allow constant surveillance for microbial invasion and can act rapidly and effectively to localize and eliminate many bacterial and fungal infections. However, by virtue of their ability to adhere to target cells and release a variety of products that are capable of both directly or indirectly causing vascular and tissue injury, phagocytes mediate host damage as well as host defense. In certain inflammatory conditions (typified by endotoxin-induced inflammation), phagocyte responses may be excessive and directly damaging to host tissue. In this section we review examples of both the beneficial and deleterious effects of neutrophils and monocytes in endotoxin-induced lung injury, emphasizing the pivotal roles that these phagocytes appear to play in both processes.

A. Beneficial Effects of Phagocytes in Endotoxin-Induced Lung Inflammation

Neutrophil Functions in Lung Host Defense

The recruitment of neutrophils and monocytes to extravascular sites in the lung is an important and fundamental part of host defense. As outlined previously, phagocytes protect the host against bacterial infection by first adhering to the vessel wall adjacent to sites of bacterial invasion and emigrating across endothelium to extravascular tissue, then phagocytizing and killing the pathogens. The process of phagocyte recruitment is dependent on the generation of cytokines (e.g., IL-1, TNF, IL-8, MCP-1) by lymphocytes, mononuclear phagocytes, and other lung cells, and requires the activation and/or induction of adherence molecules as detailed in previous sections.

Pathogen recognition and attachment to the neutrophil membrane prior to phagocytosis are mediated by Fc receptors and complement component (C3b and C3bi) receptors on the neutrophil surface (143). Three Fc receptors have been described, and bind the Fc portion of pathogen-bound IgG to monocyte or macrophages (Fc-γ RI and RIII) or neutrophils (Fc-γ RII and RIII). Complement receptors 1 (CR1) and 3 (CR3) bind particles opsonized with complement fragments C3b and C3bi, respectively. Both the binding affinity and apparent numbers of complement receptors are increased when neutrophils are activated (59,144). Despite its name, CR3 (the β_2 integrin receptor CD11b/CD18) does not require complement coating of particles for attachment, as evidenced by direct binding of *E. coli* to human macrophage CD11b/CD18 via endotoxin/lipid A in the bacterial wall (145). Once opsonized particles are bound to the phagocyte membrane by Fc or complement receptors, engulfment occurs by the localized extension of a pseudopod to form the phagocytic vacuole.

Endotoxin has variable effects on neutrophil phagocytosis. On the one hand, endotoxin exposure of isolated neutrophils in vitro has been shown to increase phagocytic activity more than twofold (146,147). In contrast, impaired phagocytosis has been observed in neutrophils isolated both from the peripheral blood and the alveoi of rats exposed to endotoxin in vivo (148). The mechanism of this impairment is unclear, since as noted above, neutrophil activation increased complement receptor binding. Of interest, an association between decreased neutrophil expression of CR3 and Fc-γ RIII (CD16) and the development of gram-negative pneumonia and sepsis has been observed in patients with both severe trauma (149) and burns (150). Evaluation of these clinical data suggest that following in vivo endotoxin exposure (gram-negative sepsis), down-regulation of specific membrane surface receptors may result in impaired bacterial phagocytosis.

Microbicidal killing is the final step of bacterial host defense and takes place within the phagosome of the neutrophil. Both oxygen-dependent (e.g., hydrogen peroxide and superoxide radicals) and oxygen-independent (e.g., lysozyme, collagenase) mechanisms are involved (143). As cytoplasmic primary (azurophil) and secondary (specific) granules fuse with the phagosome, their enzyme contents are released to kill the ingested microorganisms. Cytotoxic activity of these substances appears to be the result of oxidation and digestion of bacterial membrane phospholipids and glycoproteins.

Host defense against invasion by organisms of either low virulence or small number is the principal responsibility of the alveolar macrophage (151, 152). Major neutrophil recruitment into the alveoli occurs when the resident host defense capacity of the lung is overcome by more virulent or larger numbers of bacteria. The importance of the neutrophil can be determined when impairments arise in the cascade of events leading to emigration from bloodstream to tissue, and from phagocytosis to killing. Defects at any point in the inflammatory cascade result in impaired host defense, leading to an increased risk of recurrent infection, overwhelming sepsis, and/or death. Congenital abnormalities can result in impaired neutrophil chemotaxis (e.g., Job's syndrome), intracellular granule formation (e.g., Chediak–Higashi syndrome), and microbicidal enzyme production (e.g., myeloperoxidase deficiency, chronic granulomatous disease). One of the most severe, life-threatening neutrophil defects is found in persons with leukocyte adherence deficiency syndromes (LAD) (69,153). These syndromes result from an absence of functional neutrophil adhesion molecules or their counterstructures.

LAD type I syndrome is characterized by a profound defect in phagocyte emigration. Despite a moderate, chronic, neutrophilic leukocytosis with striking elevations of neutrophils during episodes of acute infection, neutrophils and monocytes of severely affected patients do not accumulate at sites of infection. Patients experience recurrent, nonsuppurative, soft-tissue infections (usually with *Staphylococcus* and *Pseudomonas* species) and often succumb to sepsis early in life. Antibiotic therapy alone may be ineffective, and some patients have required concurrent leukocyte transfusion to resolve the infectious episode. Tissue biopsies from sites of infection or inflammation show an absence or paucity of neutrophils and monocytes (154–156), and neutrophils and monocytes do not migrate to skin windows

(155–157). In vitro studies first demonstrated that CD11/CD18-deficient neutrophils or normal neutrophils treated with a monoclonal antibody against CD18 failed to adhere to or migrate across cultured endothelium when stimulated (158,159). The results of these experiments were consistent with the hypothesis that the profound defect found in LAD type I patients resulted from impaired phagocyte adherence to endothelium at sites of inflammation. Despite findings of impaired emigration into soft tissue, neutrophil accumulation in the lung has been documented in a person with LAD type I syndrome (141). This observation leads to the conclusion that there is an important CD18-independent adherence pathway in the pulmonary circulation.

As described above, Etzioni et al. (69) have reported a new LAD syndrome designated LAD type II. The two patients affected have a general defect in fucose metabolism and do not fucosylate proteins or lipids, resulting in the absence of SLe^X on neutrophils. Their cells fail to bind to E- or P-selectin, demonstrating that fucose is a critical determinant in selectin-mediated adherence. The phenotype of these patients is similar to those with LAD type I, showing an increased susceptibility to bacterial infections, with episodes of pneumonitis. It is unknown whether this defect results in impaired phagocyte emigration into the lung.

A similar defect in host defense against bacterial infection has been reproduced in the laboratory, using blocking monoclonal antibodies against various adhesion molecules. The incidence and severity of subcutaneous abscess formation was shown to be increased following CD18 blockade compared with control animals. Impairment was shown in response to large inocula of *S. aureus* (160) and a wide range of *E. coli* inocula (161). Interestingly, monoclonal antibody blockade of P-selectin does not appear to have any deleterious effects on host defense against subcutaneous *S. aureus* (162), subcutaneous *E. coli* (161), or intraperitoneal *E. coli* (161). Thus P-selectin-mediated adherence does not appear to be required for adequate host defense in these models. Perhaps the presence of a redundant adherence pathway mediating neutrophil rolling (e.g., L- or E-selectin) may be sufficient to allow normal neutrophil emigration in these types of bacterial infection. Animal studies examining the effects of adhesion molecule blockade in pulmonary bacterial infections have not yet been reported.

Finally, to what extent does endotoxin impair neutrophil-mediated host defense? Endotoxin (or the endotoxin-induced macrophage products TNF and IL-1) affects neutrophils in multiple ways, including (1) decreased neutrophil deformability and increased pulmonary capillary neutrophil retention (48); (2) increased CD18-dependent neutrophil adherence to endothelium in the lung (52); (3) increased surface expression of neutrophil CD11/CD18 (59–62), endothelial ICAM-1 (86,87), VCAM-1 (78,91,92), E-selectin (94,

95,102), and P-selectin (102), and decreased expression of neutrophil L-selectin (65); (4) decreased phagocytosis by alveolar neutrophils (148); and (5) decreased intraalveolar bacteriocidal activity (163,164) with associated impairment of host defense against bacterial pneumonia (163). Thus endotoxin has both stimulatory and suppressive effects on host defense.

In ARDS, the human clinical setting that often closely resembles endotoxin stimulation, blood and lung neutrophils also have been shown to have impaired migration and oxygen radical production (27). However, clinical studies have shown that overt, secondary bacterial infections complicating ARDS of noninfectious origin are rare (K.P. Steinberg and L.D. Hudson, personal communication). Thus despite laboratory evidence of several deleterious effects of endotoxin on different steps in neutrophil-mediated host defense and, to some extent, similar changes seen in ARDS, the extent to which host defense is impaired in patients is unclear.

Alveolar Macrophage Functions in Host Defense

The alveolar macrophage, the resident phagocyte of the alveolus, has primary responsibility for host defense when the lungs are invaded by organisms of either low virulence or small numbers (151,152). Alveolar macrophages can eliminate most of the small particulates (0.5 to 3 μm in diameter) and bacterial pathogens from the distal airways. These phagocytes keep the alveoli sterile through phagocytosis and phagosomal release of oxygen radicals and proteases. Phagocytosis by macrophages, much like that by neutrophils, is facilitated by pathogen opsonization with immunoglobulins and complement (165). The cytokines IL-1, TNF, and interferon-γ have been implicated in activating resting alveolar macrophages (166,167). This results in enhanced microbicidal activity, both by oxidant mechanisms [utilizing native peroxidases, and scavenged neutrophil peroxidase (168)] and enzymatic destruction (31,169). Although macrophages are perhaps less capable of generating oxygen radicals than are neutrophils or monocytes (170), their large numbers in the resting lung may account for their relatively efficient host defense.

Bacterial host defense by alveolar macrophages is not only inoculum-dependent, but also bacterial species-dependent (165). Alveolar macrophages are activated by lymphokines and become active phagocytes when challenged by *Legionella* and *Mycobacterium* species, *Pneumocystis carini*, and *Toxoplasma gondii*. They are particularly effective in clearing gram-positive organisms (e.g., *S. aureus*) without assistance from other phagocytes (171). In contrast, gram-negative, endotoxin-producing organisms such as *E. coli* and *Pseudomonas* species are minimally affected by alveolar macrophage defense mechanisms (172) and require the recruitment of neutrophils for effective eradication.

Possibly more important than their direct role as phagocytes in lung host defense, alveolar macrophages are potent effector cells, releasing cytokines (e.g., TNF, IL-1) and eicosanoids (e.g., LTB_4) (32) that regulate the adherence and emigration of circulating neutrophils as described in detail in previous sections. In addition to these chemoattractants, alveolar macrophages also release compounds that act on recruited alveolar neutrophils, stimulating their microbicidal activity. Neutrophil activating factor (NAF) (173) and enzyme releasing peptide (ERP) (174) are two such compounds that induce neutrophil oxidant release and azurophilic granule enzyme secretion, respectively.

Finally, alveolar macrophages have been implicated in the host repair process following lung injury, through the release of cell products that both induce tissue growth and inhibit further neutrophil-mediated destruction. Exudative macrophages have been observed to release several growth factors in vivo, including transforming growth factor (TGF-α, TGF-β) (175), platelet-derived growth factor (175), and fibronectin (176). In addition, a neutrophil chemotaxis inhibitor has been identified as a product released by alveolar macrophages (177). The precise mechanisms that control the release of these reparative products in vivo are not known, but the alveolar macrophage clearly has the capability to modulate lung tissue repair as well as host defense.

B. Deleterious Effects of Phagocytes in Endotoxin-Induced Lung Injury

Evidence from Animal Models

As reviewed by Brigham and Meyrick (178), a variety of animal studies have documented the pathologic and physiologic findings in the lung following exposure to intravenous endotoxin. In brief summary, endotoxin infusion in the sheep results in acute (the initial 15 to 60 min) pulmonary hypertension, increased protein-poor lung lymph flow, and hypoxemia concomitant with decreased lung compliance and increased airways resistance. Histologic observation during this period shows the accumulation, margination, and degranulation of neutrophils, with their emigration into the pulmonary interstitium. By 2 h, endothelial disruption is apparent histologically with the development of interstitial edema. Physiologically, these changes are accompanied by a return toward normal in both pulmonary vascular pressures and lung mechanics. Lung lymph flow continues to be elevated and is protein-rich, consistent with the observed endothelial cell injury, which consists of a disruption of the pulmonary alveolar–capillary barrier.

Similarly reviewed by Brigham and Meyrick (178,179), a variety of animal studies have documented a contribution of neutrophils to endotoxin-induced lung injury. When circulating neutrophil counts were reduced to less than 5% of baseline with hydroxyurea prior to endotoxin infusion in sheep, the acute pulmonary hypertension and elevated lung lymph flow were unaffected, but the alterations in lung mechanics and, more important, the late-phase increase in vascular permeability were markedly reduced (20,22). The initial acute pulmonary hypertension has been shown to be mediated by cyclooxygenase products (e.g., thromboxane A_2) (180) and therefore largely independent of direct neutrophil effects. Studies in other models of endotoxin infusion have documented a similar association between circulating neutrophil number and endotoxin-induced lung injury (17–19,21), suggesting a requirement for neutrophils to produce injury. Furthermore, canine neutrophils exposed ex vivo to endotoxin prior to reinfusion also produced a pulmonary endothelial permeability defect that was associated with neutrophil influx into the alveoli (181). Despite these associations between neutrophils and endotoxin-induced lung injury, conflicting animal data have also been reported. Neutrophil depletion with nitrogen mustard (Fig. 2) (23) or antineutrophil antibody (182) did not affect the evolution of endotoxin-induced acute lung injury as described above. A clear resolution of these conflicting data (as well as those below regarding observations of ARDS in

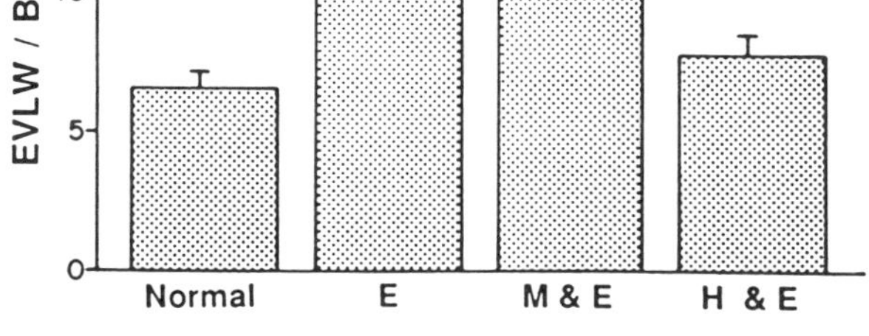

Figure 2 Extravascular lung water (EVLW)-to-body weight (BW) ratio in goats 5 h after intravenous infusion of 1 μg/kg *E. coli* endotoxin. EVLW was significantly greater in animals receiving endotoxin (E) than in normal controls. Animals depleted of neutrophils by hydroxyurea (H + E) demonstrated reduced EVLW after endotoxin infusion, whereas animals depleted of neutrophils to an equivalent degree by nitrogen mustard (M + E) developed EVLW similar to endotoxin alone. These data suggest that endotoxin-induced acute lung injury can occur in the absence of neutrophils.

septic, neutropenic humans) is lacking, and continues to remain a subject of debate.

Recently, attention has turned to the role of neutrophil–endothelial adherence in endotoxin-induced vascular injury. The hypothesis underlying these studies is that the pulmonary capillary injury caused by endotoxin is neutrophil-mediated, and that this neutrophil-mediated endothelial injury requires adherence of neutrophils to the endothelial cell. Firm adhesion of the neutrophil to the endothelial cell forms a protected microenvironment at the interface of the neutrophil and endothelial cell that is inaccessible to circulating antiproteases and antioxidants (53). Highly reactive proteases, oxidants, and phospholipase products released by activated, adherent leukocytes at the interface can react with and damage the endothelium. Inhibition of firm neutrophil adherence with a blocking monoclonal antibody would, in theory, prevent formation of the protected microenvironment and ameliorate neutrophil-mediated endothelial injury. As reviewed by Harlan et al. (53), striking protective effects against a variety of inflammatory and immune endothelial injuries have been observed in the laboratory by inhibiting CD18-dependent adherence.

Using a blocking CD18 monoclonal antibody in the context of endo-

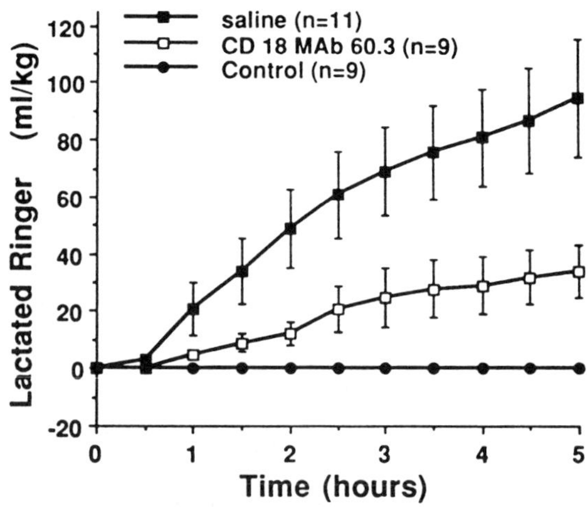

Figure 3 Cumulative fluid requirements in rabbits after intravenous infusion of 3 μg/kg *E. coli* endotoxin. Saline-treated animals required significantly more fluid to maintain cardiac output than either controls (no endotoxin) or animals treated with a blocking CD18 monoclonal antibody (60.3). These data suggest that intravenous endotoxin produces a CD18-dependent, neutrophil-mediated, diffuse endothelial injury that results in increased capillary permeability. (From Ref. 184.)

toxin-induced injury, it has been shown in rabbits that neutrophil emigration in the systemic circulation to endotoxin-soaked sponges is dependent on CD18-mediated neutrophil adherence (183). Furthermore, rabbits receiving intravenous *E. coli* endotoxin and blocking CD18 monoclonal antibody demonstrated a 60% reduction in fluid resuscitation volumes in the first 5 h of treatment for septic shock compared to controls (Fig. 3) (184). These results are consistent with a CD18-dependent, neutrophil-mediated, diffuse endothelial injury that results in increased capillary permeability. Interestingly, in the same study CD18 blockade did not affect either the endotoxin-induced leukopenia or increased pulmonary capillary permeability (Fig. 4). In a similar study, mild hemorrhagic shock (50% of baseline cardiac output for 1 h) 24 h before the initiation of endotoxic shock produced similar results. The administration of a blocking CD18 monoclonal antibody at the time of resuscitation from hemorrhagic shock again reduced the fluid resuscitation requirements following endotoxin infusion, but had no effect on the significant endotoxin-induced pulmonary capillary injury (R. K. Winn, personal communication). These data are consistent with others demonstrating that acute, endotoxin-induced leukopenia is a CD18-independent event (51,52)

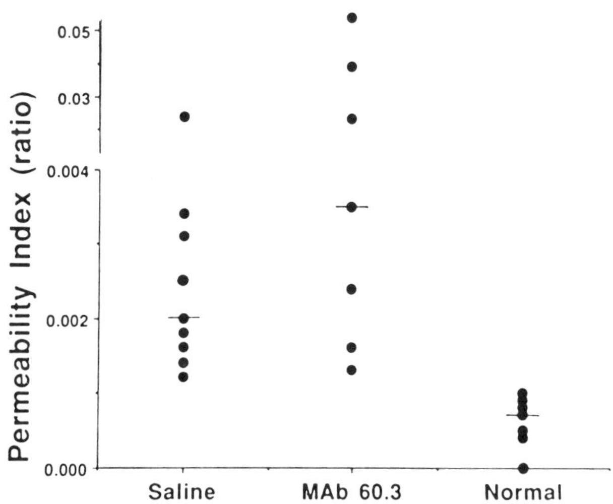

Figure 4 Lung permeability index (measured as the ratio of [^{125}I]albumin in lung lavage fluid to [^{125}I]albumin in plasma) in rabbits 5 h after intravenous infusion of 3 μg/kg *E. coli* endotoxin. Lung permeability was significantly greater in both saline-treated animals and animals receiving a blocking CD18 monoclonal antibody (60.3) than in control animals (no endotoxin). These data suggest that neutrophil-dependent, endotoxin-induced acute lung injury may be CD18-independent. (From Ref. 184.)

and suggest that neutrophil involvement in acute lung injury may be CD18-independent (139).

Endotoxin-induced neutrophil emigration into the peritoneum is also not exclusively dependent on the CD18 adherence pathway. Mileski et al. (142) demonstrated that monoclonal antibody blockade of CD18 prevented peritoneal neutrophil emigration 4 h following intraperitoneal stimulation with live *E. coli*. However, unpublished data from our lab demonstrate that neutrophil emigration 24 h following similar intraperitoneal *E. coli* inoculation is not inhibited by pretreatment with blocking CD18 monoclonal antibody. It is possible that in this case activated peritoneal macrophages release cytokines that induce neutrophil emigration by a delayed, CD18-independent pathway. A similar mechanism of cytokine release by activated alveolar macrophages may explain the existence of a CD18-independent neutrophil adherence pathway in endotoxin-induced lung injury (139,184). Precise definition of this endotoxin-induced, CD18-independent adherence pathway in the lung and peritoneum has remained elusive, since all transendothelial migration in vitro is CD18-dependent. Selectin-mediated adherence represents a potential pathway in vivo, however, as it has been shown that marked upregulation of E-selectin occurs in baboons subjected to septic shock induced by live *E. coli* (185).

Evidence from Studies in Humans

The study of the effects of endotoxemia in a controlled fashion in humans presents obvious difficulties. The infusion of very low doses of endotoxin in human volunteers has provided important information on cytokine release and more limited information on hemodynamic effects (186–189). However, for obvious reasons it is not possible to obtain information on the evolution of acute lung (or other organ) injury. Endotoxemia in humans has been shown to result in acute increases in plasma TNF and subsequently, IL-6. This is accompanied by an increase in tissue plasminogen activator and von Willebrand factor, consistent with endothelial cell activation by TNF or other cytokines. Physiologic changes associated with these observations include increased cardiac output, decreased peripheral vascular resistance, hyperthermia, and circulating leukopenia. From these data it appears that the effects of endotoxemia in humans are quite similar to those observed in experimental animals.

The sepsis syndrome allows some observations to be made on the potential contribution of endotoxin to organ injury (the lung, in particular) in humans. In this syndrome, many but not all patients have endotoxin identifiable in the blood (190,191). Gram-negative sepsis or sepsis syndrome has been shown to be a major risk factor for the development of ARDS, resulting in an 18 to 38% incidence of ARDS. Of particular interest are the recent findings on serial bronchoalveolar (BAL) from patients with ARDS from a variety of risk factors. Steinberg et al. (192) found that patients with sepsis-induced ARDS who eventually died had persistently elevated levels of BAL neutrophils and protein compared to patients with sepsis-induced ARDS who lived, or to patients with trauma-induced ARDS. ARDS mortality was also significantly greater in the sepsis-induced ARDS group (60%) compared to the trauma-induced ARDS group (29%). In addition, patients with higher numbers of alveolar macrophages in the BAL appeared to have a better prognosis. The investigators proposed that serial assessment of BAL phagocytes serves as a convenient marker of the lung inflammatory process and that continued exposure to endotoxin (i.e., sepsis syndrome) results in sustained alveolar inflammation and is associated with a poor outcome.

As discussed previously, there continues to be debate as to the precise role that neutrophils play in the development of endotoxin-induced lung injury. Although a large body of laboratory evidence supports a critical role in mediating this injury process, lung injury has been produced in animal models in the absence of circulating neutrophils (23,182). Similar conflicting data have been reported in humans with ARDS. Although neutrophils have been associated with ARDS (10–17,191), ARDS has also been documented in neutropenic patients (193–195). The histologic findings of diffuse alveolar damage and edema in these patients are identical to those seen in ARDS patients with normal neutrophil counts. However, the neutropenic patients are not normal immunocompetent hosts, in general receiving chemotherapy and/or radiation therapy for widespread malignancies. The pathogenesis of ARDS in these individuals is not known, but it would appear that neutrophils are not involved. It may be related to therapy-induced vascular damage, or from a variety of other causes, including direct lung injury by alveolar macrophages, cytokines, arachidonate metabolites, vasoactive amines, platelets, or products of the coagulation or complement cascade.

IV. Summary

Neutrophils, monocytes, and alveolar macrophages are the phagocytes principally responsible for lung host defense. The resident alveolar macrophages are effective phagocytes against invasion with limited inocula of bacterial organisms, particularly gram-positive organisms. When overwhelmed by invading organisms, especially endotoxin-producing gram-negative organisms, the resident host defense system of the lung is augmented by the recruitment of circulating neutrophils and monocytes, the latter differentiating into macrophages upon emigration into the lung. Recruitment of these circulating phagocytes into extravascular tissues is initiated by changes in leukocyte deformability, pulmonary sequestration of leukocytes, and the expression of adhesion molecules on both leukocytes and endothelial cells, resulting in enhanced leukocyte adherence to endothelium adjacent to the infectious site. This enhanced adherence is a prerequisite for normal neutrophil and monocyte emigration into extravascular tissue, and when it is impaired (e.g., the LAD syndromes) there can be disastrous infectious consequences for the host. It appears that phagocyte emigration in the pulmonary circulation may differ from that in the systemic circulation in having a significant CD18-independent pathway of adherence. Once emigrated into the lung, neutrophils and macrophages are potent effectors of host defense through phagocytosis and bacteriocidal killing of pathogens. Macrophages appear to have additional roles in regulating neutrophil adherence and chemotaxis, as well as host repair, through the release of a variety of cell products, including chemotaxins, enzymes, and growth factors. Endotoxin up-regulates certain aspects of the phagocyte response (e.g., increases expression of adhesion molecules) but may impair others (e.g., impairs neutrophil phagocytosis and oxidant release). Whether endotoxin itself is more helpful or harmful in affecting phagocyte host defense remains unresolved.

Although phagocytes are clearly critical in host defense and repair, under some circumstances they may also contribute to lung injury. Despite a significant volume of laboratory and human data collected over the past two decades, it cannot be stated with absolute certainty that endotoxin-induced acute lung injury is a phagocyte-mediated event. However, it is clear that neutrophils, monocytes, and macrophages can alter their function when stimulated with endotoxin and may interact with host tissue to produce injury in experimental models. Endotoxin or endotoxin-generated mediators induce activation or expression of adhesion molecules on both endothelial cells and circulating phagocytes, resulting in enhanced adhesiveness of the two cells. The adherent phagocytes may release oxidants and proteases that damage endothelium directly. Similarly, once emigrated, phagocytes may injure alveolar epithelium. The end result is disruption of the alveolar-capillary unit, producing acute lung dysfunction. Studies using blockade of neutrophil adherence to endothelium with monoclonal antibodies directed against adhesion molecules suggest that CD18-dependent adherence is crucial to the development of endotoxin-induced organ injury in the systemic circulation. However, endotoxin-induced, neutrophil-mediated lung injury (like neutrophil-mediated lung defense) may involve a CD18-independent pathway of adherence that has yet to be clearly defined.

Acknowledgments

This work was supported by USPHS grants HL-30542, HL-43141, and GM-42689.

References

1. Martin MA, Silverman HJ. Gram-negative sepsis and the adult respiratory distress syndrome. Clin Infect Dis 1992; 14:1213–1228.
2. Harlan J, Harker L, Striker G, Weaver L. Effects of lipopolysaccharide on human endothelial cells in culture. Thromb Res 1983; 29:15–26.
3. Harlan J, Harker L, Reidy M, Gajdusek CM, Schwartz SM, Striker GE. Lipopolysaccharide-mediated bovine endothelial cell injury in vitro. Lab Invest 1983; 48:269–274.
4. Meyrick B. Endotoxin-mediated pulmonary endothelial cell injury. Fed Proc 1986; 45:19–24.
5. Bloom RJ, Simon LM, Benitz WE. Endotoxin and pulmonary cell injury. Surg Gynecol Obstet 1988; 167:92–98.
6. Weiland JE, Davis WB, Holter JF, Mohammed JR, Dorinsky PM, Gadek JE. Lung neutrophils in the adult respiratory distress syndrome. Clinical and pathophysiologic significance. Am Rev Respir Dis 1986; 133:218–225.
7. Chang JC, Lesser M. Quantitation of leukocytes in bronchoalveolar lavage samples from rats after intravascular injection of endotoxin. Am Rev Respir Dis 1984; 129:72–75.
8. Rinaldo JE, Dauber GH, Christman J, Rogers RM. Neutrophil alveolitis following endotoxemia. Am Rev Respir Dis 1984; 130:1065–1071.
9. Fowler AA, Fischer BJ, Webster RO, Hyers TM. Lavage fluid from patients at risk for ARDS promotes neutrophil adherence. Am Rev Respir Dis 1985; 131:A137.
10. Fowler AA, Fisher BJ, Centor RM, Carchman R. Development of the adult respiratory distress syndrome: progressive alteration of neutrophil chemotactic and secretory processes. Am J Pathol 1984; 116:427–435.
11. Zimmerman GA, Renzetti AD, Hill HR. Functional and metabolic activity of granulocytes from patients with adult respiratory distress syndrome: evidence for activated neutrophils in the pulmonary circulation. Am Rev Respir Dis 1983; 127:290–300.
12. Weinberg PF, Matthay MA, Webster RO, Roskos KV, Goldstein RM, Murray JF. Biologically active products of complement in acute lung injury in patients with sepsis syndrome. Am Rev Respir Dis 1984; 130:791–796.
13. Hammerschmidt DE, Weaver LI, Hudson LD, Craddock PR, Jacob HS. Association of complement activation and elevated plasma-C5a with adult respiratory distress syndrome. Lancet 1980; 1:947–957.
14. Bachofen M, Weibel ER. Structural alterations of lung parenchyma in the adult respiratory distress syndrome. Clin Chest Med 1982; 3:35–56.
15. Elliott CG, Zimmerman GA, Orme JF, Morris AH, Mortensen JD. Granulocyte aggregation in adult respiratory distress syndrome (ARDS): serial histologic and physiologic observations. Am J Med Sci 1985; 289:70–74.
16. Coalson JJ, Hinshaw LB, Guenter CA, Berrell EL, Greenfield LF. Pathophysiologic responses of the sub-human primate in experimental septic shock. Lab Invest 1975; 32:561–569.
17. Coalson JJ, Hinshaw LB, Guenter CA. The pulmonary ultrastructure in septic shock. Exp Mol Pathol 1970; 12:84–103.

18. Shasby DM, Vanbenthuysen KM, Tate RMN, Shasby SS, McMurtry IF, Repine JE. Granulocytes mediate acute edematous lung injury in rabbits and in isolated rat lungs perfused with phorbol myristate acetate: role of oxygen radicals. Am Rev Respir Dis 1982; 125:443–447.
19. Sacks T, Moldow CF, Craddock PR, Bowers TK, Jacob HS. Oxygen radicals mediate endothelial cell damage by complement-stimulated granulocytes. J Clin Invest 1978; 60:1161–1167.
20. Heflin AC, Brigham KL. Prevention by granulocyte depletion of increased vascular permeability of sheep lung following endotoxemia. J Clin Invest 1981; 68:1253–1260.
21. Snapper JR, Bernard GR, Hinson JM, et al. Endotoxemia-induced leukopenia in sheep. Am Rev Respir Dis 1983; 127:306–309.
22. Hinson JM, Hutchison AA, Ogletree ML, Brigham KL, Snapper JR. Effect of granulocyte depletion on altered lung mechanics after endotoxemia in sheep. J Appl Physiol 1983; 55:92–99.
23. Winn R, Maunder R, Chi E, Harlan J. Neutrophil depletion does not prevent lung edema after endotoxin infusion in goats. J Appl Physiol 1987; 62:116–121.
24. Glauser FL, Fairman RP. The uncertain role of the neutrophil in increased permeability pulmonary edema. Chest 1985; 88:601–607.
25. Henson PM, Larsen GL, Webster RO, Mitchell BC, Goins AJ, Henson JE. Pulmonary microvascular alterations and injury induced by complement fragments: synergistic effect of complement activation, neutrophil sequestration and prostaglandins. Ann NY Acad Sci 1982; 384:287–300.
26. Brigham KL, Meyrick BO. Granulocyte-dependent injury of pulmonary endothelium: a case of miscommunication? Tissue Cell 1984; 16:137–155.
27. Martin TR, Pistorese BP, Hudson LD, Maunder RJ. The function of lung and blood neutrophils in patients with the Adult Respiratory Distress Syndrome: implications for the pathogenesis of lung infections. Am Rev Respir Dis 1991; 144:254–262.
28. Martin TR, Raugi G, Merritt TL, Henderson WR. Relative contribution of leukotriene B_4 to the neutrophil chemotactic activity produced by the resident human alveolar macrophage. J Clin Invest 1987; 80:1114–1124.
29. Meyrick B, Brigham KL. Acute effects of *E. coli* endotoxin on the pulmonary microcirculation of anesthestized sheep: structure:function relationships. Lab Invest 1983; 48:458–470.
30. Golde DW. Production, distribution, and fate of neutrophils. In: Williams WJ, Beutler E, Erslev AJ, Lichtman MA, eds. Hematology. New York: McGraw-Hill, 1990:795–801.
31. Golde DW, Groopman JE. Cellular kinetics of monocytes and macrophages. In: Williams WJ, Beutler E, Erslev AJ, Lichtman MA, eds. Hematology. New York: McGraw-Hill, 1990:874–878.
32. Nathan CF. Secretory products of macrophages. J Clin Invest 1987; 79:319–326.
33. Downey GP, Doherty DE, Schwab B, Elson EL, Henson PM, Worthen GS. Retention of leukocytes in capillaries: role of cell size and deformability. J Appl Physiol 1990; 69:1767–1778.

34. Ohgami M, Doerschuk CM, Gie RP, English D, Hogg JC. Monocyte kinetics in rabbits. J Appl Physiol 1991; 70:152–257.
35. Ohgami M, Doerschuk CM, Gie RP, English D, Hogg JC. Late effects of endotoxin on the accumulation and function of monocytes in rabbit lungs. Am Rev Respir Dis 1992; 146:190–195.
36. Gold DW, Finley TN, Cline MJ. The pulmonary macrophage in acute leukemia. N Engl J Med 1974; 290:875–878.
37. Van Furth RM, Diesselhoff-den Dulk MC, Mattie H. Quantitative study of the production and kinetics of mononuclear phagocytes during an acute inflammatory reaction. J Exp Med 1973; 138:1314–1330.
38. Lien DC, Wagner WW Jr, Capen RL, et al. Physiological neutrophil sequestration in the lung: visual evidence for localization in capillaries. J Appl Physiol 1987; 62:1236–1243.
39. Lien DC, Worthen GS, Capen RL, et al. Neutrophil kinetics in the pulmonary microcirculation: effects of pressure and flow in the dependent lung. Am Rev Respir Dis 1990; 141:953–959.
40. Staub BC, Schultz EL, Albertine KH. Leucocytes and pulmonary microvascular injury. Ann NY Acad Sci 1982; 384:332–343.
41. Doerschuk CM, Allard MF, Martin BA, MacKenzie A, Autor AP, Hogg JC. The marginated pool of neutrophils in the lungs of rabbits. J Appl Physiol 1987; 63:1806–1815.
42. Weibel ER. Morphometry of the Human Lung. New York: Academic Press, 1963.
43. Schmid-Schonbein GW, Shih Y, Chien S. Morphometry of human leukocytes. Blood 1980; 56:866–875.
44. Hogg JC, McLean T, Martin BA, Wiggs B. Erythrocyte transit and neutrophil concentration in the dog lung. J Appl Physiol 1988; 65:1217–1225.
45. Thommasen HV, Martin BA, Wiggs BR, Quiroga M, Baile EM, Hogg JC. Effect of pulmonary blood flow on leukocyte uptake and release by dog lung. J Appl Physiol 1984; 56:966–974.
46. Foster NK, Martyn JB, Rangno RE, Hogg JC, Pardy RL. Leukocytosis of exercise: role of cardiac output and catecholamines. J Appl Physiol 1986; 61: 2218–2223.
47. Muir AL, Cruz M, Martin BA, Thommasen HV, Belzberg A, Hogg JC. Leukocyte kinetics in the human lung: role of exercise and catecholamine. J Appl Physiol 1984; 57:711–719.
48. Erzurum SC, Downey GP, Doherty DE, Schwab B, Elson EL, Worthen GS. Mechanisms of lipopolysaccharide-induced neutrophil retention: relative contributions of adhesive and cellular mechanical properties. J Immunol 1992; 149:154–162.
49. Worthen GS, Schwab B, Elson EL, Downey GP. Mechanics of stimulated neutrophils: cell stiffening induces retention in capillaries. Science 1989; 245:183–185.
50. Doerschuk CM, Allard MF, Hogg JC. Neutrophil kinetics in rabbits during infusion of zymosan-activated plasma. J Appl Physiol 1989; 67:88–95.

51. Lundberg C, Wright SD. Relation of the CD11/CD18 family of leukocyte antigens to the transient neutropenia caused by chemoattractants. Blood 1990; 76: 1240–1245.
52. Doerschuk CM. The role of CD18-mediated adhesion in neutrophil sequestration induced by infusion of activated plasma in rabbits. Am J Respir Cell Mol Biol 1992; 7:140–148.
53. Harlan JM, Winn RK, Vedder NB, Doerschuk CM, Rice CL. In vivo models of leukocyte adherence to endothelium. In: Harlan JM, Liu DY, eds. Adhesion: Its Role in Inflammatory Disease. New York: W.H. Freeman, 1992:117–150.
54. Albelda SM, Buck CA. Integrins and other cell adhesion molecules. FASEB J 1990; 4:2868–2880.
55. Larson RS, Springer TA. Structure and function of leukocyte integrings. Immunol Rev 1990; 114:181–217.
56. Arnaout MA. Structure and function of the leukocyte adhesion molecules CD11/CD18. Blood 1990; 75:1037–1050.
57. Lasky LA. Selectins: interpreters of cell-specific carbohydrate information during inflammation. Science 1992; 258:964–969.
58. Lobb RR. Integrin-immunoglobulin superfamily interactions in endothelial-leukocyte adhesion. In: Harlan JM, Liu DY, eds. Adhesion: Its Role in Inflammatory Disease. New York: W.H. Freeman, 1992:1–18.
59. Vedder NB, Harlan JM. Increased surface expression of CD11b/CD18 (Mac-1) is not required for stimulated neutrophil adherence to endothelium. J Clin Invest 1988; 81:676–682.
60. Buyon JP, Abramson SB, Philips MR, et al. Dissociation between increased surface expression of gp165/95 and homotypic neutrophil aggregation. J Immunol 1988; 140:3156–3160.
61. Petty HR, Francis JW, Todd RF III, Petriquin PR, Boxer LA. Neutrophil C3bi receptors: formation of membrane clusters during cell triggering requires intracellular granules. J Cell Physiol 1987; 133:235–242.
62. Todd RF III, Arnaout MA, Rosin RE, Crowley CA, Peters WA, Babior BM. Subcellular localization of the large subunit of Mo1 (Mo1 alpha; formerly GP110), a surface glycoprotein associated with neutrophil adhesion. J Clin Invest 1984; 74:1280–1290.
63. Lo SK, Detmers PA, Levin SM, Wright SD. Transient adhesion of neutrophils to endothelium. J Exp Med 1989; 169:1779–1793.
64. Gallatin WM, Weissman IL, Butcher EC. A cell-surface molecule involved in organ-specific homing of lymphocytes. Nature 1983; 304:30–34.
65. Kishimoto TK, Jutila MA, Berg EL, Butcher EC. Neutrophil Mac-1 and MEL-14 adhesion protein are inversely regulated by chemotactic factors. Science 1989; 245:1238–1241.
66. Phillips ML, Nudelman E, Gaeta FCA, et al. ELAM-1 mediates cell adhesion by recognition of a carbohydrate ligand, Sialyl-Le^X. Science 1990; 250:1130–1132.
67. Walz G, Aruffo A, Kolanus W, Bevilacqua M, Seed B. Recognition by ELAM-1 of the Sialyl-Le^X determinant on myeloid and tumor cells. Science 1990; 250: 1231–1235.

68. Polley MJ, Phillips JL, Wayner E, et al. CD62 and endothelial cell-leukocyte adhesion molecule-1 (ELAM-1) recognize the same carbohydrate ligand, sialyl Lewis x. Proc Natl Acad Sci USA 1991; 88:6224–6228.
69. Etzioni A, Frydman M, Pollack S, et al. Brief report: recurrent severe infections caused by a novel leukocyte adhesion deficiency. N Engl J Med 1992; 327: 1789–1792.
70. Carlos TM, Harlan JM. Membrane proteins involved in phagocyte adherence to endothelium. Immunol Rev 1990; 114:5–28.
71. Ohmori K, Yoneda T, Ishihara G, et al. Sialyl SSEA-1 antigen as a carbohydrate marker of human natural killer cells and immature lymphoid cells. Blood 1989; 74:255–261.
72. Carlos T, Kovach N, Schwartz B, et al. Human monocytes bind to two cytokine-induced adhesive ligands on cultured human endothelial cells, ELAM-1 and VCAM-1. Blood 1991; 77:2266–2271.
73. Weller PF, Rand TH, Goelz SE, Chi-Rosso G, Lobb R. Human eosinophil adhesion to vascular endothelium mediated by binding to vascular cell adhesion molecule 1 and endothelial leukocyte adhesion molecule 1. Proc Natl Acad Sci USA 1991; 88:7430–7433.
74. Dobrina A, Menegazzi R, Carlos TM, et al. Mechanisms of eosinophil adherence to cultured vascular endothelial cells: eosinophils bind to the cytokine-induced endothelial ligand vascular cell adhesion molecule-1 via the very late activation antigen-4 integrin receptor. J Clin Invest 1991; 88:20–26.
75. Bochner BS, Luscinskas FW, Gimbrone MA Jr, et al. Adhesion of human basophils, eosinophils, and neutrophils to interleukin1-activated human vascular endothelial cells: contributions of endothelial cell adhesion molecules. J Exp Med 1991; 173:1553–1556.
76. Schwartz BR, Wayner EA, Carlos TM, Ochs HD, Harlan JM. Identification of surface proteins mediating adherence of CD11/CD18-deficient lymphoblastoid cells to cultured human endothelium. J Clin Invest 1990; 85:2019–2022.
77. Elices MJ, Osborn L, Takada Y, et al. VCAM-1 on activated endothelium interacts with the leukocyte integrin VLA-4 at a site distinct from the VLA-4/fibronectin binding site. Cell 1990; 60:577–584.
78. Osborn L, Hession C, Tizard R, et al. Direct expression cloning of vascular cell adhesion molecule 1, a cytokine-induced endothelial protein that binds lymphocytes. Cell 1989; 59:1203–1211.
79. Rice GE, Munro JM, Bevilacqua MP. Inducible cell adhesion molecule 110 (INCAM-110) is an endothelial receptor for lymphocytes: a CD11/CD18-independent adhesion mechanism. J Exp Med 1990; 171:1369–1374.
80. Anderson DC, Schmalsteig FC, Finegold MJ, et al. The severe and moderate phenotypes of heritable Mac-1, LFA-1 deficiency: their quantitative definition and relation to leukocyte dysfunction and clinical features. J Infect Dis 1985; 152:668–689.
81. Winn RK, Harlan JM. CD18-independent neutrophil and mononuclear leukocyte emigration into the peritoneum of rabbits. J Clin Invest 1993; 92:1168–1173.

82. Picker LJ, Warnock RA, Burns AR, Doerschuk CM, Berg EL, Butcher EC. The neutrophil selectin LECAM-1 presents carbohydrate ligands to the vascular selectins ELAM-1 and GMP-140. Cell 1991; 66:921–933.
83. Kuijpers TW, Hoogerwerf M, van der Laan LJW, et al. CD66 nonspecific cross-reacting antigens are involved in neutrophil adherence to cytokine-activated endothelial cells. J Cell Biol 1992; 118:457–466.
84. Lasky LA, Singer MS, Dowbenko D, et al. An endothelial ligand for L-selectin is a novel mucin-like molecule. Cell 1992; 69:927–938.
85. Dustin ML, Rothlein R, Bhan AK, Dinarello CA, Springer TA. Induction by IL-1 and interferon-gamma: tissue distribution, biochemistry, and function of a natural adherence molecule (ICAM-1). J Immunol 1986; 137:245–254.
86. Pober JS, Gimbrone MA Jr, Lapierre LA, et al. Overlapping patterns of activation of human endothelial cells by interleukin1, tumor necrosis factor, and immune interferon. J Immunol 1986; 137:1893–1896.
87. Pober JS, Bevilacqua MP, Mendrick DL, Lapierre LA, Fiers W, Gimbrone MA Jr. Two distinct monokines, interleukin 1 and tumor necrosis factor, each independently induce biosynthesis and transient expression of the same antigen on the surface of cultured human vascular endothelial cells. J Immunol 1986; 136:1680–1687.
88. Pober JS, Lapierre LA, Stolpen AH, et al. Activation of cultured human endothelial cells by recombinant lymphotoxin: comparison with tumor necrosis factor and interleukin 1 species. J Immunol 1987; 138:3319–3324.
89. Renkonen R. Regulation of intercellular adhesion molecule-1 expression on endothelial cells with correlation to lymphocyte-endothelial binding. Scand J Immunol 1989; 29:717–721.
90. Sugama Y, Tiruppathi C, Janakidevi K, Andersen TT, Fenton JW II, Malik AB. Thrombin-induced expression of endothelial P-selectin and intercellular adhesion molecule-1: a mechanism for stabilizing neutrophil adhesion. J Cell Biol 1992; 119:935–944.
91. Rice GE, Bevilacqua MP. An inducible endothelial cell surface glycoprotein mediates melanoma adhesion. Science 1989; 246:1303–1306.
92. Carlos TM, Schwartz BR, Kovach NL, et al. Vascular cell adhesion molecule-1 mediates lymphocyte adherence to cytokine-activated cultured human endothelial cells. Blood 1990; 76:965–970.
93. Masinovsky B, Urdal D, Gallatin WM. IL-4 acts synergistically with IL-1β to promote lymphocyte adhesion to microvascular endothelium by induction of vascular cell adhesion molecule-1. J Immunol 1990; 145:2886–2895.
94. Bevilacqua MP, Pober JS, Mendrick DL, Cotran RS, Gimbrone MA Jr. Identification of an inducible endothelial-leukocyte adhesion molecule. Proc Natl Acad Sci USA 1987; 84:9238–9242.
95. Pohlman TH, Stanness KA, Beatty PH, Ochs HD, Harlan JM. An endothelial cell surface factor(s) induced in vitro by lipopolysaccharide, interleukin 1, and tumor necrosis factor-alpha increases neutrophil adherence by a CD2w18-independent mechanism. J Immunol 1986; 136:4548–4553.
96. McEver RP, Beckstead JH, Moore KL, Marshall-Carlson L, Bainton DF. GMP-140, a platelet α-granule membrane protein, is also synthesized by vascular endo-

thelial cells and is localized in Weibel–Palade bodies. J Clin Invest 1989; 84: 92–99.
97. Geng JG, Bevilacqua MP, Moore KL, et al. Rapid neutrophil adhesion to activated endothelium mediated by GMP-140. Nature 1990; 343:757–760.
98. Toothil VJ, Van Mourik MA, Niewenhuis HK, Metzelaar MJ, Pearson JD. Characterization of the enhanced adhesion of neutrophil leukocytes to thrombin-stimulated endothelial cells. J Immunol 1990; 145:283–291.
99. Larsen E, Palabrica T, Sajer S, et al. PADGEM-dependent adhesion of platelets to monocytes and neutrophils is mediated by a lineage-specific carbohydrate, LNF III (CD15). Cell 1990; 63:467–474.
100. Hattori R, Hamilton KK, Fugate RD, McEver RP, Sims PJ. Stimulated secretion of endothelial von Willebrand factor is accompanied by rapid redistribution to the cell surface of the intracellular granular membrane protein GMP-140. J Biol Chem 1989; 264:7768–7771.
101. Patel KD, Zimmerman GA, Prescott SM, McEver RP, McIntyre T. Oxygen radicals induce human endothelial cells to express GMP-140. J Cell Biol 1991; 112:749–759.
102. Sander WE, Wilson RW, Ballantyne CM, Beaudet AL. Molecular cloning and analysis of in vivo expression of murine P-selectin. Blood 1992; 80:795–800.
103. Imai Y, Singer MS, Fennie C, Lasky LA, Rosen SD. Identification of a carbohydrate-based endothelial ligand for a lymphocyte homing receptor. J Cell Biol 1991; 113:1213–1221.
104. Ley K, Ceritto M, Arfors KE. Sulfated polysaccharides inhibit leukocyte rolling in rabbit mesentery venules. Am J Physiol 1991; 260:H1667–H1673.
105. Ley K, Lundgren E, Berger E, Arfors KE. Shear-dependent inhibition of granulocyte adhesion to cultured endothelium by dextran sulfate. Blood 1989; 73: 1324–1330.
106. Hernandez LA, Grisham MB, Twohig B, Arfors KE, Harlan JM, Granger DN. Role of neutrophils in ischemia-reperfusion-induced microvascular injury. Am J Physiol 1987; 238:H699–H703.
107. Zimmerman BJ, Granger DN. Reperfusion-induced leukocyte infiltration: role of elastase. Am J Physiol 1990; 259:H390–H394.
108. Atherton A, Born GVR. Quantitative investigations of the adhesiveness of circulating polymorphonuclear leukocytes to blood vessels. J Physiol (London) 1972; 222:447–474.
109. Arfors KE, Lundberg C, Lindbom L, Lundberg K, Beatty PG, Harlan JM. A monoclonal antibody to the membrane glycoprotein complex CD18 inhibits polymorphonuclear leukocyte accumulation and plasma leaking in vivo. Blood 1987; 69:338–340.
110. Perry MA, Granger DN. Role of CD11/CD18 in shear rate-dependent leukocyte–endothelial interactions in cat measenteric venules. J Clin Invest 1991; 87:1789–1804.
111. Lawrence MD, Smith CW, Eskin SG, McIntire LV. Effect of venous shear stress on CD18-mediated neutrophil adhesion to cultured endothelium. Blood 1990; 75:227–237.

112. Lawrence MB, Springer TA. Leukocytes roll on a selectin at physiologic flow rates: distinction from and prerequisite for adhesion through integrins. Cell 1991; 65:859–873.
113. Smith CW, Kishimoto TK, Abbassi O, et al. Chemotactic factors regulate lectin adhesion molecule-1 (LECAM-1)-dependent neutrophil adhesion to cytokine-stimulated endothelial cells in vitro. J Clin Invest 1991; 87:609–618.
114. Ley K, Gaehtgens P, Fennie C, Singer MS, Lasky LA, Rosen SD. Lectin-like cell adhesion molecule 1 mediates leukocyte rolling in mesenteric venules in vivo. Blood 1991; 77:2553–2555.
115. von Andrian UH, Chambers JD, McEvoy LM, Bargatze RF, Arfors KE, Butcher EC. Two-step model of leukocyte–endothelial cell interaction in inflammation: distinct roles for LECAM-1 and the leukocyte beta 2 integrins in vivo. Proc Natl Acad Sci USA 1991; 88:7538–7542.
116. Mulligan MS, Varani J, Dame MK, et al. Role of endothelial–leukocyte adhesion molecule 1 (ELAM-1) in neutrophil-mediated lung injury in rats. J Clin Invest 1991; 88:1396–1406.
117. Gundel RH, Wegner CD, Torcellini CA, et al. Endothelial–leukocyte adhesion molecule-1 mediates antigen-induced acute airway inflammation and late phase obstruction in monkeys. J Clin Invest 1991; 88:1407–1411.
118. von Andrian UH, Hansell P, Chambers JD, et al. L-selectin function is required for β_2-integrin-mediated neutrophil adhesion at physiological shear rates in vivo. Am J Physiol 1991; 263:H1034–H1044.
119. Abbassi O, Lane CL, Krater SS, et al. Canine neutrophil margination mediated by lectin adhesion molecule-1 in vitro. J Immunol 1991; 147:2107–2115.
120. Smith CW, Kishimoto TK, Abbassi O, et al. Chemotactic factors regulate lectin adhesion molecule 1 (LECAM-1)-dependent neutrophil adhesion to cytokine-stimulated endothelial cells in vitro. J Clin Invest 1991; 87:609–618.
121. Bainton DF, Miller LJ, Kishimoto TK, Springer TA. Leukocyte adhesion receptors are stored in peroxidase-negative granules of human neutrophils. J Exp Med 1987; 166:1641–1653.
122. Singer SJ. Intercellular communication and cell-cell adhesion. Science 1992; 255:1671–1677.
123. Dustin ML, Carpen O, Springer TA. Regulation of locomotion and cell–cell contact area by the LFA-1 and ICAM-1 adhesion receptors. J Immunol 1992; 148:2654–2663.
124. Huber AR, Kunkel SL, Todd RF, Weiss SJ. Regulation of transendothelial neutrophil migration by endogenous interleukin-8. Science 1991; 254:99–102.
125. Bussolino F, Camussi G, Baglioni C. Synthesis and release of platelet-activating factor by human vascular endothelial cells treated with tumor necrosis factor or interleukin-1 alpha. J Biol Chem 1988; 263:11856–11861.
126. Koyama S, Rennard SI, Leikauf GD, Robbins RA. Bronchial epithelial cells release monocyte chemotactic activity in response to smoke and endotoxin. J Immunol 1991; 147:972–979.
127. Koyama S, Rennard SI, Leikauf GD, et al. Endotoxin stimulates bronchial epithelial cells to release chemotactic factors for neutrophils; a potential mech-

anism for neutrophil recruitment, cytotoxicity, and inhibition of proliferation in bronchial inflammation. J Immunol 1991; 147:4293–4301.
128. Christman JW, Christman BW, Shepard VL, Rinaldo JE. Regulation of alveolar macrophage production of chemoattractants by leukotriene B_4 and prostaglandin E_2. Am J Respir Cell Mol Biol 1991; 5:297–304.
129. Christman JW, Petras SF, Vacek PM, Davis GS. Rat alveolar macrophage production of chemoattractants for neutrophils: response to *Escherichia coli* endotoxin. Infect Immun 1989; 57:810–816.
130. Duke SS, Bolds JM, Loyd JE, Brigham KL. Endotoxin-induced neutrophilic alveolitis and macrophage chemotaxin production in sheep. Am J Med Sci 1988; 296:381–386.
131. Issekutz AC, Morzycki W, Sadowska J. Rabbit alveolar macrophages stimulated with endotoxin and lung fragments from endotoxemic rabbits produce a leukocyte infiltration-inducing factor that lacks IL-1, TNFα, chemotactic activity. Exp Lung Res 1991; 17:803–819.
132. Rinaldo JE, Henson JE, Dauber JH, Henson PM. Role of alveolar macrophages in endotoxin-induced neutrophilic alveolitis in rats. Tissue Cell 1985; 17:461–472.
133. Snella MC. Production of a neutrophil chemotactic factor by endotoxin stimulated alveolar macrophages in vitro. Br J Exp Pathol 1986; 67:801–807.
134. Surfaro S, Berman JS. The relation between cell migration and activation in inflammation: beyond adherence. Am J Respir Cell Mol Biol 1992; 7:248–250.
135. Smith CW. Transendothelial migration. In: Harlan JM, Liu DY, eds. Adhesion: Its Role in Inflammatory Disease. New York: W.H. Freeman, 1992:83–115.
136. Hakkert BC, Kuijpers TW, Leeuwenberg JFM, Roos D. Neutrophil and monocyte adherence to and migration across monolayers of cytokine-activated endothelial cells: the contribution of CD18, ELAM-1, and VLA-4. Blood 1991; 78:2721–2726.
137. Bohnsack JF. CD11/CD18-independent neutrophil adherence to laminin is mediated by the integrin VLA-6. Blood 1992; 6:1545–1552.
138. Mulligan MS, Polley MJ, Bayer RJ, Nunn JF, Paulson JC, Ward PA. Neutrophil-dependent acute lung injury: requirement for P-selectin (GMP-140). J Clin Invest 1992; 90:1600–1607.
139. Doerschuk CM, Winn RK, Coxson HO, Harlan JM. CD18-dependent and -independent mechanisms of neutrophil emigration in the pulmonary and systemic microcirculation of rabbits. J Immunol 1990; 144:2327–2333.
140. Mulligan MS, Warren JS, Smith CW, et al. Lung injury after deposition of IgA immune complexes: requirements for CD18 and L-arginine. J Immunol 1992; 148:3086–3092.
141. Hawkins HK, Heffelfinger SC, Anderson DC. Leukocyte adhesion deficiency: clinical and postmortem observations. Pediatr Pathol 1992; 12:119–130.
142. Mileski W, Harlan J, Rice C, Winn R. *Streptococcus pneumonia*-stimulated macrophages induce neutrophils to emigrate by a CD18-independent mechanism of adherence. Circ Shock 1990; 31:259–267.

143. Klempner MS, Malech HL. Phagocytes: normal and abnormal neutrophil host defenses. In: Gorbach SL, Bartlett JG, Blacklow NR, eds. Infectious Diseases. Philadelphia: W.B. Saunders, 1992:43–62.
144. Wright SD, Meyer BC. Phorbol esters cause sequential activation and deactivation of complement receptors on polymorphonuclear leukocytes. J Immunol 1986; 136:1759–1764.
145. Wright SD, Jong MTC. Adhesion-promoting receptors on human macrophages recognize *E. coli* by binding to lipopolysaccharide. J Exp Med 1986; 164:1876–1888.
146. Bohmer RH, Trinkle LS, Staneck JL. Dose effects of LPS on neutrophils in a whole blood flow cytometric assay of phagocytosis and oxidative burst. Cytometry 1992; 13:525–531.
147. Trinkle LS, Wellhausen SR, McLeish KR. A simultaneous flow cytometric measurement of neutrophil phagocytosis and oxidative burst in whole blood. Diagn Clin Immunol 1987; 5:62–68.
148. Griswold J, Maier RV. Neutrophil phagocytosis during endotoxin-induced lung injury. J Surg Res 1988; 44:417–422.
149. White-Owen C, Alexander JW, Babcock GF. Reduced expression of neutrophil CD11b and CD16 after severe traumatic injury. J Surg Res 1992; 52:22–26.
150. Babcock GF, Alexander JW, Warden GD. Flow cytometric analysis of neutrophil subsets in thermally injured patients developing infection. Clin Immunol Immunopathol 1990; 54:117–125.
151. Onofrio JM, Toews GB, Lipscomb MF, Pierce AK. Granulocyte-alveolar macrophage interaction in the pulmonary clearance of *Staphylococcus aureus*. Am Rev Respir Dis 1983; 127:335–341.
152. Toews GB, Gross GN, Pierce AK. The relationship of inoculum size to lung bacterial clearance and phagocytic response in mice. Am Rev Respir Dis 1979; 120:559–566.
153. Arnaout M. Leukocyte adhesion molecule deficiency: its structural basis, pathophysiology and implication for modulating the inflammatory response. Immunol Rev 1990; 114:145–180.
154. Crowley CA, Curnutte JT, Rosen RE, et al. An inherited abnormality of neutrophil adhesion: its genetic transmission and its association with a missing protein. N Engl J Med 1980; 302:1163–1168.
155. Bowen TJ, Ochs HD, Altman LC, et al. Severe recurrent bacterial infections associated with defective adherence and chemotaxis in two patients with neutrophils deficient in a cell-associated glycoprotein. J Pediatr 1982; 101:932–940.
156. Waldrop TC, Anderson DC, Hallmon WW, Schmalstieg FC, Jacobs RL. Periodontal manifestations of the heritable Mac-1, LAF-1 deficiency syndrome: clinical, histopathologic, and molecular characteristics. J Periodontol 1987; 58:400–416.
157. Todd RF III, Freyer DR. The CD11/CD18 leukocyte glycoprotein deficiency. Hematol Oncol Clin North Am 1988; 2:13–31.
158. Beatty PG, Ochs HD, Rosen H, et al. Absence of monoclonal-antibody defined protein complex in a boy with abnormal leucocyte function. Lancet 1984; 1:535–537.

159. Harlan JM, Kellen PK, Senecal FM, et al. The role of neutrophil membrane glycoprotein GP-150 in neutrophil adherence to endothelium in vitro. Blood 1985; 66:167–178.
160. Sharar SR, Winn RK, Murry CE, Harlan JM, Rice CL. A CD18 monoclonal antibody increases the incidence and severity of subcutaneous abscess formation after high-dose *S. aureus* injection in rabbits. Surgery 1991; 110:213–220.
161. Talbott G, Sharar S, Paulson J, Harlan J, Winn R. CD18 but not P-selectin blockade increases incidence and severity of subcutaneous *E. coli* abscess formation. Crit Care Med 1993; 21:S228.
162. Sharar S, Sasaki S, Flaherty L, Paulson J, Harlan J, Winn R. P-selectin blockade does not impair leukocyte host defense against bacterial peritonitis or soft tissue infection in rabbits. J Immunol 1993; 151:4982–4988.
163. Nelson S, Chidiac C, Bagby G, Summer WR. Endotoxin-induced suppression of lung host defenses. J Med 1990; 21:85–103.
164. Simons RK, Maier RV, Lennard ES. Neutrophil function in a rat model of endotoxin-induced lung injury. Arch Surg 1987; 122:197–203.
165. Sibille Y, Reynolds HY. Macrophages and polymorphonuclear neutrophils in lung defense and injury. Am Rev Respir Dis 1990; 141:471–501.
166. Nathan CF, Murray HW, Weibe ME, Rubin BF. Identification of interferon-gamma as the lymphokine that activates human macrophage oxidative metabolism and antimicrobial activity. J Exp Med 1983; 158:670–689.
167. Russell S. Stimulate the phagocytes. Immunol Today 1986; 7:347–349.
168. Heifets L, Imai K, Goren MB. Expression of peroxidase-dependent iodination by macrophages ingesting neutrophil debris. J Reticuloendothel Soc 1980; 28:391–404.
169. Fels AOS, Cohn ZA. The alveolar macrophage. J Appl Physiol 1986; 60:353–369.
170. Greening AP, Lowrie DB. Extracellular release of hydrogen peroxide by human alveolar macrophages: the relationship to cigarette smoking and lower respiratory tract infections. Clin Sci 1983; 65:661–664.
171. Rehn ST, Gross GN, Pierce AK. Early bacterial clearance from murine lungs: species-dependent phagocyte responses. J Clin Invest 1980; 66:194–199.
172. Reynolds HY. Lung host defenses: a status report. Chest 1979; 75:S39–S42.
173. Pennington JE, Rossing TH, Boerth LW, Lee TH. Isolation and partial characterization of a human alveolar macrophage-derived neutrophil-activating factor. J Clin Invest 1985; 75:1230–1237.
174. MacArthur CK, Miller EJ, Cohen AB. A peptide secreted by human alveolar macrophages releases neutrophil granule contents. J Immunol 1987; 139:3456–3462.
175. Rappolee DA, Mark D, Banda MJ, Werb Z. Wound macrophages express TGF-alpha and other growth factors in vivo; analysis by mRNA phenotyping. Science 1988; 241:708–712.
176. Davis WB, Rennard SI, Bitterman PB, Crystal RG. Pulmonary oxygen toxicity. Early reversible changes in human alveolar structures induced by hypoxia. N Engl J Med 1983; 309:878–883.

177. Sibille Y, Merrill WW, Naegel GP, Care SB, Cooper JAD, Reynolds HY. Human alveolar cells release in vitro a factor(s) which inhibits neutrophil chemotaxis. Am J Respir Cell Mol Biol 1989; 1:407–416.
178. Brigham KL, Meyrick B. Endotoxin and lung injury. Am Rev Respir Dis 1986; 133:913–927.
179. Brigham KL, Meyrick B. Interactions of granulocytes with the lungs. Circ Res 1984; 54:623–635.
180. Snapper J, Hutchison A, Oglestree M, Brigham K. Effects of cyclooxygenase inhibitors on the alterations in lung mechanics caused by endotoxemia in the unanesthestized sheep. J Clin Invest 1983; 72:63–76.
181. Welsh CH, Lien DC, Worthen GS, Henson PM, Weil JV. Endotoxin-pretreated neutrophils increase pulmonary vascular permeability in dogs. J Appl Physiol 1989; 66:112–119.
182. Basadre JO, Singh H, Herndon DN, et al. Effect of antibody-mediated neutropenia on the cardiopulmonary response to endotoxemia. J Surg Res 1988; 45:266–275.
183. Price T, Beatty PG, Corpuz SR. In vivo inhibition of neutrophil function in the rabbit using monoclonal antibody to CD18. J Immunol 1987; 139:4174–4177.
184. Thomas JR, Harlan JM, Rice CL, Winn RK. Role of leukocyte CD11/CD18 complex in endotoxic and septic shock in rabbits. J Appl Physiol 1992; 73: 1510–1516.
185. Redl H, Dinges HP, Buurman WA, et al. Expression of endothelial leukocyte adhesion molecule-1 in septic but not traumatic/hypovolemic shock in the baboon. Am J Pathol 1991; 139:461–466.
186. Suffredini AF, Fromm RE, Parker MM, et al. The cardiovascular response of normal humans to the administration of endotoxin. N Engl J Med 1989; 321: 280–287.
187. van Deventer SJH, Buller HR, te Cate JW, Aarden LA, Hack CE, Sturk A. Experimental endotoxemia in humans: analysis of cytokine release and coagulation, fibrinolytic, and complement pathways. Blood 1990; 76:2520–2526.
188. Michie HR, Manogue KR, Spriggs DR, et al. Detection of circulating tumor necrosis factor after endotoxin administration. N Engl J Med 1988; 318:1481–1486.
189. Fong Y, Moldawer LL, Marano M, et al. Endotoxemia elicits increased circulating beta 2-IFN/IL-6 in man. J Immunol 1989; 142:2321–2324.
190. Danner RL, Elin RJ, Hosseini JM, Wesley RA, Reilly JM, Parrillo J. Endotoxemia in human septic shock. Chest 1991; 99:169–175.
191. Parson PE, Worthen GS, Moore EE, Tate RM, Henson PM. The association of circulating endotoxin with the development of the adult respiratory distress syndrome. Am Rev Resir Dis 1989; 140:294–301.
192. Steinberg KP, Milberg JA, Martin TR, Maunder RJ, Cockrill BA, Hudson LD. Evolution of bronchoalveolar cell populations in the adult respiratory distress syndrome. Am J Respir Crit Care Med 1994 (in press).

193. Laufe MD, Simon RH, Flint A, Keller JB. Adult respiratory distress syndrome in neutropenic patients. Am J Med 1986; 80:1022–1026.
194. Ognibene FP, Martin SE, Parker MM, et al. Adult respiratory distress syndrome in patients with severe neutropenia. N Engl J Med 1986; 315:547–551.
195. Maunder RJ, Hackman RC, Riff E, Albert RK, Springmeyer SC. Occurrence of the adult respiratory distress syndrome in neutropenic patients. Am Rev Respir Dis 1986; 133:313–316.

12

LPS and Cytokines in Lung Injury

PETER M. HENSON,
DAVID W. H. RICHES, and
G. SCOTT WORTHEN

National Jewish Center for Immunology
and Respiratory Medicine
and University of Colorado Health
Sciences Center
Denver, Colorado

DENNIS E. DOHERTY

National Jewish Center for Immunology
and Respiratory Medicine
Denver Veterans Affairs Medical Center
and University of Colorado Health
Sciences Center
Denver, Colorado

POLLY E. PARSONS

University of Colorado
Health Sciences Center
Denver, Colorado

I. Introduction

The current paradigm suggests that lipopolysaccharide (LPS) produces its many biological effects through the action of cytokines: that is, by initiating synthesis and secretion of this diverse group of proteins exhibiting potent intracellular signaling activities. In this scenario particular emphasis is given to molecules such as tumor necrosis factor (TNF-α) or interleukins 1 and 6 (IL-1 and IL-6) that are found in LPS-induced conditions and by themselves can mimic many of the effects of parenterally administered LPS. However, the groups of biologically active molecules that are released from cells in vitro and in vivo by LPS extend much more widely than those usually designated as cytokines and include a wide range of lipids, bioactive amines, mokines, and other peptides—mediators in general. Interactions between these mediators (spatially and temporally) lead to an extensive mediator network with significant consequences for migratory and sessile cells in the lung. The concept of mediator networks is now some 30 years old but is receiving new-found attention as the mediators themselves are better defined and the means to detect them and specifically block their action in vivo have become better developed. Unlike many other external inflammatory stimuli, LPS may play a unique role in such networks because it not only initiates the production and release of the mediators, it seems to participate actively in the network itself. It is this interaction between LPS and the mediator cascades that it initiates that is the subject of this chapter.

Mediators are here defined as intercellular communication molecules that have a sphere of action intermediate between neurotransmitters and hormones, the former being limited to immediately adjacent cell surface structures and the latter exhibiting their effects throughout the body. The key distinguishing difference between these groups therefore would be the persistence of the biological activity of each molecule, whether it decays spontaneously or is inhibited, inactivated, destroyed, or excreted, or if the responding cells develop a rapid refractoriness to its action. Separation of these signal molecules into such groups is quite artificial but does provide a useful conceptual framework in which to examine processes involved in local inflammatory reactions, and in this case, responses to LPS. Cytokines may be seen to fit generally into the mediator group, being distinguished primarily by being proteins and because they are synthesized and secreted in response to the initiating stimulus (i.e., are not generally stored as preformed entities ready for rapid deployment in the reaction). A key here is that even if they are relatively short acting, there is often a significant delay before they are produced. It also seems that these designations may break down somewhat when LPS gains access to the circulation since this can result in significant levels of circulating cytokines that indeed may exhibit actions at a distance, especially perhaps in the lung.

LPS has been shown, both in vitro and in vivo, to initiate production and/or release of a host of mediators from a wide variety of cells (1,14). Which of these, or combinations of these, may be involved in LPS interactions with the lung remains highly speculative. The multiple cellular sources (e.g., Ref. 1) and targets for the cytokines and mediators, the complex nature of the mediator network, and the prolonged time course of many pulmonary reactions all serve as obstacles to a detailed analysis of these processes. For example, the demonstration that endotoxemia leads to increased circulating levels of TNF-α (2) or PAF (3) (and see Chapter 9) and that antibody to the former or antagonists of the latter each are able to inhibit LPS effects on the lung (3–5) do not indicate where in the cascade (network) each mediator may act, whether in parallel or in series, or indeed, how the LPS initiates production of these mediators in the first place. A more detailed analysis, for example, to dissociate PAF and TNF-α effects (6) is required. The increasing availability (e.g., Ref. 1) of specific antagonists for given mediators and cytokines as well as the intriguing possibility of genetic manipulations of cytokine availability all suggest that these complex systems may become tractable to analysis in the near future.

As well as being capable of initiating the mediator network and a wide variety of pathophysiologic and inflammatory events, LPS is also a potentially active participant in the network itself. Thus interaction with LPS significantly alters the responsiveness of inflammatory cells to the action of a wide variety of the mediators and cytokines that are, in fact, products of the LPS reactions. Early demonstrations of this phenomenon from our group focused on production of potentially injurious toxic oxygen metabolites from both mononuclear leukocytes and neutrophils and was termed "priming" (7,8). Now extended to a number of different cell responses, shown to occur with stimuli other than LPS and perhaps mediated by many quite disparate intracellular transduction pathways, the processes of priming are considered critical for optimal inflammatory cell responses and, in addition, for both normal and abnormal consequences of the inflammation itself. Priming by LPS (or TNF) for enhanced in vivo effects on the lung has also been widely reported (e.g., Refs. 9 to 13). Most of this chapter is devoted to a discussion of the ways in which LPS interacts with the mediators and cytokines to influence inflammatory responses and their impact on the lung. However, because the LPS is presumably involved both in the initiation of the mediator cascade as well as within it, it seemed appropriate to consider briefly the target of this initial action (i.e., the original source of the mediators and cytokines) since this might also contribute to the most likely sites and targets for the secondary interactions of the LPS with the mediators themselves.

In the lung, a critical question for all issues relating to exogenous factors such as LPS is the route by which the material is encountered. Much emphasis has been placed on intravascular LPS in the context of speticemia and ARDS, whereas LPS impingement from the airways has received specific attention largely in the context of byssinosis. In either case, the cells that specifically interact with the LPS to initiate the cascade of response are not known, although a number of candidates are available. At this point it should be emphasized that LPS could conceivably initiate responses in the blood by virtue of its ability to activate the complement or possibly the coagulation system. However, it is generally held that the concentrations required for these reactions are much higher than those thought capable of initiating pathophysiologic reactions [(14) and Chapter 10] and that the initiating effect is likely to be on cells within the blood or appropriate tissues. Interaction of LPS with plasma factors and specific cell surface receptors (e.g., LPS-binding protein and CD14) are discussed in Chapter 4. Suffice it to say that there are many potential candidate cells in the lung to provide the initial mediators that can set the entire reaction in train, even without including the known responsiveness of neutrophils and monocytes to this material.

A number of different and overlapping possibilities for the initially responsive cells should be raised.

1. The most obvious of these is that LPS stimulates the circulating leukocytes. The fact that cytokine production takes time (hours) would imply that the circulating cells gradually started secreting the cytokines into the plasma. On the other hand, as discussed below, LPS also induces leukocyte retention in the pulmonary microvasculature and while localized at that site, the cells might contribute their cytokines and mediators to the local pulmonary environment. A potential problem for this scenario is that we do not really know how long an individual cell remains in the microvasculature under these conditions.

2. Increasing evidence suggests the presence of functional CD14 on endothelial cells and/or stimulation of these cells via LPS interaction with soluble CD14 (15,16). In some species endothelial cells may be especially susceptible to toxic effects resulting from LPS interaction (17–19), although the role of CD14 analogs in this process seems at this point to be unknown. In humans, however, LPS–LBP interacts with umbilical cord endothelial cells to up-regulate surface molecules involved in leukocyte adhesion, but more important in the context of this chapter, to initiate synthesis and secretion of IL-8 (16). Whether the same responses would be seen for microvascular endothelial cells in the lung remains to be determined, but if any endothelium in the body reacts in this way with a sufficiently vigorous response, the cytokines generated would be expected to impinge on the lung with their expected biological activities.

3. Some interest has been generated recently in the possibility that the so-called pulmonary intravascular macrophages (PIM) might constitute an important population of cells that may contribute to LPS responses (20, 21). As a sessile cell with great mediator-producing potential, it represents an intriguing candidate. However, the variability of PIM presence from species to species suggests that alternative sources of the mediators should also be sought. An alternative suggestion is that intravascular stimuli could induce the accumulation of monocytes in the pulmonary microvasculature, which, in that location, might mature into macrophages with properties similar to those of the so-called PIM of pigs, sheep, and so on (22).

4. Intravascular LPS might gain access to the alveolar interstitium or airspace and there interact with the pulmonary macrophages to initiate the production and release of mediators and cytokines that would then diffuse back into the circulation to initiate the types of pulmonary inflammatory reactions that would be expected from alveolar sources of phlogisitic agents. Whether LPS can cross the alveolar wall—in the normal or inflamed lung—is not clear and would be worthy of investigation. In any event, mediators produced by the LPS effects on cells in the blood or vascular lining could gain access to the macrophages and thus secondarily bring their great potential for cytokine and mediator production into the picture.

5. Finally, inhalation or local generation of LPS in the airways would be expected to impinge primarily on the pulmonary macrophages and epithelial cells. The latter have not been reported to be highly responsive to the direct action of LPS, but this may be due to the lack of intensive study. If the requirement for plasma proteins for LPS–receptor interactions is general, low concentrations of LBP in the normal lung may contribute to limited activity of LPS until the inflammatory response is under way and significant plasma exudation has occurred.

Chemokines

Although inflammatory cell accumulation at inflammatory sites may be due to the response of the leukocyte to LPS or other inciting agent, the last 10 years have provided abundant evidence that the *local* host response to bacterial products, by leading to the elaboration of new chemoattractants for leukocytes, may amplify the inflammatory response, and lead potentially to the dysregulation that seems apparent in many inflammatory responses in human disease. The recently described chemokine family of chemoattractants are ideally suited, in terms of their structure, known function, and cellular sources, to regulate the inflammatory response by acting as specific chemoattractants for particular types of leukocytes. Because different members of this family show significant cell specificity, chemokines provide for the first time a potential explanation for conditions where one migratory cell type predominates over another in an inflammatory lesion. The timing and site of elaboration of these chemokines then may regulate, at least in part, the sequential appearance of neutrophils followed by monocytes and lymphocytes. The specificity of their responses may also allow the generation of inflammatory responses in which a relatively few types of inflammatory cell are localized, as might be expected for instance in the formation of a granuloma or in the atherosclerotic plaque, where the neutrophil is poorly represented.

Chemokines can be divided into two groups, α and β, by the arrangement of the first two (of four) conserved cysteines. Members of the α-chemokine group all possess a single amino acid interposed between the first two cysteines (see Table 1). Chemokines of the α family in general induce neutrophil chemotaxis and activation through typical seven transmembrane spanning G-protein-linked receptors. IL-8 is the best characterized member of this group. Two IL-8 receptors have been identified that are 77% identical at the amino acid level (23,24). The IL-8 receptor B also appears to recognize the related α-chemokines MGSA or GRO and neutrophil-activating peptide 2 (24,25).

In β-chemokines (see Table 2) the first two conserved cysteines are adjacent. This family includes macrophage inflammatory protein-1α (MIP-1α,

Table 1 α-Chemokines: Cysteines C-X-C, Chromosome 4

Nomenclature	Source	Stimulus	Target	Effect
IL-8	Macrophages	LPS	Neutrophil	Adherence, chemotaxis, secretion
	Endothelial cells	IL-1		
	Neutrophils	TNF	Monocytes	Adherence
	T-cells		T-cells	Chemotaxis
	Fibroblasts		Keratinocyte	Proliferation
	Keratinocytes			
MGSA/Gro	Similar to IL-8	Similar to IL-8	Neutrophil	Chemotaxis, secretion
			Melanocytes	Proliferation
NAP-2	Platelets	Activation	Neutrophil	Chemotaxis, secretion
MIP-2	Macrophages	LPS	Neutrophil	Chemotaxis, secretion

MIP-1β), RANTES, and monocyte chemoattractant protein-1 (MCP-1, MCP-2, MCP-3) (26). Members of this group serve as chemoattractants for neutrophils, eosinophils, monocytes, and lymphocytes with different selectivity (27–32). An example of the remarkable selectivity possible even by closely related members of the β-chemokine family is the recent description of the ability of MIP-1α and MIP-1β to serve as a chemoattractant for T-cells. Taubin and his co-workers have shown that MIP-1α induced chemotaxis of predominantly CD8-positive T-cells, while MIP-1β induced chemotaxis of predominantly CD4-positive T-cells and also induced the corresponding

Table 2 β-Chemokines: Cysteines C-C, Chromosome 17

Nomenclature	Source	Stimulus	Target	Effect
MCP-1	Macrophages	LPS	Monocyte	Chemotaxis, activation
	Endothelial cells	IL-1		
	Neutrophils	TNF		
	Epithelial cells	IL-4		
	Fibroblasts	?IL-6		
	Keratinocytes			
RANTES	T-cells	PHA	Monocyte	Chemotaxis
	Platelets	Anti-CD3	Neutrophil	Chemotaxis
		Platelet activators	T-cells	Chemotaxis
MIP-1α	Macrophages	LPS	Neutrophil	Chemotaxis, secretion
	B-cells, T-cells	IL-1	Monocyte	Chemotaxis, activation
	Fibroblasts	TNF	$CD4^+$ T-cells	Chemotaxis
MIP-1β	Macrophages	LPS?	Monocyte	Inhibits MIP-1α
	B-cells, T-cells		$CD8^+$ T-cells	Chemotaxis

T-cell to become adherent to endothelial cells. Thus the relative expression of MIP-1α versus MIP-1β might account for the accumulation of specific T-cell subsets at an inflammatory response. To date, however, the potential regulatory role of these molecules in human disease has not been elucidated.

In contrast, the α-chemokine family has been studied more extensively and in particular, IL-8 has been found to occur in a variety of pathological circumstances. Endothelial expression of IL-8 appears to be responsible for a considerable amount of the neutrophil migration across endothelial monolayers that occurs in response to endothelial activation (33,34). LPS, acting through CD14 (in soluble form or on the endothelial surface), is a potent stimulus for this expression (15,16). In human material alone IL-8 has been shown to be present in neutrophil-rich inflammatory exudate such as that seen in the lung in ARDS (35) in idiopathic pulmonary fibrosis, in empyema fluid (36), and in chronic bronchitis (37). In some of these settings, IL-8 has been shown to be the predominant chemoattractant (37). Although not as well explored, it seems clear that the MGSA/GRO members of this family and NAP-2 may well play important roles in other circumstances as the tools to dissect their presence become more readily available. It is anticipated that these important inducers of neutrophil accumulation will be explored thoroughly, and agents to interfere with their action will be identified as novel therapeutic tools.

II. LPS–Cytokine Interactions in Neutrophil Retention, Emigration, and Pulmonary Injury

A. Mechanisms of LPS-Induced Neutrophil Retention

One of the hallmarks of systemic endotoxemia accompanying gram-negative infection, at least in the initial stages, is the disappearance of neutrophils from the bloodstream. This event has been interpreted to represent neutrophil sequestration in the lung (as well as perhaps other organs) by a process similar to that by which chemoattractants infused intravascularly induce neutrophil sequestration in the lung and lead to systemic neutropenia. However, lipopolysaccharide (LPS) is not a chemoattractant, and the mechanisms by which it interacts with cells to induce these events might indicate fundamental differences from chemoattractant-induced neutrophil sequestration.

To discern some of the fundamental mechanisms by which LPS caused leukocyte interaction with the lung (whether directly or via the generation of cytokines/mediators, we first determined the vascular site wherein neutrophils were retained (38–41). In system microcirculatory beds, neutrophil sequestration (i.e., margination) appears to occur at the postcapillary venule. In contrast, in the lung, neutrophils are retained almost exclusively within capillaries (perhaps due in part to geometric constraints and capillary size, but also probably related in an important way to the local capillary pressure). This applies to normal conditions where they form the marginating pool of neutrophils, and also in response to both intravascular and airway-directed neutrophil stimuli. In one sense it is surprising that this does not happen in other systems since the average neutrophil diameter is approximately 8 μm (42), while that of the capillary averages 5.5 to 6 μm (43). However, consistent with the notion that the low pulmonary perfusion pressure accounts for this alteration in neutrophil localization, increase of pressure and/or flow markedly diminish the retention of neutrophils within capillaries and also induce demargination of these cells from the lungs with reentry into the circulation (39). This inverse relationship between pressure and retention, described previously by Martin and colleagues (44), suggested the possibility for some novel mechanisms that might underlay the initial neutrophil retention.

B. Alteration in Neutrophil Resistance to Deformation

The neutrophil is a viscoelastic body whose cytoskeletal network is rapidly assembled in response to neutrophil stimuli. We have previously shown that stimulation of the neutrophil with chemoattractant, such as FMLP, induces a marked and extraordinarily rapid induction of neutrophil cytoskeletal assembly and marked alteration of neutrophil properties such that the cell becomes considerably more resistant to deformation. Since the size differential between the neutrophil and capillary ensures that the neutrophil must deform in order to pass through the capillary lumen, a diminished ability to deform might allow rapid and initial retention. This model of neutrophil retention appears to be the case with chemoattractants, such as FMLP, as initial retention due to FMLP was attenuated by cytochalasin D, an agent that disrupts actin cytoskeleton but not (in the hands of other workers) by monoclonal antibodies directed against the CD11/CD18 adhesion complex (see Chapter 12). To determine whether this appeared to be the case when neutrophil retention was induced by LPS, we have also examined the ability of LPS to modify the neutrophil cytoskeletal. Exposure of human neutrophils to LPS concentrations between 1 and 100 ng/mL induces reorganization of the actin cytoskeleton and net actin assembly as long as the cells are provided with a source of LPS binding protein (see Chapter 4) in the form of serum or plasma. The cells become significantly stiffer, due predominantly to actin assembly, as cytochalasin D prevents the apparent stiffness. Reorganization of the microfilament network in the neutrophil precedes the development of adhesive events, and blockade of the adhesive effects of LPS exposure on neutrophils by monoclonal antibody 60.3 directed against CD18 (although it completely blocks neutrophil adherence to surfaces) has little effect on neutrophil retention within capillary-sized pores. Furthermore, the alteration in mechanical properties, actin reorganization, and retention in capillary-sized pores were abolished by incubation with antibodies directed against the LPS-binding molecule, CD14. Furthermore, neutrophils express and are capable of up-regulating CD14 (see below).

Even if retention in capillaries based on viscoelastic properties of the cell is the initiating event (still a controversial point) the prolonged residence of neutrophils within the lung probably requires the development of adhesion-based mechanisms. We have also studied the adherence of neutrophils to endothelial cells in an attempt to discern the precise mechanisms by which LPS induces neutrophils to become adhesive (45). The first observation was that in the absence of serum or plasma, LPS is an extremely weak stimulator of neutrophil adherence. It requires approximately 10 to 100 μg/mL LPS to induce neutrophil adhesion, concentrations never seen in humans. In contrast, in the presence of serum or plasma, concentrations of LPS between 1 and 10 ng/mL induce the neutrophil to become adhesive, concentrations within an order of magnitude of those seen during sepsis. The search for the essential plasma component that allowed LPS to act in a physiologically relevant concentration was furthered by the isolation, purification, and cloning of LPS-binding protein (see Chapter 4). The addition of purified LPS-binding protein in buffer allowed LPS to induce neutrophil adhesion at relevant concentrations of LPS-binding protein in serum. Furthermore, the depletion of LPS-binding protein from rabbit plasma makes it impossible for the plasma to enhance LPS-induced neutrophil adhesion, suggesting that LBP was a necessary component of the LPS interaction. When LPS is incubated in plasma, its apparent molecular weight changes from approximately 1 million to approximately 100,000, such that many more particles of LPS are now available for interaction. Since serum is capable of extracting LPS off the surface of gram-negative bacteria (46), it seems likely that interaction between LPS and plasma components is an essential part of the host response to these extraordinarily powerful molecules. Recent observations from Tobias's group have suggested that in whole blood (i.e., with the formed elements as well) LPS can become even more active on cells (47). The mechanism for this further enhancement of LPS effects (activity now in the picogram range) is unknown but could include rapid generation of amplifying cytokines or mediators and/or even more effective presentation of the LPS to the responding cells.

As indicated above, neutrophils express and appear to be able to up-regulate the glycerophosphatidylinositol-linked glycoprotein CD14 on their surface. Antibodies directed at CD14 completely abolished neutrophil adhesion in response to LPS even in the presence of an adequate source of LPS-binding protein. Furthermore, when CD14 was removed from the surface of neutrophils using a phosphatidylinositol-specific phospholipase C, there was a dramatic reduction in the ability of LPS to induce adhesion. These data, taken together, suggest that CD14 is at least one of the LPS receptors for a complex for LPS and LBP. However, CD14 lacks any cytoplasmic component; thus the actual transducing molecule must represent another cell surface constituent. The identity of this signaling molecule remains unknown. Finally, LPS induced the neutrophil to up-regulate CD11/CD18, and most of the adhesion to endothelial cells or to more generalizable serum-coated surfaces was abolished by monoclonal antibodies directed at CD18. Thus the model for LPS interactions with the relevant cells to guide leukocyte sequestration, on the one hand, or cytokine generation, on the other, is as depicted in Fig. 1. This figure indicates the multipotential of CD14. In a membrane-bound form on neutrophils, monocytes, and macrophages, it assists in inducing adhesive events mediated in part through up-regulation of CD18, and cytoskeletal assembly, which in turn increases resistance to deformation. It is significant that there is at this point no evidence for a completely independent intracellular signaling pathway for LPS beyond the initial "receptor" interaction and perhaps the early postreceptor events. Rather, the cells respond to LPS in ways very similar to their response to chemotactic lipids, peptides, or chemokines. In contrast, soluble CD14 (right-hand panel) interacts with unknown signaling molecules on the endothelium and epithelium to induce cytokine and chemokine production.

C. Emigration

Neutrophil migration into the airspace to form a neutrophil-rich exudate within the alveolus is a hallmark of the adult respiratory distress syndrome which may follow gram-negative sepsis. The mechanisms by which neutrophil migration occurs during these events remains somewhat obscure (see above). Even if LPS induces neutrophil retention by its direct effects on the neutrophil, LPS is not a chemoattractant and hence cannot be responsible for a movement of neutrophils toward the alveolitis. It is noteworthy that although neutrophil migration has been seen in models of systemic or vascular administration of LPS (48,49) it is considerably more prominent when LPS is administered through the airspace. This appears to be the case for cytokines as well, as the airspace administration of IL-1 and TNF appears to induce considerably more neutrophil accumulation than when administered through the vascular space. However, the process of chemotaxis—directional migration in response to a gradient of stimulating ligand—still seems the most likely process to induce the emigration, even granule enzymes and eicosanoids contributing to the genesis of the adult respiratory distress syndrome.

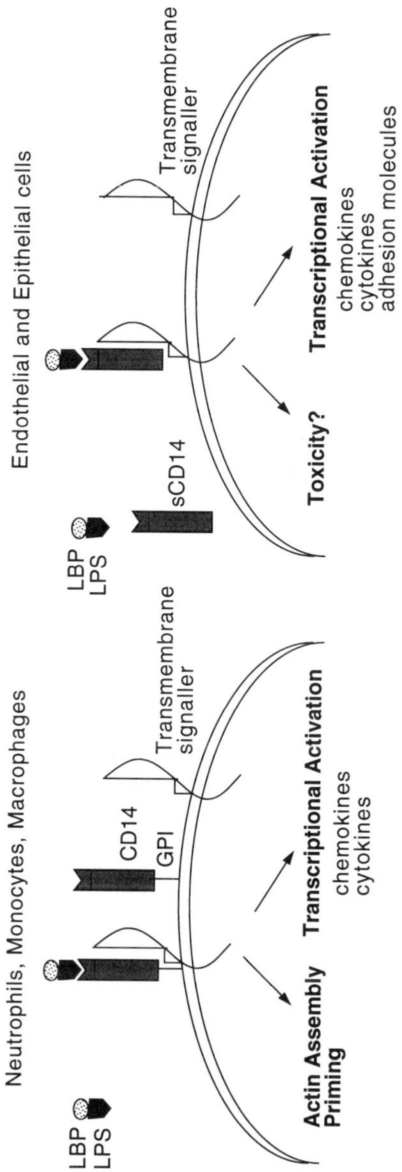

Figure 1 Hypothetical schemes for LPS-CD14-mediated activation of inflammatory and endothelial cells.

if it is helped by up-regulated adhesion molecules on the endothelial cells (50) and is influenced by increased random migratory activity of the neutrophils (i.e., nondirectional stimulation). Likely candidates for the chemotactic agents generated within the lung by LPS are members of the chemokine family.

D. Neutrophil Priming

Neutrophil activation in response to chemoattractants may appear to be an explosive, all-or-none event resulting in elaboration of toxic products—to whit, enzyme release, oxidant production, and eicosanoid release. However, it is now clear that neutrophil activation responses are considerably more tightly regulated than previously thought. Stimulation of neutrophils with typical chemoattractants that act through seven-transmembrane spanning G-protein-linked receptors, although readily able to induce calcium transients, IP-3 generation, actin assembly, and chemotaxis, is inefficient at inducing oxidant production or release of granular enzymes. However, when the neutrophil is first exposed to a variety of other stimuli, including LPS, TNF-α, GMCSF, and PAF, the neutrophil then responds to a chemoattractant with a marked increase in the production of eicosanoids, oxidants, and release of granular enzymes (8). Several experimental systems now suggest that this mechanism may be operative in models of lung injury and in humans with adult respiratory distress syndrome. In animals the infusion of low concentrations of both LPS and FMLP increased lung vascular permeability in rabbits (51)—either alone was ineffective. This model has been extended by others for use in the isolated perfused lung (52). Although none of these systems induce an exact equivalent of the adult respiratory distress syndrome in humans, it nonetheless suggests that this general notion (cell priming followed by enhanced responsiveness) may be applicable to neutrophil-mediated vascular injury. In humans we have detected the presence of both LPS as well as neutrophil stimuli such as C5a in more patients with ARDS than those who did not develop the syndrome (53). LPS here is postulated to act (1) directly on the circulating neutrophils to induce their retention; (2) to induce the synthesis of other priming agents, such as GCSF, GMCSF, TNF, and PAF (54); (3) to induce the formation of chemokines and chemotactic factors; and (4) to prime the neutrophils for enhanced synthetic and secretory responses to those chemokines and chemotactic agents. Thus in the syndrome of sepsis-induced acute lung injury, it would appear that the conditions are present to allow the primed neutrophil to be retained and migrate into the lung perhaps under the influence of chemokines and to have this primed neutrophil then respond to seven-transmembrane spanning chemokine receptors with a markedly enhanced release of oxidants

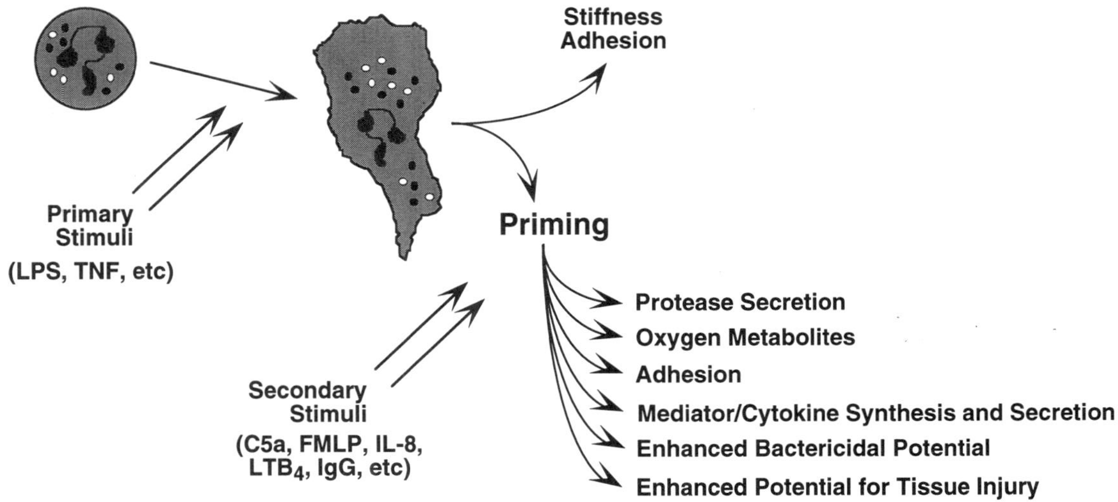

Figure 2 Neutrophil priming. Exposure of neutrophils to primary stimuli such as LPS itself or a variety of mediators and cytokines alters the responsivity of the cells to chemotactic factors or phagocytic stimuli. The combined effect normally may be to enhance extravascular activities of the cells—however, the increased production of reactive oxygen metabolites and secretion of toxic products may also contribute to pulmonary injury, for example, in ARDS.

granule enzymes and eicosanoids contributing to the genesis of the adult respiratory distress syndrome.

Despite a great deal of work, the mechanisms by which neutrophils are primed by LPS and other mediators remain obscure. Initial studies focused on the potential role for protein kinase C, IP3 generation, calcium mobilization, and receptor up-regulation, and these have not been completely ruled out as contributing causes (55,56). Nonetheless, it now appears that some other fundamental property of the neutrophils signal transduction process must be altered to allow efficient coupling between the receptor and induction of the truly activated state. Considering the recent interest in induction of protein phosphorylation, particularly tyrosine phosphorylation, induced by a variety of agents, including GMCSF, an effective neutrophil primer (57), these suggest the possibility that the priming of neutrophils for enhanced functions involve an alteration in the phosphoprotein phenotype of these cells. However, it should be noted that priming can be induced by ligands acting through a wide variety of quite different types of "receptors," including GPI-linked CD14 for LPS; seven-transmembrane G-protein-linked PAF receptors; trimeric TNF receptors, and GMCSF receptors. This observation would seem to suggest either that the priming process is different for different stimuli or that it may be evidenced late in the signal transduction systems (i.e., after convergence of the various pathways). This exciting area will almost certainly permit elucidation of the effects of LPS on neutrophils and other cells as well, and may eventually allow novel therapeutic strategies to address the genesis of lung injury.

III. LPS-Cytokine Interactions in Monocyte Retention, Emigration, and Pulmonary Injury

In general terms, the mechanisms that govern monocyte priming, retention in the pulmonary microvasculature, and emigration into the lung are similar to mechanisms regulating these processes for neutrophils. As outlined above, neutrophils can mediate tissue injury via their release of toxic agents (i.e., oxygen radicals, proteases, and cationic proteins, (58–60)). Monocytes certainly have the potential to damage tissues since they elaborate most of the same toxic substances as neutrophils (58–63). Also like neutrophils, monocytes can undergo a priming response for enhanced secretion of toxic products by a wide variety of priming agents, including LPS itself as well as mediators such as PAF, interferons, and other cytokines (64,65). However, if little is known about the underlying mechanisms for neutrophil priming, even less is known for the monocyte. LPS and FMLP induce a net assembly and reorganization of f-actin in monocytes, and in the case of LPS, this increase in f-actin occurs with the same time course as LPS-induced increases in monocyte stiffness. As with neutrophils, these processes are CD14-dependent and are also dependent on the presence of plasma and more specifically on the presence of LBP (66–68). As for neutrophils, numerous mediators, including the chemokines, induce monocyte-specific migration into the lung. However, despite several similarities in the mechanisms governing monocyte recruitment to the lung and their potential involvement in lung tissue injury, there are important differences that may contribute to the involvement of monocytes in the inflammatory/injury processes in the lung.

A. Monocyte Priming and Potential for Injury to the Lung

The role of monocytes in lung injury is unclear. Many investigators have suggested that their major function is to protect and modulate repair of the lung rather than initiating or perpetuating severe lung injury. Monocytes have been shown to contain several unique metaloproteases which are released at inflammatory sites and may be involved in the degradation of matrix proteins and presumably in the remodeling of lung architecture that occurs during the later phases of acute lung injury (69–72). The relative activity of these proteases is less than their serine protease counterparts in the neutrophil and therefore may be less likely to induce severe tissue injury and more likely perhaps to facilitate tissue remodeling. Monocytes and monocyte-derived interstitial and alveolar pulmonary macrophages also secrete chemokines such as IL-8 as well as LTB4 and may thus contribute to the continued attraction and stimulation of neutrophils (73–76). Thus the main effect of monocytes on injury, in fact, may be indirect, via the production of neutrophil priming and secondary stimuli, leading to the induction of a persistent influx of activated injurious granulocytes to areas of lung inflammation.

Acute pulmonary inflammatory reactions do not usually lead to chronic persistent lung damage; rather, they resolve to normal structure and function. Thus, despite an intense granulocytic and monocytic inflammatory response in pneumococcal pneumonia, following killing of the bacteria, complete resolution ensues. By contrast, gram-negative bacterial pneumonias often induce an equally striking inflammatory process, but here the endpoint is more likely to be pulmonary fibrosis and impaired lung function. It seems reasonable to suggest therefore that LPS could contribute to these processes, either directly or indirectly via cytokine production.

Thus an alternative role for monocytes during inflammation is one of protection or repair. As these cells mature into macrophages, they develop the capability for the synthesis and posttranslational modification of a wide variety of chemokines, cytokines, leukocyte chemotactic factors, lipidic mediators, and growth factors which may play a major role in modulating the inflammatory process and regulating the repair phases of lung injury. Monocytes and macrophages regulate connective tissue matrix protein production and deposition in the lung by mesenchymal cells (fibroblasts, myofibroblasts, and smooth muscle cells (77–81)), via their elaboration of numerous profibrogenic cytokines and growth factors (80, 82–86). Monocytes also express numerous surface receptors which orchestrate their communication with other cells during the complex process of tissue inflammation and repair (87,88), a topic that we have reviewed elsewhere (89).

Irrespective of the participation of monocytes in lung injury and/or repair, LPS- and LPS-induced cytokines/chemokines are intimately involved in the recruitment of monocytes to the lung. The content of monocytes in lung tissue is dependent on mechanisms which regulate their (1) release from the bone marrow, (2) kinetics (half-life) in the systemic circulation, (3) margination in the pulmonary vasculature, (4) migration into and through the interstitum, and (5) clearance from the alveolar and airway compartments of the lung.

Blood monocytes are derived from bone marrow stem cells (90–92). Their proliferation in, and release from, the bone marrow is primarily under the regulation of macrophage colony-stimulating factor (M-CSF), granulocyte-macrophage colony-stimulating factor (GM-CSF), and interleukin 3 (93–95), with some influence from the cytokines interleukins 1, 5, 6, and 11 (96) and interferons. The number of monocytes present in the blood during inflammatory states is dependent not only on their production in the marrow, but also on their survival in the systemic circulation. They have been shown to undergo programmed cell death (apoptosis) (97) and are cleared from the circulation predominantly in the liver by phagocytic cells. In keeping with the general theme of this chapter of LPS acting both to generate mediators and as a mediator itself, a number of LPS-generated cytokines (e.g., TNF-α and IL-1β) and also LPS itself have been shown to act on monocytes to prevent apoptosis (97), raising the possibility that during inflammation the production of these cytokines enhances monocyte survival in the blood and lung. IL-4 has been shown to enhance apoptosis of blood monocytes, a process that is antagonized by interferon-γ (98). Accordingly, the relative production, secretion, and local concentration of these cytokines, as well as the above-mentioned growth factors, during inflammatory states most certainly influences the survival of monocytes in the systemic circulation and therefore regulates their delivery, margination, and potential injurious or protective functions in the lung.

B. Retention of Monocytes in the Pulmonary Microvasculature

A marginating pool of monocytes is present in the lung under noninflammatory conditions (99–105). During endotoxemia in several animal systems, an early peripheral leukopenia develops (14, 106–109), which is accompanied by, and presumably caused by, a marked increase in the marginating pool of monocytes in the lungs (101,109,110). In vitro, LPS directly stimulates an enhanced monocyte stiffness by inducing net f-actin assembly and reorganization (101,110). Because the mean pulmonary capillary diameter (5.5 μm) (43) is smaller than that of the monocyte (8.5 μm) (111,112), we have suggested that during endotoxemia, the diminished ability of monocytes to deform and transit the narrow pulmonary capillary network leads to cell retention (113).

Others have suggested that pulmonary vascular endothelial cell deformability may also be important in allowing passage of leukocytes through the capillary network (114). LPS also induces an increased monocyte adherence to endothelial cell monolayers by direct actions on (1) on the monocyte (115) and (2) on the endothelial cell (116,117). In vitro, the effect on monocytes is more rapid than that on endothelial cells (115).

Based on these data and on our previous work, we have observed differences in the kinetics of monocyte and neutrophil retention in the lung under normal conditions and during endotoxemia (12,101,107,110,111, 118). Monocytes under normal conditions are retained in the pulmonary vasculature longer and in greater magnitude than neutrophils. That is, infused neutrophils traverse the lungs of rabbits within the first 2 to 4 h after their intravenous infusion, whereas infused monocytes require 12 to 18 h to transit the capillary network totally. The retention of unstimulated monocytes in 8-μm pores is also greater than that observed for neutrophils in vitro. This discrepancy cannot be explained totally by the differences between the mean diameter of monocytes (8.13 μm) and neutrophils (8.3 μm) (111), but was, rather, found to be due to differences in baseline stiffness (monocytes, 0.157 mdyn/μm; neutrophils, 0.05 mdyn/μm) and perhaps in the organization of f-actin in these cells. LPS induces a three- to fourfold increase in neutrophil net actin assembly compared to only a 15 to 20% increase in monocytes, yet the mean stiffness induced by LPS (10 ng/mL) is similar for both. This may be due to the fact that less assembly is required for monocytes to attain maximal stiffness due to its larger nucleus and higher baseline stiffness. However, the organizational pattern of actin in response to LPS is also different in monocytes compared to neutrophils. Another important difference between these two leukocytes is that CD18-dependent mechanisms appear to be necessary for monocyte retention in the lung during endotoxemia, whereas neutrophil retention is CD18-independent (118).

Further maintenance of monocyte margination in the lung is probably regulated by the interaction of monocyte surface adhesion molecules (the selectin and integrin families) with the immunoglobulin gene superfamily and selectin family of endothelial cell adhesion molecules (119–121). The later up-regulation and activation of adhesion molecules, first on the monocyte itself and then on the endothelial cells, would serve to keep the cells in the pulmonary microvasculature once the actin reorganization has resolved—the redevelopment of deformability being presumably required for emigration through the microvascular walls. LPS-upregulated endothelial adhesion molecules may also be involved in monocyte emigration.

A variety of cytokines and chemokines induced in cells (including mononuclear phagocytes themselves) by LPS can also alter monocyte stiffness and adhesion to endothelial cell surfaces and therefore potentiate monocyte margination in the lung. Thus, TNF-α not only induces a decrease in monocyte deformability and an increase in net f-actin assembly (unpublished observations), but also increases adhesion to endothelial cells by direct actions on both the monocyte (115) and the endothelial cell (116,117) involving up-regulation and redistribution of adhesion molecules on their surface. Monocyte margination in the rabbit lung is also rapidly increased in response to the intravenous infusion of TNF-α (unpublished observation), probably due to these TNF-α-induced increases in monocyte stiffness and adhesion to pulmonary endothelium. Monocytes themselves make TNF-α in response to LPS with significant potential, therefore, for positive-feedback amplification loops, and this too is enhanced by LBP and regulated by CD14-dependent mechanisms (68,122). LBP is an acute-phase protein whose concentration in serum increases 10-fold or greater during inflammation, possibly leading to enhanced responses to all LBP-CD14 responsive cells.

C. Emigration and Monocyte-Specific Chemoattractants

A long-standing question in understanding different manifestations of the inflammatory process is the selectivity of inflammatory cell accumulation, either as a function of time after initial insult or in certain types of lesion. Thus for many years the known chemoattractants did not show a high degree of inflammatory cell selectivity and yet in vivo inflammatory reactions were observed with only neutrophil, eosinophil, or monocyte accumulation (at least at a given time point). As mentioned in the introduction, this question has in part a potential explanation in a large number of newly described molecules (including the chemokines) which seem to show considerable cell specificity. Thus a recently described neutrophil cationic protein which is released during degranulation and phagocytosis (human neutrophil–derived chemotactic factor, CAP 37) has been shown to have monocyte-specific chemotactic activity in vitro (123). Although not yet documented in vivo, during neutrophilic inflammatory processes, the concentration of this peptide is probably increased. We have shown that instillation of this protein (or its rabbit analog) into rabbit airways leads to a significant and selective migration of rabbit blood monocytes into the lung (unpublished data). The minimal neutrophil or lymphocyte migration into the lung that occurs in response to this protein is indistinguishable from that induced by the instillation of vehicle alone. Additional monocyte-specific chemoattractants/chemokines, such as MCP-1 or MAP-1, may also play a role in regulating the influx of monocytes into the lung during inflammatory processes. These molecules would also be expected to induce retention, adhesion, and migration of monocytes into the lung, although there are no data at present to support this speculation.

A totally different source of monocyte-specific chemoattractants comes from the inflammatory process itself: namely, the digested fragments of connective tissue proteins. The enzymatic degradation products of the extracellular matrix proteins collagen (124), elastin (125), and fibronectin (126) have been shown to stimulate monocyte, but not neutrophil, migration in vitro. Generation of these biologically active monocyte chemoattractants at areas of inflammation, potentially by neutrophil enzymes such as elastase (125, 127) and collagenase (124), make them candidates for neutrophil-dependent monocyte influx into inflammatory sites. A 120-kDa thermolysin-generated fragment of fibronectin, similar in size and sequence to a neutrophil elastase-generated fibronectin fragment (127) and containing the RGDS fibroblast cell-binding domain, has chemokinetic and chemotactic activity for monocytes but not for neutrophils or lymphocytes in vitro (126). Instillation of this fragment of fibronectin into the rabbit lung induces a marked monocyte migration into the alveolar space of the lung in a neutrophil-independent fashion (100). Plasma is a major source of fibronectin; therefore, plasma leak into the lung during inflammatory processes could act as a source for the precursor fibronectin. Thus fragmented fibronectin was found in the alveolar space of rabbits with C5 fragment–induced pulmonary inflammation in normal but not neutropenic animals, and a significant proportion of the monocyte influx was abrogated by monoclonal anti-fibronectin antibodies (directed against the cell-binding domain). Accordingly, one mechanism by which monocytes are stimulated to migrate into pulmonary inflammatory lesions may be by initial neutrophil localization, priming, and secretion and the subsequent cleavage of proteins such as fibronectin into molecules which are chemotactic for monocytes.

D. Pulmonary Intravascular Macrophages

Pulmonary intravascular macrophages (PIMs) reside in the pulmonary microvasculature of certain species (cattle, sheep, pigs, goats) (20,21,128,129) and may or may not (128) be distinct from monocyte-derived macrophages that have marginated in the lung. These cells have been shown to phagocytose foreign particles, a process enhanced after LPS stimulation (21), and regulate lung microvascular reactivity (130) via their production of thromboxane (131). It is unclear whether these cells produce TNF-α, IL-1β, IL-6, or other macrophage-derived cytokines constitutively or in response to LPS, and if so, whether these molecules regulate monocyte/macrophage functions in the vascular, interstitial, or alveolar compartments of the lung. It also remains to be determined if blood monocytes, whose margination in the pulmonary microvasculature and whose state of activation is increased in response to LPS, serve similar potential functions in humans, a species that lacks PIMs (22). Even more intriguing is the possibility that monocytes might be induced to lodge in the pulmonary capillaries and adhere there following exposure to LPS or similar-acting molecules in the circulation (see above). If they then matured into macrophages in situ they might well be seen as, or at least subserve the function of, the so-called pulmonary intravascular macrophages.

IV. Functional Diversity of Macrophages in the Pulmonary Inflammatory Response

As discussed above, mononuclear phagocytes are thought to contribute actively to both the initiation and resolution of acute inflammation, as well as playing crucial roles in the perpetuation and amplification of the chronic inflammatory response (132). A principal explanation for the prominence of this single cell type in each phase of the inflammatory response is the breadth of diversity in the functional responses of macrophages. During the acute inflammatory response, resident tissue macrophages and infiltrating monocytes and differentiated monocyte-derived macrophages have been shown to secrete a variety of chemokines and cytokines which serve locally to increase leukocyte recruitment, particularly that of neutrophils, and systemically to induce the hepatic acute-phase response. In addition, mononuclear phagocytes secrete a variety of agents, including toxic reactive oxygen and nitrogen intermediates such as O_2^-, OH•, and NO• as well as tissue-degrading lysosomal hydrolases and neutral proteolytic enzymes (133). Through their combined actions, this array of toxic materials is capable of damaging both the cellular and structural elements of the lung parenchyma.

As the acute inflammatory response begins to wane, macrophages become the principal phagocytic cell type. Through their capacity to secrete several key fibroblast growth factors, such as PDGF (134) and IGF-1 and other cytokines such as TGF-β (135), this cell is recognized as a leading contender for the role of conductor of the fibrogenic phase of the inflammatory response by stimulating connective tissue matrix synthesis.

In addition to these important and perhaps sometimes apparently conflicting roles, macrophages also contribute to true resolution of the inflammatory response by recognizing and removing apoptotic granulocytes, thereby preventing potentially tissue-damaging enzymes from being released into the resolving inflammatory lesion. A major question now facing those working the in field of macrophage biology is that of delineating the mechanisms that regulate these complex aspects of macrophage functional diversity.

A. Macrophage Functional Diversity Is Determined by Cytokines

With the recognition that macrophages are capable of synthesizing and secreting a broad spectrum of cytokines and other inflammatory mediators has come the experimental demonstration that many of their functional responses are regulated in an autocrine or paracrine or intracrine fashion by specific cytokines secreted by the macrophage itself or by other juxtaposed cells. In many cases the responses are regulated by the simultaneous or sequential impingement of two (or more) cytokines and/or stimuli upon the macrophage, giving an additional (and more complex) level of control of functional diversity. As with the neutrophil blood monocyte (see above), the role of the first stimulus has frequently been referred to as that of a priming agent: a stimulus that by itself has little or no effect but which synergistically enhances the response of the cell to a second stimulus. However, it is important to emphasize that the priming processes in macrophages may be significantly different in mechanism from those in neutrophils and monocytes (e.g., by requiring protein synthesis) even though the term used and the concept are the same. However, one point in common is the inclusion of LPS along with a variety of other stimuli and cytokines as a potential priming agent. Since many responses of macrophages are enhanced by priming stimuli, it is important both to define the response enhanced and the combinations of stimuli required to induce that response. Priming agents can operate both at the level of immediate responses—for example, the ability of LPS to enhance the capacity of macrophages to produce reactive oxygen intermediates in response to PMA (7) or to enhance the release of arachidonic acid in response to zymosan particles (136)—or differentiated responses, such as the role of IFNs in priming for enhanced cytocidal activity of macrophages against tumor target cells (137) and facultative intracellular parasites such as *Leishmania* (138). Further complexity has been introduced with the recognition that the stimulatory effects of many cytokines and stimuli are often kept in check by other cytokines, which have an opposing effect on a given macrophage capacity or response. For example, the ability of IFN-γ to prime macrophages for nitric oxide synthase expression is opposed by TGF-β, while the ability of LPS to induce TNF-α expression is blocked in the presence of IL-10. Fortunately, while interactions between cytokines and stimuli with macrophages and other inflammatory and immune cells are highly complex, in vitro systems have been established to allow the reductionist approach to be applied to investigate such areas as the individual functional responses and capacities of cells, allowing the definition of cell surface structures responsible for cytokine and stimulus recognition, and to begin to elucidate the signaling pathways initiated as a consequence of receptor–ligand interaction. Because of the great potential for LPS to induce macrophage synthesis of TNF-α and IFN-β, we will focus particularly on these cytokines as autocrine regulators of macrophage functional diversity.

B. Macrophage Responses to IFNs and TNF

One of the key controlling cytokines of the inflammatory response is TNF-α. Although beyond the scope of this review, in vivo and in vitro studies have shown TNF-α to induce cachexia and shock. Most cell types are capable of recognizing and responding to TNF-α. Macrophage responses to TNF-α that are implicated in the inflammatory response include the expression of reactive oxygen intermediates, the secretion of other pro-inflammatory cytokines such as IL-8 and IL-1, and mediators such as PGE_2. Recent work in our laboratory has focused on the ubiquitous role of TNF-α in regulating two distinct capacities (maturational directions) of macrophages: the production of the fibroblast growth factor IGF-1 and the reactive nitrogen intermediate NO•. We have observed that IGF-1 synthesis is markedly stimulated by the connective tissue matrix component hyaluronic acid while NO•expression, as reflected by NO_2 accumulation in the culture supernatants, remains repressed. The mechanism of activation of IGF-1 synthesis has been shown to involve recognition of hyaluronic by CD44, which in turn stimulates the rapid accumulation of TNF-α- and IL-1β mRNA and the secretion of TNF protein into the culture medium. By a number of criteria, including the ability of anti-TNF-α antibodies to block hyaluronic acid–induced IGF-1 synthesis and of exogenous TNF-α to stimulate IGF-1 synthesis, the response to hyaluronic acid has been shown to be TNF-α-dependent.

TNF-α also plays a requisite role in NO• production, and probably in the generation of cytocidal capacity, wherein macrophages recognize and kill transformed cells. However, in this situation, additional cytokines are required to prime the macrophage to respond to TNF-α. The major category of cytokines capable of priming macrophages to produce NO• in response to TNF-α are also among the oldest in the chronology of cytokine biology, namely IFNs α and β, derived predominantly from macrophages and fibroblast, and IFN-γ derived from T-cells and NK cells. This difference in the response of the macrophage to TNF-α (depending on whether or not they are also exposed to interferons) are believed to have important implications in terms of both the regulation of macrophage functional diversity and the pleiotropic behavior of TNF-α. This diversity is not mediated by the actions of different TNF receptors since TNF receptor type I (p55) is required for both responses (139). Moreover, TNF-α exposure results in a similar level of activation of MAP1 kinase (Winston and D. W. H. Riches, unpublished observations), in both unprimed and IFN-primed macrophages, suggesting the likelihood that the response to the priming stimulus may be manifested at some level distal to MAP1 kinase. Similar conclusions have been drawn from studies using the calcium ionophore ionomycin, in which the effect of IFN priming is manifest downstream of the Ca^{2+} signal (140). Much effort is currently being directed to determine the point at which the signals generated in response to IFNs (i.e., priming) interact with the signals generated in response to ligation of the TNF receptors or whether the signals ultimately converse at the gene level.

Combinations of cytokines therefore have a profound effect on the differentiated functions and capacities of macrophages, and although the types and concentrations of cytokines and other inflammatory mediators are likely to be changing constantly during the course of an inflammatory response, it would seem appropriate to begin to consider the possibilities that overexuberant inflammatory responses, either acute or chronic, may be mediated by an inappropriate balance of pro-inflammatory and anti-inflammatory cytokines and mediators. LPS here will act both directly on the macrophage as another priming or stimulating agent and/or as an inducer of the very cytokines that induce the macrophage diversification responses. Of potential significance, discussed below, is the finding by several laboratories and in several experimental systems that IFNs, in addition to changing the cellular response to TNF-α, are also profoundly anti-inflammatory cytokines.

C. Interferons and Inflammation

IFNs α, β, and γ are secreted by a wide variety of cell types and under markedly varied conditions. Although originally observed as an antiviral response, many bacteria, such as salmonella, staphylococci, mycobacteria, rickettsia, and shigella, as well as isolated gram-negative LPS have been reported to induce IFN-α/β expression both in vivo and in vitro. IFN-α/β is also reported to be induced when fibroblasts are stimulated with PDGF. In addition, both T-cell and NK cell activation results in the secretion of IFN-γ. Thus IFNs are produced during inflammation, initiated by many types of bacteria as well as that accompanying immune and viral responses.

Experiments aimed at determining the role(s) of IFNs in the inflammatory response have shown these cytokines to antagonize both the acute and chronic inflammatory responses. Heremans et al. (141–143) have shown that the IFN-inducing LDH virus as well as systemic administration of purified

IFNs α, β, or γ significantly block the local inflammatory response initiated by injection of LPS into the footpads of mice. Similar results were obtained by Mecs et al. (144) using Sendai virus as the IFN inducer, and carrageenan as the inflammatory stimulus. A recent preliminary report also suggests that IFN-α therapy may reduce joint swelling in rheumatoid arthritis (145). The mechanism of the anti-inflammatory effect of IFN is not known, although in vitro studies have indicated that exposure of macrophages to IFNs results in a profound inhibition of the capacity of these cells to secrete prostaglandins and leukotrienes in response to stimulation with zymosan particles and calcium ionophores (146,147), suggesting that one of the effects of IFNs is to down-regulate the production of acute inflammatory mediators.

IFNs also down-regulate aspects of the reparative phase on the inflammatory response. In a study of the effects of IFN inducers such as poly [I:C] and purified IFN-γ on the development of lung fibrosis following intratracheal instillation of bleomycin into mice, Hyde et al. (148) and Granstein et al. (149) observed a profound inhibition of lung fibrosis following injury. While the mechanism of this effect is not known, IFNs have been observed in vitro to antagonize the growth factor–dependent proliferation of mesenchymal cells and to block the synthesis of collagen by these cells. For example, Granstein et al. (150) have reported that recombinant IFN-γ reduced the collagen content of argon laser–induced wounds and promoted an increase in the acid mucopolysaccharide content of healing wounds. Thus the presence of IFNs appears to influence the characteristics of the inflammatory response. How this effect is mediated and how much can be attributed to effects on the macrophage need to be addressed.

D. Influence of Cytokines on Macrophage Production of Nitric Oxide

The past five years has led to the increasing recognition of nitric oxide as a powerful mediator of many diverse mammalian responses and culminated in its recognition in 1992 by the journal *Science* as the "molecule of the year". Nitric oxide is synthesized during the oxidation of L-arginine to L-citrulline by nitric oxide synthase (NOS). Now recognized to comprise a family of closely related enzymes, NOS can be broadly divided into two groups: (1) the constitutive NOSs expressed by endothelial cells, neuronal cells, and smooth muscle cells, and (2) the inducible NOS expressed mainly by macrophages. Although the mechanism of NO• synthesis is apparently identical for both its NOS isoforms, their mechanism of activation differ in that the cNOS isoforms are activated in response to changes in $[Ca^{2+}]_i$ in a calmodulin-dependent fashion, whereas the iNOS is Ca^{2+} independent. Another significant difference is that iNOS is expressed at much higher levels than cNOS, and consequently, NO• production is greater and more sustained in cells expressing iNOS compared to cells expressing cNOS.

As mentioned earlier, NO• has been identified as a key mediator of cellular cytostasis and parasite elimination by macrophages activated with IFNγ and LPS, probably through its ability to bind and inactivate FeS-containing enzymes such as aconitase, an enzyme of the tricarboxylic acid cycle (151). Recent studies have begun to focus on the role of NO• in the inflammatory response. The general approach applied to both in vitro and in vivo studies has been to investigate the effects of inhibitors of NO• synthesis such as NG-monomethyl-L-arginine and nitro-L-arginine on a given response. The results from these studies have shown that NO• exerts multiple effects on the inflammatory response. However, given the diversity of cell types capable of synthesizing the molecules, the variety of conditions leading to its formation as well as its short half-life (i.e., the likelihood that its actions will be manifest locally), different results have been obtained according to the model studied. For example, in a murine model of acute endotoxemia, NO• has been suggested to protect the liver from endotoxin-mediated injury (152), whereas in an immune complex model of vascular injury in rat lung, NO• has been suggested to be a principal mediator of injury (153). This effect may be mediated in part by the ability of NO• to increase vascular permeability as has been shown in a model of carrageenin-induced edema formation in rats (154). Clearly, more studies are required to delineate the effects of NO• on epithelial and endothelial barriers and the consequence to tissue injury and repair and to identify the principal cell types responsible for NO• production in vivo.

V. LPS-Cytokine Interactions in ARDS

Many of the LPS–cytokine interactions proposed in the foregoing sections of this discussion are likely to occur in the development of acute lung injury (ARDS) in humans. In studies of patients at risk for and with ARDS, we, and others, have demonstrated that endotoxin is present in the blood (155,156) and appears to be associated with the development of ARDS (53). Circulating LPS would be expected to stimulate the release of cytokines and secretion of chemokines both systemically and within the lung. Several recent studies in patients strongly suggest that this occurs.

Alveolar macrophages stimulated with LPS, TNF-α, or IL-1β express mRNA for IL-8 within 60 min (157). The chemokine IL-8 is a likely candidate for attracting the large number of neutrophils found in the airspaces of patients with ARDS. In early studies we described a neutrophil chemotactic factor in bronchoalveolar lavage (BAL) from patients with established ARDS that had characteristics very similar to those of what is now known as IL-8 (158). In a recent study IL-8 was measured in BAL from patients with

ARDS and shown to exhibit an association with mortality (159). Even more intriguing is the finding by Donnelly that IL-8 is present within munites or hours in the BAL of patients at risk for ARDS and that the IL-8 concentration may predict which patients will ultimately develop the syndrome (35). The source of the measured IL-8 may be alveolar macrophages, as alveolar macrophages from patients who developed ARDS (even at these very early time points) had greater immunocytochemical staining than did macrophages from those patients who did not develop the syndrome (35). IL-8 was measured in plasma samples obtained at the same time that the lavage was performed and was found to be undetectable in some patients, again suggesting (but not proving) a pulmonary origin, even if the precipitating condition was peripheral to the lung. There was no association between plasma IL-8 levels and the development of ARDS. Similarly, in a study of 47 patients with sepsis, while IL-8 was present in the blood of 42 of them (160) there were no differences between the levels in patients who developed ARDS compared with those who did not, even though the absolute levels did correlate with the severity of sepsis.

In addition to effects on neutrophils (see above) it has been suggested that LPS might stimulate pulmonary cells (including macrophages) to generate chemokines and cytokines and thus initiate or enhance the condition. Donnelly's observations are important in this regard since they would suggest that if LPS was indeed involved in this very early response, it would have had to be (1) liberated from some site (presumably the gut), (2) cross the blood–lung barrier, and (3) stimulate cytokine production—all in a period of less than an hour. This seems highly unlikely and suggests that if LPS is involved in these very early events, it probably acts by liberation of some rapidly acting mediators that can readily cross into the airspaces. On the other hand, the potential for LPS and the cytokines it induces acting later in the development of the syndrome remains high.

LPS-stimulated alveolar macrophages can also release IL-1β, which has also been implicated in neutrophil accumulation in the lung in experimental animals (161). IL-1β levels are increased in the BAL of patients with ARDS (162), and alveolar macrophages from patients with ARDS release more IL-1β at baseline and after LPS stimulation than alveolar macrophages from normal subjects or other patients with lung disease (163). Whether this too is a very early response of macrophages from patients at risk for ARDS is at present unknown.

Many studies have attempted to detect an association between circulating cytokines and the development of ARDS, both in the hope of identifying those patients at risk for the syndrome who will actually develop ARDS and to provide some indication of likely pathogenesis. The presumed hematogenous route of lung damage in patients whose inciting factors are distant to the lung have added impetus to these investigations. Many of these cytokines are released from cells (including neutrophils as well as mononuclear phagocytes in the blood and lung) following exposure to LPS. Reports of the presence of TNF-α in both septic and ARDS patients have been variable (164–166). In our own studies on circulating TNF-α we found that its levels varied significantly with the assay methods used, and although it was certainly found in the plasma of patients at risk for ARDS, we were unable to demonstrate any association with development of the syndrome (167). In like fashion, IL-1β has also been detected in the serum of patients with sepsis syndrome in two studies, although in a single study of patients at risk for and with ARDS, the levels were not clearly different from those measured in normal individuals (162,164,166).

These discrepancies may also be explained by variability in available assay methods (168). IL-6 levels were shown to be increased in the circulation of patients with sepsis (166,169). Levels were higher in patients with gram-negative sepsis than in those with gram-positive sepsis (169), and there was an association between the absolute level of IL-6 early in the course of sepsis and subsequent mortality. There was also an increase in mortality in those patients with persistently elevated IL-6 levels (169). By contrast, a recent preliminary study of septic patients found no demonstrated association between levels of IL-6 and the development of ARDS (170). Overall, these studies have to be considered at best as no more than suggestive. However, if defined ARDS is seen as no more than the extreme manifestation of injurious processes that impinge upon the lung in virtually all cases where the known risk factors occur, the cytokines found in the circulation may be considered to be of critical importance to setting up these injurious processes. In this context, when LPS is present it would be expected to play a significant role, both directly as a cytokine mimic in its own right and/or as a contributor to the network of cytokines and mediators produced.

Acknowledgments

These studies were carried out in the F. L. Bryant, Jr. Research Laboratory for the study of mechanisms of lung disease and were supported by a SCOR in ARDS (HL-40784), a SCOR in ILD (HL-27353), GM-48211, HL-34303, HL-50319, Veteran's Administration Merit Review, and Clinical Investigator Awards.

References

1. Giroir B, Johnson J, Brown T, Allen G, Beutler B. The tissue distribution of tumor necrosis factor biosynthesis during endotoxemia. J Clin Invest 1992; 90:693–698.

2. Michie HR, Manogue KR, Spriggs DR, et al. Detection of circulating tumor necrosis factor after endotoxin administration. N Engl J Med 1988; 318:1481–1486.
3. Chang S, Feddersen C, Henson P, Voelkel N. Platelet-activating factor mediates hemodynamic changes and lung injury in endotoxin-treated rats. J Clin Invest 1987; 79:1498–509.
4. Beutler B, Milsard IW, Ceramic A. Passive immunization against cachectin/tumor necrosis factor protects mice from the lethal effect of endotoxin. Science 1985; 229:869–871.
5. Mathison JC, Wolfson E, Ulevitch RJ. Participation of tumor necrosis factor in the mediation of gram negative bacterial lipopolysaccharide induced injury in rabbits. J Clin Invest 1988; 81:1925–1937.
6. Chang S, Ohara N, Kuo G, Voelkel N. Tumor necrosis factor-induced injury is not mediated by platelet-activating factor. Am J Physiol 1989; 257:L232–L239.
7. Pabst M, Johnston JRB. Increased production of superoxide anion by macrophages exposed in vitro to muramyl dipeptide or lipopolysaccharide. J Exp Med 1980; 151:101–114.
8. Guthrie L, McPhail L, Henson P, Johnston R Jr. Priming of neutrophils for enhanced release of oxygen metabolites by bacterial lipopolysaccharide. Evidence for increased activity of the superoxide-producing enzyme. J Exp Med 1984; 160:1656–1671.
9. Rabinovici R, Yeh C, Hillegass L, et al. Role of complement in endotoxin/platelet-activating factor-induced lung injury. J Immunol 1992; 149:1744–1750.
10. Rabinovici R, Esser KM, Lysko PG, et al. Priming by platelet-activating factor of endotoxin-induced lung injury and cardiovascular shock. Circ Res 1991; 69: 12–25.
11. Johnson A, Hocking D, Ferro T. Tumor necrosis factor-alpha primes pulmonary hemodynamic response to *N*-formyl-L-methionyl-L-leucyl-L-phenylalanine. Am J Physiol 1991; 261:H996–H1004.
12. Worthen G, Haslett C, Rees A, Gumbay R, Henson J, Henson P. Neutrophil-mediated pulmonary vascular injury. Synergistic effect of trace amounts of lipopolysaccharide and neutrophil stimuli on vascular permeability and neutrophil sequestration in the lung. Am Rev Respir Dis 1987; 136:19–28.
13. Heuer H, Letts G, Meade C. Tumor necrosis factor (TNF) and endotoxin prime effects of PAF in vivo. J Lipid Mediators 1990; 2:S101–S108.
14. Morrison DC, Ulevitch RJ. The effects of bacterial endotoxin on host mediation systems. Am J Pathol 1978; 93:527–617.
15. Frey EA, Miller DS, Jahr TG, et al. Soluble CD14 participates in the response of cells to lipopolysaccharide. J Exp Med 1992; 176:1665–1671.
16. Pugin J, Schurer-Maly CC, Leturcq D, Moriarty A, Ulevitch RJ, Tobias PS. Lipopolysaccharide activation of human endothelial and epithelial cells is mediated by lipopolysaccharide-binding protein and soluble CD14. Proc Natl Acad Sci USA 1993; 90:2744–2748.

17. Brigham K, Meyrick B, Berry JLC, Repine J. Antioxidants protect cultured bovine lung endothelial cells from injury by endotoxin. J Appl Physiol 1987; 63:840–850.
18. Meyrick B, Johnson J, Brigham K. Endotoxin-induced pulmonary endothelial injury. Prog Clin Biol Res 1989; 308:91–99.
19. Pohlman T, Harlan J. Human endothelial cell response to lipopolysaccharide, interleukin-1, and tumor necrosis factor is regulated by protein synthesis. Cell Immunol 1989; 119:41–52.
20. Rybicka K, Daly BDT, Migliore JJ, Norman JC. Intravascular macrophages in normal calf lung: an electron microscopic study. Am J Anat 1974; 139:353–368.
21. Warner AE, Brain JD. Intravascular pulmonary macrophages: a novel cell removes particles from blood. Am J Physiol 1986; 250:R278–R732.
22. Warner A. Chest 1993 (Suppl). In press.
23. Holmes WE, Lee J, Kuang W-U, Rice GC, Wood WI. Structure and functional expression of a human interleukin-8 receptor. Science 1991; 253:1278.
24. Murphy PM, Tiffany HL. Cloning of complementary DNA encoding a functional interleukin-8 receptor. Science 1991; 253:1280.
25. Lee J, Horuk R, Rice GC, Bennett GL, Camerato T, Wood WI. Characterization of two high affinity human interleukin-8 receptors. J Biol Chem 1992; 267: 16283.
26. Schall TJ. Biology of the RANTES/SIS cytokine family. Cytokine 1991; 3:165.
27. Schall TJ, Bacon K, Toy KJ, Goeddel DV. Selective attraction of monocytes and T lymphocytes of the memory phenotype by cytokine RANTES. Nature 1990; 347:669.
28. Wolpe SD, Davatelis G, Sherry B, et al. Macrophages secrete a novel heparin-binding protein with inflammatory and neutrophil chemokinetic properties. J Exp Med 1988; 167:570.
29. Van Damme J, Proost P, Lenaerts J-P, Opdenakker G. Structural and functional identification of two human, tumor-derived monocyte chemotactic proteins (MCP-2 and MCP-3) belonging to the chemokine family. J Exp Med 1992; 178:59.
30. Kameyoshi Y, Dorschner A, Mallet AI, Christophers E, Schroder J-M. Cytokine RANTES released by thrombin-stimulated platelets is a potent attractant for human eosinophils. J Exp Med 1992; 176:587.
31. Matsushima K, Oppenheim JJ. Interleukin 8 and MCAF: Novel inflammatory cytokines inducible by IL-1 and TNF. Cytokine 1989; 1:2.
32. Oppenheim JJ, Zachariae COC, Mukaida N, Matsushima K. Properties of the novel proinflammatory suspergene "intercrine" cytokine family. Annu Rev Immunol 1991; 9:617.
33. Kuijpers TW, Hakkert BC, Hart MH, Roos D. Neutrophil migration across monolayers of cytokine-prestimulated endothelial cells: a role for platelet-activating factor and IL-8. J Cell Biol 1992; 117:565–572.
34. Huber AR, Kunkel SL, Todd RF III, Weiss SJ. Regulation of transendothelial neutrophil migration by endogenous interleukin-8. Science 1991; 254:99–102.

35. Donnelly SC, Strieter RM, Kunkel SL, et al. Interleukin-8 and development of adult respiratory distress syndrome in at-risk patient groups. Lancet 1993; 341: 643–647.
36. Broaddus VC, Hébert CA, Vitangcol RV, et al. Interleukin-8 is a major neutrophil chemotactic factor in pleural liquid of patients with emphysema. Am Rev Respir Dis 1992; 146:825–830.
37. Richman-Eisenstat JBY, Jorens PG, Hébert CA, Ueki I, Nadel JA. Interleukin-8: An important chemoattractant in sputum of patients with chronic inflammatory airway diseases. Am Physiol Soc (Lung Cell Mol Physiol 8) 1993; 264:L413–L418.
38. Lien DC, Worthen GS, Capen RL, et al. Neutrophil kinetics in the pulmonary microcirculation: effects of pressure and flow in the dependent lung. Am Rev Respir Dis 1990; 141:953–959.
39. Lien DC, Capen RL, Hanson WL, et al. The acute inflammatory response in the canine lung: direct observation of neutrophil behavior in response to C5 fragments. 1991. Submitted.
40. Lien DC, Henson PM, Capen RL, et al. Neutrophil kinetics in the pulmonary microcirculation during acute inflammation. Lab Invest 1991; 65:145–159.
41. Downey GP, Worthen GS, Henson PM, Hyde DM. Neutrophil sequestration and migration in localized pulmonary inflammation: capillary localization and migration across the intraalveolar septum. Am Rev Respir Dis 1993; 147:168–176.
42. Schmid-Shonbein GW, Shih YY, Chien S. Morphometry of human leukocytes. Blood 1985; 56:866.
43. Guntheroth WG, Lachter DL, Kawabori I. Pulmonary microcirculation: tubules rather than sheet and post. J Appl Physiol 1982; 53:510–515.
44. Martin BA, Wright JL, Thommasen H, Hogg JC. Effect of pulmonary blood flow on the exchange between the circulating and marginating pool of polymorphonuclear leukocytes in dog lungs. J Clin Invest 1982; 69:1277.
45. Worthen GS, Avdi N, Vukajlovich S, Tobias PS. Neutrophil adherernce induced by lipopolysaccharide in vitro: role of plasma-mediated lipopolysaccharide disaggregation. J Clin Invest 1993; 90:2526–2535.
46. Tesh VL, Morrison DC. The interaction of Escherichia coli with normal human serum: factors affecting the capacity of serum to mediate lipopolysaccharide release. Microb Pathol 1988; 4:175–187.
47. Tobias PS. Chest 1993 (Suppl). In press.
48. Cardozo C, Edelman J, Jagirdar J, Lesser M. Lipopolysaccharide-induced pulmonary vascular sequestration of polymorphonuclear leukocytes is complement independent. Am Rev Respir Dis 1991; 144:173–178.
49. Rinaldo JE, Moore SA, Lee RE, Dauber JH. Shift in subfractions of rate alveolar macrophages in vivo during endotoxin-induced alveolitis. J Leukoc Biol 1987; 42:230–238.
50. Luscinskas FW, Cybulsky MI, Kiely JM, Peckins CS, Davis VM, Gimbrone MA Jr. Cytokine-activated human endothelial monolayers support enhanced neutrophil transmigration via a mechanism involving both endothelial–leukocyte

adhesion molecule-1 and intercellular adhesion molecule-1. J Immunol 1991; 146:1617–1625.
51. Worthen GS, Haslett C, Rees AJ, Gumbay RS, Henson JE, Henson PM. Neutrophil pulmonary vascular injury: synergistic effects of trace amounts of lipopolysaccharide and neutrophil stimuli on vascular permeability and neutrophil sequestration in the lung. Am Rev Respir Dis 1987; 136:19–28.
52. Anderson BO, Poggetti RS, Shanley PF, et al. Primed neutrophils injure rat lung through a platelet-activating factor-dependent mechanism. J Surg Res 1991; 50:510–514.
53. Parsons PE, Worthen GS, Moore EE, Tate RM, Henson PM. The association of circulating endotoxin with the development of the adult respiratory distress syndrome. Am Rev Respir Dis 1989; 140:294–301.
54. Taveira da Silva AM, Kaulbach HC, Chuidian FS, Lambert DR, Suffredini AF, Danner RL. Brief report: shock and multiple-organ dysfunction after self-administration of Salmonella endotoxin. N Engl J Med 1993; 328:1457–1460.
55. Yasui K, Becker EL, Shaafi RI. Lipopolysaccharide and serum cause the translocation of G-protein to the membrane and prime neutrophils via CD14. Biochem Biophys Res Commun 1992; 183:1280–1286.
56. Mege JL, Gomez-Cambronero J, Molski TF, Becker EL, Shaafi RI. Effect of granulocyte-macrophage colony-stimulating factor on superoxide production in cytoplasts and intact human neutrophils: role of protein kinase and G-proteins. J Leukoc Biol 1989; 46:161–168.
57. Gomez-Cambronero J, Huang C-K, Gomez-Cambronero TM, Waterman WH, Becker EL, Shaafi RI. Granulocyte-macrophage colony-stimulating factor-induced protein tyrosine phosphorylation of microtubule-associated protein kinase in human neutrophils. Proc Natl Acad Sci USA 1992; 89:7551–7555.
58. Sibille Y, Reynolds HY. Macrophages and polymorphonuclear neutrophils in lung defense and injury. Am Rev Respir Dis 1990; 141:471–501.
59. Henson PM, Johnston RB Jr. Tissue injury in inflammation: oxidants, proteinases, and cationic proteins. J Clin Invest 1987; 79:669–674.
60. Cochrane CG, Spragg RG, Schraufstatter IU, Revak SD, Hyslop PA, Hinshaw DB. Oxidant and protease effectors in acute inflammation. In: van Furth RF, ed. Mononuclear Phagocytes. Dordrecht, The Netherlands: Martinus Nijhoff, 1985:777–783.
61. Fantone JC. Role of oxygen-derived free radicals and metabolites in leukocyte-dependent inflammatory reactions. Am J Pathol 1982; 107:397–418.
62. Hinman LM, Stevens CA, Matthay RA, Gee JL. Elastase and lysozyme activities in human alveolar macrophages: effects of smoking. Am Rev Respir Dis 1980; 121:263–271.
63. Werb Z, Gordon S. Elastase secretion by stimulated macrophages: characterization and regulation. J Exp Med 1975; 142:361–377.
64. Pabst MJ, Hedegaard HB, Johnston RB. Cultured human monocytes require exposure to bacterial products to maintain an optimal oxygen radical response. J Immunol 1982; 128:123.
65. Pabst MJ, Aida Y. Neutrophil responses to LPS: effect of adherence on triggering and priming of the respiratory burst. J Immunol 1991; 146:1271–1276.

66. Doherty DE, Leadbeater J, Zagarella L. LPS-induced monocyte retention in model capillaries: role of plasma and the LPS receptor CD14. Am Rev Respir Dis 1993; 147:A245.
67. Doherty DE, Tobias PS, Ulevitch RJ, Worthen GS. The role of LPS-binding protein and CD14 in LPS-induced monocyte adherence. Am Rev Respir Dis 1992; 145:A666.
68. Schumann RR, Leong SR, Flaggs GW, et al. Structure and function of lipopolysaccharide binding protein. Science 1990; 249:1429.
69. Shapiro SD, Campbell EJ, Welgus HG, Senior RM. Elastin degradation by mononuclear phagocytes. Ann NY Acad Sci 1991; 624:95–98.
70. Senior RM, Connolly NL, Cury JD, Welgus HG, Campbell EJ. Elastin degradation by human alveolar macrophages: a prominent role of metalloproteinase activity. Am Rev Respir Dis 1989; 139:1251–1256.
71. Campbell EJ, Senior RM, Welgus HG. Extracellular matrix injury during lung inflammation. Chest 1987; 92:161–167.
72. Albin RJ, Senior RM, Welgus HG, Connolly NL, Campbell EJ. Human alveolar macrophages secreted an inhibitor of metalloproteinase elastase. Am Rev Respir Dis 1987; 135:1281–1285.
73. Hunninghake GW, Gallin JI, Fauci AS. Immunologic reactivity of the lung: the in vivo and in vitro generation of a neutrophil chemotactic factor by alveolar macrophages. Am Rev Respir Dis 1978; 117:15–23.
74. Hunninghake GW, Gadek JE, Lawley JJ, Crystal RG. Mechanisms of neutrophil accumulation in the lungs of patients with idiopathic pulmonary fibrosis. J Clin Invest 1981; 68:259–269.
75. Carré PC, Mortenson RL, King TE Jr, Nobel PW, Sable CL, Riches DWH. Increased expression of the interleukin-8 gene by alveolar macrophages in idiopathic pulmonary fibrosis: a potential mechanism for the recruitment and activation of neutrophils in lung fibrosis. J Clin Invest 1991; 88:1802–1810.
76. Kunkel SL, Standiford T, Kasahara K, Streiter RM. IL-8: The major neutrophil chemotactic factor in the lung. Exp Lung Res 1991; 17:17–23.
77. Libbey P, Wyler DJ, Janicka MW, Dinarello CA. Differential effects of interleukin-1 on growth of human fibroblasts and vascular smooth muscle cells. Aterosclerosis 1985; 5:186.
78. Martin BM, Gimbrone MA, Unanue ER, Cotran RS. Stimulation of nonlymphoid mesenchymal cell proliferation by a macrophage-derived growth factor. J Immunol 1981; 126:1510.
79. Hunt TK, Knighton DR, Thakral KK, Goodson WH, Andrews WS. Studies on inflammation and wound healing: angiogenesis and collagen synthesis stimulated in vivo by resident and activated wound macrophages. Surgery 1984; 96: 48.
80. Rappolee DA, Mark D, Bander MJ, Werb Z. Wound macrophages express TGF-β and other growth factors in vivo: analysis by mRNA phenotyping. Science 1988; 241:708.
81. Crystal RG, Bitterman PB, Rennard SI, Hance AJ, Keogh BA. Interstitial lung diseases of unknown cause: disorders characterized by chronic inflammation

of the lower respiratory tract (first of two parts). N Engl J Med 1984; 310:154–156.
82. Lebovitch SJ, Ross R. A macrophage-dependent factor that stimulates the proliferation of fibroblasts in vitro. Am J Pathol 1976; 84:501.
83. Shimokado K, Raines EW, Madtes DK, Barret TB, Benditt EP, Ross R. A significant part of macrophage-derived growth factor consists of two forms of PDGF. Cell 1985; 43:277.
84. Associan RK, Fleurdelys BE, Stevenson HC, et al. Expression and secretion of type beta transforming growth factor by activated human macrophges. Proc Natl Acad Sci USA 1987; 84:6020.
85. Khalil N, Bereznay O, Sporn M, Greenberg AH. Macrophage production of transforming growth factor beta and fibroblast collagen synthesis in chronic pulmonary inflammation. J Exp Med 1989; 170:727.
86. Khalil N, O'Connor R, Unruh H, Warren P, Kemp A, Greenberg A. Enhanced expression and imunohistochemical distribution of transforming growth factor-beta in idiopathic pulmonary fibrosis. Chest 1991; 99:65S–66S.
87. Berridge MJ. Molecular basis of communication within the cell. Sci Am 1985; 253:142–152.
88. Bitterman PB, Rennard SI, Hunninghake GW, Crystal RG. Human alveolar macrophage growth factor for fibroblasts: regulation and partial characterization. J Clin Invest 1982; 70:806–822.
89. Doherty DE, Worthen GS, Henson PM. Inflammation in the interstitial lung disease. In: Schwarz MI, King TE Jr, eds. Interstitial Lung Disease. Philadelphia: Mosby-Year Book, 1992:23–43.
90. Volkman A, Gowans JL. Production of macrophages in the rat. Br J Exp Pathol 1965; 46:50–57.
91. Virolainen M. Hematopoetic origin of macrophages as studied by chromosome markers in mice. J Exp Med 1968; 127:943–951.
92. Wulff HR, Sparlevohn S. The origin of mononuclear cells in human skin windows. Acta Pathol Microbiol Scand 1966; 68:401.
93. Morstyn G, Burgess AW. Hemopoietic growth factors: a review. Cancer Res 1988; 48:5624–5637.
94. Metcalf D, Morstyn G. Colony-stimulating factors: general biology. In: DeVita VT Jr, Hellman S, Rosenberg SA, eds. Biologic Therapy of Cancer. Philadelphia: JB Lippincott, 1991:417–444.
95. Groopman JE, Molina J-M, Scadden DT. Hematopoietic growth factors: biology and clinical applications. N Engl J Med 1989; 321:1449–1459.
96. Moore MAS. Clinical implications of positive and negative hematopoietic stem cell regulators. Blood 1991; 78:1–19.
97. Mangan DF, Welch GR, Wahl SM. Lipopolysaccharide, tumor necrosis factor-α, and IL-1β prevent programmed cell death (apoptosis) in human peripheral blood monocytes. J Immunol 1991; 146:1541.1546.
98. Mangan DF, Robertson B, Wahl SM. IL-4 enhances programmed cell death (apoptosis) in simulated human monocytes. J Immunol 1992; 148:1812–1816.

99. Doherty DE, Downey G, Worthen GS, Haslett C, Henson PM. Monocyte retention and migration in pulmonary inflammation: requirement for neutrophils. Lab Invest 1988; 59:200–213.
100. Doherty DE, Henson PM, Clark RAF. Fibronectin fragments containing the RGDS cell-binding domain mediate monocyte migration into the rabbit lung: a potential mechanism for C5 fragment-induced monocyte lung accumulation. J Clin Invest 1990; 86:1065–1075.
101. Doherty DE, Worthen GS. Lipopolysaccharide-induced monocyte lung retention in the rabbit: the role of the CD11/CD18 leukocyte adhesion complex and cell stiffness. American Thoracic Society Meeting, May 1991, Anaheim, CA. Am Rev Respir Dis 1991; 143:A583.
102. Doherty DE, Hirose N, Cherniack RM. Monocyte lung accumulation during the acute phase of bleomycin-induced lung inflammation in the rabbit: noninvasive assessment by external scintigraphy. Lab Invest 1992; 66:231–242.
103. Doerschuk CM, Downey GP, Doherty DE, et al. Leukocyte and platelet margination within the microvasculature of rabbit lungs. J Appl Physiol 1990; 68:1956–1961.
104. Ohgami M, Doerschuk CM, Gie RP, English D, Hogg JC. Monocyte kinetics in rabbits. J Appl Physiol 1991; 70:152–157.
105. Van Furth R, Sluiter W. Distribution of blood monocytes between a marginating and a circulating pool. J Exp Med 1986; 163:474–479.
106. Coalson JJ, Hinshaw LB, Guenter CA. The pulmonary ultrastructure in septic shock. Exp Mol Pathol 1970; 12:84.
107. Haslett C, Worthen GS, Giclas PC, Morrison DC, Henson JE, Henson PM. The pulmonary vascular sequestration of neutrophils in endotoxemia is initiated by an effect of endotoxin on the neutrophil in the rabbit. Am Rev Respir Dis 1987; 136:9–18.
108. Kane MA, May JE, Frank MM. Interactions of the classical and alternate complement pathway with endotoxin lipopolysaccharide effect on platelets and blood coagulation. J Clin Invest 1973; 52:370.
109. Meyrick B, Brigham KL. Repeated *Escherichia coli* endotoxin-induced pulmonary inflammation causes chronic pulmonary hypertension in sheep: structural and functional changes. Lab Invest 1986; 55:164–176.
110. Doherty DE, Schawb B, Elson E, Worthen GS. Lipopolysaccharide (LPS)-induced monocyte lung retention: the role of cell stiffness. Am Rev Respir Dis 1990; 141:A642.
111. Downey GP, Doherty DE, Schwab B III, Elson EL, Henson PM, Worthen GS. Retention of leukocytes in capillaries: role of cell size and deformability. J Appl Physiol 1990; 69:1767–1778.
112. Schmidt-Schonbein GW, Shih YY, Chien S. Morphometry of human leukocytes. Blood 1985; 56:866–876.
113. Doherty DE. Chest 1993 (Suppl). In press.
114. Valberg PA, Meyrick B, Brain JD, Brigham KL. Phagocytic and motile properties of endothelial cells measured magnetometrically: effects of endotoxin. Tissue Cell 1988; 20:345–354.

115. Doherty DE, Zagarella L, Henson PM, Worthen GS. Lipopolysaccharide stimulates monocyte adherence by effects on both the monocyte and the endothelial cell. J Immunol 1989; 143:3673–3679.
116. Bevilaqua MP, Pober JS, Wheeler ME, Cotran RS, Gimbrone MA. Interleukin-1 acts on cultured vascular endothelium to increase the adhesion of polymorphonuclear leukocytes, monocytes, and related leukocyte cell lines. J Clin Invest 1985; 76:2003–2011.
117. Bevilaqua MP, Pober JS, Wheeler ME, Cotran RS, Gimbrone MA. IL-1 activation of vascular endothelium: effects on procoagulant activity and leukocyte adhesion. Am J Pathol 1985; 121:394.
118. Erzurum SC, Downey GP, Doherty DE, Schwab B III, Elson EL, Worthen GS. Mechanisms of lipopolysaccharide-induced neutrophil retention: relative contributions of adhesive and cellular mechanical properties. J Infect 1992; 149:154–162.
119. Osborn L. Leukocyte adhesion to endothelium in inflammation. Cell 1990; 62:3–6.
120. Wallis WJ, Beatty PG, Ochs HD, Harlan JM. Human monocyte adherence to cultured vascular endothelium: monoclonal antibody-defined mechanisms. J Infect 1985; 135:2323–2330.
121. Mentzer Sj, Crimmins MAV, Burakoff SJ, Faller DV. Alpha and beta subunits of the LFA-1 membrane molecule are involved in human monocyte–endothelial cell adhesion. J Cell Physiol 1987; 130:410–415.
122. Wright SD, Ramos RA, Tobias PS, Ulevitch RJ, Mathison JC. CD14, a receptor for complexes of lipopolysaccharide (LPS) and LPS-binding protein. Science 1990; 249:1431–1433.
123. Pereira HA, Shafer WM, Pohl J, Martin LE, Spitznagel JK. CAP37, a human neutrophil-derived chemotactic factor with monocyte specific activity. J Clin Invest 1990; 85:1468–1476.
124. Postlewaite AE, Kang AH. Collagen- and collagen peptide-induced chemotaxis of human blood monocytes. J Exp Med 1976; 143:1299–1307.
125. Senior RM, Griffen GL, Mecham RP. Chemotactic activity of elastin-derived peptides. J Clin Invest 1980; 66:859–862.
126. Clark RAF, Wickner NE, Doherty DE, Norris DA. Cryptic chemotactic activity of fibronectin for human monocytes resides in the 120k Da fibroblastic cell-binding fragment. J Biol Chem 1988; 263:12115–12123.
127. McDonald JA, Kelley DG. Degradation of fibronectin by human leukocyte elastase. J Biol Chem 1980; 255:8848–8858.
128. Winkler GC. Pulmonary intravascular macrophages in domestic animal species: review of structural and functional properties. Am J Anat 1988; 181:217–234.
129. Wheeldon EG, Hansen-Flaschen J. Intravascular macrophages in the sheep lung. J Leukoc Biol 1986; 40:657–661.
130. Albertine KH, Staub NC. Vascular tracers alter hemodynamics and airway pressure in anesthetized sheep. Microvasc Res 1986; 32:279–288.
131. Miyamoto K, Schultz E, Heath T, Mitchell M, Albertine K, Staub NC. Pulmonary intravascular macrophages and hemodynamic effects of liposomes in sheep. J Appl Physiol 1988; 64:1143–115.

132. Riches DWH. The multiple roles of macrophages in wound healing. In: Clark RAF, Henson PM, eds. The Molecular and Cellular Biology of Wound Repair. New York: Plenum Press, 1988:213–239.
133. Riches DWH, Channon JY, Leslie CC, Henson PM. Receptor-mediated signal transduction in mononuclear phagocytes. Prog Allergy 1988; 42:65–122.
134. Laszlo DJ, Henson PM, Weinstein L, et al. Development of functional diversity in mouse macrophages. Mutual exclusion of two phenotypic states. Am J Pathol 1993. In press.
135. Noble PW, Henson PM, Carre PC, Riches DWH. TGFβ primes macrophages to express inflammatory gene products in response to particulate stimuli by an autocrine/paracrine mechanism. J Immunol 1993. In press.
136. Aderem A, Cohen D, Wright S, Cohn Z. Bacterial lipopolysaccharides prime macrophages for enhanced release of arachidonic acid metabolites. J Exp Med 1986; 164:165–179.
137. Adams DO, Hamilton TA. The cell biology of macrophage activation. Annu Rev Immunol 1984; 2:283–318.
138. Green SJ, Meltzer MS, Hibbs JBJ, Nacy CA. Activated macrophages destroy intracellular *Leishmania major* amastigotes by an L-arginine-dependent killing mechanism. J Immunol 1990; 144:278–283.
139. Lake F, Noble P, Henson P, Riches D. Functional switching of macrophage responses to TNFα by interferons. Implications for the pleiotropic activities of TNFα. 1993. Submitted for publication.
140. Underwood GA, Riches DWH. Transmembrane-mediated changes in $[Ca^{2+}]_i$ are involved in the signalling pathway leading to macrophage cytocidal differentiation. Implications of localized changes in $[Ca^{2+}]_i$ and of IFN-priming on Ca^{2+} utilization. Mol Biol Cell 1992; 3:335–347.
141. Heremans H, Dijkmans R, Sobis H, Vandekerckhove F, Billiau A. Regulation by interferons of the local inflammatory response to bacterial lipopolysaccharide. J Immunol 1987; 138:4175–4179.
142. Heremans H, Billiau A, Coutelier JP, De SP. The inhibition of endotoxin-induced local inflammation by LDH virus or LDH virus-infected tumors is mediated by interferon. Proc Soc Exp Biol Med 1987; 185:6–15.
143. Heremans H, Billiau A. The potential role of interferons and interferon antagonists in inflammatory disease. Drugs 1989; 38:957–972.
144. Mecs I, Koltai M. In vivo hyporesponsiveness induced by Sendai virus in CFLP mice. Acta Virol (Praha) 1982; 26:346–352.
145. Shiozawa S, Shiozawa K, Kita M, Kishida T, Fujita T, Imura S. A preliminary study on the effect of alpha-interferon treatment on the joint inflammation and serum calcium in rheumatoid arthritis. Br J Rheumatol 1992; 31:405–408.
146. Boraschi D, Censini S, Bartalini M, et al. Interferon inhibits prostaglandin biosynthesis in macrophages: effects on arachidonic acid metabolism. J Immunol 1984; 132:1987–1992.
147. Boraschi D, Censini S, Bartalini M, Tagliabue A. Regulation of arachidonic acid metabolism in macrophages by immune and nonimmune interferons. J Immunol 1985; 135:502–505.

148. Hyde D, Henderson T, Giri S, Tyler N, Stovall M. Effect of murine gamma interferon on the cellular responses to bleomycin in mice. Exp Lung Res 1988; 14:687–704.
149. Granstein R, Deak M, Jacques S, et al. The systemic administration of gamma-interferon inhibits collagen synthesis and acute inflammation in a murine skin wounding model. J Invest Dermatol 1989; 93:18–27.
150. Granstein R, Flotte T, Amento E. Interferons and collagen production. J Invest Dermatol 1990; 95:75S–80S.
151. Drapier JC, Hibbs JBJ. Murine cytotoxic activated macrophages inhibit aconitase in tumor cells. Inhibition involves the iron-sulfur prosthetic group and is reversible. J Clin Invest 1986; 78:790–797.
152. Harbrecht BG, Billiar TR, Stadler J, et al. Nitric oxide synthesis serves to reduce hepatic damage during acute murine endotoxemia. Crit Care Med 1992; 20:1568–1574.
153. Mulligan MS, Hevel JM, Marletta MA, Ward PA. Tissue injury caused by deposition of immune complexes is L-arginine dependent. Proc Natl Acad Sci USA 1991; 88:6338–6342.
154. Ialenti A, Ianaro A, Moncada S, Di RM. Modulation of acute inflammation by endogenous nitric oxide. Eur J Pharmacol 1992; 211:177–182.
155. Brandtzaeg P, Kierulf P, Gaustad P, et al. Plasma endotoxin as a predictor of multiple organ failure and death in meningococcal disease. J Infect Dis 1989; 159:195–204.
156. Donner RL, Elin RJ, Hossein SM, Wesley RA, Reilly JM, Parillo JE. Endotoxemia in human septic shock. Chest 1991; 99:169–175.
157. Standiford TJ, Kunkel SL, Rolfe MW, Evanoff HL, Allan RM, Strieter RM. Regulation of human alveolar macrophage and blood monocyte-derived interleukin-8 by prostaglandin E_2 and dexamethasone. Am J Respir Cell Mol Biol 1992; 6:675–681.
158. Parsons PE, Fowler AA, Hyers TM, Henson PM. Chemotactic activity in bronchoalveolar lavage fluid from patients with adult respiratory distress syndrome. Am Rev Respir Dis 1985; 141:98–103.
159. Miller EJ, Cohen AB, Nagao S. Elevated levels of NAP-1/interleukin-8 are present in the airspaces of patients with the adult respiratory distress syndrome and are associated with increased mortality. Am Rev Respir Dis 1992; 146: 427–432.
160. Hack CE, Hart M, van Schijndel RJ, et al. Interleukin-8 in sepsis: relation to shock and inflammatory mediators. Infect Immun 1992; 60:2835–2842.
161. Leff J, Baer J, Kirkman J, Bodman M, Ostro M, Repine J. Post-injury treatment with liposome-encapsulated prostaglandin E_1 (Lip-PGE_1 or TLC C-53), but not free PGE_1, decreases acute edematous lung injury (ARDS) in rats given interleukin-1 intratracheally. Am Rev Respir Dis 1992; 145:A845.
162. Siler TM, Swierkosz JE, Hyers TM, Fowler AA, Webster RO. Immunoreactive interleukin-1 in bronchoalveolar lavage fluid of high-risk patients with ARDS. Exp Lung Res 1989; 15:881–894.
163. Jacobs RF, Tabor DR, Burks AW, Campbell GD. Elevated interleukin-1 release by human alveolar macrophages during the adult respiratory distress syndrome. Am Rev Respir Dis 1989; 140:1686–1692.

164. Girardin E, Graw GE, Dayer JM, Roux-Lombard P, Group TJS, Lambert PH. Tumor necrosis factor and interleukin-1 in the serum of children with severe infectious purpura. N Engl J Med 1988; 319:397–400.
165. Marks JD, Marks CB, Luce JM, Montgomery AB, Turner J, Metz CA].F. Plasma tumor necrosis factor in patients with septic shock: mortality rate, incidence of ARDS, and effects of methylprednisolone administration. Am Rev Respir Dis 1990; 141:94–97.
166. Waage A, Brondtzag P, Halstensen A. The complex pattern of cytokines in serum from patients with meningococcal septic shock. Association between interleukin-6, interleukin-1 and the fatal outcome. J Exp Med 1989; 169:333–338.
167. Parsons PE, Moore FA, Moore EE, Ikle' DN, Henson PM, Worthen GS. Studies on the role of tumor necrosis factor in adult respiratory distress syndrome. Am Rev Respir Dis 1992; 146:694–700.
168. Herzyk DJ, Wewers MD. ELISA detection of IL-1β in human sera needs independent confirmation. Am Rev Respir Dis 1993; 147:139–142.
169. Fisher CJ, Opal SM, Dhainaut J-F. Influence of an anti-tumor necrosis factor monoclonal antibody on cytokine levels in patients with sepsis. Crit Care Med 1993; 21:318–327.
170. Campagna AC, Adelman D, Szaflarski N, Cheng P, Matthay MA. The predictive value of interleukin-6 (IL-6) levels in the serum of sepsis patients. Am Rev Respir Dis 1993; 147:A97.

13

Endotoxin-Dependent Cytokine Networks

STEVEN L. KUNKEL, STEPHEN W. CHENSUE, THEODORE J. STANDIFORD, and ROBERT M. STRIETER

University of Michigan Medical School
Ann Arbor, Michigan

I. Introduction

Increasing scientific evidence supports the concept that many of the pathophysiologic manifestations of pulmonary diseases are due to the expression of a variety of polypeptide mediators known as cytokines (1). These inflammatory mediators are not limited to a particular set of polypeptides but represent a large collection of heterogeneous factors, which are exemplified by the interferons, interleukins, tumor necrosis factors, growth factors, and colony-stimulating factors. Historically, these polypeptides were thought to be synthesized only by leukocytes, but studies now have demonstrated that all nucleated cells are potential sources of cytokines. While cytokines do indeed represent a heterogeneous group of polypeptide mediators, they share a number of common parameters. Cytokines often possess overlapping activities, they are pleiotropic, they exert different effects at different concentrations, they can either synergize or antagonize the effects of other cytokines, they are regulated in a complex manner, and they often function via cytokine cascades. The latter aspect of cytokine biology is important in establishing cell-to-cell communication circuits during the evolution of an inflammatory response (2–4). This process is paramount in the normal orchestration of inflammation, as the expression of an early response cytokine can activate a particular cell to express a more distal cytokine, which may be important in maintaining the inflammatory response. Thus cytokines can activate a variety of immune and nonimmune cells, as well as cause specific inflammatory cells to proliferate, differentiate, and undergo chemotaxis and phagocytosis.

In addition, cytokine cascades are probably involved in the generation of a localized inflammatory response, as well as mediating the progression of systemic life-threatening inflammatory reactions. Under physiologic conditions, cytokines are important in regulating the activity of immune/inflammatory cells, endothelial cells, and stromal cells. However, when produced aberrantly or in an exaggerated manner, these cytokines can exert toxic effects. For example, there is strong evidence which suggests that the clinical manifestations of sepsis and adult respiratory distress syndrome (ARDS) are due to the unregulated production of certain cytokines, especially interleukin-1 (IL-1) and tumor necrosis factor (TNF) (5–10). The role of TNF as an early response mediator of septic shock has been addressed in both clinical and experimental models (11,12).

Clinical studies examining TNF levels in sera from patients with meningitis-induced septicemia found a high correlation with the circulating levels of this cytokine and the mortality associated with this disease (6–8,10). In addition, initial experimental studies assessing the effects of systemic TNF challenge found that this mediator could mimic the pathophysiologic effects of lipopolysaccharide (LPS) or live gram-negative bacterial challenge. In these experimental studies the systemic administration of TNF resulted in fever, intraalveolar hemorrhage, edema, increased levels of circulating catecholamines, and multiorgan dysfunction (13–16). In further support of a role of TNF in sepsis, the administration of neutralizing antibody against TNF was found to markedly attenuate the pathophysiologic effects of LPS (16). Additional experimental studies demonstrated that neutralizing TNF activity in a mouse model of endotoxemia could significantly augment the survival of the LPS-treated animals (11). Subsequent studies demonstrated that treatment of baboons with a monoclonal antibody directed against human TNF was found to protect against a subsequent challenge with live *Escherichia coli* (12). These animal studies are of interest in that much of the subsequent pathology normally associated with gram-negative bacteremia was significantly reduced with in vivo anti-TNF antibody treatment. The importance of these findings are further underscored by the observations that pathologic changes not directly induced by TNF alone were ameliorated with the neutralizing antibody. Lipopolysaccharide-induced tumor necrosis factor was probably activating a cytokine cascade in vivo, which in turn was generating more distal mediators/cytokines, and the latter mediators were then exerting many of the actual pathologic effects. These observations were important in defining specific LPS-derived early response cytokines, which are able to initiate cytokine networks via their interactions with a number of immune/inflammatory and nonimmune cell populations.

Presently, a number of phase II and III clinical trials are in process to determine the efficacy of anti-TNF monoclonal antibodies in the treatment of sepsis syndrome. These double-blind trials will have endpoints that include survival at 28 days and prevention or reversal of shock and organ failure (17). Much of the strategy behind the development of potential therapies targeting TNF in patients with sepsis has taken into account the role of LPS/TNF in establishing in vivo pathologic cytokine networks. In this chapter we focus on the expression of cytokine genes in response to lipopolysaccharide and assess the subsequent networks which are likely to play a role in the pathophysiology of LPS-dependent inflammation.

II. Pathologic Effects of Experimental Endotoxemia

The pathophysiologic effect of LPS are clearly mediated via the release of soluble polypeptide cytokines derived from mononuclear phagocytic cells. These early response cytokines include, but are not limited to, IL-1, TNF, and IL-6 (18–20). While these early signals alter the normal physiology of the host, they also orchestrate a chain of events that may result in life-threatening consequences. A number of experimental models have assessed the hypothesis that the effects of LPS are secondary to the generation of host-derived soluble mediators. For example, the administration of recombinant TNF to experimental animals caused injury to a number of organs, including kidneys, lungs, adrenal glands, and small intestines. The latter organ was particularly sensitive to the effects of TNF, and the pathology was similar to that caused by LPS. Changes in the small intestine were probably due to diminished blood flow as a consequence of vasoconstriction. Histologically, the pattern of injury in the small bowel was compatible with ischemic changes, as the tips of the intestinal villi were necrotic and the submucosal intestinal endothelial cells demonstrated blebbing and damaged endothelial cells (15, 16).

Subsequent production of TNF in response to the infusion of LPS in the foregoing experimental models was rapid and transient in character. Kinetic profiles of LPS-induced TNF production was demonstrated to occur over a wide concentration range of LPS. Mice challenged with different concentrations of LPS (2 mg to 10 μg per animal) were found to generate biologically active TNF by 1 h and return to basal levels by 4 h. A similar pattern of expression for LPS-induced IL-1 was also observed in this model, as IL-1 was detected by 1 h post-LPS exposure and the levels of IL-1 then persisted for approximately 16 h. The expression of steady-state TNF mRNA was also found to be extremely rapid and transient. The levels of TNF mRNA were markedly increased by 15 min post-LPS infusion and reached a peak in steady-state levels by 30 min. By 60 min post-LPS administration, the levels of TNF mRNA were no longer detectable. These data support the concept that TNF is indeed an early response cytokine which is generated rapidly after LPS challenge. Studies assessing the regulatory mechanisms that control TNF expression at the cellular and molecular levels have shown that dexamethasone can suppress LPS-induced TNF expression. However, the efficacy of dexamethasone was demonstrated only if this anti-inflammatory steroid was given prior to LPS exposure. The delay in treatment with dexamethasone for 20 min after LPS challenge had no suppressive effects on the expression of TNF. This information again supports the role of TNF as a polypeptide that is transcribed and translated rapidly after LPS stimulation.

III. LPS-Induced Cytokine Networks in the Liver

As discussed above, the rapid expression of specific cytokines and the delayed pathology induced by these mediators post LPS challenge suggests that a cascade of events may be important in dictating the pathologic response in endotoxemia. In acute, systemic pathologic states, such as endotoxemia, fixed tissue macrophages probably play an important role in the initiation of this cascade (21). Recent investigations have demonstrated the in vivo contribution of fixed macrophages, such as Kupffer cells, to the production of specific cytokines known to be important in endotoxin-mediated events. The expression of antigenic IL-1α, IL-1β, and TNF in the liver were assessed by immunohistochemical techniques at 1, 3, and 6 h post-LPS challenge. The immunohistochemical localization of all three cytokines was clearly restricted to the sinusoidal lining cells and was not observed in bile ductal cells, endothelial cells, or hepatocytes. The staining pattern of these sinusoidal cells for IL-1α, IL-1β, and TNF were consistent with sessile macrophages in the liver (Kupffer cell), as these cells were also immunoreactive for the mouse macrophage antigen Mac-1. The staining pattern for IL-1α, IL-1β, and TNF was observed only in LPS-challenged animals, as the sinusoidal lining cells in control animals were not reactive.

The kinetics in the appearance of antigenic IL-1α, IL-1β, and TNF by the Kupffer cells demonstrated a rapid expression pattern. As shown in Fig. 1, antigenic TNF was observed by 1 h post-LPS challenge followed by the expression of antigenic IL-1β and then IL-1α. The production of antigenic TNF (1 h) was rapidly followed by the expression of antigenic IL-1β (2 to 3 h) and then antigenic IL-1α (3 to 6 h). The specific expression of antigenic

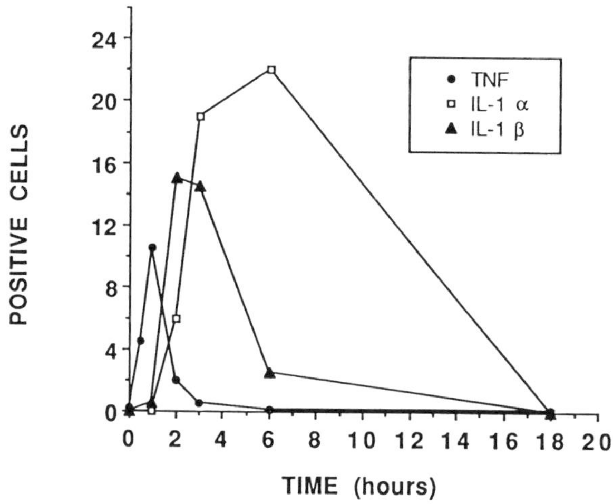

Figure 1 Kinetics in the expression of antigenic TNF, IL-1α, and IL-1β by mouse Kupffer cells postinfusion of LPS. Data are expressed as number of antigen-positive cells per high-power field. Positive cells were localized via immunohistochemical techniques.

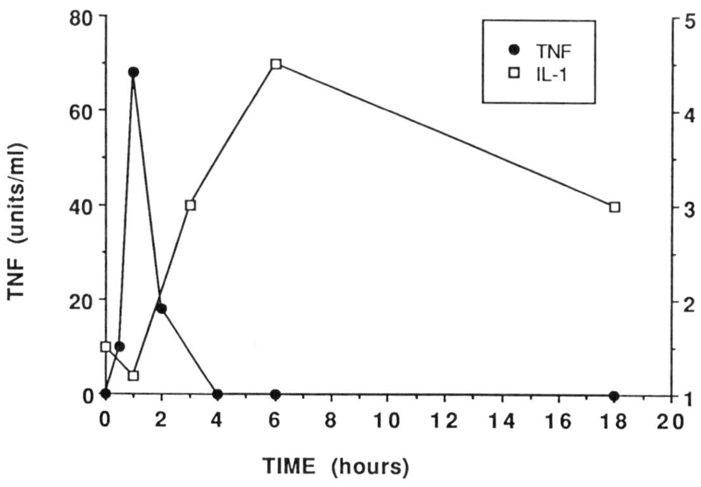

Figure 2 Kinetics in the expression of TNF and IL-1 in the sera of mice treated in vivo with LPS.

TNF and IL-1 by Kupffer cells mimics the kinetic pattern in the production of biologically active TNF and IL-1 found in the circulating sera (Fig. 2).

Previous studies have demonstrated that corticosteroids can suppress the production of IL-1 and TNF both in vivo and in vitro (22,23). Thus the effects of this immunosuppressive steroid was assessed on the production of Kupffer cell–derived cytokines. Experimental animals treated with 4 mg/kg of dexamethasone 3 h prior to endotoxin challenge were studied for the expression of antigenic TNF, IL-1α, and IL-1β. While the production of both antigenic TNF and IL-1α were reduced by 60 to 80%, IL-1α was only reduced by approximately 30%. Serum levels for biologically active TNF accurately reflected the suppressive effects of dexamethasone treatment, as systemic TNF levels were reduced by 75%. However, IL-1 activity was reduced by only about 40%. Since endotoxemia is known to induce endogenous cortisol production, and both TNF and IL-1 can cause the expression of cortisol in vivo, studies were conducted to define the relationship between cytokine production and cortisol production (Fig. 3). An in vivo LPS challenge caused a rise in corticosterone levels by 1 hour and were maximal by 6 h. The kinetics in corticosterone production was similar to the kinetics of IL-1 activity. This observation may aid in explaining the relative resistance of IL-1 production to the effects of anti-inflammatory steroids, as this cytokine may physiologically be expressed in a cortisol-rich environment.

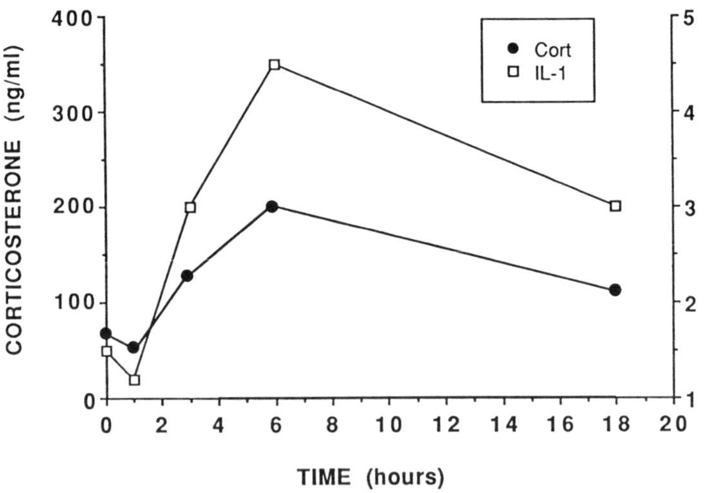

Figure 3 Kinetics in the concentration of corticosterone and IL-1 in the sera of mice following in vivo challenge with LPS.

IV. Kupffer Cell–Hepatocyte Interactions

The presence of neutrophils in the liver is associated with a number of pathologic disorders, including ischemia–reperfusion and multiple-organ failure (24,25). The latter state may be observed subsequent to endotoxemia-induced sepsis. The hypothesis that TNF mediates the liver injury and neutrophil influx which occur after exposure of experimental animals to LPS is supported by studies showing that neutralization of circulating TNF with specific antisera attenuated liver injury in LPS-challenged galactosamine-sensitized mice (26). In addition, studies have shown that antibodies to TNF can block much of the liver injury, as determined by reduction of plasma AST activity and total plasma bilirubin concentration, normally due to LPS treatment of mice. Depletion of neutrophils also affords protection against LPS-induced hepatotoxicity. Thus data suggest that LPS-dependent TNF production may be networking with cells in the liver and aid in the recruitment of neutrophils to this organ.

Based on the observations that both IL-1 and TNF can be expressed by the resident, fixed macrophages in the liver, in vitro experiments were designed to assess the ability of IL-1 and/or TNF to induce the production of the neutrophil chemotactic cytokine interleukin-8 (IL-8) by hepatocytes. These studies were performed because nonimmune cells have been identified as an important source of IL-8 (27–29). This chemotactic cytokine or chemokine has been identified as a low-molecular-weight polypeptide that possesses activity for the movement of neutrophils at n*M* concentrations. Interleukin-8 is a novel chemotactic polypeptide that belongs to a supergene family of inflammatory mediators (30,31). This supergene family is characterized by the configuration of two of the four cysteine amino acids being separated by one addition amino acid (C-X-C family). This amino acid motif is conserved throughout the supergene family members. In general terms, members of this polypeptide family have relative specificity for the recruitment of neutrophils. Although a number of polypeptide mediators are susceptible to physiochemical degradation, IL-8 is relatively resistant to changes in pH, temperature, and mild enzymatic treatment. The stability of IL-8 and the expression of this chemotactic cytokine by both immune/inflammatory and nonimmune cells make this factor a leading candidate in mediating the pathologic processes characterized by significant neutrophil infiltrates (32).

The importance of IL-8 derived by nontraditional inflammatory cells has been studied in the context of organ-based inflammation. Initial investigations have demonstrated that human hepatocyte cell lines challenged with either IL-1 or TNF, but not LPS or IL-6, could express high levels of a neutrophil chemotactic factor. By 2 to 4 h post-TNF or IL-1 stimulation, human hepatoma cell lines began to express a neutrophil chemotactic activity that could be detected in the conditioned media. This neutrophil chemotactic activity continued to rise over the next 20 to 30 h poststimulation with IL-1 or TNF. Hepatoma cells treated with LPS did not generate a neutrophil chemotactic factor over the defined study period. The chemotactic activity of the hepatoma cell line—conditioned media, treated with specific antibodies directed against IL-8, were found to possess reduced levels of neutrophil chemotactic activity. Northern blot analysis of RNA isolated from the stimulated cells demonstrated a rapid rise in the expression of steady-state IL-8 mRNA. Within the first hour postchallenge, the hepatocytes initiated transcription for IL-8, which peaked by approximately 4 h (Fig. 4). The expression of steady-state IL-8 mRNA continued to be evident as long as IL-1 or TNF was in the media. Once the cells were washed and the agonists were no longer present, IL-8 mRNA decayed in a relatively rapid manner. In addition to the kinetic expression of steady-state IL-8 mRNA, the hepatocytes responded to either IL-1 or TNF in a dose-dependent manner. Both IL-1 and TNF in the pg/mL range were effective in inducing the expression of IL-8 mRNA. Interestingly, hepatocyte cell lines challenged with TNF plus cycloheximide resulted in the superinduction of IL-8 mRNA, suggesting that the expression of IL-8 mRNA may be regulated via intracellular repressor proteins.

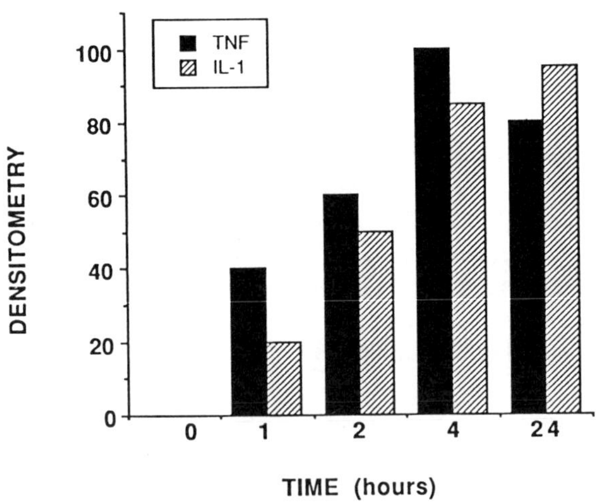

Figure 4 Kinetics in the expression of steady-state IL-8 mRNA synthesized by hepatocyte cell lines in response to IL-1 and TNF. The steady-state levels of IL-8 mRNA are represented by bar graphs of the laser densitometry scans from Northern blots of the hepatocytes.

The next area of investigation was directed at determining the potential for cytokine networking to occur in a particular organ. Since Kupffer cells were identified as a rich source of IL-1 and TNF, and both IL-1 and TNF were potent in inducing the production of IL-8 from hepatocytes, studies were initiated to determine cytokine communication links between Kupffer cells and hepatocytes. To determine the ability of Kupffer cells to secrete soluble factors that could induce IL-8 production from hepatocytes, conditioned media from LPS-stimulated Kupffer cells were overlaid on human hepatocytes. While control medium alone did not cause an increase in the expression of hepatocyte steady-state IL-8 mRNA, the conditioned media from the LPS-stimulated Kupffer cells did elicit a significant expression in IL-8 mRNA. As shown in Fig. 5, conditioned media from both LPS-treated monocytes or Kupffer cells could induce steady-state IL-8 mRNA from cultured hepatocytes by 4 to 6 h. The levels of hepatocyte-derived IL-8 mRNA continued to be detected by 24 h postchallenge. To determine the role of IL-1 and/or TNF in mediating the expression of hepatocyte-derived IL-8, IL-1 and TNF neutralization studies were conducted on the Kupffer cell–conditioned media. Preincubation with neutralizing antibody to IL-1 and TNF effectively reduced the ability of the LPS-conditioned media to express hepatocyte-derived IL-8.

In the scenario above, a primary challenge, such as LPS, can serve as an effective signal for the fixed Kupffer cells, which can respond by generating

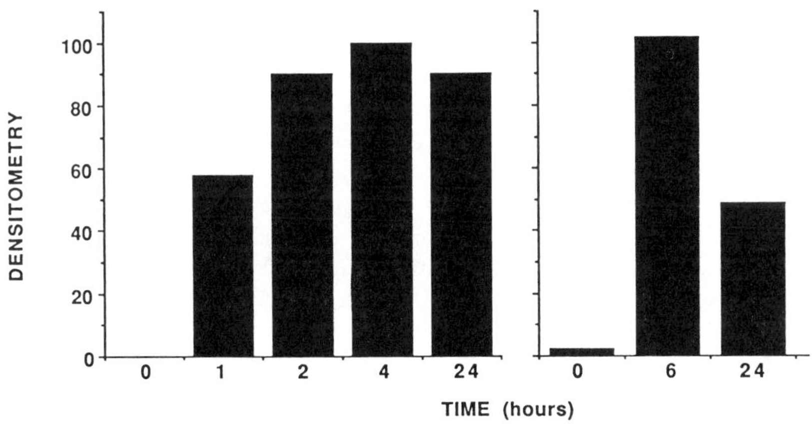

Figure 5 Kinetics in the expression of steady-state IL-8 mRNA by normal human hepatocytes in response to LPS-conditioned media from human monocyte and Kupffer cells. The steady-state levels of IL-8 mRNA are represented by bar graphs of the laser densitometry from Northern blots of the hepatocytes.

sufficient levels of IL-1 and TNF necessary to activate resident hepatocytes. The resident hepatocytes, in turn, respond to the IL-1 and TNF generated by the Kupffer cells and generate IL-8. The production of IL-8 serves as a potent chemotactic factor for the elicitation of neutrophils into the liver. These observations are important in that they describe a unique cell-to-cell communication system for the recruitment of granulocytes to an organ which is often involved in multiorgan failure due to fulminate sepsis.

V. LPS-Dependent Cytokine Networks in the Lungs

The sequestration of neutrophils is not unique to the liver or any particular organ, but is a characteristic associated with a variety of acute inflammatory reactions in all organs. However, in pathologic processes associated with endotoxemia and sepsis, the lungs are often identified as susceptible target organs. The presence of neutrophils in the interstitium or alveolar space is due to the production of chemotactic factors that establish a gradient from the vasculature to the site of inflammation. Although various peptides, polypeptides, and lipids have been identified as lung-derived chemotactic factors, none of these have specificity for recruiting a particular leukocyte population. In addition, the pulmonary origin of chemotactic factors has historically been an enigma. While substantial information has been generated regarding alveolar macrophages as a source of chemotactic factors, recent studies have identified other cellular constituents of the alveolar capillary wall as important sources of chemotactic factors. Both pulmonary fibroblasts and pulmonary type-II-like epithelial cells can express the chemotactic cytokine IL-8 in a stimulus-specific pattern (2,3). These lung cells respond in a manner similar to that of hepatocytes when challenged with particular stimulants. While alveolar macrophages have the ability to generate a number of cytokines in response to LPS, the noninflammatory cells in the lung are generally unresponsive to this activating agent. The participation of these cells in an inflammatory event appears to be dependent on cytokine networking, a repeating scenario that is probably common to all organs.

The activation of this pathway in the lungs may be especially important in disease states characterized by bacteremia, endotoxemia, sepsis, or septic shock, as the lungs are particularly vulnerable to these pathologic states. As with the Kupffer cell in the sinusoidal lining of the liver, the alveolar and interstitial lung macrophages are important in generating early response signals (IL-1 and TNF) when challenged with LPS. These early response cytokines can then act in either an autocrine or paracrine fashion to stimulate the production of more distal mediators of inflammation by the nonimmune cells found in the lung. To test this hypothesis, pulmonary fibroblasts were challenged with LPS, TNF, or IL-1 and the conditioned fibroblast media assessed for neutrophil chemotactic activity. Both IL-1 (IL-1β) and TNF caused a significant increase in the production of a neutrophil chemotactic factor by 4 h postchallenge, which peaked by 8 h and continued to be expressed for the next 20 h. In these studies IL-1β was a more potent agonist than TNF for the production of the neutrophil chemotactic factor. However, LPS over a wide concentration range did not induce pulmonary fibroblasts to produce a neutrophil chemotactic factor over the 24-h study period. These observations led to subsequent investigations to determine if the chemotactic activity was due, at least in part, to IL-8. Pulmonary fibroblasts were challenged with IL-1β (20 ng/mL), TNF (20 ng/mL), or LPS (1 μg/mL) and the expression of steady-state IL-8 mRNA was determined in a kinetic fashion. IL-1β and TNF were both effective in causing the expression of IL-8 mRNA in a rapid and prolonged fashion, while LPS did not cause the expression of this chemokine gene. By 1 h poststimulation, IL-1β and TNF induced detectable levels of IL-8 mRNA that persisted for 24 h. Steady-state IL-8 mRNA levels were also found to be elevated in a dose-dependent manner in response to either IL-1β or TNF but not to LPS. Concentrations of IL-1β and TNF in the pg/mL range could cause a significant increase in the expression of IL-8 mRNA from the pulmonary fibroblasts. These same findings were also observed at the translational level, as both IL-1β and TNF, over a wide dose-response range, could increase the production of antigenic IL-8. However, the production of antigenic IL-8 by LPS-stimulated pulmonary fibroblasts was not significantly elevated over that of control cells.

In the experiments above, signal specificity for the generation of IL-8 by pulmonary fibroblasts was demonstrated. Both IL-1 and TNF stimulation of the pulmonary fibroblasts resulted in the expression of IL-8 mRNA, antigenic IL-8 protein, and biologically active IL-8, while LPS was not efficacious in this system. Since the resident alveolar macrophage is susceptible to LPS stimulation with the rapid generation of IL-1 and TNF, studies were designed to determine if alveolar macrophage-derived soluble mediators could influence the production of pulmonary fibroblast–derived IL-8. In this study normal human alveolar macrophages were challenged with 1 μg/mL of LPS for 24 h and the macrophage supernatant was collected. The supernatant was then overlaid on cultured fibroblasts in the presence or absence of neutralizing TNF or IL-1β antibody. The pulmonary fibroblast were placed back in culture for an additional 8 h and then total fibroblast RNA isolated. The addition of LPS-stimulated alveolar macrophage–conditioned media to the cultured pulmonary fibroblasts resulted in significant steady-state expression of pulmonary fibroblast-derived IL-8. However, the LPS-stimulated alveolar macrophage–conditioned media treated with either neutralizing antibody to IL-1β or TNF resulted in an approximate 40% reduction in steady-state IL-8 mRNA expression. Additional studies using a combination of neutralizating antibodies to both IL-1 and TNF resulted in over a 90% inhibition of fibroblast-derived IL-8 mRNA.

To further address the networking between alveolar macrophages and pulmonary fibroblasts, a co-culture system using both normal human alveolar macrophages and normal lung fibroblasts was utilized. These cells were cultured together in the absence or presence of 1 μg/mL LPS. At 1, 4, 8, and 24 h following the addition of LPS the cells were processed for immunohistochemical localization of antigenic IL-8. Upon the addition of LPS, the alveolar macrophages rapidly expressed antigenic IL-8, which was observed clearly by 4 h. However, at the 4 h time point, the expression of antigenic IL-8 by the lung fibroblasts in the co-culture system was low. By 8 h, significant levels of antigenic IL-8 could be detected by the fibroblasts. Finally, by 24 h, all of the pulmonary fibroblasts were found to be expressing large levels of antigenic IL-8. At this point, over 85% of the fibroblasts were positive for IL-8. These studies support the role of alveolar macrophages in aiding the granulocyte recruitment process to the lung interstitium and alveolar space via cytokine networking with lung resident cells. This resident lung macrophage can respond to a number of initiating agents and rapidly generate early response cytokines. These mediators are important in orchestrating the inflammatory response by activating inflammatory cells directly, as well as serving a paracrine role in networking with resident normal tissue. The latter activity elicits an effector response from the surrounding normal cells and augments the initiation and maintenance response of the inflammatory reaction.

VI. LPS-Dependent Bone Marrow–Derived Cytokines

The early nonspecific host response to an LPS challenge is complex and involves both a cellular and a humoral response. Paramount to the success of this early response is the recruitment of neutrophils from the vasculature to the site of inflammation. As the inflammatory event is initiated and maintained, a substantial increase in the absolute number of circulating neutrophils is often observed, with a disproportionate increase in the number of immature (band) neutrophils. Thus studies were initiated to address LPS-dependent, bone marrow–derived mononuclear cell production of the neutrophil chemotactic factor IL-8 (33). Bone marrow was removed from normal volunteers via the posterior iliac crest and the mononuclear fraction isolated by density-gradient centrifugation. The bone marrow–derived mononuclear cells responded to LPS in a time-dependent manner, with the maximum expression of IL-8 (4.5 ng/mL) occurring at 24 h postchallenge. These levels of IL-8 were approximately 12-fold lower than the levels observed for LPS-treated peripheral blood mononuclear cells. The quantities of IL-8 produced by the bone marrow mononuclear cells were compatible with the levels expressed by LPS-stimulated blood neutrophils, suggesting that the myeloid precursors could be the source of the IL-8. Kinetic studies of steady-state IL-8 mRNA expressed by LPS-treated bone marrow mononuclear cells demonstrated a rapid rise in de novo IL-8 mRNA. LPS stimulation resulted in the appearance of IL-8 mRNA by 30 min and a plateau by 90 min. In the presence of a protein synthesis inhibitor, the expression pattern for bone marrow-derived mononuclear cell IL-8 mRNA was superinduced, suggesting that IL-8 transcription is under the influence of an intracellular regulatory protein.

The ability of a bone marrow-derived cell to express IL-8 may have important consequences for the role of this cytokine in infectious processes. Recent investigations have demonstrated that the systemic administration of IL-8 to experimental animals can induce a leukocytosis and bandemia in these animals (34). It may be hypothesized that the neutrophilia commonly associated with infectious diseases may be a result of IL-8 interacting at the level of the marrow and inducing the release of granulocytes from the marrow sinusoids.

VII. Summary

The information above lends support to the generation of LPS-derived cytokines during the initiation and maintenance of an inflammatory response. These polypeptide mediators of inflammation are necessary to orchestrate the response by serving as key cell-to-cell communication signals. One of the interesting scenarios of inflammation, which aids in establishing these communication signals, is the generation of cytokine networks. While cytokine networks are important to various phases of the inflammatory response, they are clearly one of the major mechanisms for the elicitation, accumulation, and activation of specific leukocyte populations to a defined area of inflammation. The migration of inflammatory leukocytes from the vascular compartment to the interstitium is a complex process that is dependent on a number of cytokine-initiated events. These include cytokine-induced "rolling" of neutrophils, adherence, followed by directed migration to chemotactic factors, and the final activation of the cells during their trafficking to the inflammatory site. The sequence of events that drives these steps probably includes the induction of early response cytokines, such as IL-1 and TNF, which can network with other local, resident cells. This networking leads to the expression of further cytokines which are needed to maintain the inflammatory response. One of the cytokines produced during this networking process is the chemotactic cytokine IL-8, which is important to the continued leukocyte recruitment process. The studies cited above also point to the critical role of both fixed, resident macrophages and resident non-immune cells that comprise normal tissue. These cells are probably involved in truly relevant cellular communications, which are ultimately responsible for the long-term maintenance of inflammation by regulating continued leukocyte recruitment.

Acknowledgments

These studies were supported in part by NIH grants HL-02401, HL-31963, and HL-31237, and by the Tobacco Research Council.

References

1. Kunkel SL, Standiford TJ, Kasahara K, Strieter RM. Interleukin-8 (IL-8): the major neutrophil chemotactic factor in the lung. Exp Lung Res 1991; 17:17–23.
2. Rolfe MW, Kunkel SL, Standiford TJ, et al. Pulmonary fibroblast expression of interleukin-8: a model for alveolar macrophage-derived cytokine networking. Am J Respir Cell Mol Biol 1991; 5:493–501.
3. Standiford TJ, Kunkel SL, Basha MA, et al. Interleukin-8 gene expression by a pulmonary epithelial cell line: a model for cytokine networks in the lung. J Clin Invest 1990; 86:1945–1953.
4. Rolfe MW, Kunkel SL, Standiford TJ, et al. Expression and regulation of human pulmonary fibroblast-serived monocyte chemotactic peptide-1. Am J Physiol 1992; 263:235–247.
5. Cohen J, Exley AR, Buurman W, et al. Monoclonal antibody to TNF in severe septic shock. Lancet 1990; 335:1275–1277.
6. Damas P, Reuter A, Gysen P, Demonty J, Lamy M, Franchimont P. Tumor necrosis factor and interleukin-1 serum levels during severe sepsis in humans. Crit Care Med 1989; 17:975–978.
7. Debets JMH, Kampmeijer R, van der Linden MPMH. Plasma tumor necrosis factor and mortality in critically ill septic patients. Crit Care Med 1989; 17:489–494.
8. Girardin E, Grau GE, Dayer JM. Tumor necrosis factor and interleukin-1 in the serum of children with severe infectious purpura. N Engl J Med 1988; 319: 397–400.
9. Michie HR, Manogue KR, Spriggs DR, Revhaug A, O'Dwyer WC, Dinarello CA. Detection of circulating tumor necrosis factor after endotoxin administration. N Engl J Med 1988; 318:1481–1486.
10. Waage A, Halstensen A, Espevik T. Association between TNF in serum and fatal outcome in patients with meningococcal disease. Lancet 1987; i:355–357.
11. Beutler B, Milsark IW, Cerami A. Passive immunization against cachectin/tumor necrosis factor protects mice from lethal effect of endotoxin. Science 1985; 229:869–871.

12. Tracey KJ, Fong Y, Hesse DG, et al. Anti cachectin/TNF monoclonal anatibodies prevent septic shock during lethal bacteraemia. Nature 1987; 330:662–664.
13. Mathison JC, Wolfson E, Ulevitch RJ. Participation of tumor necrosis factor in the mediation of gram negative bacterial lipopolysaccharide-induced injury in rabbits. J Clin Invest 1988; 81:1925–1937.
14. Remick DG, Strieter RM, Lynch JP, Nguyen D, Eskandari M, Kunkel SL. In vivo dynamics of murine tumor necrosis factor alpha gene expression: kinetics of dexamethasone-induced suppression. Lab Invest 1989; 60:766–771.
15. Remick DG, Kunkel RG, Larrick JW, Kunkel SL. Acute in vivo effects of human recombinant tumor necrosis factor. Lab Invest 1987; 56:583–590.
16. Remick DG, Strieter RM, Eskandari MK, et al. Role of tumor necrosis factor-alpha in lipopolysaccharide-induced pathologic alterations. Am J Pathol 1990; 136:49–60.
17. Pennington J. TNF: therapeutic target in patients with sepsis. ASM News 1992; 58:479–482.
18. Le J, Vilcek J. Tumor necrosis factor and interleukin-1: cytokines with multiple overlapping biological activities. Lab Invest 1987; 56:234–248.
19. Ulich TR, Guo K, Del Castillo J. Endotoxin-induced cytokine gene expression in vivo. I. Expression of tumor necrosis factor mRNA in visceral organs under physiologic conditions and during endotoxemia. Am J Pathol 1989; 134:11–14.
20. Tracey KJ, Beutler B, Lowry SF, et al. Shock and tissue injury induced by recombinant human cachectin. Science 1986; 234:470–474.
21. Chensue SW, Terebuh PD, Remick DG, Scales WE, Kunkel SL. In vivo biologic and immunohistochemical analysis of interleukin-1 alpha, beta, and tumor necrosis factor during experimental endotoxemia. Am J Pathol 1991; 138:395–402.
22. Lew W, Oppenheim JJ, Matsushima K. Analysis of the suppression of IL-1 alpha, and IL-1 beta production in human peripheral blood mononuclear adherent cells by a glucocorticoid hormone. J Immunol 1988; 140:1895–1902.
23. Zuckerman SH, Shellhaas J, Butler LD. Differential regulation of lipopolysaccharide induced interleukin 1 and tumor necrosis factor synthesis: effects of endogenous and exogenous glucocorticoids and the role of the pituitary-adrenal axis. Eur J Immunol 1989; 19:301–305.
24. Colletti LM, Remick DG, Burtch GD, Kunkel SL, Strieter RM, Campbell DA. Role of tumor necrosis factor-alpha in the pathophysiologic alterations after hepatic ischemia/reperfusion injury in the rat. J Clin Invest 1990; 85:1936–1943.
25. Waage A, Halstensen A, Shalaby R, Brandtzaeg P, Kierulf P, Espevik T. Local production of tumor necrosis factor, interleukin-1, and interleukin-6 in meningococcal meningitis. Relation to the inflammatory response. J Exp Med 1989; 170:1859–1867.
26. Tiegs G, Wolter M, Wendel A. Tumor necrosis factor is a terminal mediator in galactosamine/endotoxin-induced hepatitis in mice. Biochem Pharm 1989; 38:627–631.
27. Elner VM, Strieter RM, Elner SG, Baggiolini M, Lindley I, Kunkel SL. Neutrophil chemotactic factor (IL-8) expression by cytokine treated retinal pigmented epithelial cells. Am J Pathol 1990; 136:745–750.

28. Strieter RM, Kunkel SL, Showell HJ, et al. Endothelial cell gene expression of a neutrophil chemotactic factor by TNF, LPS, and IL-1. Science 1989; 243: 1467–1469.
29. Thornton AJ, Strieter RM, Lindley I, Baggiolini M, Kunkel SL. Cytokine-induced gene expression of a neutrophil chemotactic factor/IL-8 by human hepatocytes. J Immunol 1990; 144:2609–2613.
30. Oppenheim JJ, Zachariae COC, Mukaida N, Matsushima K. Properties of the novel proinflammatory supergene "intercrine" cytokine family. Annu Rev Immunol 1991; 9:617–648.
31. Matsushima K. Oppenheim JJ. Interleukin-8 and MCAF: novel inflammatory cytokines inducible by IL-1 and TNF. Cytokine 1989; 1:2–13.
32. Baggiolini M, Walz A, Kunkel SL. Neutrophil-activating peptide-1/interleukin-8, a novel cytokine that activates neutrophils. J Clin Invest 1989; 84:1045–1049.
33. Dibb CR, Strieter RM, Burdick M, Kunkel SL. Expression of interleukin-8 by lipopolysaccharide-stimulated bone marrow–derived mononuclear cells. Infect Immun 1992; 60:3052–3058.
34. Van Damme J, Van Beeumen J, Opdenakker G, Billiau A. A novel NH terminal sequence-characterized human monokine possessing neutrophil chemotactic, skin reactive, and granulocytosis-promoting activity. J Exp Med 1988; 167:1364–1376.

14

Sepsis, Multiple-Organ Dysfunction Syndrome, and Adult Respiratory Distress Syndrome in Humans

JAMES E. McFEELY and LEONARD D. HUDSON

University of Washington
Seattle, Washington

I. Introduction

The relationships between endotoxin, sepsis syndrome, multiple-organ dysfunction syndrome (MODS), and the adult respiratory distress syndrome (ARDS) have been studied extensively over the past decade. With increased use of therapies resulting in immunosuppression, more patients are being admitted to the intensive care unit (ICU) with endotoxin-induced sepsis syndrome, ARDS, and MODS. Despite the improvements in ICU care, many patients are still dying of sepsis, ARDS, and MODS. In our institution, the fatality rate for ARDS over the past decade has slowly declined from approximately 60% to 40% (1). MODS remains the major cause of death in these patients, as it has for the past decade (2,3). Patients who develop ARDS have a higher mortality rate than that of patients with similar clinical risk factors who do not develop ARDS (2). In this chapter we define and give an overview of these clinical syndromes, including discussions of their clinical epidemiology, presentation, pathogenesis, and treatment. More extensive discussions of many of these topics are covered elsewhere in this volume.

II. Definitions

A. Sepsis Syndrome

Common definitions for some terms are well accepted, such as infection, bacteremia, and endotoxemia. However, there has been some confusion in the literature regarding the definition of sepsis and the sepsis syndrome. Over the years a number of terms have been used to describe the clinical scenario that clinicians recognize as sepsis. Early definitions of sepsis described it as a putrefactive process with a systemic response. Terms such as "hot sepsis" and "cold sepsis" then developed, describing systemic responses to infection. These terms are no longer commonly used, as we have come to understand that they represent a variable response to a common stimulus. Understanding of research studies performed during the 1970s and early 1980s was hampered by a lack of a common definition of sepsis (4–8).

In the late 1970s an attempt was made to identify patients in the surgical intensive care unit who had sepsis syndrome: the clinical picture or syndrome that clinicians recognize as being associated with serious infections with major systemic findings (9). "Clinical sepsis," as defined, clearly included patients who all clinicians would agree did have the sepsis syndrome (9). In this series of consecutive patients seen in a surgical intensive care unit, 42 of 153 were classified as having sepsis; of these, 22 of 42 had positive blood cultures, and 20 of 42 had negative cultures. The latter group were referred to as "nonbacteremic clinical sepsis" (9). The diagnosis of sepsis was made on the basis of clinical criteria such as "overt signs of occult sepsis" (hypoxia, hyper-bilirubinemia, renal failure, thrombocytopenia, abnormal mentation, gastric bleeding, and transient hypotension) in addition to clinical and laboratory findings and a knowledge of the original disease process (9). Unfortunately, this study suffered from a lack of rigorous specific criteria defining the group of patients with sepsis. This lack of specificity made comparisons between this and other studies difficult.

In the early 1980s a definition of sepsis syndrome designed for clinical research purposes was proposed which defined arbitrary but specific criteria involving two primary elements: (1) serious infection or inflammation, and (2) a deleterious systemic response associated with the inflammation. The specific definition included the following: (1) any two factors suggesting serious infection of (a) temperature $<35°C$ or $>39°C$, (b) WBC < 2000 or $> 12{,}000$ cells/mm^3, or 20% band forms, (c) positive blood cultures, or (d) documented source of infection; plus (2) any one deleterious systemic response of (a) unexplained systolic blood pressure <90, (b) systemic vascular resistance (SVR) < 800, or (c) unexplained metabolic acidosis (10,11). This definition with minor modifications was used subsequently by several investigators in clinical trials. Since then other definitions have evolved, including modifications of criteria reflecting evidence of infection and signs of inflammation plus a broader list of organ system abnormalities seen with sepsis (12).

A consensus definition has recently been developed that allows for comparisons between research studies (13). This definition is based on clinical data easily obtained from patients. All patients must have evidence of infection, although positive culture results are not necessary. In addition, they must have systemic manifestations of a response to the infection, including fever or hypothermia, tachypnea, and tachycardia. They also must have evidence of impaired organ function or perfusion, as evidenced by one of the following: altered mental status, oliguria, hypoxemia, or elevated plasma lactate (14,15). Although broad, this definition does include the more commonly seen sequelae of sepsis and allows for earlier recognition of patients with the syndrome than does a definition requiring positive culture results. Using this definition in a study of steroid treatment of sepsis, Bone and colleagues (16) found a similar overall mortality rate and incidence of ARDS as in other published reports using earlier definitions (6,7,17). It may not be necessary, however, to have culture-positive evidence of infection to have sepsis syndrome. Patients without evidence of infection who have similar organ system manifestations and dysfunction to those with proven infection have matching outcomes and complications (18,19). This process of systemic inflammation without evidence of infection can occur in patients with a wide variety of disorders, including pancreatitis, multiple trauma, ischemia, and hemorrhagic shock. Systemic inflammatory response syndrome (SIRS) is the preferred term used to describe this group of patients with similar response to diffuse inflammation (13). This shared response may reflect a mediator pathway common to a number of different insults.

B. ARDS

ARDS was initially described by Ashbaugh and colleagues as a picture of marked hypoxemia (defined as a widened A-a DO_2), diffuse alveolar infiltrates, and low lung compliance, similar to that seen in infants with hyaline membrane disease (20). Their report of 12 cases highlighted the variable causes of ARDS (trauma, 7; infection, 3; pancreatitis, 1; drug overdose, 1), and the overall common clinical presentation of the illness regardless of the inciting event. They speculated even then on the possible common etiology, saying, "In view of the similar response of the lung to a variety of stimuli, a common mechanism of injury may be postulated" (20). They made measurements of surface tension as a measure of surfactant function and commented on the utility of PEEP in the ventilatory management of ARDS.

Several other terms have been applied to the same syndrome. Certain names reflect the clinical situations in which patients may develop ARDS, such as "shock lung," "posttraumatic pulmonary insufficiency," and "DaNang lung." Other names reflect pathological descriptions of the disorder, including "diffuse alveolar damage," "alveolar capillary leak syndrome," and "increased permeability edema." What is now generally agreed is that ARDS represents one end of a spectrum of acute lung injury (21–24). This spectrum of lung injury can be roughly quantitated using the lung injury score described by Murray and colleagues (24). This scoring system contains components describing chest x-ray abnormalities, a hypoxemia score (PaO_2/FIO_2), a measure of respiratory mechanics (compliance), and the need for PEEP to support oxygenation. For research purposes a definition including some threshold level of physiological and clinical effects has been developed to compare clinically similar groups of patients. Criteria for the diagnosis of ARDS that are widely used are included in Table 1.

Most definitions of ARDS include the minimum of:

1. A severe degree of hypoxemia
2. Chest radiographic abnormalities, usually diffuse (25)
3. Clinical setting known to be associated with ARDS
4. An attempt to exclude other causes of the findings, particularly heart failure

Common criteria defining the syndrome include the following: (1) PaO_2/FIO_2 ratio < 150 (10), (2) bilateral alveolar-filling infiltrates (in some definitions requiring the involvement of all four quadrants) (11,19,25–27), (3) pulmonary artery wedge pressure (PAWP) less than 18 mmHg, and (4) no other cause to explain the foregoing values (2,11). Other recent intervention trials in ARDS have another definition: (1) $PaO_2 < 65$ mmHg while breathing $FIO_2 > 0.4$ or a ratio of $PaO_2/PAO_2 < 0.3$, (2) bilateral diffuse infiltrates, and (3) PAWP < 18 (18,28). Other common blood gas criteria have included $PaO_2/FIO_2 < 200$ (26), or $PaO_2/PAO_2 \leqslant 0.20$ (27). These definitions all describe the group of patients with low-vascular-pressure pulmonary edema commonly thought to have ARDS. Hopefully, the American-European Consensus Conference definition of ARDS (Table 1) will become the standard used in all future research studies, so that all clinical trials will utilize comparable patient groups (29).

Recently, ARDS has been understood to be the pulmonary manifestation of the multiple-organ dysfunction caused by SIRS. SIRS may be thought of as a malignant form of intravascular inflammation leading to endothelial injury in several organs of the body, thus resulting in multiple-organ dysfunction syndrome (MODS). When SIRS is initiated by infection, it is termed sepsis syndrome. Inflammation is usually a beneficial process, representing an essential part of the repair process when the body is injured by any of several different mechanisms, including endotoxin-induced damage. When

Table 1 Clinical Trials ARDS Definitions

Study	Hypoxemia	CXR	Setting	Exclusion
Sprung, 1984 (119)	$PaO_2/FIO_2 < 160$	Diffuse bilateral pulmonary infiltrates	Sepsis	PCWP < 15
Montgomery, 1985 (2)	$PaO_2/FIO_2 < 150$	Diffuse infiltrates in all lung fields	Known ARDS risk factors	PCWP $\leqslant 18$
Bone, 1987 (18)	$PaO_2 < 70$ with $FiO_2 \geqslant 0.4$	Bilateral pulmonary infiltrates	Clinical evidence of infection	PCWP < 18
Bernard, 1987 (28)	$PaO_2 < 70$ with $FIO_2 \geqslant 0.4$ or $PaO_2/PAO_2 \leqslant 0.3$	Bilateral diffuse pulmonary infiltrates compatible with pulmonary edema	Illness known to be associated with ARDS	PCWP < 18
Bone, 1989 (170)	$PaO_2/FIO_2 \leqslant 150$ without PEEP, or $PaO_2/FIO_2 \leqslant 200$ with any PEEP	Bilateral diffuse infiltrates	Trauma, surgery, sepsis	PCWP < 15
Parsons, 1989 (62)	$PaO_2/PAO_2 < 0.25$	Bilateral infiltrates	Known risk factors for ARDS	PCWP < 18, static compliance < 50 mL/cm H_2O
American-European Consensus Conference on ARDS, 1994 (29)	$PaO_2/FIO_2 \leqslant 150$ regardless of PEEP	Bilateral infiltrates on frontal CXR	–	PCWP $\leqslant 18$ or no evidence of left atrial hypertension

normal control mechanisms fail, however, inflammation may cause further injury, resulting in MODS.

C. MODS

MODS, also referred to as multiple-organ system failure (MOSF) and multiple-organ failure syndrome, has become increasingly prevalent over the last 15 to 20 years as our ability to manage single-organ system failure in the ICU has improved. The initial description of "sequential systems failure" after rupture of abdominal aortic aneurysms occurred in 1973 (30). In an editorial in 1975, Baue was the first to describe the common clinical scenario referred to as "multiple progressive or sequential systems failure" (31). Patients who probably would have died a few years ago from single-organ dysfunction, are now surviving their initial insult, only to develop the sequelae of activation of immune and inflammatory stimuli affecting multiple-organ systems. Treatment for single-organ dysfunction often is not directed at the inflammatory stimuli causing multiple-organ failure, allowing persistent damage to other organ systems.

MODS is a clinical syndrome of severe dysfunction of several organ systems occurring either in succession or concomitantly after a generalized insult to the patient. MODS may result from a wide variety of initial insults, most commonly sepsis, trauma, or other causes of persistent hypoperfusion (32–34). As with the study of sepsis syndrome and ARDS, early definitions varied from study to study, making comparisons difficult (2,19). There is still no universal agreement on the definitions of organ failure; however, the development of the APACHE II scoring system, which attempts to describe severity of illness, has given some common ground from which research on the epidemiology and prognosis of MODS may be based (35). A general definition of MODS might be the development of severe organ dysfunction in at least two organ systems lasting at least 24 to 48 hours. A more specific definition awaits consensus regarding definitions of organ failure for each system of the body and an improved understanding of the interrelationship between severity of dysfunction of individual organs and overall morbidity and mortality.

III. Clinical Epidemiology

A. Sepsis Syndrome

In hospitalized patients, the incidence of bacteremia and Gram-negative bacteremia in particular has increased dramatically over the past 40 years. Between the years 1951 and 1974, the rate of hospital admission for Gram-negative bacteremia rose from 0.75 of 1000 admissions to 13 of 1000 admissions (6,36). The number of cases annually during the 1970s has been estimated to be between 70,000 and 300,000 (37,38). Approximately 170,000 admissions for Gram-negative sepsis occurred during 1987 (39). The mortality from Gram-negative bacteremia is estimated to be 25 to 32% (40,41). This increase in the frequency of Gram-negative bacteremia is probably due to a number of different factors. There are many more patients at risk as a result of increasing immunosuppression resulting from therapy for an ever-larger spectrum of disease. In addition, there may be an increase in diagnosis of bacteremia or in the frequency of reporting of bacteremia as our surveillance methods improve and physicians become more aware of the possibility of sepsis.

B. ARDS

Of the group of patients with Gram-negative infections, a subset will develop ARDS. This is thought to occur in about 18 to 42% of such patients (15, 42–46). The likelihood of developing ARDS from sepsis varies depending on whether there is associated hypotension (43,44). In addition, there is a trend toward increasing incidence of ARDS with Gram-negative sepsis as opposed to Gram-positive sepsis (44). In a recent study of methylprednisolone in sepsis, 25% of patients enrolled developed ARDS (15). Patients who develop ARDS in association with sepsis have a much higher mortality rate than that of those who do not develop ARDS (42–44). In addition, patients with Gram-positive sepsis and fungal sepsis may also develop ARDS (42,44).

Development of ARDS is associated with a number of different conditions in addition to Gram-negative sepsis, although sepsis is the most common risk factor. In a prospective study at our institution, Pepe et al. identified several risk factors for the development of ARDS (11). In addition to the sepsis syndrome, these included aspiration of gastric contents, lung contusion, multiple transfusions, and multiple fractures (11). We have reevaluated the clinical risks of developing ARDS, to provide the information necessary for appropriate development of intervention trials in patients at risk. In this study 695 medical and surgical patients were identified meeting criteria for seven clinical risks associated with ARDS. These patients were prospectively followed for development of ARDS and eventual outcome. ARDS occurred in 179 of the 695 patients (26%), with the highest incidence occurring in the group of patients with sepsis syndrome (47). The incidence of ARDS increased as the number of risk factors increased in both the trauma and sepsis groups of patients. The time to onset of ARDS from identification of risks was rapid: 32% of sepsis patients who developed ARDS from sepsis did so in the first 12 h and 54% within 24 h. Those developing ARDS from trauma did so more slowly: 16% by 12 h, 29% within the first 24 h (47). Endotoxin itself also predisposes patients to develop ARDS. In a study of patients with septic shock, those with measurable levels of circulating endotoxin were twice as likely to develop ARDS (48). Circulating endotoxin also predisposed these patients to MODS (48).

C. MODS

The data available on the incidence and outcome of MODS vary with the differing definitions of MODS, the types of patients studied (medical versus surgical) and the duration of organ dysfunction (2,19,32,49–52). Sepsis has been identified as the most frequent predisposing factor to MODS (2,19,50). In the largest study to date, Knaus et al. prospectively evaluated the outcome of 5815 ICU admissions at 13 medical centers around the country. In this mixed population of medical and surgical patients, 49% (2843 patients) met at least one of the criteria for organ failure lasting 1 day, and 15% (891 patients) met criteria for failure of more than two organ systems (51). The two major risk factors for development of MODS identified in this study were diagnosis of sepsis or infection at admission to the ICU, and patient age < 65 years (51). In addition, patients in this study with high Apache II scores were more likely to have MODS (51). Like the majority of other studies of prognosis, this group found that the greater the number of organ systems affected and the greater the length of time of disease, the higher the likelihood of death. Single-organ dysfunction lasting more than one day carried a risk of in-hospital mortality of 40%. Failure of two organ systems for more than 24 h was associated with a hospital mortality rate of 60%, and failure of three or more organ systems for more than 72 h was almost uniformly fatal (51).

IV. Clinical Presentation

A. Sepsis Syndrome

The clinical presentation of a patient with sepsis is variable. Patients present with some or all of the following findings: fever or hypothermia, tachycardia, hypotension associated with a low systemic vascular resistance, mental status changes, thrombocytopenia, disseminated intravascular coagulation, hypoxemia, increased liver function tests (AST, ALT, bilirubin), oliguria, or hypoxemic respiratory failure. If a pulmonary artery (PA) catheter is placed, findings usually include an elevated PA pressure, a relatively "normal" or high cardiac output, and a low systemic vascular resistance. Sepsis syndrome is more likely to occur in immunosuppressed individuals, particularly those infected with *Pseudomonas* or *Escherichia coli*, although any bacterial species may be the source in a given patient. The more common sites of infection include the urinary tract, the lung, and the gastrointestinal tract. Less common sites of origin of sepsis include the central nervous system and the skin, although again, these may be the source in any given patient.

B. ARDS

Individuals with ARDS usually present with relatively abrupt onset of severe dyspnea associated with arterial hypoxemia. Approximately one-half of patients with sepsis syndrome who develop ARDS do so within the first 24 h (47). They often do not respond to supplemental oxygen as a result of shunt. The chest radiograph commonly shows diffuse infiltrates, although CT scanning often will reveal a patchy distribution of the infiltrates, worse in the dependent portions of the lung (53–56).

C. MODS

Patients presenting with either ARDS or the sepsis syndrome may go on to develop MODS over the course of their hospitalization as additional organ systems are affected by either the initial event, intercurrent infections, or other insults. As defined, sepsis syndrome may be synonymous with MODS. Organ dysfunction typically evolves at a highly variable rate. This reflects individual organ response rates to a simultaneous insult affecting all organ systems and tissues, including, in particular, the endothelium and the immune system. While the lungs are usually the first organs to fail, there is no common pattern of progression of subsequent organ dysfunction. There is no typical course of development or progression of MODS (52); therefore, patients at risk for MODS should be followed closely for development of the syndrome.

Those patients who develop ARDS following insults other than sepsis (e.g., trauma, aspiration, drug overdose, multiple transfusions) often go on to develop sepsis during their time in the ICU (2). These patients often progress to MODS as a result of sepsis and have a very poor prognosis as a result. Sepsis syndrome with associated MODS is the leading cause of death in patients with ARDS (2). In our most recent analysis of deaths in patients who developed ARDS, sepsis and multiple-organ failure continue to be the leading cause of death, regardless of the initial insult leading to development of ARDS (3).

V. Pathogenesis

The biological effects of endotoxin are dealt with in detail elsewhere in this book. It is clear from these descriptions that the pathogenesis of the sepsis syndrome related to Gram-negative infection is closely linked to the effects of bacterial endotoxin (57). Endotoxin has multiple direct and indirect deleterious effects through its effect on mediators of inflammation. It is known to damage lung endothelial cells leading to increased permeability (58). In a sheep model, endotoxin has been shown to increase vascular permeability, decrease lung compliance, increase PA pressures, and pulmonary vascular resistance without changes in left-sided pressures, all alterations similar to those seen in humans with sepsis syndrome and ARDS (59–61). Circulating endotoxin acting with other toxic metabolites, such as complement fragments, may directly increase the risk of developing ARDS (57,62). In one study of 98 patients at risk for development of ARDS, 74% of those who went on to develop ARDS had detectable levels of endotoxin in their blood, as opposed to 22% of those who did not go on to develop ARDS (62). Endothelial cell damage extends beyond the lung in the case of MODS. The diffuse organ damage is probably a result of either direct or indirect damage caused in part by endotoxin (63–65). Endotoxin itself may also cause accumulation of leukocytes in the pulmonary vasculature (66). In addition, endotoxin upregulates the expression of the endothelial adhesion molecule ICAM-1 and increases its affinity for leukocytes, adding to the accumulation of these cells in affected organs (67–69).

Endotoxin causes further damage by activating a number of other inflammatory mediators. Endotoxin activates the complement cascade producing C5a, which in turn induces granulocyte aggregation and generation of oxygen radicals (70–74). Endotoxin activates the blood coagulation and fibrinolytic cascades in patients with ARDS (75,76). It is also known to cause release of many pro-inflammatory cytokines: in particular, tumor necrosis factor (TNF) (77–81). TNF increases pulmonary vascular permeability, causes pulmonary hemorrhage and inflammation, and is associated with damage to other organ systems (82–89). This and other cytokines, such as interleukin-1 (IL-1), interleukin-6, and platelet activating factor, are possibly responsible for many of the clinical features of MODS, ARDS, and sepsis syndrome (77,90–94). A similar cascade of events may be stimulated by lipoteichoic acid from the cell walls of gram-positive bacteria and by mannan from fungi (95).

In addition to the direct effects of endotoxin, secondary organ damage can exacerbate shock and mortality after sepsis or trauma. Hepatic reticuloendothelial cell dysfunction has been implicated in contributing to other host defense abnormalities in the setting of sepsis-related ARDS and MODS (96–98). Preexisting liver damage and/or altered hepatic flow, either via hypotension or vasoactive drugs, may negatively affect the ability to clear endotoxin in severe infections (99). Hypotension and edema formation may add to the direct effect of endotoxin on the gut to promote translocation of organisms and endotoxin, which may cause secondary damage (100,101). Thus via both direct and indirect mechanisms, endotoxin could be responsible for a majority of the clinical findings in sepsis syndrome.

VI. Treatment of Sepsis Syndrome and ARDS

Given the overall high mortality associated with sepsis syndrome, ARDS, and MODS, a great deal of attention has been given to possible therapies directed at the mediators of organ dysfunction. Several chapters in this book deal extensively with specific therapeutics. We attempt to provide a broad overview here.

A. Sepsis Syndrome

Traditionally, therapeutic interventions in sepsis have been based on identification of the source of infection, with drainage of a closed space if necessary, rapid institution of appropriate antibiotics, and support of blood pressure and tissue oxygen delivery. These supportive measures continue to be critical interventions; however, a succession of newer modalities are being evaluated which are directed at the specific mediators of sepsis and sepsis syndrome.

Hemodynamic Support

Patients with sepsis syndrome and ARDS have an increased tissue demand for oxygen (102). It has been argued that these patients also have a deficiency in ability to extract oxygen and that oxygen consumption is dependent on oxygen supply (103,104). Several investigators have argued that augmentation of cardiac output to improve oxygen delivery will enhance survival in sepsis syndrome (105–110). Others, however, argue that the finding of supply dependence is an artifact of the measurement of oxygen consumption by arterial–mixed venous oxygen difference and that the supply dependency point in humans is not physiologically relevant to the values seen in patients with sepsis or ARDS (111–114). Although this discussion continues, it does seem reasonable to augment oxygen delivery both by increasing cardiac output and by increasing blood oxygen content when it can be done without deleterious consequences for the patient.

Fluids

ARDS from sepsis is generally associated with a relative deficiency of peripheral tissue oxygenation. This results from a combination of decreased O_2 delivery (from decreased PaO_2 and maldistribution of blood flow), increased metabolic demand for O_2, and a defect in peripheral oxygen extraction (102). A current controversy exists regarding the optimal intravascular volume (as defined by PCWP or extravascular lung water) in this setting (115,116). Some argue for relatively low intravascular volumes to protect the alveoli from further flooding and worsening of the hypoxemia. Keeping patients "dry" is thought to hasten weaning from mechanical ventilation and to maximize PaO_2 allowing for minimal FIO_2 (117). Arguments against attempting to maintain low wedge pressures include the very real concern for decreased peripheral tissue perfusion, increased need for inotropic agents, and the potential for increased morbidity resulting from multiorgan dysfunction due to decreased tissue perfusion (105). The argument for maintenance of relatively high intravascular volumes is that maintenance of vital organ perfusion should be paramount, and that delays in weaning from ventilation would be minimal with judicious use of volume expansion. Those who disagree point out that high intravascular pressures increase alveolar flooding and worsen hypoxemia, thereby worsening oxygen delivery. We feel there is a middle ground that maintains adequate organ perfusion while not delaying eventual weaning from the ventilator. Fluids are given based on a determination of adequacy of tissue perfusion using clinical indicators (such as mentation, urine output, and—if a pulmonary artery (PA) catheter is in place—arteriovenous O_2 difference and O_2 extraction ratios). Tissue perfusion is reassessed frequently, in particular, after any adjustment of pressors or ventilatory support (e.g., PEEP). Diuresis is not attempted until the blood pressure is stable and pressors are minimized.

Blood Pressure Support

The hypotension seen in patients with sepsis/MODS is commonly the result of low filling pressures, inadequate cardiac output, and low SVR. This causes inadequate O_2 delivery to tissues, resulting in tissue hypoxia and acidosis. If fluids are insufficient, current practice includes the use of pressors/ionotropic agents to augment cardiac output as well as SVR in order to maintain adequate tissue perfusion. Dopamine is generally administered first, in doses intended to support renal perfusion. Should this be inadequate, the dose of dopamine may be increased into the range known to stimulate α- and β-adrenergic receptors. In general, use of pressors beyond the "renal dose" of dopamine requires the use of a PA catheter and often an arterial line for monitoring. Again, clinical indicators of organ perfusion are used to assess adequacy of therapy. In severe cases of sepsis/MODS, a single pressor often is inadequate to maintain blood pressure. Dobutamine, norepinephrine, or a pure alpha agonist such as neosynephrine are then added (depending on the relative abnormalities in systemic resistance and cardiac output), titrating the dose to optimal blood pressure, cardiac output, and distal O_2 delivery. If possible, the dopamine is reduced at that point back to "renal" doses. Assessment of the need for both pressors and invasive monitoring must be an ongoing process, as both are known to be associated with iatrogenic complications. Both the indwelling monitors and the pressors should be discontinued as soon as the patient's condition is stable enough to tolerate their removal.

Corticosteroids

The use of high-dose corticosteroids was one of the first pharmacologic therapies aimed at improving the outcome of sepsis and MODS to be studied in controlled trials. Experimental evidence suggested that the increased microvascular permeability induced by the administration of endotoxin to animals could be prevented by giving methylprednisolone. In one early randomized clinical trial of patients with septic shock, 33 of 86 patients given placebo died, while 9 of 86 given one or two doses of steroids died (118); the study design of this investigation, however, has been criticized. Since 1984, there have been a number of large multicenter trials of steroids in sepsis and ARDS. A summary of the entry criteria and results of these studies appears in Table 2. Several studies were designed to evaluate the role of corticosteroids in prevention/reversal of shock or ARDS and to determine the effect of steroids on short-term mortality (16,45,46,119). As shown in Table 2, no benefit from steroid treatment was noted in terms of prevention of disease or improved mortality, with an apparent increase in ARDS incidence associated with the use of steroids (16,18,45). In a separate trial, steroids were used in a group of "high-risk" surgical and trauma patients as a means of preventing the development of ARDS (120). In this group, treatment with steroids did not prevent the development of ARDS (120). Three other large studies have been conducted using steroids to treat patients with sepsis syndrome and ARDS (18,28,45). Again in none of these studies were steroids effective in reversing ARDS or in improving mortality (18,28,45). In fact, one of the studies found an increase in mortality in the steroid group compared to placebo (18). Overall, there does not appear to be any role for the use of corticosteroids in the management of sepsis syndrome or ARDS at the present time, at least relatively early in the course of the disease. There is anecdotal evidence that steroids may have a role late in the course of ARDS in those patients who have persistent ongoing inflammation (121–123). This hypothesis has yet to be tested in a large randomized trial.

Monoclonal Antibodies

In addition to steroids, the use of newer agents developed through cell and molecular biology techniques are being evaluated for the treatment of sepsis and ARDS (124–126). In addition to monoclonal antibodies, various small molecules, receptor antagonists, soluble receptors (acting as competitive inhibitors), and combinations of agents are now reaching clinical trials. Agents other than monoclonal antibodies are of interest because antibodies carry a risk of immunogenicity, whereas small molecules, because of their size, do not. Theoretically, they may be used more than once with no increased risk to the patient. Because of their relative ease of development, however, mono-

Table 2 Clinical Trials of Steroids in Sepsis

Trial	Definition of sepsis	# PTS	Dose	Endpoints	Results	p^a
Schumer, 1976 (118)	Septic history, falling blood pressure, positive blood cultures	172	Placebo vs. MPSS 30 mg/kg vs. Dex. 6 mg/kg	Mortality Complications	38% vs. 10% 6% vs. 6%	<0.001 NS
Sprung, 1987 (119)	SBP < 90 or 50 mmHg < baseline; altered mental status or oliguria (<20 mL/h); continued low BP despite 500 mL NS; bacteremia or identified source	59	Placebo vs. MPSS 30 mg/kg vs. Dex. 6 mg/kg	Reversal of shock Development of ARDS Survival to discharge	0/16 vs. 4/21 vs. 7/22 2/13 vs. 3/13 vs. 5/16 11/16 vs. 17/22 vs. 16/21	NS NS NS
Bone, 1987 (16)	Clinical evidence of infection; T > 38.3 or < 35.6; HR > 90; RR > 20 while breathing spontaneously; at least one of the following: Mental status changes PaO_2 < 75 on room air Elevated lactate Urine output < 30 mL/h or < 0.5 mL/kg/h for 1 h Septic shock defined as SBP < 90 or drop of 40 mmHg for 12 h	382	Placebo vs. MPSS 30 mg/kg q 6 h × 4 doses	Prevention of shock Reversal of shock 14-day mortality	45/112 vs. 51/112 83/144 vs. 85/130 48/190 vs. 65/191	NS NS 0.06

Bone, 1987 (18)	Same as Bone, 1987 (16)	304	Placebo vs. MPSS 30 mg/kg q 6 h × 4 doses	Prevention of ARDS Reversal of ARDS 14-day mortality	38/152 vs. 50/152 23/38 vs. 15/50 8/38 vs. 26/50	NS 0.005 0.004
VA, 1987 (46)	Clinical suspicion of sepsis and at least four of the following: Chills or T > 38.9 or < 35.5 RR > 28 or PaO_2 < 32 HR >100 Supine systolic pressure < 90 WBC < 3500 or > 15,000 or < 35% or > 85% PMNs or bands > 20% Platelets < 100,000 Primary focus of infection or surgery within 48 h	223	Placebo vs. MPSS 30 mg/kg followed by 5 mg/kg/h × 9 h	14-day mortality	24/11 vs. 23/112	NS
Luce, 1988 (45)	Clinical suspicion *and* T elevation within 4 h *and* decrease systolic pressure 20 mmHg from baseline Or patient with known onset of T > 38.5 or < 35.5 and SBP < 90	172	Placebo vs. MPSS 30 mg/kg q 6 h × 4 doses	Development of ARDS Resolution of ARDS Mortality	14/37 vs. 13/38 2/14 vs. 4/13 20/37 vs. 22/38	NS NS NS

[a]NS, not significant.

clonal antibodies have been the most extensively studied. A controlled clinical trial of a polyclonal antiserum directed at the lipid A moiety (the portion of the endotoxin molecule responsible for its biologic effect) of *E. coli*, demonstrating significant improvements in survival, generated enthusiasm for the development and use of monoclonal antibodies aimed at endotoxin (127). Since then there have been two multicenter trials of monoclonal antibodies directed against lipid A, using antibodies called HA1A and E5 (128,129). Unfortunately, the results of these trials have been less dramatic and call into question the utility of these antibodies (130,131). In the study of the antibody E5, increased survival was seen in only a small subset of patients with gram-negative sepsis who were not in refractory shock (128). Although subset analysis of the treatment groups does show an improvement in mortality in those patients with true gram-negative bacteremia, the problem remains the possibility of identifying this subset of patients. In addition, there have been several other criticisms of the published HA-1A study, regarding the role of inadequate antibiotic therapy in the placebo group affecting the mortality rate in the multivariate analysis, as well as the wide variability in efficacy based on study site (the entire effect was seen at sites with a greater than 50% mortality in the placebo group in the subset of patients with gram-negative bacteremia; in centers with less than 50% mortality in the placebo patients, no effect was seen) (130). At this point, the questionable efficacy, exorbitant cost of the therapy, and difficulty with identifying patients likely to benefit from the treatment preclude the use of these agents in many patients with sepsis syndrome (130). Further studies are ongoing that may settle the issue in the near future.

Other monoclonal antibodies have been developed which are promising. An antibody directed against C5a des arg prevented septic primates from developing hypoxemia, increased lung water, or death (132). Recent lab experience has shown that the end-organ effects of sepsis may be prevented by administration of anti-TNF antibody (133). Antibodies against TNF and soluble TNF receptor are undergoing trials. Antibodies directed against the adhesion molecule CD11b/CD18 are also currently in a phase I trial in trauma patients at risk for MODS and ARDS. In a phase II trial of IL-1 receptor antagonist (IL-1Ra) a decreased 28-day mortality was noted in the group receiving IL-1Ra. Preliminary results of a phase III trial showed a statistically significant improvement in survival in a subset of patients with sepsis who were at the highest risk of death but not in the group as a whole (134).

Nonsteroidals

The use of nonsteroidal anti-inflammatory drugs to inhibit the effects of arachidonic acid cascade are also being investigated (135). In a dog model of *E. coli* sepsis, ibuprofen was able to reverse the deleterious hemodynamic and metabolic effects seen in septic shock (136). In other animal studies of sepsis-induced lung injury, ibuprofen was able to attenuate the early alveolar–capillary permeability change and decrease neutrophil migration into the alveoli as well as decrease alveolar macrophage TNF production (137,138). In one animal study, pigs made septic by infusion of *Pseudomonas aeruginosa*, then given ibuprofen, maintained normal systemic arterial pressure, improved their oxygenation, and decreased shunting as well as pulmonary edema (135). There is currently an ongoing multicenter clinical trial of ibuprofen in sepsis and ARDS that will assess efficacy in humans.

Other Experimental Strategies

Many other therapeutic measures have been suggested to treat sepsis and ARDS, including such things as *N*-acetylcysteine and other antioxidants, pentoxifylline, hemofiltration, anticoagulants, and fibrinolytic agents (139–143). There are several ongoing clinical trials, with results expected in the near future.

Although there is much optimism about the potential benefit of these new agents, several factors should cause us to pause and consider the problems associated with their development and possible use. While the animal model work is very promising, the clinical situation in humans is much more complicated. This may account for some of the variability in response seen in the human trials. Humans with sepsis syndrome, ARDS, and MODS develop their illness from a variety of causes and at variable rates. Even within one particular subset of patients, considerable heterogeneity exists in host factors and underlying clinical conditions. In addition to these host factors, the inflammatory cascade itself has a built-in redundancy that may thwart attempts to shut off the inflammatory process by blocking one mediator, depending on the specific mediator being addressed and its position in the inflammatory cycle.

Clinical trials will need to be completed carefully and successfully before these agents can be put into general use. These trials are difficult to organize and perform. Problems include carrying out a trial on an inappropriate patient population, proper timing of the intervention in an individual patient's illness, difficulties with dosing and duration of therapy, and selection of an appropriate endpoint for study. Finally, the reality that funding for clinical trials is difficult to obtain and that studies are expensive to perform suggests that one negative study for an agent might preclude further testing of what might otherwise be an effective therapeutic intervention. It may be that agents that affect the pathophysiology but do not have dramatic effects on mortality alone might be beneficial when given in combination.

B. ARDS

Ventilatory Management

Despite improvements in the understanding of the pathogenesis of ARDS, management currently is almost completely supportive. As with the treatment of patients with sepsis, the first step is to treat the illness that caused the patient to develop ARDS. The initial step in treatment of the patient with ARDS/MODS is resuscitation and stabilization. Once this is accomplished, the illness causing the ARDS/MODS must be identified and dealt with. The management of ARDS can be broadly divided into pharmacological treatment and ventilatory management. Both of these categories are addressed in detail elsewhere in this book.

The goals of ventilatory management of the patient of ARDS have changed little since the initial description by Ashbaugh and colleagues (20). These goals consist of adequate oxygenation of the blood while minimizing lung injury and are achieved by a combination of oxygen therapy, mechanical ventilation, and PEEP (144). Mechanical ventilation traditionally has consisted of volume-limited ventilation with PEEP at normal inspiratory-expiratory (I:E) ratios (145). When this has been inadequate to meet patient needs, a number of other strategies have been attempted, including pressure control ventilation (146,147), inverse ratio ventilation (146,148–150), extracorporeal techniques (151), intravascular gas exchanger (152–154), high-frequency ventilation (155–157), and changes in position of the patient (158). These methods all have their proponents but have not been validated extensively and therefore are not recommended for general use. However, in any patient failing conventional therapy, these newer modalities may be utilized, provided that there is careful monitoring and a recognized goal/endpoint to the change in therapy. Recent experimental evidence in animal models has provided convincing evidence that mechanical ventilation, especially when high pressures and volumes are used, may aggravate lung injury or even induce lung injury in relatively spared regions of the lung (159–162). There are new strategies for mechanical ventilation that have been proposed to limit this lung injury, such as "permissive hypercapnia," although they have yet to be tested in controlled clinical trials (163). For now the consensus remains that maintenance of adequate tissue oxygen delivery while minimizing barotrauma and oxygen toxicity are the mainstays of ventilatory management.

Pharmacologic Management

The pharmacologic management of ARDS is in its infancy (164,165). Although a great deal has been learned about the inflammatory process that occurs in ARDS, as yet no effective therapy has been developed to treat it. There is a great deal of overlap in the antiinflammatory therapy of both sepsis syndrome and ARDS (124). As with sepsis syndrome, corticosteroids were thought to be beneficial in treatment of ARDS. Unfortunately, no prospective clinical trial has shown a benefit in terms of prevention of or outcome from ARDS (45,120). The success of exogenously administered surfactant in neonatal respiratory distress syndrome has led many investigators to interest in its use in ARDS (166–168). There are several problems, however, that may limit the feasibility and effectiveness of surfactant in adults with ARDS. IRDS is a disease marked by a pure deficiency of surfactant, whereas the dysfunctional surfactant in ARDS is probably caused by the inflammatory pulmonary edema. ARDS may require several doses of surfactant and may require large doses that may compromise gas exchange even further. There are currently ongoing clinical trials involving the use of both animal and artificially produced surfactant products in humans. Vasodilators have been utilized in ARDS patients in an attempt to increase oxygen delivery while minimizing pulmonary intravascular pressures. Prostaglandin E_1 (PGE_1), a pure pulmonary vasodilator, has been tested in ARDS patients (169–171). While a single center study suggested a marginal advantage in survival, this was not confirmed in a larger multicenter controlled trial (169, 170). However, PGE_1 did reduce pulmonary vascular resistance and increase cardiac output and might be warranted for these goals in selected patients (171). Additional pharmacologic therapy for ARDS patients primarily supports organ systems other than the lungs. Inotropic agents are often used to improve oxygen delivery while minimizing intravascular pressures, in an attempt to reduce development of lung edema (115,117). There are some data to suggest that anticoagulants and fibrinolytic agents may be useful in ARDS. Animal model data suggest that these agents might prevent the intravascular and extravascular activation of the clotting cascade seen in ARDS, but they have not been studied in humans, in part because of their obvious potential risk (141,143,172).

C. MODS

There is no specific therapy at this time for patients with multiple-organ failure other than specific support for individual organ dysfunction (e.g., dialysis) and the general supportive measures mentioned for sepsis syndrome and ARDS. Mortality from multiple-organ failure remains very high. Improvements in mortality must await new therapeutic interventions effectively controlling the malignant inflammatory response responsible for the sequential organ failure seen in patients with sepsis syndrome and ARDS.

VII. Conclusions

The patient with sepsis, ARDS, or MODS is one of the greatest management challenges in the intensive care unit. Care of these patients requires meticulous attention to detail and frequent reevaluation by a well-trained, coordinated team of caregivers to optimize their chance of survival. The future holds the promise of improved therapeutic interventions directed against the mediators of sepsis with resultant improvement in patient survival.

Acknowledgments

This work was supported by NIH grants HL-30542 and HL-07287 and by the Parker B. Francis Foundation.

References

1. Mitchell DR, Milberg JA, Steinberg KP, Hudson LD. Trends in adult respiratory distress syndrome (ARDS) fatality rates from 1983–1992 (abstr). Am Rev Respir Dis 1993; 147:A348.
2. Montgomery AB, Stager MA, Carrico CJ, Hudson LD. Causes of mortality in patients with the adult respiratory distress syndrome. Am Rev Respir Dis 1985; 132:485–489.
3. Steinberg KP, McHugh LG, Hudson LD. Causes of mortality in patients with the adult respiratory distress syndrome (ARDS): an update (abstr). Am Rev Respir Dis 1993; 147:A347.
4. Miller SI, Wallace RJJ, Musher DM, Septimus EJ, Kohl S, Baughn RE. Hypoglycemia as a manifestation of sepsis. Am J Med 1980; 68:649–654.
5. Maclean LD, Mulligan WG, McLean AP, Duff JH. Patterns of septic shock in man: a detailed study of 56 patients. Ann Surg 1967; 166:543–562.
6. Kreger BE, Craven DE, Carling PC, McCabe WR. Gram-negative bacteremia. III. Reassessment of etiology, epidemiology, and ecology in 612 patients. Am J Med 1980; 68:332–343.
7. Kreger BE, Craven DE, McCabe WR. Gram-negative bacteremia. IV. Reevaluation of clinical features and treatment in 612 patients. Am J Med 1980; 68: 344–355.
8. Christy JH. Treatment of gram-negative shock. Am J Med 1971; 50:77–88.
9. Meakins JL, Wicklund B, Forse RA, et al. The surgical intensive care unit: current concepts in infection. Surg Clin North Am 1980; 60:117–132.
10. Pepe PE, Hudson LD, Carrico CJ. Early application of positive end-expiratory pressure in patients at risk for the adult respiratory distress syndrome. N Engl J Med 1984; 311:281–286.
11. Pepe PE, Potkin RT, Reus DH, Hudson LD, Carrico CJ. Clinical predictors of the adult respiratory distress syndrome. Am J Surg 1982; 144:124–130.

12. Bone RC. Toward an epidemiology and natural history of SIRS (systemic inflammatory response syndrome). JAMA 1992; 268:3452–3455.
13. Bone RC, Balk RA, Cerra FB, et al. Definitions for sepsis and organ failure and guidelines for the use of innovative therapies in sepsis. The ACCP/SCCM Consensus Conference Committee. American College of Chest Physicians/ Society of Critical Care Medicine. Chest 1992; 101:1644–1655.
14. Balk RA, Bone RC. The septic syndrome. Definition and clinical implications. Crit Care Clin 1989; 5:1–8.
15. Bone RC, Fisher CJJ, Clemmer TP, Slotman GJ, Metz CA, Balk RA. Sepsis syndrome: a valid clinical entity. Methylprednisolone Severe Sepsis Study Group. Crit Care Med 1989; 17:389–393.
16. Bone RC, Fisher CJ, Clemmer TP, Slotman GS, Metz CA, Balk RA, and the Methylprednisone Severe Sepsis Study Group. A controlled clinical trial of high-dose methylprednisolone in the treatment of severe sepsis and septic shock. N Engl J Med 1987; 317:653–658.
17. Abraham E, Shoemaker WC, Bland RD. Sequential cardiorespiratory patterns in septic shock. Crit Care Med 1983; 11:799–803.
18. Bone RC, Fisher CJJ, Clemmer TP, Slotman GJ, Metz CA. Early methylprednisolone treatment for septic syndrome and the adult respiratory distress syndrome [published erratum appears in Chest 1988; 94(2):448]. Chest 1987; 92: 1032–1036.
19. Bell RC, Coalson JL, Smith JD, Johanson WG. Multiple organ system failure and infection in adult respiratory distress syndrome. Ann Intern Med 1983; 99: 293–298.
20. Ashbaugh DG, Bigelow DB, Petty TL, Levine BE. Acute respiratory distress in adults. Lancet 1967; 2:319–323.
21. Rocker DM, Pearson D, Wiseman MS, Shale DJ. Diagnostic criteria for the adult respiratory distress syndrome: time for reappraisal. Lancet 1989; i:120–123.
22. Rinaldo JE. The prognosis of the adult respiratory distress syndrome; inappropriate pessimism? Chest 1986; 96:470–471.
23. Petty T. ARDS: Refinement of concept and redefinition. Am Rev Respir Dis 1988; 138:724.
24. Murray J, Matthay M, Luce J, Flick M. An expanded definition of the adult respiratory distress syndrome. Am Rev Respir Dis 1988; 138:720–723.
25. Aberle D, Brown K. Radiologic considerations in the adult respiratory distress syndrome. Clin Chest Med 1990; 11:737–754.
26. Calindrino FS, Anderson DJ, Mintun MA, Schuster DP. Pulmonary vascular permeability during the adult respiratory distress syndrome: a positron emission tomographic study. Am Rev Respir Dis 1988; 138:421–428.
27. Fowler AA, Hamman RF, Good JT, et al. Adult respiratory distress syndrome: Risk with common predispositions. Ann Intern Med 1983; 98:593–597.
28. Bernard GR, Luce JM, Sprung CL, et al. High-dose corticosteroids in patients with the adult respiratory distress syndrome. N Engl J Med 1987; 317:1565–1570.
29. Bernard GR, Artigas A, Brigham KL, et al., and the Consensus Committee. Report on the American-European Consensus Conference on Acute Respiratory

Distress Syndrome. Definitions, mechanisms, relevant outcomes, and clinical trial coordination. Am J Respir Crit Care Med 1994; 149:818–824.
30. Tilney NL, Bailey GE, Morgan AP. Sequential systems failure after rupture of abdominal aortic aneurisms: An unsolved problem in postoperative care. Ann Surg 1973; 178:117–122.
31. Baue A. Multiple, progressive or sequential systems failure: A syndrome of the 1970's. Arch Surg 1975; 110:779–781.
32. Goris RJA, Boekhorst TPA, Nuyinck JKS, Bimbrere JSF. Multiple-organ failure: Generalized autodestructive inflammation. Arch Surg 1985; 120:1109–1115.
33. Pinsky MR, Matuschak GM. Multiple systems organ failure: failure of host defense homeostasis. Crit Care Clin 1989; 5:199–220.
34. Carrico CJ, Meakins JL, Marshall JC, Fry D, Maier RV. Multiple-organ-failure syndrome. Arch Surg 1986; 121:196–208.
35. Knaus WA, Draper EA, Wagner DP, Zimmerman JE. APACHE II: A severity of disease classification system. Crit Care Med 1985; 13:818–829.
36. McCabe WR, Jackson GC. Gram-negative bacteremia. I. Etiology and ecology. Arch Intern Med 1962; 110:847–855.
37. McCabe W, Kreger B, Johns M. Type-specific and cross-reactive antibodies in gram-negative bacteremia. N Engl J Med 1972; 287:261–267.
38. Wolff S, Bennett J. Gram-negative rod bacteremia. N Engl J Med 1974; 291: 733–734.
39. Martin MA, Silverman HJ. Gram-negative sepsis and the adult respiratory distress syndrome. Clin Infect Dis 1992; 14:1213–1228.
40. Wenzel RP. The mortality of hospital-acquired bloodstream infections: Need for a new vital statistic? Int J Epidemiol 1988; 17:225–227.
41. Martin CM, Cuomo AJ, Geraghty MJ, Zager JR, Mandes TC. Gram-negative rod bacteremia. J Infect Dis 1969; 119:506–517.
42. Martin MA, Wenzel RP, Gorelick KJ, and the Xoma Sepsis Study Group. Prospective U.S. study of Gram-negative bacterial sepsis: Natural history in the 1980's. Circ Shock 1990; 31:245.
43. Kaplan RL, Sahn SA, Petty TL. Incidence and outcome of the respiratory distress syndrome in Gram-negative sepsis. Arch Intern Med 1979; 139:867–869.
44. Fein AM, Lippmann M, Holtzman H, Eliraz A, Goldberg SK. The risk factors, incidence, and prognosis of ARDS following septicemia. Chest 1983; 83:40–42.
45. Luce J, Montgomery A, Marks J, Turner J, Metz C, Murray J. Ineffectiveness of high-dose methylprednisolone in preventing parenchymal lung injury and improving mortality in patients with septic shock. Am Rev Respir Dis 1988; 138:62–68.
46. Veterans Administration Systemic Sepsis Cooperative Study Group. Effect of high-dose glucocorticoid therapy on mortality in patients with clinical signs of systemic sepsis. N Engl J Med 1987; 317:659–665.
47. Hudson LD, Milberg JA, Anardi D, Maunder RJ. Clinical risks for development of the Adult Respiratory Distress Syndrome. 1993. Submitted.
48. Danner RL, Elin RJ, Hosseini JM, Wesley RA, Reilly JM, Parillo JE. Endotoxemia in human septic shock. Chest 1991; 99:169–175.

49. Pine RW, Wertz MJ, Lennard ES, Dellinger EP, Carrico CJ, Minshew BH. Determinants of organ malfunction or death in patients with intra-abdominal sepsis. A discriminant analysis. Arch Surg 1983; 118:242–249.
50. Fry DE, Pearlson L, Fulton RL, Polk HC. Multiple system organ failure: The role of uncontrolled infection. Arch Surg 1980; 115:242–249.
51. Knaus WA, Wagner DP. Multiple systems organ failure: Epidemiology and prognosis. Crit Care Clin 1989; 5:221–232.
52. Cockrill BA, Hudson LD, Kubilis PS, Maunder RJ. Organ system dysfunction in patients with multiple trauma and sepsis. Personal communication, 1993.
53. Gattinoni L, Pelosi P, Pesenti A, et al. CT scan in ARDS: Clinical and physiopathological insights. Acta Anaesthesiol Scand 1991; 35:87–96.
54. Gattinoni L, Pesenti A, Bombino M, et al. Relationships between lung computed tomographic density, gas exchange, and PEEP in acute respiratory failure. Anesthesiology 1988; 69:824–832.
55. Maunder R, Shuman W, McHugh J, Marglin S, Butler J. Preservation of normal lung regions in the adult respiratory distress syndrome. JAMA 1986; 255:2463–2465.
56. Stark P, Greene R, Kott MM, Hall T, Vanderslice L. CT-findings in ARDS. Radiologe 1987; 27:367–369.
57. Brigham KL, Meyrick B. Endotoxin and lung injury. Am Rev Respir Dis 1986; 133:913–927.
58. Meyrick BO, Ryan US, Brigham KL. Direct effects of *E. coli* endotoxin on structure and permeability of pulmonary endothelial monolayers and the endothelial layer of intimal explants. Am J Pathol 1986; 122:140–151.
59. Snapper JR, Hutchinson AA, Ogletree ML, Brigham KL. Effects of cyclooxygenase inhibitors on the alterations in lung mechanics caused by endotoxemia in unanesthetized sheep. J Clin Invest 1983; 72:63–76.
60. Brigham K, Bowers R, Haynes J. Increased sheep lung vascular permeability caused by *Escherichia coli* endotoxin. Circ Res 1979; 45:292–297.
61. Esbenshade AM, Newman JH, Lams PM, Jolles H, Brigham KL. Respiratory failure after endotoxin infusion in sheep: Lung mechanics and lung fluid balance. J Appl Physiol 1982; 53:967–976.
62. Parsons P, Worthen G, Moore E, Tate R, Hensen P. The association of circulating endotoxin with the development of the adult respiratory distress syndrome. Am Rev Respir Dis 1989; 140:294–301.
63. Foulis AK, Murray WR, Galloway D, et al. Endotoxaemia and complement activation in acute pancreatitis in man. Gut 1982; 23:656–661.
64. Kreuzfelder E, Joka T, Keinecke HO, et al. Adult respiratory distress syndrome as a specific manifestation of a general permeability defect in trauma patients. Am Rev Respir Dis 1988; 137:95–99.
65. Brandtzaeg P, Kierulf P, Gaustad P, et al. Plasma endotoxin as a predictor of multiple organ failure and death in systemic meningococcal disease. J Infect Dis 1989; 159:195–204.
66. Meyrick B, Brigham KL. Acute effects of *Escherichia coli* endotoxin on the pulmonary microcirculation of anesthetized sheep: Structure–function relationships. Lab Invest 1983; 48:458–470.

67. Pober JS, Lapierre LA, Stolpen AH, et al. Activation of cultured human endothelial cells by recombinant lymphotoxin: Comparison with tumor necrosis factor and interleukin 1. J Immunol 1987; 138:3319–3324.
68. Pober JS, Bevilacqua MP, Mendrick DL, Lapierre LA, Gimbrone MA. Two distinct monokines, interleukin-1 and tumor necrosis factor, each independently induce biosynthesis and transient expression of the same antigen on the surface of cultured human vascular endothelial cells. J Immunol 1986; 136:1680–1687.
69. Pober JS, Gimbrone MA, Lapierre LA, et al. Overlapping patterns of activation of human endothelial cells by interleukin-1, tumor necrosis factor, and immune interferon. J Immunol 1986; 137:1893–1896.
70. Sacks T, Moldow CF, Craddock PR, Bowers TK. Oxygen radicals mediate endothelial cell damage by complement-stimulated granulocytes. An *in vitro* model of immune vascular damage. J Clin Invest 1978; 61:1161–1167.
71. Craddock PR, Hammerschmidt A, White JG, Dalmasso AP, Jacob HS. Complement (C5a)-induced granulocyte aggregation in vitro: A possible mechanism of complement-induced leukostasis and leukopenia. J Clin Invest 1977; 60:260–264.
72. Hosea SF, Brown E, Hammer C, Frank M. Role of complement activation in a model of adult respiratory distress syndrome. J Clin Invest 1980; 66:375–382.
73. Stevens JH, Raffin TA. Adult respiratory distress syndrome. I. Etiology and mechanisms. Postgrad Med J 1984; 60:505–513.
74. Morrison DC, Kline LF. Activation of the classical and properdin pathways of complement by bacterial lipopolysaccharides. J Immunol 1977; 118:362–368.
75. Carvalho ACA. Blood alterations in ARDS. In: Zapol WM, Falke KJ, eds. Acute Respiratory Failure. New York: Marcel Dekker, 1985:303–346.
76. Hyers TM, Fowler AA. Adult respiratory distress syndrome: causes, morbidity, and mortality. Fed Proc 1986; 45:25–29.
77. Zentella A, Manogue K, Cerami A. The role of cachectin/TNF and other cytokines in sepsis. Prog Clin Biol Res 1991; 367:9–24.
78. Roten R, Markert M, Feihl F, Schaller MD, Tagan MC, Perret C. Plasma levels of tumor necrosis factor in the adult respiratory distress syndrome. Am Rev Respir Dis 1991; 143:590–592.
79. Parsons PE, Moore FA, Moore EE, Ikl'e DH, Henson PM, Worthen GS. Studies on the role of tumor necrosis factor in adult respiratory distress syndrome. Am Rev Respir Dis 1992; 146:694–700.
80. Hyers TM, Tricomi SM, Dettenmeier PA, Fowler AA. Tumor necrosis factor levels in serum and bronchoalveolar lavage fluid of patients with the adult respiratory distress syndrome. Am Rev Respir Dis 1991; 144:268–271.
81. Neuhof H. Actions and interactions of mediator systems and mediators in the pathogenesis of ARDS and multiorgan failure. Acta Anaesthesiol Scand Suppl 1991; 95:7–13.
82. Tracey KJ, Beutler B, Lowry SF, et al. Shock and tissue injury induced by recombinant human cachectin. Science 1986; 234:470–474.
83. Stephens KE, Ishizaka A, Larrick JW, Raffin TA. Tumor necrosis factor causes increased pulmonary permeability and edema: Comparison to septic acute lung injury. Am Rev Respir Dis 1988; 137:1364–1370.

84. Goldblum SE. Tumor necrosis factor-a augments pulmonary arterial transendothelial albumin flux in-vitro. Am J Physiol 1990; 258:L57–L67.
85. Bellomo R. The cytokine network in the critically ill. Anaesth Intensive Care 1992; 20:288–302.
86. Zhang D. [Tumour necrotic factor (TNF) in the pathogenesis of liver necrosis in viral hepatitis and strategy for its prevention and treatment]. Chung Hua I Hsueh Tsa Chih 1990; 70:438–441.
87. Wada H, Ohiwa M, Kaneko T, et al. Plasma level of tumor necrosis factor in disseminated intravascular coagulation. Am J Hematol 1991; 37:147–151.
88. Aderka D. Role of tumor necrosis factor in the pathogenesis of intravascular coagulopathy of sepsis: Potential new therapeutic implications. Isr J Med Sci 1991; 27:52–60.
89. Bertani T, Abbate M, Zoja C, et al. Tumor necrosis factor induces glomerular damage in the rabbit. Am J Pathol 1989; 134:419–430.
90. Suter PM, Suter S, Girardin E, Roux LP, Grau GE, Dayer JM. High bronchoalveolar levels of tumor necrosis factor and its inhibitors, interleukin-1, interferon, and elastase, in patients with adult respiratory distress syndrome after trauma, shock, or sepsis. Am Rev Respir Dis 1992; 145:1016–1022.
91. Antonelli M, Bufi M, De Blasi RA, et al. Detection of leukotrienes B4, C4 and of their isomers in arterial, mixed venous blood and bronchoalveolar lavage fluid from ARDS patients. Intensive Care Med 1989; 15:296–301.
92. Hseuh W, Gonzalez-Crussi F, Arroyave JL. Platelet-activating factor: An endogenous mediator for bowel necrosis in endotoxemia. FASEB J 1987; 1: 403–405.
93. Petrak RA, Balk RA, Bone RC. Prostaglandins, cyclo-oxygenase inhibitors, and thromboxane inhibitors in the pathogenesis of multiple systems organ failure. Crit Care Clin 1989; 5:303–314.
94. Wang J, Dunn MJ. Platelet-activating factor mediates endotoxin-induced acute renal insufficiency in rats. Am J Physiol 1987; 253:F1283–F1289.
95. Marshall JC, Christou NV, Meakins JL. Immunomodulation by altered gastrointestinal tract flora: the effects of orally administered, killed *Staphylococcus epidermidis, Candida,* and *Pseudomonas* on systemic immune responses. Arch Surg 1988; 123:1465–1469.
96. Matuschak GM, Rinaldo JE. Organ interactions in the adult respiratory distress syndrome during sepsis: Role of the liver in host defense. Chest 1988; 94: 400–406.
97. Schwartz DB, Bone RC, Balk RA, Szidon JP. Hepatic dysfunction in the adult respiratory distress syndrome. Chest 1989; 95:871–875.
98. Matuschak GM, Martin DJ. Influence of end-stage liver failure on survival during multiple systems organ failure. Transplant Proc 1987; 40–46.
99. Wolter J, Liehr M, Grun M. Hepatic clearance of endotoxins: Differences in arterial and portal venous infusions. J Reticuloendothel Soc 1978; 23:145–152.
100. Deitch EA, Berg R, Specian R. Endotoxin promotes the translocation of bacteria from the gut. Arch Surg 1987; 122:185–190.
101. Moore FA, Moore EE, Poggetti R, et al. Gut bacterial translocation via the portal vein: A clinical perspective with major torso trauma. J Trauma 1991; 31:629–636.

102. Astiz M, Rachow E, Falk J, Kaufman B, Weil M. Oxygen delivery and consumption in patients with hyperdynamic septic shock. Crit Care Med 1987; 15: 26–28.
103. Duff J, Groves A, Mclean A, LaPointe R, McLean L. Defective oxygen consumption in septic shock. Surg Gynecol Obstet 1969; 128:1051–1060.
104. Danek S, Lynch J, Weg J, Dantzker D. The dependence of oxygen uptake in oxygen delivery in the adult respiratory distress syndrome. Am Rev Respir Dis 1980; 122:387–395.
105. Tuchschmidt J, Fried J, Astiz M, Rackow E. Elevation of cardiac output and oxygen delivery improves outcome in septic shock. Chest 1992; 102:216–220.
106. Tuchschmidt J, Fried J, Swinney R, Sharma O. Early hemodynamic correlates of survival in patients with septic shock. Crit Care Med 1989; 17:719–723.
107. Weg JG. Oxygen transport in adult respiratory distress syndrome and other acute circulatory problems: relationship of oxygen delivery and oxygen consumption. Crit Care Med 1991; 19:650–657.
108. Shoemaker WC, Appel PL, Kram HB, Waxman K, Lee TS. Prospective trial of supranormal values of survivors as therapeutic goals in high-risk surgical patients. Chest 1988; 94:1176–1186.
109. Edwards J, Brown G, Nightengale P, Slater R, Fargher B. Use of survivors' cardiorespiratory values as therapeutic goals in high risk patients. Crit Care med 1989; 17:1098–1103.
110. Yu M, Levy MM, Smith P, Takiguchi SA, Miyasaki A, Myers SA. Effect of maximizing oxygen delivery on morbidity and mortality rates in critically ill patients: A prospective, randomized, controlled trial. Crit Care Med 1993; 21: 830–838.
111. Archie J. Mathematic coupling of data. A common source of error. Ann Surg 1981; 193:296–303.
112. Snyder J, Carroll G. Tissue oxygenation: A physiological approach to a clinical problem. Curr Probl Surg 1989; 19:652–719.
113. Annat G, Viale J-P, Percival C, Froment M, Motlin J. Oxygen delivery and uptake in the adult respiratory distress syndrome. Am Rev Respir Dis 1989; 133:999–1001.
114. Carlile P, Gray B. Effect of opposite changes in cardiac output and arterial PO2 on the relationship between mixed venous PO2 and oxygen transport. Am Rev Respir Dis 1989; 140:891–896.
115. Mitchell JP, Schuller D, Calandrino FS, Schuster DP. Improved outcome based on fluid management in critically ill patients requiring pulmonary artery catheterization. Am Rev Respir Dis 1992; 145:990–998.
116. Hudson LD. Fluid management strategy in acute lung injury. Am Rev Respir Dis 1992; 145:988–989.
117. Broadus VC, Berthiaume Y, Biondi JW, Matthay MA. Hemodynamic management of the adult respiratory distress syndrome. J Intensive Care Med 1987; 2:190–213.
118. Schumer W. Steroids in the treatment of clinical septic shock. Ann Surg 1976; 184:333–341.

119. Sprung CL, Caralis PV, Marcial EH, et al. The effects of high-dose corticosteroids in patients with septic shock. A prospective, controlled study. N Engl J Med 1984; 311:1137–1143.
120. Weigelt J, Norcross J, Borman K, Snyder W. Early steroid therapy for respiratory failure. Arch Surg 1985; 120:536–539.
121. Meduri GU, Belenchia JM, Estes RJ, Wunderink RG, el Torky M, Leeper KVJ. Fibroproliferative phase of ARDS. Clinical findings and effects of corticosteroids. Chest 1991; 100:943–952.
122. Hooper RG, Kearl RA. Established ARDS treated with a sustained course of adrenocortical steroids. Chest 1990; 97:138–143.
123. Ashbaugh DG, Maier RV. Idiopathic pulmonary fibrosis in adult respiratory distress syndrome. Arch Surg 1985; 120:530–535.
124. Giroir BP. Mediators of septic shock: New approaches for interrupting the endogenous inflammatory cascade. Crit Care Med 1993; 21:780–789.
125. St John RC, Dorinsky PM. Immunologic therapy for ARDS, septic shock and multiple organ failure. Chest 1993; 103:932–943.
126. Dinarello CA, Gelfand JA, Wolff SM. Anticytokine strategies in the treatment of the systemic inflammatory response syndrome. JAMA 1993; 269: 1829–1835.
127. Ziegler EJ, McCutchan JA, Fierer J, et al. Treatment of gram-negative bacteremia and shock with human antiserum to a mutant *Escherichia coli*. N Engl J Med 1982; 307:1225–1230.
128. Greenman RL, Schein RM, Martin WA, et al. A controlled clinical trial of E5 murine monoclonal IgM antibody to endotoxin in the treatment of gram-negative sepsis. The XOMA Sepsis Study Group. JAMA 1991; 266:1097–1102.
129. Ziegler EJ, Fisher CJJ, Sprung CL, et al. Treatment of gram-negative bacteremia and septic shock with HA-1A human monoclonal antibody against endotoxin. A randomized, double-blind, placebo-controlled trial. The HA-1A Sepsis Study Group. N Engl J Med 1991; 324:429–436.
130. Warren H, Danner D, Munford R. Anti-endotoxin monoclonal antibodies. N Engl J Med 1992; 326:1153–1157.
131. Wolff S. Monoclonal antibodies and the treatment of gram-negative bacteremia and shock. N Engl J Med 1991; 324:486–488.
132. Stevens JH, O'Hanley P, Shapiro JM, et al. Effects of anti-C5a antibodies on the adult respiratory distress syndrome in septic primates. J Clin Invest 1986; 77:1812–1816.
133. Tracey KJ, Fong Y, Hesse DG, et al. Anti-cachectin/TNF monoclonal antibodies prevent septic shock during lethal bacteraemia. Nature 1987; 330:662–664.
134. Fisher CJ, Dhainaut JF, Pribble J, Knaus W, Catalano MA. A study evaluating the efficacy of human recombinant interleukin-1 receptor antagonists in the treatment of patients with sepsis syndrome: preliminary results from a phase III multicenter trial (abstr). 13th International Symposium on Intensive Care and Emergency Medicine, Brussels, Belgium. New York: Springer-Verlag, 1993.

135. Kopolovic R, Thraikill KM, Martin DT, Carey LC, Cloutier CT. A critical comparison of the hematologic, cardiovascular, and pulmonary response to steroids and nonsteroidal anti-inflammatory drugs in a model of sepsis and adult respiratory distress syndrome. Surgery 1986; 100:679–690.
136. Coran AG, Drongowski RA, Paik JJ, Remick DG. Ibuprofen intervention in canine septic shock: Reduction of pathophysiology without decreased cytokines. J Surg Res 1992; 53:272–279.
137. Jenkins JK, Carey PD, Byrne K, Sugerman HJ, Fowler AA. Sepsis-induced lung injury and the effects of ibuprofen pretreatment. Analysis of early alveolar events via repetitive bronchoalveolar lavage. Am Rev Respir Dis 1991; 143:155–161.
138. Leeper WSK, Fisher BJ, Sugerman HJ, Fowler AA. Pharmacologic reduction in tumor necrosis factor activity of pulmonary alveolar macrophages. Am J Respir Cell Mol Biol 1993; 8:169–175.
139. Lechner AJ, Rouben LR, Potthoff LH, Tredway TL, Matuschak GM. Effects of pentoxifylline on tumor necrosis factor production and survival during lethal *E. coli* sepsis vs. disseminated candidiasis with fungal septic shock. Circ Shock 1993; 39:306–315.
140. Sable CA, Wispelwey B. Pharmacologic interventions aimed at preventing the biologic effects of endotoxin. Infect Dis Clin North Am 1991; 5:883–898.
141. Taylor FBJ, Chang A, Esmon CT, D'Angelo A, Vigano DS, Blick KE. Protein C prevents the coagulopathic and lethal effects of *Escherichia coli* infusion in the baboon. J Clin Invest 1987; 79:918–925.
142. Bernard GR, Lucht WD, Niedermeyer ME, Snapper JR, Ogletree ML, Brigham KL. Effect of *N*-acetylcysteine on the pulmonary response to endotoxin in the awake sheep and upon *in vitro* granulocyte function. J Clin Invest 1984; 73: 1772–1784.
143. Hardaway RM, Williams CH. A new treatment for traumatic shock and ARDS. Resuscitation 1990; 19:61–76.
144. Stoller JK, Kacmarek RM. Ventilatory strategies in the management of the adult respiratory distress syndrome. Clin Chest Med 1990; 11:755–772.
145. Lee R, Balk R, Bone R. Ventilatory support in the management of septic patients. Crit Care Clin 1989; 5:157–175.
146. Lain DC, DiBenedetto R, Morris SL, Van NA, Saulters R, Causey D. Pressure control inverse ratio ventilation as a method to reduce peak inspiratory pressure and provide adequate ventilation and oxygenation. Chest 1989; 95: 1081–1088.
147. Chan K, Abraham E. Effects of inverse ratio ventilation on cardiorespiratory parameters in severe respiratory failure. Chest 1992; 102:1556–1561.
148. Tharratt R, Allen R, Albertons T. Pressure controlled inverse ratio ventilation in severe adult respiratory failure. Chest 1988; 94:755–762.
149. Anderson J. Ventilatory strategy in catastrophic lung disease: Inversed ratio ventilation (IRV) and combined high frequency ventilation (CHFV). Acta Anaesthesiol Scand 1989; 33(suppl 90):145–148.
150. Marcy TW, Marini JJ. Inverse ratio ventilation in ARDS. Rationale and implementation. Chest 1991; 100:494–504.

151. Suchyta M, Elliott C, Colby T, Rasmussen B, Morris A, Jensen R. Open lung biopsy does not correlate with pulmonary function after the adult respiratory distress syndrome. Chest 1991; 99:1232–1237.
152. von Segesser LK, Weiss BM, Pasic M, et al. Temporary lung support using an intravascular gas exchanger. Thorac Cardiovasc Surg 1992; 40:121–125.
153. Jurmann MJ, Demertzis S, Schaefers HJ, Wahlers T, Haverich A. Intravascular oxygenation for advanced respiratory failure. ASAIO J 1992; 38:120–124.
154. High KM, Snider MT, Richard R, et al. Clinical trials of an intravenous oxygenator in patients with adult respiratory distress syndrome. Anesthesiology 1992; 77:856–863.
155. Carlon GC, Howland WS, Groeger JS, Ray CJ, Miodownik S. Role of high-frequency jet ventilation in the management of respiratory failure. Crit Care Med 1984; 12:777–779.
156. Holzapfel L, Robert D, Perrin F, Gaussorgues P, Giudicelli DP. Comparison of high-frequency jet ventilation to conventional ventilation in adults with respiratory distress syndrome. Intensive Care Med 1987; 13:100–105.
157. Borg UR, Stoklosa JC, Siegel JH, et al. Prospective evaluation of combined high-frequency ventilation in post-traumatic patients with adult respiratory distress syndrome refractory to optimized conventional ventilatory management. Crit Care Med 1989; 17:1129–1142.
158. Langer M, Mascheroni D, Marcolin R. The prone position in ARDS. Chest 1988; 94:103–107.
159. Bowton DL, Kong DL. High tidal volume ventilation produces increased lung water in oleic acid–injured rabbit lungs. Crit Care Med 1989; 17:908–911.
160. Cilley RE, Wang JY, Coran AG. Lung injury produced by moderate lung overinflation in rats. J Pediatr Surg 1993; 28:488–493.
161. Hickling KG. Ventilatory management of ARDS: Can it affect the outcome? Intensive Care Med 1990; 16:219–226.
162. Tsuno K, Prato P, Kolobow T. Acute lung injury from mechanical ventilation at moderately high airway pressures. J Appl Physiol 1990; 68:956–961.
163. Hickling K, Henderson S, Jackson R. Low mortality associated with low volume pressure limited ventilation with permissive hypercapnia in severe adult respiratory distress syndrome. Intensive Care Med 1990; 16:372–377.
164. Goldstein G, Luce JM. Pharmacologic treatment of the adult respiratory distress syndrome. Clin Chest Med 1990; 11:773–787.
165. Maunder R, Hudson L. Pharmacologic strategies for treating the adult respiratory distress syndrome. Respir Care 1990; 35:241–246.
166. Holm BA, Matalon S. Role of pulmonary surfactant in the development and treatment of adult respiratory distress syndrome. Anesth Analg 1989; 69:805–818.
167. Enhorning G. Surfactant replacement in adult respiratory distress syndrome. Am Rev Respir Dis 1989; 140:281–283.
168. Lewis JF, Jobe AH. Surfactant and the adult respiratory distress syndrome. Am Rev Respir Dis 1993; 147:218–233.
169. Holcroft J, Vassr M, Weber C. Prostaglandin E1 and survival in patients with the adult respiratory distress syndrome. Ann Surg 1986; 203:371–378.

170. Bone RC, Slotman G, Maunder R, et al. Randomized double-blind, multicenter study of prostaglandin E1 in patients with the adult respiratory distress syndrome. Prostaglandin E1 Study Group. Chest 1989; 96:114–119.
171. Silverman HJ, Slotman G, Bone RC, et al. Effects of prostaglandin E1 on oxygen delivery and consumption in patients with the adult respiratory distress syndrome. Results from the prostaglandin E1 multicenter trial. The Prostaglandin E1 Study Group. Chest 1990; 98:405–410.
172. Hardaway RM, Williams CH, Marvasti M. Prevention of ARDS with urokinase (abstr). J Trauma 1988; 28:1087.

15

Immunological Therapy of Endotoxemia

ARTHUR P. WHEELER

Vanderbilt University Medical Center
Nashville, Tennessee

I. Introduction

Sepsis syndrome is the clinical constellation of altered temperature, tachycardia, and increased minute ventilation accompanied by a major organ system failure (pulmonary, renal, hepatic, circulatory, hematologic, neurologic). Sepsis syndrome occurs more than 400,000 times each year in the United States, taking 100,000 to 200,000 lives and costing \$5 to 10 billion (1–3). Interestingly, one-half of patients with the syndrome have no demonstrable infection, and numerous noninfectious disorders mimic this condition (4). Hence investigators have proposed renaming "sepsis syndrome" the "systemic inflammatory response syndrome" (SIRS), to emphasize that not all patients have infection as the etiology of their illness and that in large part, host response is the critical determinant of severity (5).

Clinically significant lung injury occurs in more than half of all patients developing the sepsis syndrome, and its most severe form, the adult respiratory distress syndrome (ARDS) defines a subset of septic patients with a poor prognosis. For practical purposes it is difficult and probably counterproductive to segregate ARDS from milder lung injury associated with sepsis syndrome because of shared pathophysiology.

Although SIRS may result from noninfectious causes or any infectious agent, gram-negative bacteria constitute the single most common cause, accounting for one-third of all cases (1,4,6,7). Endotoxin is the integral cell-wall lipopolysaccharide (LPS) responsible for gram-negative sepsis. Patients at risk of developing SIRS often share risk factors for endotoxemia, notably the following:

Gram-negative bacteremia
Hemorrhagic shock (10)
Cardiopulmonary arrest
Bowel ischemia (11–13)
Profound hypoxemia (14)
Ionizing radiation (15)
Graft versus host disease (16)
Hyperthermia (17,18)
Burns (19,20)
Multiple trauma (21)
Gastroenteritis (22,23)
Excessive vasoconstrictor agents (24)
Decompression sickness (25)

As the diverse nature of these predisposing conditions suggests, endotoxemia can and does occur in the absence of bacteremia. Endotoxin may be shed by reproducing bacteria; LPS from bacteria in feces may leak directly across the bowel wall (8–25); and LPS may be precipitously liberated by treatment with bactericidal antibiotics (26). Unfortunately, a poor correlation exists between bacteremia and endotoxemia and an unacceptably high rate of false positive and false negative tests prevents use of currently available assays for the clinical detection of endotoxin (8,27–31). These limitations effectively preclude early identification of patients who might benefit from antiendotoxins (32–34).

Plasma LPS levels correlate well with illness severity and mortality in some forms of gram-negative sepsis (35–38), but endotoxin concentrations do not tell the entire story. Only 30 to 40% of patients with SIRS have endotoxemia (8,9,39,40), and even though plasma LPS levels in volunteers given endotoxin transiently approach those of septic patients, severe illness does not develop. Furthermore, even transient LPS exposure rapidly incites inflammation, causing cellular injury and physiological changes within minutes (41–44).

Living or dead bacteria, endotoxin, and lipid A (the major active component of endotoxin) are all comparably toxic but do not directly cause damage (45). Most changes of SIRS result, instead, from the rapid production of a cascade of endogenous mediators, including tumor necrosis factor (TNF-α), interleukin-1 (IL-1), and interleukin-6 (IL-6). Hence an understanding of the actions of these inflammatory proteins, collectively known as cytokines, is essential.

Even the development of sophisticated support technology and use of investigational drugs has not had a dramatic impact on the mortality rates of SIRS and ARDS (4,46–49). Therefore, immunologic antagonism of endotoxin and cytokines has become a fertile area of research.

II. Antibodies to Antagonize Endotoxin

Immunological antagonism of endotoxin is not as simple as producing one antibody that binds and neutralizes all varieties of LPS. The effectiveness of immunotherapy is substantially influenced by endotoxins structure. Endotoxins outermost "O antigen" region is composed of oligosaccharide chains that impart a "smooth" appearance to bacterial colonies. The O region is the most antigenic and most variable portion of the LPS molecule, differing by species, strain, and bacterial growth rate (50–52). Because the O region is highly antigenic, when "smooth" bacterial strains or endotoxins are used as immunogens, the vast majority of antibody produced is O-region specific (53–55). It is relatively easy to produce affine, high-titer, neutralizing antibodies to oligosaccharides, but unfortunately, such antibodies reliably protect against challenge only with homologous LPS or bacteria, making them clinically useless (56–66).

The core region has a nearly constant structure among gram-negative bacteria and links the O-region to the toxic innermost, lipid A portion of LPS (52,67–70). "Rough" bacterial mutants exist that lack the enzymes necessary to polymerize or attach oligosaccharides to the core. Discovery of the rough mutants, which have an exposed core region, raised hopes that a widely cross-reactive core-directed antibody could be engineered, and most efforts in the field have focused on production of anticore antibodies.

Lipid A can reproduce the effects of the entire LPS molecule (52,71). Preservation of lipid A's structure is critical to its toxicity; cleavage of even a single bond within lipid A can transform it into a nontoxic compound (72–74). Although anti-lipid A antibodies exhibit nearly universal binding to a variety of endotoxins, they do not prevent the effects of LPS (52,56,57,71, 75–77).

III. Endogenous Mediators of Endotoxin: TNF, IL-1, and IL-6

Endotoxin serves as a critical trigger to induce production of endogenous inflammatory mediators. Because of the transient nature of endotoxemia and the absence of endotoxemia in the majority of SIRS victims, attention has been drawn to the antagonism of the common early downstream mediators of inflammation tumor necrosis factor, interleukin-1, and interleukin-6.

Tumor necrosis factor α (TNF-α), also known as cachectin, is a highly conserved inflammatory protein produced predominately by blood monocytes and tissue macrophages (78–82). The trimeric, active form, of TNF-α is rapidly produced after LPS stimulation by a gene locus on chromosome 6 (83–85). Two TNF receptors have been cloned and are present on all cells except erythrocytes (86–90). Within minutes of appropriate stimulation, TNF-α messenger RNA is generated and is rapidly translated into the mature protein (91–95). After a discrete intravenous stimulus, TNF-α levels typically peak within 1 to 3 h and wane within 4 to 6 h; however, prolonged low-level TNF-α production has been reported.

TNF-α's actions mimic the sepsis syndrome and include altered hemodynamics, lung mechanics, and oxygenation. TNF also causes fever, cachexia, leukocyte and coagulation activation, and the prouction of acute-phase reactants (78,79,96–100) and is directly cytolytic for some cells (101,102). Neutrophils accentuate the injury caused by TNF-α, but even profound neutropenia does not preclude the development of TNF-α-induced changes (96). TNF-α's effects depend not only on its concentration but also on its location. For example, TNF may be markedly elevated in a single anatomical compartment without producing deleterious effects at other sites (103–105).

In many respects IL-1 mimics the effects of TNF-α (106,107), but in addition IL-1 indirectly causes T-cell proliferation. Alone and together, TNF-α and IL-1 have potent effects on neutrophil adhesion and activation (108,109). Monocytes, macrophages, and endothelial cells are the predominant cell sources of the two forms of IL-1 (IL-1α and IL-1β), both of which cause identical actions by binding to a shared receptor (110,111). IL-1β represents more than 90% of both the IL-1 mRNA and active protein secreted in vitro by human monocytes following LPS challenge and is the predominate form of IL-1 circulating in sepsis.

IL-6 is a protein produced by endothelial cells, T-lymphocytes, lung fibroblasts, and monocytes, induced by TNF-α and IL-1 (112). IL-6 causes hepatic production of acute-phase reactants and reduces albumin synthesis (113,114), while simultaneously stimulating B-lymphocytes to augment immunoglobulin production (115). IL-6 is also a pyrogen (116), promotes expression of IL-2 and IL-2 receptors (117), cytolytic T-cell differentiation (118), and stimulates thrombopoiesis (119) by binding to a specific receptor (120). Despite these numerous biological actions, IL-6's role in sepsis is unknown (121,122). Like TNF and IL-1, IL-6 expression is not due exclusively to sepsis (123–126).

IV. Cytokine Interactions in Inflammation

Cytokines exhibit complex behavior. A single agent may have contrasting actions on different target cells (101,102,127–131), and because cytokines functions may be redundant or counterregulatory, it is often difficult to distinguish the compound responsible for a given action (116,132). For example, in vivo and in vitro, TNF-α stimulates its own production (132) and that of IL-1 (133,134) and IL-6 (135–137). IL-1 also induces itself, TNF-α, and IL-6 (138,139). In contrast, IL-6 does not cause TNF-α or IL-1 production and may down-regulate TNF and IL-1. Cytokine effects may overlap or act synergistically, as is the case with TNF-α- and IL-1β-induced shock (140–143), with IL-1β- and IL-6-induced thymocyte stimulation (144) and immunoglobulin production (145–147), and with the inflammatory effects of TNF-α and interferon gamma (147–149).

Recognizing the synergistic and amplifying interactions of cytokines is important because it suggests that inhibition of one cytokine could interrupt the inflammatory cascade. This suspicion has been borne out in vivo where TNF-α antibodies dramatically reduce IL-1 and IL-6 production (150, 151). TNF-α is also synergistic with LPS in causing injury in vivo (152). One cytokine's actions may also be inhibited by another, as when IL-6 or IL-1 down-regulate inflammatory cytokine production (153,154), or when IL-1 exposure blunts the action of TNF-α or LPS (155,156).

Cytokines are not entirely detrimental, and some cytokines are necessary for normal immune function. For example, the removal of TNF-α dramatically reduces the ability to control local infection (157–161), and pretreatment with IL-1 or TNF may augment local bacterial defenses (155,156). At high concentrations or when released into the circulation, however, cytokines may become deleterious.

V. Evidence That Cytokines Are Involved in Sepsis

A. Administration of Recombinant Cytokines Mimics Endotoxemia

The infusion of recombinant TNF-α or IL-1β results in physiological alterations identical to those seen in SIRS and sepsis-associated lung injury (162–165). In rodents, dogs, and rabbits, recombinant TNF-α causes hemoconcentration and hypovolemic shock by diffusely increasing vascular permeability (166–170). In sheep and swine TNF-α has comparable effects to LPS on hemodynamics and extravascular lung water but acts more rapidly (171). In sheep, TNF-α infusion causes pulmonary hypertension, hypoxemia, leukopenia, and the generation of prostanoid mediators (172,173). Increased vascular permeability suggested by the protein-rich lymph flow seen in TNF-α-treated sheep has now been confirmed in rodents (174–176). TNF-α also causes perivascular edema (177) and pulmonary sequestration of granulocytes, reduces lung compliance, and increases airway resistance and reactivity (173,178–180). Like endotoxin, many of TNF's effects are attenuated by cyclooxygenase inhibitors (181–183).

In humans and animals, TNF-α causes fever, leukopenia, stress hormone release, and acute-phase reactant production (141,184–190). In humans, exogenously administered TNF-α (35 to 200 $\mu g/m^2$) has a half-life of approximately 20 min (184,186,187) and results in serum concentrations similar to those of sepsis patients (2.5 to 80 ng/mL) (184–190). Chills, rigors, and fever are nearly universal during TNF infusion (190). High doses of TNF cause hypotension, an effect exacerbated by concurrent administration of IFN-γ (170). TNF-α stimulates IL-6 production within minutes, and peak IL-6 levels are achieved within 2 to 3 h (191). Concurrent administration of IL-1 (141,151,192) or IFN-γ (193) may accelerate TNF-induced injury or death; however, TNF-α is unique in that it is the only cytokine that by itself is known to produce shock and death (167).

In rabbits, large intravenous doses of IL-1β increase cardiac output, reduce systemic vascular resistance, and cause hypotension (141,142,194). Doses of IL-1β and TNF-α that are each alone not sufficient to alter hemodynamics result in a profound shock state when given together (141). Like LPS and TNF-α, IL-1's effects can be altered by cyclooxygenase inhibitors (141). In human volunteers given an agent that induces IL-1β production, etiocholanolone, fever, leukocytosis, and increased acute-phase reactant production occur (195). When humans are given high doses (ca. 1 $\mu g/kg$) of IL-1, cyclooxygenase product generation (196) and nitric oxide release (197,198) result in hypotension.

B. Cytokines in Experimental and Clinical Sepsis

Cytokines are released following toxic challenge and there is a rough correlation between serum cytokine level, severity of sepsis, and mortality. Cytokine detection cannot be regarded exclusively as a marker of infection, however (92,199), since circulating TNF-α has been detected in patients with renal allograft rejection (200,201), heart failure (202), neoplasia (203), thermal burns, and some skin diseases (137). TNF-α, IL-1, and IL-6 are not restricted to the circulation and have been recovered from localized sites of inflammation such as bronchoalveolar lavage fluid of patients with ARDS (204–207), joint fluid of patients with rheumatoid arthritis (208–210), and cerebrospinal fluid of patients with meningitis and head trauma (211–216).

When considered collectively, studies indicate that in sepsis, persistent or profound elevations of TNF-α, IL-1β, and IL-6 are associated with more severe illness and higher mortality. Most experimental studies demonstrate that limited intravenous infusions of bacteria or their toxins causes a brief sequential rise in TNF-α, IL-1β, and IL-6 (81,93–95); however, in some experimental models (94) and in humans with septic shock, TNF-α levels may remain elevated for days (204,217–221).

Plasma TNF-α and IL-1β increase in LPS-challenged animals and humans, but TNF-α elevations are usually of greater magnitude and occur faster (peak levels at 90 min) than IL-1β, where peak concentrations occur at 180 min (92,222–224). Studies of LPS infusion in human volunteers have demonstrated abnormal physiology concurrent with rises in TNF but not IL-1 levels in plasma (225). Specific splanchnic organ production of TNF-α has also been observed during experimental human endotoxemia (226,227).

TNF-α is detectable in 25 to 33% of all SIRS victims (199,218), and TNF-α, IL-1, and LPS levels are more frequently elevated in septic patients with shock. A certain prognostic relationship between toxin levels and outcome has not been established, but persistent elevations of TNF-α and IL-1 and high TNF-α levels are associated with greater likelihood of developing shock (228,229). Bacteremic ICU patients essentially all have measurable plasma levels of TNF-α and IL-6, and nonsurvivors have higher TNF-α and IL-6 levels than those of survivors (230). Cytokine elevations are not confined to gram-negative infections. Both gram-positive and gram-negative sepsis cause cytokine release (199,218,228,229); however, some infections, such as meningococcemia (211,212,231), are more likely to cause profound TNF-α elevations, especially in fatal cases.

Among burn patients, a crude relationship exists between TNF-α levels and sepsis, where 11 of 16 (69%) of patients with clinical episodes of sepsis had detectable TNF-α levels, whereas only 9 of 27 (33%) without sepsis had detectable TNF-α (218). TNF elevations among these patients were brief but repetitive, and the mortality in patients with detectable TNF-α exceeded 70% versus only a 33% mortality rate in patients without detectable TNF-α.

In sepsis, TNF-α levels correlate with the severity of pulmonary dysfunction (218), illness scores (211,212,218,232), level of plasma fibrinogen, and mortality rate (199,218,231). A positive correlation between plasma TNF-α concentration, sepsis severity score, and mortality has been reported in patients with bacteremia (232), where TNF-α levels >200 pg/mL were associated with near-certain death, while levels <200 pg/mL were associated with a 36% mortality rate. Death was roughly twice as common among 43 ICU patients with elevated TNF-α levels (73% mortality, 8/11) compared to patients with undetectable TNF-α levels (34% mortality, 11/32). In a separate study of a small group of patients with septic shock and elevated TNF-α levels, two main groups could be discerned, those with persistent TNF elevations, who uniformly died, and those with declining levels, who had a low (20%) mortality rate (220).

In contrast, other studies (230,233,234) indicate that the duration of cytokine presence is more critical than absolute level. Substantial elevations of cytokines at the onset of sepsis may be associated with death, but in humans the duration of cytokine circulation is probably more important (230, 233,234). These data are similar to those in animal models, in which TNF-α levels are more likely to remain elevated in nonsurvivors (94). TNF-α and IL-6 levels are elevated in septic shock patients compared to nonseptic hypotensive controls (221). Although there was no correlation between initial or peak TNF-α or IL-6 level and outcome, patients with plasma TNF levels > 150 pg/mL persisting for 24 h have excessively high mortality rates.

The relationship between IL-1 level and mortality is less clear. Although IL-1β is increased in about 20% of patients with gram-negative sepsis (231), neither the absolute IL-1 level nor its duration correlate well with mortality (228,229,232). Plasma IL-6 concentrations are elevated in the overwhelming majority of patients with sepsis syndrome and correlate with elevated plasma lactate levels, tachycardia, hypotension, and thrombocytopenia (212,235, 236). Neither the type of bacteria nor the site of infection appear to have a significant impact on IL-6 levels, although plasma IL-6 concentrations in septic shock patients are sometimes 100-fold greater than those of normotensive septic patients (237).

Waage has reported an association between IL-1β and IL-6 levels and a fatal outcome and suggested that TNF-α alone may not be sufficient to cause death (211,212). In this study, high levels of IL-6 combined with elevated TNF-α levels were associated with a fatal outcome. Elevations of IL-1β occurred exclusively in patients with high levels of IL-6 and TNF-α and predicted a rapidly fatal outcome. IL-6 levels are detectable in the majority of patients with gram-negative septic shock early in their course but rapidly decline. In contrast to the findings of Dofferhoff (230), IL-6 levels had such a poor predictive value, they were clinically useless (234). In patients with malaria, a correlation between serum concentrations of TNF-α and IL-6 and the density of parasitemia (235) exists and there is a correlation between TNF-α plasma level and mortality (236). Dofferhoff (230) has also found a correlation between IL-6 level and outcome. Marano noted that the initial serum levels of TNF-α were elevated and remained elevated for 10 days in 13 nonsurvivors compared to lower initial levels, which declined in 34 survivors of gram-negative sepsis (237). Marano reported similar findings in 43 burn victims, where the incidence of detectable TNF-α in the peripheral circulation predicted mortality (238). IL-6 levels are elevated in survivors and nonsurvivors of sepsis (264), but a correlation can be observed between admission IL-6 level and mortality in shock and nonshock patients.

VI. Therapeutic Strategies

A. Active Immunization to Endotoxin

Despite its effectiveness in experimental animals, active immunization of humans with LPS is not practical because (1) even low doses of LPS are extremely toxic, (2) there is significant delay in producing antibodies following immunization, and (3) without repeated injections antibody titer rapidly wanes. Therefore, essentially all immunological efforts have focused on passive immunization initially using crude antisera, later followed by purified polyclonal antibody preparations, monoclonal antibodies, and antibody fragments.

B. Passive Immunizations with Endotoxin Antibodies

Studies as early as 1896 (239) demonstrated a protective effect of crude antisera and purified antibodies against endotoxins; however, the mechanism of protection was uncertain and remains obscure even today. There is little evidence to demonstrate direct neutralizing effects of antibody on LPS. Similarly, LPS clearance is not the mechanism since homologous LPS is more rapidly cleared in the presence of antibody, but the same is not true for heterologous bacteria or endotoxin (55,58,240). There is also little evidence to support enhanced complement-mediated bacterial lysis, toxin or bacteria opsonization, or nonspecific immune stimulation as mechanisms of action (55,241,242). Because of the prospect of cross-protection to numerous different endotoxins, most investigators have studied antibody preparations directed at the core region of rough-mutant bacteria. *Escherichia coli 0111:B4 J5* and *Salmonella minnesota Re 595* rough mutants have been the most widely used immunogens in this regard.

Use of Anticore Antibodies

Active and passive immunization with anticore antibodies can prevent *bacteria-induced* physiological changes (45,55–57,240–248). Similarly, passive immunization with anticore antibodies can prevent fever, the Shwartzman reaction, and other physiological changes of *endotoxin* and *hemorrhagic shock* and are accepted therapy in veterinary medicine (10,12–15,22,23,55, 58,66,75–77,233,240–246,249–259). Occasionally, such antibodies can have dramatic effects on the hemodynamic abnormalities of sepsis (22,258,259). Despite these successes anticore antibodies have failed in a number of models of endotoxemia, bacteremia (66,260,261), and experimental gram-negative pneumonia (57,60,262–264).

Further support for the concept of anticore antibody protection comes from clinical data. Analysis of plasma from patients with gram-negative bacteremia indicates that those with naturally occurring high titers of anticore antibodies present at the onset of sepsis have superior survival rates (265,266). Finally, clinical studies using human polyclonal antisera suggest that survival is increased by passive immunization with rough-mutant antisera (267,268), implying that such antibodies can circumvent the stearic interference of the oligosaccharide region (55). Although cross-reactive, the magnitude of the protection offered by anticore antibodies is substantially less than that offered by homologous O antigen antibodies (266) and anticore/lipid A antibodies have failed in some settings (43,269–271).

Impact of Anticore Antibody Class and Titer on Efficacy

Controversy surrounds the ideal source and antibody isotype to treat endotoxemia. In experimental animals, IgM (252,272) and IgG antibody fractions are effective (240,242,270) in treating some aspects of endotoxemia, but in humans, conflicting data exist. Studies using hemagglutination assays suggest that the presence of high titers of naturally occurring antibodies to the core regions of *S. minnesota Re* mutants correlates with survival in septic humans, but that (266) the titer of anti-O region antibodies does not. Further study (273) using immunofluorescence techniques suggests that (1) survival correlates best with the titer of IgG O-specific antibody to the patient's infecting organism; (2) oligosaccharide-specific IgM antibody levels do not correlate with outcome; and (3) high titers of anticore antibodies of all isotypes correlate with survival. In patients with *P. aeruginosa* bacteremia, the titer of strain-specific antibodies is the best predictor of outcome, followed by IgM, then IgG anticore antibodies (265). Of note is that the vast majority of anti-LPS antibody produced during convalesce from human gram-negative sepsis is of the IgG class (45,245,249).

The controversy over the relative importance of antibody classes was fueled further by reports that there were no differences between the levels of IgG and IgM antibodies to rough-mutant bacteria, or IgG class antibodies to lipid A, among surviving and fatal cases of human gram-negative bacteremia. In fact, the titers of IgM antibodies to lipid A were higher in nonsurvivors (7,274). Several reports detail lower efficacy for monoclonal antibodies than for polyclonal antibodies (53,252,275,276).

Prospective Studies of Human Polyclonal Core Antiserum

In a 1978 human study, J5 antiserum roughly halved the mortality rate among 136 patients with gram-negative bacteremia (26% in controls versus 14% in treated patients). In the subset with shock, mortality was reduced from 89% among controls to 29% among J5 recipients (277). A larger (304-patient) randomized, double-blind study of human sepsis using human sera rich in core antibodies or control serum was published in 1982 (267). The 212 patients (70% of the total population) comprising the group of primary analysis contained 191 patients with gram-negative bacteremia and 21 patients with a gram-negative site of infection but sterile blood cultures. A mortality rate of 39% (42/109) in controls and 22% (23/103) among bacteremic recipients of antisera ($p < 0.05$) was encouraging, and benefits were also seen in patients with shock, who had a reduction in mortality from 77% (30/39) among controls to 44% (18/41) among treated patients ($p < 0.05$).

Because of a high observed incidence of gram-negative bacteremia, the applicability of these results to more heterogeneous groups of sepsis syndrome patients has been questioned. Other problems of this study include a poorly defined time point for determining mortality, and the exclusion of late deaths occurring as the result of multiple-organ system failure from sepsis mortality calculations.

This 1982 study (267) spurred several prophylactic trials of similar antisera in patients at high risk of developing gram-negative infections. McCutchan (278) studied neutropenic patients receiving chemotherapy who received either human control antisera or that rich in *E. coli J5* antibodies. Prophylactic antisera did not reduce the risk of infection (8/59 infections among controls and 8/52 in treated patients), bacteremia (12% in controls versus 15% in the treated group), or death [4% (2/59 deaths) in controls compared to 8% (4/52 deaths) in treated patients]. There was a reduction in the incidence of fever among a single subgroup of patients that was of questionable clinical importance. There are several potential explanations for failure of this intervention, including a small sample size and low incidence of infection, making it difficult to demonstrate a treatment effect. Moreover, because only a single dose of antisera was given at the initiation of chemotherapy, by the time neutropenia developed (approximately 1 week) it is unlikely that significant amounts of antibody remained. Wolff (279) has also demonstrated a reduction in the incidence of fever among a subgroup of neutropenic patients receiving prophylactic J5 antiserum, but similarly, did not observe a survival benefit.

Baumgartner (268) conducted a trial of J5 antibodies obtained by plasmapheresis administered prophylactically to surgical patients deemed to be at high risk for sepsis. Antibodies were routinely administered every 5 days and were given more often if sepsis developed. Although treatment did not prevent gram-negative infections, it reduced the risk of developing septic shock from 11% (15/136) among placebo-treated patients to 5% (6/126) in plasma recipients ($p < 0.05$), and reduced mortality from 7% (9/136) among controls to 2% (2/126) among treated patients ($p < 0.05$). The significance of this study has been questioned because the beneficial effects were derived from a subgroup of patients undergoing abdominal surgery in whom mortality was reduced from 16% (13/83) among controls to 3% (2/71) among treated patients. It is also difficult to compare the groups in this trial with respect to severity of illness because little detail about the patient populations was provided. In neither the Baumgartner (268) nor Zeigler (267) studies did the titer of anticore antibodies correlate with survival.

Failure to observe benefits at interim analysis resulted in early termination of a randomized double-blind placebo-controlled study of J5 immune plasma in 70 children with meningococcemia conducted by Girardin (280).

The effects of purified intravenous human globulin preparations have also been studied in sepsis. Calandra (7) tested an intravenous human IgG preparation in patients with gram-negative septic shock. Treatment with 200 mg/kg of intravenous J5-rich IgG (J5 IVIG) did not reduce the incidence of organ system failures or the mortality rate in the treated group [15/30 died (50% mortality) versus 20/41 deaths (49% mortality) among the group given control IgG]. A probable explanation for failure was the low titer of specific anti-J5 antibody in the J5 IVIG preparation. Other potential explanations include the possibilities that patients with septic shock cannot benefit from an antiendotoxin antibody or that IgG isotype antibodies are ineffective.

Studies by Duswald (281), Glinz (282), Rommelshein (283), and Class (284) have used various combinations of poly-specific IgG, IgM-enriched IgG, or IgG enriched with high levels of IgM antibodies. Patient populations have varied, but each study showed some benefit on the surrogate endpoints of fever duration, antibiotic usage, shortened ICU time, decreased risk of nosocomial pneumonia, and a decreased need for respiratory therapy treatments. None of these studies showed an improvement in mortality. The most recent trial utilizing a polyclonal human intravenous immunoglobulin preparation in septic shock (285) demonstrated a reduction in mortality from 32% (9/28) among patients receiving placebo to 4% (1/27) in patients given immunoglobulin daily for 3 days. Reduction in mortality was accompanied by immunological and physiological improvement.

Studies of Monoclonal Anticore/Lipid A Antibodies

The emergence of monoclonal antibody technology and recognition of the human acquired immunodeficiency syndrome (AIDS) substantially shaped the course of antiendotoxin development. The prospect of producing a pure, pathogen-free, specific antiendotoxin antibody in unlimited quantities using monoclonal technology in vitro was inviting. At least eight different monoclonal antibodies to various determinants on endotoxin have been produced and tested (286–290). Two of these compounds, E5 and HA1A, have undergone human testing (286,287).

E5

Unfortunately, there is a dearth of evidence to support the efficacy of E5, a murine monoclonal IgM antibody to *E. coli* J5 LPS. Although initially reported to be protective when used in combination with antibiotic therapy in bacteremic rodents (287), subsequent studies have had difficulty demonstrating substantial protection using E5 in vitro or in sheep or mice (43,291, 292).

For example, in awake sheep with chronic lung lymph fistulas, hemodynamic and ariway responses have been monitored before and after infusion of endotoxin in the presence or absence of E5. Neither pretreatment of animals with a (2 mg/kg) infusion of E5 4 h before LPS (0.75 μg/kg) challenge or mixing the antibody and LPS in vitro before administration prevented changes of endotoxemia except for a modest blunting of the early pulmonary hypertension. Control animals given only antibody showed no change in any measured parameter, while animals receiving LPS alone exhibited a typical endotoxin response. In all animals receiving endotoxin, lung compliance rapidly declined by 50% and was accompanied by a sixfold increase in airflow resistance. Accompanying the lung mechanics abnormalities were pulmonary hypertension, leukopenia, and widening of the A-aDO_2. Similar results were seen when lower doses of LPS (0.075, 0.05, and 0.005 μg/kg) were infused 1 h after antibody (2 mg/kg) administration (unpublished observations). In vitro, incubation of LPS with E5 in molar ratios from $1{:}10^3$ to $1{:}10^6$ for 30 min at 37°C did not reduce LPS-induced limulus lysate activation under serum-free or serum-containing conditions (43).

In two phase 3 randomized blinded human studies (293–295) using E5 in sepsis, conflicting results have been reported. In the first trial, patients received either two doses of E5 (2 mg/kg) or isotonic saline placebo 24 h apart. Of 486 enrolled patients, 468 were considered to be evaluable. Gram-negative infection was seen in 316 of 468 patients (68%) and gram-negative bacteremia in 171 of 468 (37%). Shock occurred in 56% (262/468) of the total group and in 57% of the patients with gram-negative infection (179/316). Among the total study population, treated patients had a mortality rate of 31% (167/242) compared to 37% (143/226) among placebo-treated patients (p = ns). Similarly, there was no difference in mortality between treated patients with gram-negative bacteremia 33% (31/94) versus 42% (32/77) among placebo recipients. Even when the group of patients with gram-negative bacteremia was subdivided into patients with or without shock, no treatment effect was observed. For patients with gram-negative bacteremia without shock, mortality rates were 4 in 44 (9%) among treated patients and 5 in 27 (18.5%) among placebo recipients. Among patients without gram-negative bacteremia there was also no difference in mortality: 44 of 148 (30%) in E5 recipients versus 51 of 149 (34%) among placebo-treated patients.

The only subgroup for which this study claimed benefit were patients with nonbacteremic gram-negative infections who were not in shock. The mortality was 24 of 72 (33%) among treated patients in this group, and 28 of 65 (43%) among placebo patients ($p < 0.05$) after adjustment for covariates. Although no physiologic data were presented to suggest an early effect of treatment, the resolution rate of organ system failures was reported. A total of 66 organ-failure reversals was seen in the 137 patients without shock.

Among treated patients, 19 of 53 (54%) of the organ system failures resolved compared to 8 of 31 (26%) in the placebo group ($p < 0.05$ by chi-square).

Much criticism has been leveled at the E5 study because of the choice of isotonic saline as a placebo instead of an irrelevant monoclonal IgM antibody and because of low APACHE scores in the subset of patients in whom these scores were reported. The study population was quite ill, however, with a 50% incidence of shock, a high prevalence of multiple-organ failure, and a high mortality rate. The prevalence of organ failures in the study and placebo groups at the time of study entry was well balanced.

Even if not effective, E5 appears to be safe with only a 1.6% incidence of allergic reactions despite the development of human anti-mouse antibodies (HAMA) in almost half of the E5 recipients. The authors of this study concluded that E5 was safe and effective in gram-negative sepsis, a conclusion not widely accepted since E5 benefited only a small subset of patients and there is no clear and compelling rationale as to why the treatment effect should be limited to the subgroup shown to benefit.

A second phase 3 study of E5 was undertaken in patients with suspected gram-negative infection without shock (296,297). Of 811 enrolled patients, 530 (65%) had gram-negative infection and 139 of these had organ system failures, including 52 cases of adult respiratory distress syndrome, 73 cases of acute renal failure, and 40 cases of disseminated intravascular coagulation. There was no survival benefit in patients with gram-negative infection. [Thirty-day survival was 70% in the treated group and 74% in the placebo group (p = ns).] Among the subgroup of 139 patients with organ system failures, survival was 59% among E5 recipients and 53% in patients receiving placebo (p = ns). Organ failures reversed in 48% of the treated patients and in 25% of the placebo group; however, this effect failed to reach statistical significance. In the subgroup of patients with gram-negative infection and acute renal failure, 63% reversed the renal failure if treated and 29% recovered renal function if given placebo ($p < 0.05$).

E5 was again relatively safe, with a 2.6% incidence of adverse reactions (four cases of anaphylaxis, five episodes of rash) and a 48% incidence of HAMA formation compared to 12% in placebo-treated patients. As in the first trial, there was no clear-cut benefit of E5 treatment. Furthermore, the subgroup claimed to have benefited in the first clinical study did not benefit from drug treatment. Based on data presented to date, E5 does not appear to exhibit a substantial biological effect in septic humans

HA1A

Like E5, few published data exist to support the effectiveness of HA1A, a chimeric human-mouse monoclonal IgM class antibody directed against *E. coli* J5. In early studies, HA1A contained in mouse ascitic-hybridoma fluid blocked the dermal Schwartzman reaction in rabbits and reduced the mortality of gram-negative bacteremia in mice (286). Subsequent studies have failed to show protective effects. Baumgartner reported that HA1A fails to block gram-negative bacteremia and LPS-induced death in mice and fails to prevent the Schwartzman reaction in rabbits (269). Wheeler has also shown failure of HA1A in a lethal murine model of intraperitoneal endotoxemia (292). Recent studies in chronically instrumented sheep have failed to show protective effects of HA1A against endotoxin-induced physiological changes (291). Disappointingly, in vitro, HA1A has also failed to prevent LPS-induced generation of cytokines (269).

A phase 3 randomized, blinded clinical trial of HA1A in patients with sepsis syndrome has been completed in which patients received 100 mg of HA1A or human serum albumin, then were followed until death, 28 days, or study withdrawal (298). A total of 543 patients, 262 given HA1A and 281 given albumin, were enrolled. Overall, 58% of patients (317/543) had gram-negative infection and 37% (200/543) had gram-negative bacteremia. There was no difference in mortality at 28 days between the HA1A- (39%) and albumin-treated (43%) groups (p = ns). In the 37% of patients with gram-negative bacteremia, a statistically significant reduction in mortality, from 49% in the placebo group to 30% in the HA1A-treated group, was reported; however, the preadjustment p value for this analysis was not provided. Survival benefits appeared to extend to the smaller subgroup of bacteremic patients with shock, in whom mortality was 57% in placebo group and 33% among treated patients.

Two of 291 patients had self-limited adverse reactions to HA1A. None of the 116 treated patients tested for antichimeric antibody formation had such antibodies found, but the method used to look for this response was not reported. Much criticism has been leveled at this study (299,300). Most critics have suggested that the most appropriate placebo would be an identical quantity of an irrelevant monoclonal antibody of the same species and isotype. Despite claims that the two groups in the HA1A study were similar, the two populations have clear differences. The APACHE score was more than 2 points higher in the placebo group than in the treated group, and for each organ system malfunction for which there was a difference in incidence between groups (ARDS, acute renal failure, and hepatic failure), the difference favored the HA1A-treated group. There was also a striking difference in the incidence of *Pseudomonas* bacteremia between groups favoring the HA1A-treated group. It is not clear from published data that appropriate corrections for imbalances between treatment groups at entry were performed, and because of imbalance of organ failures and incidence of *Pseudomonas* bacteremia, the results of this study are open to question.

Some attempts have been made to compare the 1991 study to the 1982 study of Ziegler et al. (267), despite the obvious problems in comparing mortality statistics among groups with different entry characteristics treated 10 years apart. Although it is true that the mortality rates in the antibody-treated groups are similar in both studies, the same is not true for the placebo groups. For patients with gram-negative bacteremia, the 1991 study has a higher placebo group mortality rate that any other comparable subgroup in either study. In addition, although the percentage improvement in mortality was similar in the two studies, the patient populations clearly were not. The 1982 study had a 10-fold higher incidence of neutropenia, and higher rates of *Pseudomonas* bacteremia, neoplasia, and liver disease, suggesting that the 1991 trial studied less-ill patients but found similar mortality rates. Because of the unusually high mortality rate in the placebo arm and imbalances in prevalence of organ system failures at entry, the conclusions of this study are open to question.

In the past several years, the role of cytokines in SIRS has evolved to be critical. If antiendotoxin antibodies are effective, one mechanism of action could be inhibition of cytokine generation. DeGroote (301) and Baumgartner (269) have shown that E5 and HA1A fail to inhibit cytokine generation in vitro and in vivo. In contrast, Wortel (302) has reported a reduction in TNF-α and IL-6 levels in endotoxemic patients treated with HA1A.

VII. Cytokine Antibodies

A. Tumor Necrosis Factor Antibodies

Pretreatment with a *polyclonal* antiserum to TNF-α prevents LPS and bacteria-induced physiologic changes and death; however, only a narrow therapeutic window exists (170,303,304). Also, *monoclonal* TNF-α antibodies (305) given before or up to 30 min after uniformly lethal *E. coli* challenge dramatically improve 72-h survival rates to near 80%; however, delay in treatment of only 2.5 h reduces survival to 25%. Survival has not been as good when using a *Pseudomonas* challenge in immunocompetent mice, and no improvement in survival is seen in immunosuppressed mice challenged with *Klebsiella.* Interestingly, in these studies serum TNF-α levels were highest in *E. coli*–challenged animals and were 10-fold greater than those previously reported in LPS-challenged mice (306), rabbits (170), or baboons (192). In these studies TNF antibodies did not impair clearance of bacteria from blood or tissue. Similar treatment effects have been demonstrated in rats with *Pseudomonas bacteremia* (307,308). In addition to blocking the acute septic effects of TNF-α, TNF antibodies may prevent the chronic wasting effects of TNF-α (309).

Contrasting results have been shown when comparing peritonital and intravenous models of endotoxin challenge. Peritonitis models have consistently failed to demonstrate the benefit of anti-TNF-α treatment (310–312). Zaneti has shown that mice challenged with lethal intravenous doses of *E. coli* LPS have prompt and profound but brief TNF-α elevations and are dramatically protected by an anti-TNF antibody (94). In contrast, following infected mucin-clot peritonitis, TNF-α antibodies are ineffective in reducing cytokine levels or improving survival (94), and the profile of cytokine release is dramatically different. Bagby (310) has published similar results reporting low plasma levels of TNF-α in rats with peritonitis and lack of protection using anti-TNF-α antibodies.

In baboons, TNF-α monoclonal Fab2 antibody fragments given 2 h prior to lethal *E. coli* challenge (192) prevent shock and organ failure, but treatment 1 h before challenge is not beneficial. From these experiments it was discovered that the Fab2 antibody abolished detectable serum TNF and attenuated IL-1β and IL-6 production by more than 95% (156). The quenching effect of TNF-α antibodies on IL-1 and IL-6 production has subsequently been confirmed in mice (94,313).

Studies by Hinshaw have investigated the utility of anti-TNF-α after the onset of sepsis (95). Anesthetized baboons given a lethal intravenous dose of *E. coli* were treated 30 min later with TNF antibody or placebo. All treated baboons survived 7 days without evidence of organ failure, while all baboons receiving placebo died within 34 h. Physiologic changes of tachycardia, tachypnea, and decreases in blood pressure were observed in both groups but were more marked in the placebo group. TNF-α antibody had no effect on the incidence of leukopenia, thrombocytopenia, or hypofibrinogemia, but lactate and creatinine elevations were more marked in the placebo group. Histological evidence of organ damage was also attenuated by antibody, an effect confirmed in another model by Grau (314).

After demonstrating beneficial effects in primates (315), Exley reported preliminary observations of a phase I study of a murine monoclonal TNF-α antibody (CB0006) in 14 patients with septic shock (316). Although the antibody was free of immediate toxicity and was associated with a rise in mean arterial pressure, mortality remained high. Absence of a control group prevents meaningful interpretation of the physiological and survival data. Using the same TNF-α antibody, Vincent has reported (317) minimal reductions in heart rate, increases in left ventricular stroke work index, and transient improvements in gas exchange in a group of 10 patients with septic shock. Again the lack of a control group precludes interpretation of the data except to say that there does not appear to be major immediate toxicity associated with the antibody.

Essentially no information is available regarding the effect of other cytokine antibodies in human sepsis; however, animal data indicate that IL-6 antibodies may abrogate the effects of endotoxin infusion. Pretreatment of mice with IL-6 antibodies prevents death from intraperitoneal *E. coli* or intravenous TNF-α (313). In this study, IL-6-antibody-treated animals had increased circulating TNF-α levels, suggesting that IL-6 exerts negative feedback on TNF-α production. In these studies, anti-TNF-α antibody resulted in a 70% decrease in IL-6 levels, providing further evidence of the complex interaction between IL-6 and TNF-α.

The use of cytokine antibodies in general and TNF antibodies specifically should be approached with caution. Even though investigator concerns that removal of the effects of TNF-α might hamper the host clearance of bacteria from the circulation have proven unfounded (94,95,192,305), there are emerging data to suggest that in some models of infection anti-TNF-α antibodies may worsen outcome (161,318,319).

VIII. Circulating Cytokine Inhibitors and Receptor Antagonists

Soluble inhibitors have been described for IL-1 (320–329), IL-2 (330), IL-6 (331), and TNF-α (139,327). These proteins are structurally similar to either the cytokine itself or the extracellular domain of the cytokine receptor. The largest amount of data has been generated using the IL-1 receptor antagonist (IL-1ra), originally named the IL-1 inhibitor, a compound initially isolated from children with juvenile arthritis (320–323). This protein binds to the IL-1 receptor, blocking access and activation by IL-1 (320–322,326). IL-1ra has now been isolated from human monocytes and alveolar macrophages (329) and has been cloned (324,325).

Pretreatment with IL-1ra prevents death in rabbits given lethal intravenous doses of LPS (194), *E. coli* (328), or IL-1, but IL-1ra does not prevent all the physiologic changes of sepsis. Sepsis-induced reductions in systemic vascular resistance result in hypotension in both treated and untreated animals but occur to a lesser degree in IL-1ra-treated animals and reverses more quickly. IL-1ra blunts declines in peripheral leukocyte count and pulmonary accumulation of neutrophils, but has no effect on the occurrence of thrombocytopenia. Because receptor blockade is the mechanism of action, serum TNF-α and IL-1 levels are not reduced by IL-1ra (194,328). IL-1ra also prevents lethal *Klebsiella* sepsis in newborn rats (332) and *E. coli* sepsis in baboons. IL-1ra is also effective when given up to 2 h after the onset of lethal *E. coli* bacteremia (194).

A recent open label trial of IL-1ra in septic humans reported a dose-dependent reduction in mortality from 44% in placebo recipients to 16% in patients treated with 133 mg/h of IL-1ra (333). Even though substantial reductions in mortality were reported, nonblinded trials must be considered preliminary until a full-scale randomized double-blind placebo controlled trial of the drug is completed. One potential problem in using the IL-1ra clinically may be the dose requirements for the drug. In vitro and in vivo 50- to 500-fold molar excesses of IL-1ra have been required to observe protective effects (194,327,328) and the effects of the drug are very dose dependent (194,333). Such huge doses of drug may preclude economical clinical use.

IL-1ra also has potent anti-inflammatory effects in nonsepsis models. For example, IL-1ra reduces the inflammation of immune complex colitis (334,335), meningitis (336), adjuvant-induced arthritis, and blocks the soporific, febrile, and chemoattractant effects of IL-1 (337). In vitro the IL-1ra also blocks IL-1-induced production of IL-1, TNF, and IL-6 from human peripheral blood monocytes.

The IL-1 inhibitor described above differs from the soluble IL-1 inhibitor reported by Giri, which has biochemical characteristics of the extracellular binding domain of the IL-1 receptor (338). Antibodies to the murine IL-1 receptor have also now been produced and are reported to reduce the severity of IL-1-induced and radiation-induced injury and death (339–341).

Two soluble receptors (27 and 30 kDa) for TNF-α have been isolated from human urine (331,342). The human soluble TNF-α receptor represents a shed portion of the cell surface receptor (343). Seckinger and Engelmann described 27- and 33-kDa TNF-α inhibitors which when incubated with TNF-α blocked binding to the TNF-α receptor (344,345). Heller further characterized the shed portion of the TNF-α receptor (89). Administration of the soluble TNF-α receptor is now reported to prevent death in mice from LPS-induced shock; however, such results are preliminary.

Novick has reported a soluble extracellular fragment of the IL-6 receptor which functions as a soluble IL-6 inhibitor (346). To date there are no data to support a solutary role for the IL-6 inhibitor in sepsis.

IX. Advantages and Limitations of Antibodies in Sepsis

Antiendotoxin antibodies have several theoretical potential advantages over other pharmacological agents in sepsis: they may directly antagonize LPS, a potent trigger of the inflammatory cascade; unlike antibiotics, antiendotoxins may be useful in nonbacteremic endotoxemia; prophylactic use in high-risk patients is possible; and polyclonal antiendotoxins have been shown to be safe and effective in animals.

Unfortunately, there are many disadvantages of antiendotoxins. Only 30 to 40% of patients with SIRS have endotoxemia, suggesting that antiendotoxins will fail to benefit two-thirds of treated patients. Furthermore, because of the structural diversity of endotoxin, even anti-LPS antibodies recognizing the core region will bind to various endotoxins with different affinities (and presumably not at all to some).

It is of concern that the mechanism by which antiendotoxin antibodies afford protection is unknown. There is little evidence to indicate that neutralization of endotoxin or enhanced circulatory clearance of LPS is the mechanism of action of currently produced antiendotoxins.

At present the critical site within the organism for the action of antiendotoxins is also unknown. Because of their high molecular weight, IgM class antibodies are confined predominately to the circulation. Restriction of anti-LPS antibodies to the intravascular space alone is probably neither crucial or desirable. Lower-molecular-weight antibodies or antibody fragments capable of gaining access to a larger volume of distribution will probably prove more effective in neutralizing LPS closer to its source.

Even if antigen–antibody binding is complete, avid, and irreversible, it is possible that antibodies will not sequester or inactivate toxins sufficiently to prevent inflammation, and LPS-antibody complexes may themselves be toxic. Because of the speed with which inflammation occurs, any antibody given after the onset of SIRS will be less than ideal. Furthermore, the patient population developing SIRS often has underlying fatal diseases, making it unrealistic to expect an antiendotoxin or anticytokine antibody to improve long-term survival.

Antibodies are large and complex immunogenic molecules. The presence of Fc fragments gives antibodies the ability to activate complement and bind to the surface of mononuclear cells, stimulating immediate allergic and long-term immunologic reactions. Because of the possibility of allergic reactions upon rechallenge, repeated dosing of antiendotoxins will probably not be tried. Finally, the cost of monoclonal antibody therapy is huge. With a condition as common as SIRS and its associated lung injury, financial considerations may prove to be the limiting factor preventing widespread antibody use.

X. Summary

Immunological therapy of SIRS is in its infancy. The two major therapeutic strategies are the use of cross-reactive antibodies to common determinants on endotoxin and antibodies or antagonists to host-generated cytokines. Antisera and polyclonal antiendotoxins can bind and neutralize endotoxin in vitro and in vitro; however, high-affinity, widely cross-reactive, neutralizing monoclonal endotoxin antibodies have not yet been produced. Because of the risk of transfusing pooled human blood products, it is unlikely that a human polyclonal antiendotoxin will emerge (59,61).

Unlike polyclonal antisera, data supporting the effectiveness of monoclonal antiendotoxins is flimsy, occasionally contradictory, and in dispute.

Evidence of improved survival in three large clinical trials of monoclonal antiendotoxins is weak. Extensive subset analysis was used to show the benefit of E5, and these findings were not replicated in a second clinical trial. Lack of supporting preclinical data and failure to replicate early results makes the efficacy of E5 questionable. HA1A also leaves much to be desired, and its practical value is unclear. Even if the HA1A data are taken at face value, two-thirds of patients did not benefit from treatment. Imbalances in numbers of organ failures and distribution of bacteriologic isolates favoring the treated group make the reported conclusions suspect. Because patients who might benefit from antiendotoxins cannot be identified prospectively, most treated patients will receive these drugs, exposing them unnecessarily to untoward reactions and substantial expense.

Because of the marginal results in human clinical trials and contradictory animal data, definitive clinical studies should be completed before antiendotoxins are accepted. If the effects of antiendotoxins cannot be replicated, these drugs should not be used in humans. It is probable that antiendotoxins will rapidly be superseded by the use of anticytokine agents that are effective regardless of etiology.

Unfortunately, at this time there are few data showing that cytokine antagonists curtail established sepsis. In animals, the type of inflammatory challenge (endotoxin versus bacteria) and its location (intravenous versus peritoneal) are critical determinants of the effectiveness of cytokine antibodies/antagonists. Very narrowly defined patient populations may be needed to demonstrate the effectiveness of cytokine antagonists. Early studies with cytokine antibodies have been disappointing and like LPS antagonists, logistical problems in administration, less than universal efficacy, and high cost may prove problematic.

Cytokines have beneficial effects in appropriate local physiologic concentrations, and abolishing their action may prove dangerous by impairing the local immune response. To maximize the safety and efficacy of immunological therapy in sepsis, inexpensive, avidly binding and neutralizing low-molecular-weight compounds with large volumes of distribution and short half-lives are most desirable. Safety and financial considerations dictate that use of these drugs cannot be indiscriminate. For each antagonist, carefully conducted large, multicenter, randomized double-blind, placebo controlled trials will be required to demonstrate safety and efficacy. Because of the complexity of cytokine interactions, it would not be surprising that any single cytokine antagonist might fail to treat effectively human sepsis syndrome requiring a "combined chemotherapy" strategy for success.

Acknowledgment

This work was supported by SCOR Grant No. HL-19153 from the National Institutes of Health.

References

1. Centers for Disease Control. MMWR 1990; 39:31–34.
2. Balk RA, Bone RC. The septic syndrome: definition and clinical implications. Crit Care Clin 1989; 5:1–8.
3. Bone RC, Fisher CJ, Clemmer TP, Slotman GJ, Metz CA, Balk RA. Sepsis syndrome: a valid clinical entity. Crit Care Med 1989; 17:389–393.
4. Bone RC, Fisher CJ, Clemmer TP, et al. A controlled clinical trial of high-dose methylprednisolone in the treatment of severe sepsis and septic shock. N Engl J Med 1987; 317:653–658.
5. ACCP/SCCM Consensus Conference Committee. Definitions for sepsis and organ failure and guidelines for the use of innovative therapies in sepsis. Chest 1992; 101:1644–1655.
6. Ispahani P, Perason NJ, Greenwood D. An analysis of community and hospital acquired bacteremia in a large teaching hospital in the United Kingdom. Q J Med 1987; 241:427–440.
7. Calandra T, Glauser MP, Schellekens J, Verhoef J, and the Swiss–Dutch Immunoglobulin Study Group. Treatment of gram-negative septic shock with human IgG antibody to *Escherichia coli* J5: a prospective, double blind, randomized trial. J Infect Dis 1988; 158:312–319.
8. Dofferhoff ASM, Bom VJJ, van Ingen J, et al. Patterns of cytokines, plasma endotoxin, and acute phase proteins during the treatment of severe sepsis in humans. In: Sturk A, van Deventer SJH, Wouter ten Cate J, Büller HR, Thijs LG, Levin J, eds. Bacterial Endotoxins: Cytokine Mediators and New Therapies of Severe Sepsis in Humans. New York: Wiley-Liss, 1991:43–54.
9. Schlag G, Redl H, Dinges HP, Davies J. Sources of endotoxin in the posttraumatic setting. In: Sturk A, van Deventer SJH, Wouter ten Cate J, Büller HR, Thijs LG, Levin J, eds. Bacterial Endotoxins: Cytokine Mediators and New Therapies of Severe Sepsis in Humans. New York: Wiley-Liss, 1991; 121–134.
10. Gaffin SL, Grinberg Z, Abraham C, Birkhan J, Shechter Y. Protection against hemorrhagic shock in the cat by human plasma containing endotoxin-specific antibodies. J Surg Res 1981; 31:18–21.
11. Papa M, Halperin Z, Rubinstein E. The effect of ischemia of the dog's colon on transmural migration of bacteria and endotoxin. J Surg Res 1983; 35:264.
12. Gathiram P, Gaffin SL, Wells MT, Brock Utne JG. Superior mesenteric artery occlusion shock in cats: modification of the endotoxemia by anti-lipopolysaccharide antibodies (anti-LPS). Circ Shock 1986; 19:231–237.
13. Zanotti AM, Gaffin SL. Prophylaxis of superior mesenteric artery occlusion shock in rabbits by anti-lipopolysaccharide (anti-LPS antibodies). J Surg Res 1985; 38:113–115.
14. Gaffin SL, Brock-Utne JG, Zanotta A. Hypoxia-induced endotoxemia in primates: role of the raticulo-endothelial system function and anti-lipopolysaccharide plasma. Aviat Space Environ Med 1986; 57:1044–1048.
15. Gaffin SL, Wells M, Jordan JP. Anti-lipopolysaccharide toxin therapy for whole body x-irradiation overdose. Br J Radiol 1985; 58:881–884.

16. Walker RI, Ledney G, Galley G. Septic endotoxemia in radiation injury and graft-versus-host disease. Radiat Res 1975; 62:242–249.
17. Graber C, Reinhold R, Bremen J, Harley R. Fatal heatstroke. JAMA 1971; 216:1195–1196.
18. Neuman F, Gaffin S, Epstein Y, Magazanik H, Shapiro Y. Changes in membrane permeability to endotoxin in heat stroke. Isr Soc Intern Med Proc Herzliya, 1978.
19. Hammer-Hodges D, Woodruff P, Cuevas P. Role of intraintestinal gram-negative bacterial flora on response to major burn. Surg Gynecol Obstet 1974; 138: 599.
20. Cuevas P, Ishiyama N, Koizumi S, Woodruff P, Kaufmann A, Fine J. Role of endotoxemia of intestinal origin in early death from large burns. Surg Gynecol Obstet 1974; 138:725–731.
21. Michaeli D. Medicine on the battlefield. J R Soc Med 1979; 72:370–373.
22. Gaffin SL, Baker B, DuPreez J, et al. Prophylaxis and therapy with anti-endotoxin hyperimmune serum against gastroenteritis and endotoxemia in horses. Proc Congr Am Assoc Eq Pract 1982; 28:335–340.
23. Wessels BC, Gaffin SL, Wells MT. Circulating plasma endotoxin (LPS) concentration in healthy and hemorrhagic enteric dogs: anti-endotoxin immunotherapy in hemorrhagic enteric endotoxemia. J Am Anim Hosp Assoc 1987; 23:291–296.
24. Cuvas P, Fine J. Production of fatal endotoxin shock by vasoactivate substances. Gastroenterology 1973; 64:285–291.
25. Daily O, Uddin S, Walker RI, Spaur W. Endotoxemia during hyperbaric stress. Undersea Biomed Res 1978; 5(suppl):37–38.
26. Shenep JL, Morgan K. Kinetics of endotoxin release during antibiotic therapy for experimental gram-negative bacterial sepsis. J Infect Dis 1984; 150:380.
27. Buller HR, Ten Cate JW, Sturk A, Thomas LL. Validity of the endotoxin assay in post surgical patients. In: Ten Cate JW, Buller HR, Sturk A, Levin J, eds. Bacterial Endotoxins: Biomedical Significance and Detection with the Limulus Amebocyte Lysate Test. New York: Alan R. Liss, 1985:425–428.
28. Dolan SA, Riegle L, Berzofsky R, Cooperstock M. Clinical evaluation of the plasma chromogenic limulus assay. In: Watson SW, Levin J, Novitsky T, eds. Detection of Bacterial Endotoxins with the Limulus Amebocyte Lysate Test. New York: Alan R. Liss, 1987:405–416.
29. Pearson FC, Dubczak J, Weary M, Bruszer G, Donohue G. Detection of endotoxemia in plasma of patients with gram-negative bacterial sepsis by the *Limulus* amebocyte lysate assay. J Clin Microbiol 1985; 21:865.
30. McCartney AC, Piotrowicz BI, Edlin SE, Ledingham IM. Evaluation of the chromogenic LAL assay in septic shock. In: Watson SW, Levin J, Novitsky T, eds. Detection of Bacterial Endotoxins with the Limulus Amebocyte Lysate Test. New York: Alan R. Liss, 1987:459–463.
31. Cooperstock MS, Tucker RP, Baublio JV. Possible pathogenic role of endotoxin in Reye's syndrome. Lancet 1975; 1:1272.
32. van Deventer SJH. Endotoxemia: an early predictor of septicemia in febrile patients. Lancet 1988; 1:605–608.

33. Sturk A, Janssen ME, Muylaert FR, Thomas JK, ten Cate JW. Endotoxin testing in blood. In: Watson SW, Levin J, Novitsky T, eds. Detection of Bacterial Endotoxins with the Limulus Amebocyte Lysate Test. New York: Alan R. Liss, 1987:371–385.
34. van Deventer SJH, Buller HR, ten Cate JW, Aarden LA, Hack E, Sturk A. Experimental endotoxemia in humans. Analysis of cytokine release and coagulation, fibrinolytic and complement pathways. Blood. In press.
35. Brandtzag P, Kieruli P, Gaustad P, et al. Plasma endotoxin as a predictor of multiple organ failure and death in systemic meningococcal disease. J Infect Dis 1989; 159:195–204.
36. Brandtzag P, Mollnes TE, Kierrulf P. Complement activation and endotoxin levels in systemic meningococcal disease. J Infect Dis 1989; 160:58–65.
37. Brandtzag P, Sandset PM, Joo GB, Ovstebo R, Abildgaard U, Kieruli P. The quantitative association of plasma endotoxin, anti-thrombin, protein C, extrinsic pathway inhibitor and fibriopeptide A in systemic meningoccoccal disease. Thromb Res 1989; 55:459–470.
38. Brandtzag P, Ovstebo RM, Bruscetto B, Kierulf P. Plasminogen activator inhibitor 1 and 2, alpha 2 anti-plasmin, plasminogen and endotoxin levels in systemic meningoccocal disease. Thromb Res 1990; 57:271–278.
39. van Deventer SJH, Pauw W, ten Cate JW, Janssen ME, Buller HR, Sturk A. Clinical evaluation in febrile patients of an optimized endotoxin assay in blood. Prog Clin Biol Res 1987; 231:489–499.
40. Danner RL, Elin RJ, Hosseini JM, Wesley RA, Reilly JM, Parillo JE. Endotoxemia in human septic shock. Chest 1991; 99:169–175.
41. van Deventer SJH, Hack CE, Wolbink CE, et al. Endotoxin-induced neutrophil activation: the role of complement revisited. In: Sturk A, van Deventer SJH, Wouter ten Cate J, Büller HR, Thijs LG, Levin J, eds. Bacterial Endotoxins: Cytokine Mediators and New Therapies of Severe Sepsis in Humans. New York: Wiley-Liss, 1991:101–109.
42. Meyrick B, Hoover R, Jones M, Berry LC, Brigham KL. In vitro effects of endotoxin on bovine and sheep lung microvascular and pulmonary artery endothelial cells. J Cell Physiol 1989; 138:165–174.
43. Wheeler AP, Hardie WD, Bernard G. Studies of an anti-endotoxin antibody in preventing the physiologic changes of endotoxemia in awake sheep. Am Rev Respir Dis 1990; 142:775–781.
44. Michie HR, Manogue KR, Spriggs DR, et al. Detection of circulating tumor necrosis factor after endotoxin administration. N Engl J Med 1988; 318:1481–1486.
45. Young LS, Stevens P, Ingram J. Functional role of antibody against "core" glycolipid of Enterobacteriaceae. J Clin Invest 1975; 56:850–861.
46. Bone RC, Fisher CJ, Clemmer TP. Early methylprednisolone treatment for septic syndrome and adult respiratory distress syndrome. Chest 1987; 932:1032–1036.
47. Neugebauer E, Schirren J, Lorenz W. High dose steroids and sepsis. N Engl J Med 1988; 318:514–518.
48. Veterans Administration Systemic Sepsis Cooperative Study Group. Effect of high dose glucocorticoid therapy on mortality in patients with signs of systemic sepsis. N Engl J Med 1987; 317:659–665.

49. DeMaria A, Craven DE, Heffernan JJ, McIntosh TK, Grindlinger GA, McCabe WR. Naloxone versus placebo in the treatment of septic shock. Lancet 1985; 1: 1363–1365.
50. Goldman RC, Leive L. Heterogeneity of antigenic-side-chain length in lipopolysaccharide from *Escherichia coli* 0111 and *Salmonella typhimurium* LT2. Eur J Biochem 1980; 107:145–153.
51. Munford RS, Hall CL, Rick PD. Size heterogentiy of *Salmonella typhimurium* lipopolysaccharides in outer membranes and culture supernatant membrane fragments. J Bacteriol 1980; 144:630–640.
52. Luderitz O, Galanos C, Lehmann V, et al. Lipid A: chemical structure and biological activity. J Infect Dis 1973; 128:S17–S29.
53. Gigliotti E, Shenep JL. Failure of monoclonal antibodies to core glycolipid to bind intact smooth strains of *Escherichia coli.* J Infect Dis 1985; 151:1005–1011.
54. Johns MA, Bruins SC, McCabe WR. Immunizations with R mutants of *Salmonella minnesota.* II. Serological response to lipid A and the lipopolysaccharide of Re mutant. Infect Immunol 1977; 17:9–15.
55. Chedid L, Parant M, Parant F, Boyer F. A proposed mechanism for natural immunity of enterobacterial pathogens. J Immunol 1968; 100:292–301.
56. Bruins SC, Stumacher R, Johns MA, McCabe WR. Immunization with R mutants of *Salmonella minnesota.* III. Comparison of the protective effect of immunization of lipid A and the Re mutant. Infect Immun 1977; 17:16–20.
57. McCabe WR, Bruins SC, Craven DE, Johns M. Cross reactive antigens: their potential for immunization-induced immunity to gram-negative bacteria. J Infect Dis 1977; 136:S161–S166.
58. Tate WJ, Douglas H, Braude AE. Protection against lethality of *E. coli* endotoxin with "O" antiserum. Ann NY Acad Sci 1966; 133:746–762.
59. McCabe WR, Greeley A. Immunization with R mutants of *S. minnesota.* I. Protection against challenge with heterologus gram negative bacilli. J Immunol 1972; 108:601–610.
60. Pennington JE, Menkes E. Type-specific vs. cross-protective vaccination for gram-negative bacterial pneumonia. J Infect Dis 1981; 144:599–603.
61. Pennington JE, Hickey WF, Blackwood LL, Arnaut MA. Active immunization with lipopolysaccharide *Pseudomonas* antigen for chronic *Pseudomonas* bronchopneumonia in guinea pigs. J Clin Invest 1981; 68:1140–1148.
62. Wolberg G, Dewitt CW. Mouse virulence of K(L) antigen containing strains of *Escherichia coli.* J Bacteriol 1969; 100:730–737.
63. Markley K, Smallman E. Protection by vaccination against *Pseudomonas* infection after thermal injury. J Bacteriol 1968; 96:867–874.
64. McCabe WR, DeMaria A, Berberich H, Johns MA. Immunization with rough mutants of *Salmonella minnesota*: protective activity of IgM and IgG antibody to the R595 (Re chemotype mutant. J Infect Dis 1988; 158:291–300.
65. Davis D. Brown K. Douglas H, Take W, Braude AI. Prevention of death from endotoxin with antiserum. J Immunol 1969; 102:563–567.
66. Ng A, Chen CLH, Chang CM, Nowotny A. Relationship of structure to function in bacterial endotoxins: serologically cross-reactive components and their effect on protection of mice against some gram-negative organisms. J Gen Microbiol 1976; 94:107–116.

67. Pollack M, Chia JKS, Koles NL, Miller M, Guelde G. Specificity and cross-reactivity of monoclonal antibodies reactive with the core and lipid A regions of bacterial lipopolysaccharide. J Infect Dis 1989; 159:168–188.
68. Fuller NA, Ming-Chi W, Wilkinson RG, Heath EC. The biosynthesis of cell wall lipopolysaccharide in *Escherichia coli.* J Biol Chem 1973; 248:7938–7950.
69. Jansson PE, Lindberg AA, Lindberg B, Wollin R. Structural studies on the hexose region of the core in lipopolysaccharides from enterobacteriaceae. Eur J Biochem 1981; 115:571.
70. Retschel ET, Wollenweber HW, Russa R, Brade H, Zahringer U. Concepts of the chemical structure of lipid A. Rev Infect Dis 1984; 6:432–438.
71. Galanos C, Luderitz O, Westphal O. Preparation and properties of antisera against the lipid A component of bacterial lipopolysaccharides. Eur J Biochem 1971; 24:116–122.
72. Golenbock DT, Will JA, Raetz CR, Proctor RA. Lipid X ameliorates pulmonary hypertension and protects sheep from death due to endotoxin. Infect Immun 1987; 55:2471–2476.
73. Golenbock DT, Ebert S, Will JA, Proctor RA. Elimination and tissue distribution of the monosaccharide lipid A precursor, lipid X in mice and sheep. Antimicrob Agents Chemother 1988; 32:37–41.
74. Golenbock DT, Leggett JE, Rasmussen P, Craig WA, Raetz CR, Proctor RA. Lipid X protects mice against fatal *Escherichia coli* infection. Infect Immun 1988; 56:779–784.
75. Hodgin L, Drevis J. Effect of active and immunizations with lipid A and *Salmonella minnesota* Re 595 on gram-negative infections in mice. Infection 1976; 4:5–13.
76. Mullan WA, Newsome PM, Cunnington PG, Palmer GH, Wilson ME. Protection against gram-negative infections with anti-serum to lipid A form *Salmonella minnesota* R595. Infect Immun 1974; 10:1195–1201.
77. Reitschel ETh, Galanos C. Lipid A antiserum-mediated protection against lipopolysaccharide and lipid A–induced fever and skin necrosis. Infect Immun 1977; 15:34–49.
78. Beutler B, Greenwald D, Hulmes JD, et al. Identity of tumor necrosis factor and the macrophage-secreted factor cachectin. Nature 1986; 316:552–554.
79. Beutler B, Cerami A. Cachectin: more than a tumor necrosis factor. N Engl J Med 1986; 316:379–385.
80. Sherry B, Cerami A. Cachectin/tumor necrosis factor exerts endocrine, paracrine, and autocrine control of inflammatory responses. J Cell Biol 1988; 107: 1269–1277.
81. Beutler BA, Milsark IW, Cerami A. Cachectin/tumor necrosis factor, production, distribution and metabolic fate in vivo. J Immunol 1985; 135:3972–3977.
82. Wang AM, Creasey AA, Ladner MB, Lin LS, Strickler J, van Arsdell JN, Yamamoto R, Mark DF. Molecular cloning of the complementary DNA for human tumor necrosis factor. Science 1985; 228:149–156.
83. Jones EY, Stuart DI, Walker NPC. Structure of tumor necrosis factor. Nature 1989; 338:225–228.
84. Eck MJ, Sprang SR. The structure of tumor necrosis factor at 2.6 Å resolution. J Biol Chem 1989; 264:17595–17605.

85. Tavernier J, van Ostade, Hauuquier G, et al. Conserved residues of tumor necrosis factor and lymphotoxin structure constitute the framework of the trimeric structures. FEBS Lett 1989; 257:315–318.
86. Schall TJ, Lewis M, Koller KJ, et al. Molecular cloning and expression of a receptor for human tumor necrosis factor. Cell 1990; 61:361–370.
87. Loetscher H, Pan YCE, Lahm HW, et al. Molecular cloning and expression of the human 55 kd tumor necrosis factor receptor. Cell 1990; 61:351–359.
88. Shepard VL, Abdolrasulnia R. Isolation of a receptor for tumor necrosis factor from human lung tissue. Cytokine 1980; 1:96.
89. Heller RA, Song K, Onasch MA, Fisher WH, Chang D, Ringold GM. Complementary DNA cloning of a receptor for tumor necrosis factor and demonstration of a shed form of the receptor. Proc Natl Acad Sci USA 1990; 87: 6151–6155.
90. Gray P, Barrett K, Chantry D, Turner M, Feldman M. Cloning of human tumor necrosis factor (TNF) receptor cDNA and expression of recombinant soluble TNF-binding protein. Proc Natl Acad Sci USA 1990; 87:7380–7384.
91. Ulich TR, Guo K, del Castillo J. Endotoxin-induced cytokine gene expression in vivo. I. Expression of tumor necrosis factor mRNA in visceral organs under physiologic conditions and during endotoxemia. Am J Pathol 1989; 134:11–14.
92. Hesse DG, Tracey KJ, Fong Y, et al. Cytokine appearance in human endotoxemia and primate bacteremia. Surg Gynecol Obstet 1988; 166:147–154.
93. Kelly N, Cross A. Non-lethal and lethal production of cytokines (TNF and IL-6) in response to bacterial LPS. In: Dinarello CA, Kluger MJ, Powanda MC, Oppenheim JJ, eds. The Physiological and Pathological Effects of Cytokines. New York: Wiley-Liss, 1990:147–152.
94. Zaneti G, Heumann D, Gerain J, et al. Cytokine production after intravenous or peritoneal gram-negative bacterial challenge in mice. J Immunol 1992; 148: 1890–1897.
95. Hinshaw LB, Tekamp-Olson P, Chang ACK, et al. Survival of primates in LD100 septic shock following therapy with antibody to tumor necrosis factor (TNFα). Circ Shock 1990; 30:279–292.
96. Stephens KE, Ishizaka A, Wu Z, Larrick JW, Raffin TA. Granulocyte depletion prevents tumor necrosis factor-mediated acute lung injury in guinea pigs. Am Rev Respir Dis 1988; 138:1300–1307.
97. Ferrari-Baliviera E, Mealy K, Smith RJ, Wilmore DW. Tumor necrosis factor induces adult respiratory distress syndrome in rats. Arch Surg 1989; 124: 1400–1405.
98. Kuei JH, Tashkin DP, Figlin RA. Pulmonary toxicity of recombinant human tumor necrosis factor. Chest 1989; 96:334–338.
99. Aderka D. Role of tumor necrosis factor in the pathogenesis of intravascular coagulopathy of sepsis: potential new therapeutic implications. Isr J Med Sci 1991; 27:12–60.
100. Gamble JR, Harlan JM, Klebanoff SJ, Vadas MA. Stimulation of the adherence of neutrophils to umbilical vein endothelium by human recombinant tumor necrosis factor. Proc Natl Acad Sci USA 1985; 82:8667–8671.
101. Flick DA, Gifford GE. Comparison of in vivo cell cytotoxic assays for tumor necrosis factor. J Immunol Meth 1984; 68:167–175.

102. Hay H, Cohen J. Studies on the specificity of the L929 cell bioassay for the measurement of tumor necrosis factor. J Clin Lab Immunol 1989; 29:151–155.
103. Oliff A, Defeo-Jones D, Boyer M, et al. Tumors secreting human TNF/cachectin induce cachexia in mice. Cell 1987; 50:555–563.
104. Tracey KJ, Morgello S, Koplin B, et al. Metabolic effects of cachectin/tumor necrosis factor are modified by site of production. Cachectin/tumor necrosis factor-secreting tumor in skeletal muscle induces chronic cachexia, while implantation in brain induces predominantly acute anorexia. J Clin Invest 1990; 86:2014–2024.
105. Martinet Y, Yamauchi K, Crystal RG. Differential expression of the tumor necrosis factor/cachectin gene by blood and lung mononuclear phagocytes. Am Rev Respir Dis 1988; 138:659–665.
106. Dinarello CA, Cannon JG, Wolff SM. New concepts on the pathogenesis of fever. Rev Infect Dis 1988; 10:168–175.
107. Dinarello CA. The proinflammatory cytokines interleukin-1 and tumour necrosis factor and treatment of the septic shock syndrome. J Infect Dis 1991; 163:1177–1184.
108. Pohlman TH, Stanness KA, Beatty PG, Ochs HD, Harlan JM. An endothelial cell surface factor(s) induced in vitro by lipopolysaccharide, interleukin 1 and tumor necrosis factor increase neutrophil adherence by a CDW18 dependent mechanism. J Immunol 1986; 136:4548–4553.
109. Bevilacqua MP, Pober JS, Wheeler ME, Cotran RS, Gimbrone MA. Interleukin 1 acts on cultured human vascular endothelium to increase the adhesion of polymorphonuclear leukocytes, monocytes and related leukocyte cell lines. J Clin Invest 1985; 76:2003–2011.
110. March CJ, Mosley B, Larsen A, et al. Cloning, sequence and expression of two distinct human interleukin-1 complementary DNAs. Nature 1985; 315:641–643.
111. Sims JE, March C, Cosman D, et al. cDNA expression cloning of the IL-1β receptor, a member of the immunoglobulin superfamily. Science 1988; 243:585–589.
112. Van Snick J. Interleukin 6: an overview. Annu Rev Immunol 1990; 8:253–278.
113. Gauldie J, Richards C, Harnish M. Interferon β_2/BSF-2 shares identity with monocyte derived hepatocyte stimulating factor and regulates the major acute phase response in liver cells. Proc Natl Acad Sci USA 1988; 9:7252–7255.
114. Geiger T, Andus T, Klapproth J, Hirano T, Kishimoto T, Heinrich PC. Induction of rat acute phase proteins by interleukin 6 in vivo. Eur J Immunol 1988; 18:717–722.
115. Sehgal PB, May LT, Tamm I, Vilcek J. Human β_2 interferon and B-cell differentiation factor BSF-2 are identical. Science 1987; 235:731–732.
116. Helle M, Brakenhoff JPJ, DeGroot ER, Aarden LA. Interleukin 6 is involved in interleukin 1 induced activities. Eur J Immunol 1988; 18:957–959.
117. Hirano T, Taga T, Nakano N, et al. Purification to homogeneity and characterization of human B-cell differentiation factor (BCDF, BSF p-2). Proc Natl Acad Sci USA 1985; 82:5490–5494.

118. Okada M, Kitahara M, Kishimoto S, Matsuda T, Hirano T, Kishimoto T. IL-6/BSF2 functions as a killer helper factor in the in vitro induction of cytotoxic T-cells. J Immunol 1988; 141:1543–1549.
119. Hill RJ, Warren MK, Levin J. Stimulation of thrombopoesis in mice by human recombinant interleukin 6. J Clin Invest 1990; 85:1242–1247.
120. Yamasaki K, Taga T, Hiratas Y, et al. Cloning and expression of the human interleukin-6 receptor. Science 1988; 243:825–828.
121. Wong GG, Clark SC. Multiple actions of interleukin-6 within a cytokine network. Immunol Today 1988; 137–139.
122. Le J, Vilcek J. Interleukin 6: a multifunctional cytokine regulating immune reactions and the acute phase protein response. Lab Invest 1989; 61:588–602.
123. Maury DPJ, Teppo AM. Raised serum levels of cachectin/tumor necrosis factor alpha in renal allograft rejection. J Exp Med 1987; 166:1132–1137.
124. Grossman RM, Kreuger J, Yourish D, et al. Interleukin 6 is expressed in high levels in sporiatic skin and stimulates proliferation of cultured human keratinocytes. Proc Natl Acad Sci USA 1988; 86:6367–6371.
125. Shenkin A, Fraser WD, Series J, Winstanley FP et al. The serum interleukin 6 response to elective surgery. Lymphokine Res 1989; 8:123–127.
126. McClain CJ, Cohen D, Phillips R, Miller B, Talwalkar R, Young B. Increased plasma interleukin 6 levels in head injury patients. In: Dinarello CA, Kluger MJ, Powanda MC, Oppenheim JJ, eds. The Physiological and Pathological Effects of Cytokines. New York: Wiley-Liss, 1990:61–67.
127. Elias JA. Tumor necrosis factor interacts with interleukin 1 and interferons to inhibit fibroblast proliferation via fibroblast prostaglandin-dependent and independent mechanisms. Am Rev Respir Dis 1988; 138:652–658.
128. Sugarman BJ, Aggarwal BB, Hass PE, Figari IS, Palladino MA, Shepard HM. Recombinant tumor necrosis factor α: effects on proliferation of normal and transformed cells in vitro. Science 1985; 230:943–945.
129. Vilcek J, Palombella VJ, Henriksen-DeStefano D, et al. Fibroblast growth enhancing activity of tumor necrosis factor and its relationship to other polypeptide growth factors. J Exp Med 1986; 163:632–638.
130. Sugarman BJ, Lewis GD, Eesalu TE, Aggarwal BB, Shepard M. Effects of growth factors on the antiproliferative activity of tumor necrosis factor. Cancer Res 1987; 47:780–786.
131. Sato N, Goto T, Haranaka K, et al. Actions of tumor necrosis factor on cultured vascular endothelial cells: morphologic modulation, growth inhibition and cytotoxicity. J Natl Cancer Inst 1986; 76:1113–1121.
132. Oppenhein JJ, Matsushima K, Yoshimura T, Leonard EJ, Neta R. Relationship between interleukin-1, tumor necrosis factor and a neutrophil attracting peptide. Agents Actions 1989; 26:134–140.
133. Dinarello CA, Cannon JG, Wolff SM, et al. Tumor necrosis factor (cachectin) is an endogenous pyrogen and induces production of interleukin-1. J Exp Med 1986; 163:1433–1450.
134. Nawroth PP, Bank I, Handley D, Cassimeris J, Chess L, Stern D. Tumor necrosis factor/cachectin interacts with endothelial cell receptors to induce release of interleukin-1. J Exp Med 1986; 163:1363–1375.

135. Van Damme J, Opdenakker G, Simpson RJ, et al. Identification of the human 26 kD protein interferon β2 (IFN β2) as a growth factor induced by interleukin 1 and tumor necrosis factor. J Exp Med 1987; 165:914.
136. McIntosh JK, Jablons DM, Mule JJ, et al. In vivo induction of IL-6 by administration of exogenous cytokines and detection de novo serum levels of IL-6 in tumor bearing mice. J Immunol 1989; 143:162–167.
137. Brouckaert P, Spriggs DR, Demetri G, Kufe DW, Fiers W. Circulating IL-6 during a continuous infusion of tumor necrosis factor and interferon gamma. J Exp Med 1989; 169:2257–2262.
138. Warner SJC, Auger KR, Libby P. Interleukin-1 induces interleukin-1: recombinant human interleukin-1 induces interleukin-1 production by adult human vascular endothelial cells. J Immunol 1987; 139:1911–1917.
139. Dinarello CA, Ikejima T, Warner SJC, et al. Interleukin-1 induces interleukin-1: induction of circulating interleukin-1 in rabbits in vivo and in human mononuclear cells in vitro. J Immunol 1987; 139:1902–1910.
140. Dinarelo CA. Interleukin 1 and its biologically related cytokines. Adv Immunol 1989; 44:153–205.
141. Okusawa S, Gelfand JA, Ikejima T, Connolly RJ, Dinarello CA. Interleukin-1 induces a shock-like state in rabbits. J Clin Invest 1988; 81:1162–1172.
142. Weinberg JR, Wright DJM, Guz A. Interleukin-1 and tumor necrosis factor cause hypotension in the conscious rabbit. Clin Sci 1988; 75:251–255.
143. Waage A, Espevik T. Interleukin-1 potentiates the lethal effect of tumor necrosis factor a/cachetin in mice. J Exp Med 1988; 167:1987–1992.
144. Elias JA, Trinchieri G, Beck JM, et al. A synergistic interaction of IL-6 and IL-1 mediates the thymocyte stimulating activity produced by recombinant IL-1 stimulated fibroblasts. J Immunol 1989; 142:509–514.
145. Kunimoto DY, Nordan RP, Striber W. IL-6 is a potent cofactor of IL-1 in IgM synthesis and of IL-5 in IgA synthesis. J Immunol 1989; 143:2230–2235.
146. Elias JA, Lentz V. Interleukin 1 and tumor necrosis factor synergistically stimulate fibroblast IL-6 production and stabilize IL-6 mRNA. J Immunol 1990; 145:161–166.
147. Heremans H, van Damme J, Dillen C, Dijkmans R, Billiau A. Interferon gamma a mediator of lethal lipopolysaccharide-induced Shwartzman-like shock reactions in mice. J Exp Med 1990; 171:1853–1869.
148. Heinzel FP. The role of IFN gamma in experimental endotoxemia. J Immunol 1990; 145:2920–2924.
149. Kiner PA, Marek F, Rodges G, Lin PF, Warr G, Desiderio J. Induction of tumor necrosis factor IFN gamma and acute lethality in mice by toxic and non-toxic forms of lipid A. J Immunol 1988; 141:870–874.
150. Fong Y, Tracey KJ, Moldawer LL, et al. Antibodies to cachetin/tumor necrosis factor reduce interleukin-1β and interleukin-6 appearance during lethal bacteremia. J Exp Med 1989; 170:1627–1635.
151. Brennan FM, Jackson A, Chantry D, Maini R, Feldman M. Inhibitory effect of TNFα antibodies on synovial cell interleukin-1 production in rheumatoid arthritis. Eur J Immunol 1992; 22:1907–1912.

152. Rothstein JL, Schreiber H. Synergy between tumor necrosis factor and bacterial products causes hemorrhagic necrosis and lethal shock in normal mice. Proc Natl Acad Sci USA 1988; 85:607–611.
153. Moriarity JL, Starnes HF. Transforming growth factor β reduces lethal effects of *E. coli* infection. In: Dinarello CA, Kluger MJ, Powanda MC, Oppenheim JJ, eds. The Physiological and Pathological Effects of Cytokines. New York: Wiley-Liss, 1990:153–158.
154. Espevik T, Figari IS, Shalaby MR, et al. Inhibition of cytokine production by cyclosporin A and transforming growth factor β. J Exp Med 1987; 166:571–576.
155. Cross AS, Sadoff JC, Kelly N, Bernton E, Gemski F. Pretreatment with recombinant murine tumor necrosis factor α/cachectin and murine interleukin 1α protects mice from lethal bacterial infection. J Exp Med 1989; 169:2021–2027.
156. Sheppard BC, Norton JA. Tumor necrosis factor and interleukin-1 protection against the lethal effects of tumor necrosis factor. Surgery 1991; 109: 698–705.
157. Nelson S, Bagby GJ, Baintaon BG, Wilson LA, Thompson JJ, Summer WR. Compartmentalization of intraalveolar and systemic lipopolysaccharide-induced tumor necrosis factor and the pulmonary inflammatory response. J Infect Dis 1989; 159:189–194.
158. Nelson S. LPS induced inhibition of lung TNF. In: Dinarello CA, Kluger MJ, Powanda MC, Oppenheim JJ, eds. The Physiological and Pathological Effects of Cytokines. New York: Wiley-Liss, 1990:141–146.
159. Kindler V, Sappino AP, Grau GE, Piquet PF, Vassal P. The inducing role of tumor necrosis factor in the development of bactericidial granulomas during BCG infection. Cell 1989; 56:731–740.
160. Titus RG, Sherry B, Cerami A. Tumor necrosis factor plays a protective role in experimental murine cutaneous leshmaniasis. J Exp Med 1989; 170:2097–2104.
161. Nakane A, Minagawa, T. Kato K. Endogenous tumor necrosis factor (cachectin) is essential to host resistance against *Listeria monocytogenes* infection. Infect Immun 1988; 56:2563–2569.
162. Suffredini AF, Fromm RE, Parker MM, et al. The cardiovascular response of normal humans to the administration of endotoxin. N Engl J Med 1989; 321:280–287.
163. Parker MM, Parrillo JE. Septic shock: hemodynamics and pathogenesis. JAMA 1983; 250:3324–3327.
164. Parker MM, Shelhamer JH, Bacharach SL, et al. Profound but reversible myocardial depression in patients with septic shock. Ann Intern Med 1984; 100:483–490.
165. Shirmer WJ, Schirmer JM, Fry DE. Recombinant human tumor necrosis factor produces hemodynamic changes characteristics of sepsis and endotoxemia. Arch Surg 1989; 124:445–448.
166. Remick DG, Kunkel R, Larrick JW, Kunkel SL. Acute in vivo effects of human recombinant tumor necrosis factor. Lab Invest 1987; 56:583–590.

167. Tracey KJ, Beutler B, Lowry SF, et al. Shock and tissue injury induced by recombinant human cachetin. Science 1986; 234:470–476.
168. Tracey KJ, Lowry SF, Fahey TJ, et al. Cachectin/tumor necrosis factor induces lethal shock and stress hormone release in the dog. Surg Gynecol Obstet 1987; 164:415–422.
169. Natanson C, Eichenholz PW, Danner RL, Eichacker PQ. Endotoxin and tumor necrosis factor challenges in dogs simulate the cardiovascular profile of human septic shock. J Exp Med 1989; 169:823–832.
170. Mathison JC, Wolfson E, Ulevitch RJ. Participation of tumor necrosis factor in the mediation of bacterial lipopolysaccharide-induced injury in rabbits. J Clin Invest 1988; 81:1925–1937.
171. Leighton TA, Averbook AW, Klein SR, Bongard FS. Time course of cardiopulmonary effects tumor necrosis factor and endotoxin are similar. Am Surg 1991; 57:836–842.
172. Johnson J, Meyrick B, Jesmok G, Brigham KL. Human recombinant tumor necrosis factor α induction mimics endotoxemia in awake sheep. J Appl Physiol 1989; 66:1448–1454.
173. Wheeler AP, Jesmock G, Brigham KL. Tumor necrosis factor's effects on lung mechanics, gas exchange and airway reactivity in sheep. J Appl Physiol 1990; 68:2542–2549.
174. Stephens KE, Ishizaka A, Larrick JW, Raffin TA. Tumor necrosis factor causes increased pulmonary permeability and edema. Am Rev Respir Dis 1988; 137: 1364–1370.
175. Goldblum SE, Sun WL. Tumor necrosis factor-α augments pulmonary arterial transendothelial albumin flux in vitro. Am J Physiol 1990; 258:157–167.
176. Goldblum SE, Jay M, Yoneda K, Cohen D, McClain CJ, Gillespie M. 1987, Monokine-induced acute lung injury in rabbits. J Appl Physiol 1987; 63:2093–2100.
177. Johnson J, Brigham KL, Jesmok G, Meyrick B. Morphologic changes in lungs of anesthesized sheep following intravenous infusion of recombinant tumor necrosis factor α. Am Rev Respir Dis 1991; 144:179–186.
178. Kips J, Tavernier J, Pauwels RA. Tumor necrosis factor causes bronchial hyperresponsiveness in rats. Am Rev Respir Dis 1992; 145:332–336.
179. Redl H, Schlag G, Lamche H. TNF and LPS-induced changes of lung vascular permeability: studies in unanesthetized sheep. Circ Shock 1990; 31:183–192.
180. Kreil EA, Greene E, Fitzgibbon C, Robinson DR, Zapol WM. Effects of recombinant human tumor necrosis factor-alpha, lymphotoxin and *E. coli* lipopolysaccharide on hemodynamics, lung microvascular permeability and eicosanoid synthesis in anesthetized sheep. Circ Res 1989; 65:502–514.
181. Wheeler AP, Hardie W, Bernard GR. The role of cyclooxygenase products in lung injury induced by tumor necrosis factor. Am Rev Respir Dis 1992; 145: 632–639.
182. Kettelhut IC, Fiers W, Goldberg AL. The toxic effects of tumor necrosis factor in vivo and their prevention by cyclooxygenase inhibitors. Proc Natl Acad Sci USA 1987; 84:4273–4277.
183. Evans DA, Jacobs DO, Revhaug A, Wilmore DW. The effects of tumor necrosis factor and their selective inhibitionby ibuprofen. Ann Surg 1989; 209:312–321.

184. Blick M, Sherwin SA, Rosenblum M, Gutterman J. Phase I study of recombinant tumor necrosis factor in cancer patients. Cancer Res 1987; 47:2986–2989.
185. Warren RS, Starnes HF, Gabrilove JL, Oettgen HF, Brennan MF. The acute metabolic effects of tumor necrosis factor administration in humans. Arch Surg 1987; 122:1396–1400.
186. Starnes HF, Warren RS, Jeevanandam M, Gabrilove JL, Larchian W, Oettgen HF. Tumor necrosis factor and the acute metabolic response to tissue injury in man. J Clin Invest 1988; 82:1321–1325.
187. Chapman PB, Lester TJ, Casper ES, et al. Clinical pharmacology of recombinant human tumor necrosis factor in patients with advanced cancer. J Clin Oncol 1987; 5:1942–1951.
188. Michie HR, Spriggs DR, Manouge KR, et al. TNF and endotoxin induce similar metabolic responses in human beings. Surgery 1988; 104:280–286.
189. van der Poll T, Buller HR, Ten Cate H, et al. Activation of coagulation after administration of tumor necrosis factor to normal subjects. N Engl J Med 1990; 322:1622–1627.
190. Demetri GD, Spriggs DR, Sherman ML, Arthur KA, Imamura K, Kufe DW. A phase I trial of recombinant human tumor necrosis factor and interferon gamma: effects of combination cytokine administration in vivo. J Clin Oncol 1989; 7:1545–1553.
191. Jablons DM, Mule JJ, McIntosh JK, et al. IL-6/IFNβ2 as a circulating hormone: induction by cytokine administration in humans. J Immunol 1989; 142:1542–1547.
192. Tracey KJ, Fong JY, Hesse DG, et al. Anti-cachectin/TNFα monoclonal antibodies prevent septic shock during lethal bacteremia. Nature 1987; 330:662–667.
193. Talmadge JE, Bowersox O, Tribble H, Lel SH, Shepard HM, Ligitt D. Toxicity of tumor necrosis factor is synergistic with gamma interferon and can be reduced with cyclooxygenase inhibitors. Am J Pathol 1987; 128:410–414.
194. Ohlsson K, Bjork P, Bergenfeldt M, Hageman R, Thompson RC. Interleukin-1 receptor antagonist reduces mortality from endotoxin shock. Nature 1990; 343:550–552.
195. Watters JM, Bessey PQ, Dinarello CA, Wolff SM, Wilmore DW. The induction of interleukin-1 in humans and its metabolic effects. Surgery 1985; 98: 298–306.
196. Smith J, Urba W, Steis R, et al. A phase I trial of interleukin 1 alpha (IL-1α) alone and in combination with indomethacin. Lymphokine Res 1990; 9:568A.
197. Beasley D. Interleukin 1 and endotoxin activate soluble guanylate cyclase in vascular smooth muscle. Am J Physiol 1990; 259:R38–R44.
198. Smith J, Urba W, Steis R. Interleukin 1 alpha: results of a phase I toxicity and immunomodulatory trial. Am Soc Clin Oncol 1990; 9:717.
199. Debets JM, Kampmeijer R, van der Linden MP, Buurman WA, van der Linden CJ. Plasma tumor necrosis factor and mortality in critically ill septic patients. Crit Care Med 1989; 17:489–494.
200. Maury CPJ, Teppo AM. Raised serum levels of cachetin/tumor necrosis factor α in renal allograft rejection. J Exp Med 1987; 166:1132–1137.

201. Imagawa DK, Millis JM, Olthoff KM, et al. The role of tumor necrosis factor in allograft rejection. Transplantation 1990; 50:219–225.
202. Levine B, Kalman J, Mayer L, Fillit HM, Packer M. Elevated circulating levels of tumor necrosis factor in severe chronic heart failure. N Engl J Med 1990; 323:236–241.
203. Balkwell F, Burke R, Talbot D, et al. Evidence for tumor necrosis factor/cachetin production in cancer. Lancet 1987; 2:1229–1232.
204. Millar AB, Singer M, Meager A, Foley NM, Johnson NM, Rook GAW. Tumor necrosis factor in bronchopulmonary secretions of patients with adult respiratory distress syndrome. Lancet 1989; 2:712–714.
205. Hyers TM, Tricomi SM, Dettenmeier PA, Fowler AA. Tumor necrosis factor levels in serum and bronchoalveolar lavage fluid in patients with the adult respiratory distress syndrome. Am Rev Respir Dis 1991; 144:268–271.
206. Suter PM, Suter S, Girardin E, Roux-Lombard P, Grau GE, Dayer JM. High bronchoalveolar levels of tumor necrosis factor and its inhibitors, interleukin-1, interferon and elastase in patients with adult respiratory distress syndrome after trauma shock or sepsis. Am Rev Respir Dis 1992; 145:1016–1022.
207. Roten R, Markert M, Feihl F, Schaller MD, Tagan MC, Perret C. Plasma levels of tumor necrosis factor in the adult respiratory distress syndrome. Am Rev Respir Dis 1991; 143:590–592.
208. Hirano T, Matsuda T, Turner M, et al. Excessive production of interleukin 6/B cell stimulating factor-2 in rheumatoid arthritis. Eur J Immunol 1988; 18:1797–1801.
209. Swaak AJ, van Rooyen A, Nieuwenhuis E, Aarden LA. Interleukin 6 (IL6) in synovial fluid and serum of patients with rheumatic disease. Scand J Rheumatol 1988; 17:469–474.
210. Gosset P, Perez T, Lassale P, et al. Increased TNFα secretion by alveolar macrophages from patients with rheumatoid arthritis. Am Rev Respir Dis 1991; 143:593–597.
211. Waage A, Halstensen A, Espevik T. Association between tumor necrosis factor in serum and fatal outcome in patients with meningococcal disease. Lancet 1989; 1:355.
212. Waage AP, Brandtzaeg P, Halstensen A, Kierulf P, Espevik T. The complex pattern of cytokines in serum from patients with meningococcal septic shock. J Exp Med 1989; 169:333–339.
213. Leist TP, Frei K, Kam-Hansen S, Zinkernagal RM, Fontana A. Tumor necrosis factor α in cerebrospinal fluid during bacterial but not viral meningitis. J Exp Med 1990; 167:1743–1746.
214. Waage A, Halstensen A, Shalaby R, Brandtzeg P, Kierluf P, Espevik T. Local production of tumor necrosis factor α, interleukin 1 and interleukin 6 in meningicococcal meningitis. J Exp Med 1989; 170:1859–1867.
215. McClain CJ, Cohen D, Ott L, Dinarello CA, Young B. Ventricular fluid interleukin 1 activity in patients with head injury. J Lab Clin Med 1987; 110:48–53.
216. Mustafa MM, Ramilo O, Llorens XS, Olsen KD. Mangness RR, McCracken GH. Cerebrospinal fluid prostaglandins, interleukin 1β and tumor necrosis factor in bacterial meningitis. Am J Dis Child 1990; 144:883–887.

217. Hinshaw L, Olson P, Kuo G. Efficacy of post-treatment with anti-TNF monoclonal antibody in preventing the pathophysiology and lethality of sepsis in the baboon (abstr). Circ Shock 1989; 27:362.
218. Marks JD, Marks CB, Luce JM, Montgomery AB, Turner J, Murray JF. Plasma tumor necrosis factor in patients with septic shock: mortality rate, incidence of adult respiratory distress syndrome. Am Rev Respir Dis 1990; 141:94–97.
219. Swiss-Dutch J5 Immunoglobulin Study Group. Prognostic values of tumor necrosis factor/cachetin, interleukin-1, interferon-α, and interferon-gamma in the serum of patients with septic shock. J Infect Dis 1990; 161:982–987.
220. Baud L, Cadranel J, Offenstadt G, et al. Tumor necrosis factor and septic shock (letter). Crit Care Med 1990; 18:349–350.
221. Pinsky MR, Vincent JL, Kahn RJ, Schandene L, Dupont E, Content J. Inflammatory mediators of septic shock in man. Am Rev Respir Dis 1990; 141: A512.
222. Cannon JG, Thompkins RG, Gelfand JA, et al. Circulating interleukin-1 and tumor necrosis factor in septic shock and experimental endotoxin fever. J Infect Dis 1990; 161:79–84.
223. Fong Y, Moldawer LL, Marano M, et al. Endotoxemia elicits increased circulating β2 IFN/IL-6 in man. J Immunol 1989; 142:2321–2324.
224. Mackensen A, Galanos C, Englehardt R. Treatment of cancer patients with endotoxin induces release of endogenous cytokines. Pathobiology 1991; 59: 264–267.
225. Michie HR, Manogue KR, Spriggs DR, et al. Detection of circulating tumor necrosis factor after endotoxin administration. N Engl J Med 1988; 318:1481–1485.
226. Fong Y, Marano MA, Moldawer LL, et al. Splanchnic monokine and metabolic response to endotoxin in man. Surg Forum 1988; 39:21–23.
227. Fong Y, Marano MA, Moldawer LL, Wei H, Calvano SE, Kennedy JS. The acute splanchnic and peripheral tissue metabolic response to endotoxin in humans. J Clin Invest 1990; 85:1896–1904.
228. Casey L, Simpson S, Balk R, et al. Tumor necrosis factor in patients with septic syndrome. FASEB J 1989; 3:A635.
229. Casey L, Balk R, Simpson S, et al. Plasma-tumor necrosis factor, interleukin 1 β and endotoxin in patients with sepsis. In: Dinarello CA, Kluger MJ, Powanda MC, Oppenheim JJ, eds. The Physiological and Pathological Effects of Cytokines. New York: Wiley-Liss, 1990:37–42.
230. Dofferhoff AS, Bom VJJ, De Vries-Hospers HG, et al. Patterns of cytokines, plasma endotoxin, plasminogen activator inhibitor, and acute phase proteins during the treatment of severe sepsis in humans. Crit Care Med 1992; 20:185–192.
231. Girardin E, Grau GE, Dayer JM, Roux-Lombard P, the J5 study group and Lambert PH. Tumor necrosis factor and interleukin-1 in the serum of children with severe infectious purpura. N Engl J Med 1989; 319:397–402.
232. Damas P, Reuter A, Gysen P, Demonty J, Lamy M, Franchimont P. Tumor necrosis factor and interleukin-1 serum levels during severe sepsis in humans. Crit Care Med 1989; 17:975–978.

233. Calandra T, Baumgartner JD, Grau GE, et al., and the Swiss–Dutch J5 Study Group. Prognostic values of tumor necrosis factor/cachectin, interleukin 1, alpha interferon and gamma interferon in the serum of patients with septic shock. J Infect Dis 1990; 161:982–989.
234. Calandra T, Gerain J, Heumann D, Baumgartner JD, Glauser MP. High circulating levels of interleukin 6 in patients with septic shock: evolution during sepsis, prognostic value and interplay with other cytokines. Am J Med 1991; 91:23–29.
235. Kern P, Hemmer CJ, Gruss HJ, Dietrich M. Elevated tumor necrosis factor α and interleukin-6 serum levels as markers for complicated plasmodium falciparum malaria. Am J Med 1989; 87:139–143.
236. Grau GE, Taylor TE, Molyneux ME, et al. Tumor necrosis factor and disease severity in children with falciparium malaria. N Engl J Med 1989; 320: 1586–1591.
237. Marano MA, Fong Y, Maldawe LL, et al. Serum cachetin/tumor necrosis factor in critically ill patients with burns correlates with infection and mortality. Surg Gynecol Obstet 1990; 170:32–38.
238. Hack CE, De Groot ER, Felt-Bersma RJ, et al. Increased plasma level of interleukin-6 in sepsis. Blood 1989; 74:1704–1710.
239. Pfeiffer R, Kolle W. Uber die speccifiscle Immunutatsreactin der Typhusbacillii. Z Hyg Infektionskr 1896; 21:203–214.
240. Braude AI, Douglas H. Passive immunization against the local Schwartzman reaction. J Immunol 1972; 108:505–512.
241. Ziegler EJ, Douglas H, Sherman JE, Davis CE, Braude AI. Treatment of *E. coli* and *Klebsiella* bacteremia in agranulocytic animals with antiserum to a UDP-GAL-epimerase-deficient mutant. J Immunol 1973; 111:433–438.
242. Braude AI, Douglas H, Davis CE. Treatment and prevention of intravascular coagulation with antiserum to endotoxin. J Infect Dis 1973; 128:S157–S164.
243. Dunn D, Furgeson R. Immunotherapy of gram-negative bacterial sepsis: enhanced survival in a guinea pig model by use of rabbit antiserum to *Escherichia coli* J5. Surgery 1982; 92:212–219.
244. McCabe WR. Immunization with R mutants of *S. minnesota.* I. Protection against challenge with heterologus gram-negative bacilli. J Immunol 1972; 108:601–610.
245. Young LS, Stevens PR. Cross-protective immunity to gram negative bacilli: studies with core glycolipid of *Salmonella minnesota* and antigens of *Streptococcus pneumoniae.* J Infect Dis 1977; 136(suppl):S174–S180.
246. Marks MI, Ziegler EJ, Douglas H, Corbeil LB, Braude AI. Induction of immunity against lethal *Hemophilus influenza* type b infection by *Escherichia coli* core lipopolysaccharide. J Clin Invest 1982; 69:742–749.
247. Braude AI, Ziegler EJ, Douglas H, McCuthcen JA. Antibody to cell wall glycolipid of gram negative bacteria: induction of immunity to bacteria and endotoxin. J Infect Dis 1977; 136:S167–S173.
248. McCabe WR, Greely A, DiGenio T, Johns MA. Humoral immunity to type-specific and cross-reactive antigens of gram-negative bacilli. J Infect Dis 1973; 128:S284–S289.

249. Mutharia LM, Crockford G, Bogard WC, Hancock RE. Monoclonal antibodies specific for *E. coli* J5 lipopolysaccharide: cross-reaction with other gram-negative bacterial species. Infect Immun 1984; 45:631–636.
250. Brown KR, Douglas H, Braude AI. Prevention of death from endotoxin with antiserum. II. Elimination of risk of anaphylaxis to endotoxin. J Immunol 1971; 106:324–333.
251. Freedman H. Passive transfer of protection against lethality of homologus and heterologus endotoxins. Proc Soc Exp Biol Med 1959; 102:504–506.
252. Kim YB, Watson DW. Modification of host responses to bacterial endotoxins. II. Passive transfer of immunity to containing 19S antibodies. J Exp Med 1965; 121:751–759.
253. Ziegler EJ, Douglas H, Braude A. Human antiserum for prevention of the local Shwartzman reaction and death from bacterial lipopolysaccharide. J Clin Invest 1973; 52:3236–3238.
254. Braude AE. Endotoxic immunity. Adv Intern Med 1980; 26:427–445.
255. Greisman S, Hornick R. Mechanism of endotoxin tolerance with special reference to man. J Infect Dis 1973; 128:S265–S276.
256. Lachman E, Samkar D, Pitsoe SB, et al. Septic abortion in rats: protection by human lipopolysaccharide-specific globulin. S Afr J Sci 1985; 81:525–526.
257. Wessels BC, Gaffin SL. Anti-endotoxin immunotherapy for canine parvovirus endotoxemia. J Small Anim Pract 1986; 27:609–612.
258. Lachman E, Pitsoe SB, Gaffin SL. Anti-lipopolysaccharide immunotherapy in management of septic shock of obstetric and gynecological origin. Lancet 1984; 1:981–983.
259. Gaffin SL, Lachman E. The use of anti-lipopolysaccharide (anti-LPS) antibodies in the management of septic shock. S Afr Med 1984; 65:158–161.
260. Greisman SE, Dubut JB, Woodward CL. Experimental gram-negative bacterial sepsis: reevaluation of the ability of rough mutant antisera to protect mice. Proc Soc Exp Biol Med 1978; 158:482–490.
261. van Dijk WC, Verbrugh HA, Van Erne-Van der Tol ME, Peters R, Verhoef J. *Escherichia coli* antibodies in opsonization and protection against infection. J Med Microbiol 1981; 14:381–389.
262. Pennington JE, Menkes E. Type specific vs. cross-protection vaccination for gram-negative bacterial pneumonia. J Infect Dis 1981; 144:599–603.
263. Pennington JE. Lipopolysaccharide *Pseudomonas* vaccine: efficacy against pulmonary infection with *Pseudomonas aeruginosa*. J Infect Dis 1979; 140: 73–80.
264. Pennington JE. Mechanism for pulmonary protection by lipopolysaccharide *Pseudomonas* vaccine. J Infect Dis 1980; 142:191–198.
265. Pollack M, Huang AI, Prescott RK, et al. Enhanced survival in *Pseudomonas aeruginosa* septicemia associated with high-levels of circulating antibody to *Escherichia coli* endotoxin core. J Clin Invest 1983; 72:1874–1881.
266. McCabe WR, Kreger BE, Johns M. Type specific and cross-reactive antibodies in gram-negative bacteremia. N Engl J Med 1972; 287:261–267.
267. Ziegler EJ, McCutchan JA, Fierer J, et al. Treatment of gram-negative bacteremia and shock with human antiserum to a mutant *Escherichia coli*. N Engl J Med 1982; 307:1225–1230.

268. Baumgartner JD, Glauser MP, McCutchan JA, et al. Prevention of gram-negative shock and death in surgical patients by antibody to endotoxin core glycolipid. Lancet 1985; 2:59–63.
269. Baumgartner JD, Heuman D, Weinbreck P, Grau GE, Glauser MP. Association between protective efficacy of anti-lipopolysaccharide (LPS) antibodies and suppression of PS-induced tumor necrosis factor α and interleukin 6. J Exp Med 1990; 171:889–896.
270. Greisman SE, Young EJ, Dubut B. Mechanisms of endotoxin tolerence. VIII. Specificity of serum transfer. J Immunol 1973; 111:1349–1360.
271. Baumgartner JD, Glauser MP. Immunotherapy of gram-negative septic shock. In: Vincent JL, ed. Update in Intensive Care and Emergency Medicine. New York: Springer Verlag, 1990:109–120.
272. McCabe WR, De Maria A, Berberich H, Johns MA. Immunization with rough mutants of *Salmonella minnesota*: protective activity of IgM and IgG antibody to the 595 (Re chemotype) mutant. J Infect Dis 1988; 158:291.
273. Zinner S, McCabe WR. Effects of IgM and IgG antibody in patients with bacteremia due to gram-negative bacilli. J Infect Dis 1976; 133:37.
274. Baumgartner JD, Heumann D, Glauser MP. Effects of vaccination of humans with E. coli: J5 bacteria on antibodies (Ab) to core LPS (cLPS) and immunoglobulin (IgG) levels. In: Nelson JD, Grossi C, eds. Proceedings of the 28th Interscience Congress on Antimicrobial Agents and Chemotherapy. Washington, DC: American Society for Microbiology, 1988:145.
275. Dunn DL, Bogard WC, Cerra FB. Efficacy of type specific and cross-reactive murine monoclonal antibodies directed against endotoxin during experimental sepsis. Surgery 1985; 98:283–290.
276. Gorelick KJ, Harkonen S, Rosendorf LR, Wilson K, Greenberg RN. Clinical investigation of antilipid A monoclonal antibody XOMEN-E5 in patients with gram negative sepsis (abstr). Crit Care Med 1988; 16:398.
277. Ziegler EJ, McCutchan JA, Braude AI. Clinical trial of core glycolipid antibody in gram-negative bacteremia. Trans Assoc Am Phys 1978; 91:253–258.
278. McCutchan JA, Wolf JL, Ziegler EJ, Braude AI. Ineffectiveness of single dose human anti-serum to core glycolipid (*E. coli* J5) for prophylaxis of bacteremic gram-negative infections in patients with prolonged neutropenia. Schweiz Med Wochenschr 1983; 113(suppl 14):40–45.
279. Wolff JL, McCutchan JA, Ziegler EJ, Braude AI. Prophylactic antibody to core lipopolysaccharide in neutropenia. In: Nelson JD, Grossi C, eds. Proceedings of the 17th Interscience Congress on Antimicrobial Agents and Chemotherapy. Washington, DC: American Society for Microbiology, 1975:439.
280. Girardin E, Grau G, Dayer J, Roux-Lombard P, the J5 Study Group, Lambert PH. Tumor necrosis and interleukin 1 in serum of children with severe infectious purpura. N Engl J Med 1988; 319:397–400.
281. Duswald KH, Muller K, Seifert J, Ring J. Wirksamkeit von I.V. Gammaglobulin gegen bakterielle infektionen chirurgisher Patienten. Munch Med Wochenschr 1980; 122:832–836.
282. Glinz W, Grob PJ, Nydegger UE. Polyvalent immunoglobulins for prophylaxis of bacterial infections in patients following multiple trauma. Intensive Care Med 1985; 11:288–294.

283. Rommelshein K. Prophylaktische Einsatz von Pentaglobin in der Traumapatienten. Anasthesiol Intensivther Notfallmed 1989; 24:162–166.
284. Class I, Schorer R. Adjuvante Therapie mit Pseudomonas Immunoglobulin. Anasthesiol Intensivther Notfallmed 1989; 24:167–171.
285. Schedel I, Dreikhausen U, Nentwig B, et al. Treatment of gram-negative septic shock with an immunoglobulin preparation: a prospective, randomized clinical trial. Crit Care Med 1991; 19:1104–1113.
286. Teng NNH, Kaplan HS, Hebert JM, et al. Protection against gram-negative bacteremia in endotoxemia with human monoclonal IgM antibodies. Proc Natl Acad Sci USA 1985; 82:1790–1794.
287. Young LS. Functional activity of monoclonal antibodies against lipopolysaccharide (LPS) antigens of gram-negative bacilli (abstr). Clin Res 1984; 32: 518A.
288. Appelmelk BJ, Verweij-von vught AMJJ, Maaskant JJ, Schouten WF, Thijs LG, Maclaren DM. 1987. Monoclonal antibodies detecting novel structures in the core region of *Salmonella minnesota* lipopolysaccharide. FEMS Microbiol Lett 1987; 40:71–74.
289. Young LS, Gascon R, Alam S, Bermeudez LE. Monoclonal antibodies for treatment of gram-negative infections. Rev Infect Dis 1989; 11(S7):1564–1571.
290. Kirkland TN, Colwell DE, Michalek SM, McGhee JR, Ziegler EJ. Analysis of the fine specificity and cross reactivity of monoclonal anti-lipid A antibodies. J Immunol 1986; 137:3614–3619.
291. Chen TY, Greene E, Black K, Frostell CG, Robinson DR, Zapol WR. Protective effects of anti-O polysaccharide and anti-lipid A (HA1-A) monoclonal antibodies on the ovine pulmonary circulation. Am Rev Respir Dis 1992; 145: A417.
292. Wheeler AP, Wheeler LA, Morris P, Bernard Gr. Failure of endotoxin antibodies in murine endotoxemia. American College of Chest Physicians Scientific Assembly. Chest 1992; 102(suppl):155S.
293. Gorelick K, Jacobs R, Chmel H, Trenholme G, Greenman R, and the XOMA Sepsis Study Group. Efficacy results of a randomized multi-center trial of E5 anti-endotoxin antibody in patients with gram-negative sepsis. 29th International Conference on Antimicrobial Agents and Chemotherapy. Abstract 2. 1990:101.
294. Gorelick K. Scannon PJ, Hannigan J, Wedel N, Ackerman SK. Randomized placebo controlled study of E5 monoclonal antiendotoxin antibody. In: Larrick J, Borrebaeck C, eds. Therapeutic Monoclonal Antibodies. New York: Stratton, 1990:253–264.
295. Greenman RL, Schein RM, Martin MA, et al., and the XOMA Sepsis Study Group. A controlled clinical trial of E5 murine monoclonal IgM antibody to endotoxin in the treatment of gram-negative sepsis. JAMA 1991; 266:1097–1102.
296. SCRIP (World Pharmaceutical Newsletter), Oct. 11, 1991:22.
297. Swarz JR., Goldman Sachs Investment Research. Centocor, Nov. 11, 1991.
298. Ziegler EJ, and the HA-1A Sepsis Study Group. Treatment of gram-negative bacteremia and septic shock with HA-1A human monoclonal antibodies against endotoxin: a randomized double blind placebo controlled trial. N Engl J Med 1991; 324:429–436.

299. The HA1-A monoclonal antibody for gram-negative sepsis: letters. N Engl J Med 1991; 325:279–283.
300. Warren HS, Danner RL, Munford RS. Sounding board: anti-endotoxin monoclonal antibodies. N Engl J Med 1992; 326:1153–1157.
301. deGroote MA, Martin MA, Densen P, Pfaller MA, Wenzel RP. Plasma tumor necrosis factor levels in patients with presumed sepsis: results in those treated with anti-lipid A antibody versus placebo. JAMA 1989; 262:249–251.
302. Wortel CH, Sprung CL, Van Deventer SJH, Lubbers MJ, Ten Cate JW. 1990. Anti-endotoxin treatment with HA1A: possible mechanism of beneficial effects in patients with gram-negative septicemia. Proceedings of the International Congress for Infectious Disease, Montreal, 1990:206.
303. Beutler B, Milsarek IW, Cerami AC. Passive immunization against cachetin/tumor necrosis factor protects mice from lethal effect of endotoxin. Science 1985; 229:869–871.
304. Grau GE, Vesin C, DeGroote D, et al. Prevention of human TNF-induced cutaneous Shwartzman reaction and acute mortality in mice treated with anti-human TNF monoclonal antibodies. Clin Exp Immunol 1991; 84:411–414.
305. Silva AT, Bayston KF, Cohen J. Prophylactic and therapeutic effects of a monoclonal antibody to tumor necrosis factor-alpha in experimental gram-negative shock. J Infect Dis 1990; 162:421–427.
306. Sheehan KCF, Ruddle NH, Schreiber RD. Generation and characterization of hampster monoclonal antibodies that neutralize murine tumor necrosis factor. J Immunol 1989; 142:3884–3893.
307. Opal SM, Cross AS, Kelly NM, Sadoff JC, Bodmer MW, Palardy JE. Efficacy of a monoclonal antibody directed against tumor necrosis factor in protecting neutropenic rats from lethal infection with *Pseudomonas areuginosa.* J Infect Dis 1990; 161:1148–1152.
308. Opal SM, Cross AS, Sadoff JC, et al. Efficacy of antilipopolysaccharide and anti-tumor necrosis factor monoclonal antibodies in a neutropenic rat model of *Pseudomonas* sepsis. J Clin Invest 1991; 88:885–890.
309. Sherry BA, Gel,in J, Fong Y, et al. Anti-chectin/tumor necrosis factor antibodies attenuate development of cachexia in tumor models. FASEB J 1989; 3:1956–1962.
310. Bagby GJ, Plessala KJ, Wilson LA, Thompson JJ, Nelson S. Divergent efficacy of antibody to tumor necrosis factor α in intravascular and peritonitis models of sepsis. J Infect Dis 1991; 163:83.
311. Echtenacher B, Falk W, Mannel DN, Krammer PH. Requirement of endogenous tumor necrosis factor/cachectin for recovery from experimental peritonitis. J Immunol 1990; 145:3762–3766.
312. Evans GF, Snyder YM, Butler LD, Zukerman SH. Differential expression of interleukin 1 and tumor necrosis factor in murine septic shock models. Circ Shock 1989; 29:279–286.
313. Starnes HF, Pearce MK, Rewari A, Yim JH, Zou JC, Abrams JC. Anti-IL-6 monoclonal antibodies protect against lethal *Escherichia coli* infection and lethal tumor necrosis factor α challenge in mice. J Immunol 1990; 145:4185–4191.

314. Grau G, Fajardo LF, Piguet PF, Allet B, Lambert PH, Vassalli P. Tumor necrosis factor (cachectin) as an essential mediator in murine cerebral malaria. Science 1987; 237:210.
315. Exley AR, Buurman W, Bodmer M, Bohen J. Monoclonal antibody (Mab) to recombinant human tumor necrosis factor (rhTNF) in the prophylaxis and treatment of endotoxic shock in *Cynomolgus* monkeys (abstr). Clin Sci 1989; 76:50.
316. Exley AR, Cohen J, Buurman W, et al. Monoclonal antibodies to TNFα in severe septic shock (letter). Lancet 1990; 335:1275–1277.
317. Vincent JL, Bakker J, Marecaux G, Schandene L, Kahn R, Dupont E. Administration of anti-TNF antibody improves left ventricular function in septic shock patients: results of a pilot study. Chest 1992; 101:810–815.
318. Blanchard DK, Djeu JY, Klein TW, Friedman H, Stewart WE II. Protective effects of tumor necrosis factor in experimental *Legionella pneumophila* infections of mice via activation of PMN function. J Leukoc Biol 1988; 43:429–435.
319. Nelson S, Summer WR, Bagby GJ. LPS-induced inhibition of lung TNF and host defenses. Prog Leukoc Biol 1990; 10B:141–146.
320. Seckinger P, Lowenthal JW, Williamson K, Dayer JM, MacDonald HR. A urine inhibitor of interleukin-1 activity that blocks ligand binding. J Immunol 1987; 139:1546–1549.
321. Seckinger P, Williamson K, Balavione JF, et al. A urine inhibitor of interleukin-1 activity affects both interleukin 1α and 1β but not tumor necrosis factor α. J Immunol 1987; 87:1541–1545.
322. Arend WP, Welgus HG, Thompson RC, Eisenberg SP. Biological properties of recombinant human monocyte derived interleukin 1 receptor antagonist. J Clin Invest 1990; 85:1694–1699.
323. Prieur AM, Kaufman MT, Griscelli C, Dayer JM. Specific interleukin-1 inhibitor in serum and urine of children with systemic juvenile chronic arthritis. Lancet 1987; 2:1240–1242.
324. Eisenberg SP, Evans RJ, Arend WP, et al. Primary structure and functional expression from complementary interleukin-1 receptor antagonist. Nature 1990; 343:341–346.
325. Carter DB, Deibel MR, Dunn CJ, et al. Purification, cloning, expression and biological characterization of an interleukin-1 receptor antagonist protein. Nature 1990; 344:633–638.
326. Hannum CH, Wilcox CJ, Arend WP, et al. Interleukin-1 receptor antagonist activity of a human interleukin-1 inhibitor. Nature 1990; 343:336–342.
327. Arend WP, Welgus HG, Thompson RC, Eisenberg SP. Biological properties of recombinant human monocyte-derived interleukin 1 receptor antagonist. J Clin Invest 1990; 85:1694–1697.
328. Wakabayashi G, Gelfand JA, Burke JF, Thompson RC, Dinarello CA. A specific receptor antagonist for interleukin-1 prevents *Escherichia coli*-induced shock in rabbits. FASEB J 1991; 5:338–343.
329. Galve-de Rochemonteix B, Nicod LP, Junod AF, Dayer JM. Characterization of a specific 20–25 kD interleukin 1 inhibitor from cultured human lung macrophages. Am J Respir Cell Mol Biol 1990; 3:355–361.

330. Robb RJ, Kutny RM. Structure–function relationships for the IL-1 receptor system: analysis of the sequence and ligand-binding properties of soluble Tac protein. J Immunol 1987; 139:855–862.
331. Novick D, Engelmann H, Wallach D, Rubenstein M. Soluble cytokine receptors are present in normal human urine. J Exp Med 1989; 170:1409–1414.
332. Mancilla J, Garcia P, Dinarello CA. Either interleukin 1 or the interleukin-1 receptor antagonist prevent lethal sepsis in newborn rats. Clin Res. In press.
333. Fischer CJ, Slotman GJ, Pribble J, Stiles D, Catalano M, and the IL-1ra Sepsis Study Group. Interleukin-1 receptor antagonist (IL1-ra) reduces mortality in patients with sepsis syndrome. Chest, Nov 1991.
334. Cominelli F, Nast CC, Clark BD, et al. Interleukin 1 gene expression synthesis and effect of specific IL-1 receptor blockade in rabbit immune complex colitis. J Clin Invest 1990; 86:972–980.
335. Cominelli F, Llerena R, Clark BD, Nast CC, Thompson RC, Dinarello CA. In vivo anti-inflammatory properties of recombinant interleukin 1 receptor antagonist (IL-1ra). Lymphokine Res 1990; 9:597–598.
336. Ramilo O, Saez-Llorens X, Mertsola J, et al. Tumor necrosis factor alpha/cachectin and interleukin 1 beta initiate meningeal inflammation. J Exp Med 1990; 172:497–507.
337. Cybulsky ML, Kolditz IG, Movat H. The role of IL-1 in neutrophil leukocyte emigration induced by endotoxin. Am J Pathol 1986; 124:367.
338. Giri JG, Newton RC, Horuk R. Identification of soluble interleukin-1 binding protein in cell-free supernatants: evidence for soluble interleukin-1 receptor. J Biol Chem 1990; 265:17416–17419.
339. Gershelwald JE, Fong YM, Fahey TJ. Interleukin 1 receptor blockade attenuates the host inflammatory response. Proc Natl Acad Sci USA 1990; 87:4966–4970.
340. Neta R, Vogel SN, Plocinski JM. In vivo modulation with anti-Il-1 receptor (p80) antibody of the response to IL-1. Blood 1990; 76:57–62.
341. McIntyre KW, Stepan GJ, Kolinsky KD, et al. Inhibitor of interleukin 1 (IL-1) binding and bioactivity in vitro and modulation of acute inflammation in vivo by IL-1 receptor antagonist and anti-IL-1 receptor monoclonal antibody. J Exp Med 1991; 173:931–939.
342. Engleman H, Novick D, Wallach D. Two tumor necrosis factor binding proteins purified from human urine. J Biol Chem 1990; 170:1531–1536.
343. Porteau F, Nathan C. Shedding of tumor necrosis factor receptors by activated human neutrophils. J Exp Med 1990; 172:599–607.
344. Seckinger P, Isaaz S, Dayer JM. Purification and biologic characterization of a specific tumor necrosis factor α inhibitor. J Biol Chem 1989; 20:11966–11973.
345. Engelmann H, Aderka D, Rubinstein M, Rotman D, Wallach D. A tumor necrosis factor-binding protein purified to homogeneity from human urine protects cells from tumor necrosis factor toxicity. J Biol Chem 1989; 20:11974–11980.
346. Novick D, Engelman H, Wallach D, Rubinstein M. Soluble cytokine-receptors are present in normal human urine. In: Dinarello CA, Kluger MJ, Powanda MC, Oppenheim JJ, eds. The Physiological and Pathological Effects of Cytokines. New York: Wiley-Liss, 1990.

16

Surfactant as Therapy for ARDS

ROGER G. SPRAGG and ROBERT M. SMITH

San Diego VA Medical Center
San Diego, California

I. Lung Injury in the Septic Patient

As described in Chapter 15, patients with systemic sepsis are at risk for developing adult respiratory distress syndrome (ARDS). In fact, of the various predisposing causes of ARDS, sepsis is perhaps the most common. Although infection involving the lung may be responsible for a portion of these cases, it is clear that many other cases of ARDS occur in the setting of a distant infection. Thus a patient with a deep infection of the foot or a patient in the postoperative period with infection apparently limited to the peritoneal space may develop diffuse pulmonary infiltrates, reduced pulmonary compliance, and severely disordered gas exchange. The lung edema seen in these patients is unexplained by elevations of pulmonary vascular pressures, and thus is a nonhydrostatic (noncardiogenic) pulmonary edema.

A plethora of mediators that may initiate or perpetuate the adult respiratory distress syndrome has been identified. Polymorphonuclear leukocytes (PMNs) are usually predominant in both bronchoalveolar lavage fluid (BAL) and the lung parenchyma of patients with ARDS (1–3), although the syndrome is reported in neutropenic patients (4). A variety of mediators may attract PMN from the systemic circulation into the lung (5). These include the macrophage products interleukins 1 and 8 (IL-1, IL-8), platelet derived growth factor (PDGF) and platelet activating factor (PAF), epithelial cell products (PDGF, IL-8, and chemotactic lipoxygenases), and endothelial cell products (IL-1 and IL-8). BAL recovered from patients with ARDS reveals additional substances chemotactic for PMN. These include LTB_4, fibrinopeptide B, oxidized α_1-proteinase inhibitor (α_1-PI), and bacterial products. Events in the lung microvasculature provide a second mechanism of inflammatory cell recruitment, and include activation of complement and expression of adhesion molecules.

Once recruited into the lung, PMNs produce a variety of substances that are likely to participate in acute lung injury. Stimulated PMNs, as well as alveolar macrophages, may produce superoxide radical ($O_2^{\cdot -}$) by the single electron reduction of O_2 catalyzed by a membrane-bound NADPH-dependent oxidoreductase (6). Superoxide radical dismutates to form hydrogen peroxide (H_2O_2). In the presence of transition metals, these products may produce hydroxyl radical ($HO^{\cdot}$). In addition, H_2O_2 may react with chloride ions in the presence of myeloperoxidase to form hypochlorous acid (7). Thus a variety of highly reactive oxygen intermediates (ROIs) may be formed by phagocytic cells in the lower airway.

ROIs may react with proteins in the lower airway, including α_1-PI, the major inhibitor of neutrophil elastase in the lower airway. The α_1-PI is rendered inactive when a methionine in the active site of the inhibitor is oxidized (8). Consequently, elastase from neutrophils in the lower airways may act unopposed to cause proteolytic injury of lung extracellular proteins. The concept of protease–antiprotease imbalance in the lower airway may be of major pathogenetic significance in acute lung injury, as active elastase and/or elastase complexed to α_1-PI has been recovered in BAL from patients with ARDS (9–11). In addition to neutrophil elastase, other neutrophil proteases (including collagenase and cathepsin B) and macrophage proteases (including macrophage elastase and collagenase) may be present and active in the lower airway. Thus in the septic patient, both oxidative and proteolytic mechanisms may act unopposed in the lung and thereby perturb those systems, including the lung surfactant system, that operate to ensure adequate pulmonary function.

The initial changes seen on histologic examination of the lung early in the course of ARDS include interstitial and alveolar edema with evidence of destruction of type I alveolar epithelial cells and patchy denudation of alveolar basement membrane (3). Polymorphonuclear leukocytes are seen in the interstitium and in alveoli. Endothelial cells appear swollen and vacuolated, but endothelial cell death is not a feature of the acute histopathologic picture. After approximately 1 week, a subacute or chronic stage evolves that is characterized by thickening of alveolar septi, proliferation and infiltration of interstitial cells, and exuberant interstitial and possibly alveolar fibrosis (3). Patients who experience both acute and chronic phases of ARDS may show little change in the chest radiograph. Nevertheless, histopathologic studies suggest a marked transition over a period of several weeks, from a predominantly edematous lung to a predominantly fibrotic lung (12).

Physiologic alterations that accompany these pathologic changes include markedly decreased pulmonary compliance, with decreased functional residual capacity, airflow limitation (perhaps due to edema of small airways), and increased resistance to blood flow. Hypoxemia is profound and is due predominantly to shunt. During mechanical ventilation, dead-space ventilation is greatly increased, and during the later stages of illness, p_aCO_2 increases, consistent with greatly diminished alveolar ventilation.

Investigators who initially provided formal description of the adult respiratory distress syndrome questioned whether abnormality of the lung surfactant system might contribute to the pathophysiology they were observing (13). Subsequently, formal analyses of surfactant composition and function in patients with ARDS have disclosed consistent abnormalities. These observations have occurred coincident with experience in the clinical use of surfactant replacement for neonates at risk for infant respiratory distress syndrome (IRDS) or who have already developed IRDS. Although in both groups there is deficiency of surfactant production, in the latter group investigators have demonstrated many of the same markers of acute inflammation that are found in the lungs of patients with ARDS (14). The beneficial effect of exogenous surfactant in treating infants with established IRDS (15) raises the possibility that adults—with a functional if not absolute deficiency of lung surfactant—might also benefit from similar treatments. Several brief reports support the value of this intervention, and thus it is appropriate to understand the surfactant system of the lung, how it may be altered in the septic patient with ARDS, and to review the evidence that exogenous surfactant may provide therapeutic benefit.

II. Composition and Function of Lung Surfactant

A. Composition of Lung Surfactant

Lung surfactant is a complex of lipids and apoproteins that is distributed over the air–liquid interface of the pulmonary alveoli and distal airways. Surfactant minimizes alveolar wall surface tension, thereby stabilizing alveolar volume during tidal breathing and preventing the movement of fluid into the alveolus (16). It may also modulate the function of alveolar macrophages (17).

Surfactant is synthesized in the alveolar type II cell, stored in lamellar bodies of that cell, and secreted into the alveolar hypophase, in which it undergoes complex conformational changes. Current understanding suggests that the structure identified in the alveolar hypophase by electron microscopy as tubular myelin is a transitional state between the lamellar body and dispersed surfactant. The latter is likely to exist as small and large aggregates that have different activities (18) and as a monolayer covering the air–liquid interface.

There is no uniform definition of lung surfactant. Rather, a variety of methods exist for purifying material with surfactant activity from lungs, from fluid that has washed the airspace of the lung, or from amniotic fluid. These methods include differential and density-gradient centrifugation, lipid extraction, and chromatographic purification. To the extent that methods differ, different compositional and functional results may be obtained. Nevertheless, composition of lung surfactant in healthy adult humans and animals is very constant, although it may change in disease states (19).

Surfactant isolated by endobronchial lavage followed by differential and density-gradient centrifugation consists of approximately 90% lipid (of which more than 85% is phospholipid) and 10% surfactant associated apoproteins. The lipid composition of lung surfactant from adults is well described (Table 1) (16,19,20). Of note, phosphatidylcholine (PC) accounts for 70 to 80% of the phospholipids, and of this 60 to 70% is dipalmitoyl phosphatidylcholine (DPPC). Approximately 93 to 95% of the phospholipid acyl fatty acids (of bovine lung surfactant from adult animals) are fully saturated (21).

Several surfactant apoproteins are currently recognized. Surfactant apoprotein A (SP-A) is a group of relatively acidic glycoproteins of M_r =

Table 1 Composition of Lung Surfactant from Normal Subjects

Element	Percent[a]
Total phospholipid	100
Phosphatidylcholine (PC)	76.3 ± 2.1
Phosphatidylglycerol (PG)	11.6 ± 1.2
Phosphatidylinositol (PI)	3.9 ± 0.8
Phosphatidylserine (PS)	1.8 ± 0.3
Phosphatidyethanolamine (PE)	3.3 ± 0.5
Sphingomyelin (SM)	1.5 ± 0.4

Source: Data from Ref. 71.

[a]Values are the means ± SEM of determinations organic extracts of crude surfactant pellets from 16 to 29 patients.

30,000 to 35,000. SP-A is synthesized in alveolar type II cells and Clara cells and is secreted into the airway as a multimeric protein of approximately 650 kDa. The protein contains an N-terminal collagen-like sequence, a lipid-binding domain, and a carboxy-terminal lectin-like region. SP-A complexes with PC and may promote formation of tubular myelin structures from newly secreted phospholipids (22). Alveolar type II cells express a high-affinity receptor for SP-A (23), and it is likely that SP-A plays a significant role in phospholipid uptake and reutilization by type II cells (24,25). Finally, SP-A enhances the ability of hydrophobic apoproteins to facilitate adsorption of phospholipids to the gas–liquid interface (26).

SP-B and SP-C are highly conserved, markedly hydrophobic proteins rich in leucine and valine that are derived from precursor proteins. SP-B is a nonglycosylated protein of M_r = 7500 to 10,000 synthesized in type II pneumocytes and Clara cells and present extracellularly predominantly as a homeodimer. With the finding that SP-B is the apoprotein that is, in greatest part, responsible for surface-tension-lowering properties and prevention of collapse of pulmonary alveoli (27–29), peptides derived from the structure of SP-B and simplified analogs of SP-B have been constructed and structure–function correlates sought (30). Certain of these peptides (KL_4) that contain repeating sequences of four uncharged amino acids (leucine) separated by a positively charged amino acid (lysine), in dispersions with DPPC, dioleylphosphatidylglycerol, and palmitic acid (KL_4S), reduce surface tension in vitro and in vivo comparable to native surfactant. Substitution of hydrophobic residues or acidic residues for the basic R or K residues greatly diminished activity of the dispersions. The SP-B-like peptides are found by tryptophan fluorescence to partition within the phospholipid layer in contact with both polar head groups and acyl side chains (29). Raman spectroscopy, performed with these peptides in dispersions of DPPC and POPG, indicates interactions occurring between peptides and phospholipids, leading to a denser head group packing order than demonstrated with phospholipid alone, providing increased intermolecular and intrachain ordering of the phospholipid layer (30). These data support a concept that electrostatic charges of the intermittent hydrophilic residues of SP-B interact with phospholipid head groups and the stretches of hydrophobic residues interact with acyl side chains of phospholipids to induce lateral intermolecular order. These interactions contribute markedly to the ability of the phospholipid monolayer to resist surface tension and prevent alveolar collapse.

SP-C has an M_r value of 3000 to 6000 and has a remarkable terminal amino acid sequence composed of 23 exclusively hydrophobic amino acids that are believed to form a trans-membrane helix (31). Additionally, it is a lipopeptide containing two palmitoylated cysteines that provide additional hydrophobic sites (32).

Combining SP-B or SP-C with DPPC and PG results in a surfactant with apparently full biophysical properties, including rapid adsorption rates and excellent surface-tension-lowering properties (33–38). The apparent clinical efficacy of surfactants devoid of SP-A [e.g., calf lung surfactant extract (39), Surfactant-TA (40), or porcine lung surfactant (41)] in the treatment of neonatal respiratory distress syndrome is consistent with the proposal that SP-B and SP-C are sufficient for apparently full biophysical function of lung surfactant.

Finally, SP-D is a recently recognized surfactant apoprotein that is a multimeric complex of disulfide-bonded trimers comprised of identical 43-kDa collagenous glycoprotein subunits (42). The amino acid sequences and cDNA sequences of the lung surfactant apoproteins are known, and the genes for human SP-A, SP-B, and SP-C have been cloned. Current investigations focus on the function and metabolism of these proteins.

B. Metabolism of Lung Surfactant

Surfactant metabolism is complex and includes formation of lipid components from endogenous sources and from reutilization of previously synthesized material. Surfactant PC synthesized in the endoplasmic reticulum of the type II cell is packaged in lamellar bodies that are secreted into the alveolus. These bodies, which also contain apoproteins, rupture with subsequent unwinding of phospholipid contents to form tubular myelin structures. These structures are rich in SP-A and are the probable precursor for the monolayer film adsorbed to the gas–liquid interface. This film is continually being expanded and compressed during tidal breathing, and it is likely that the addition of newly synthesized surfactant to the film and the loss of surfactant from this film are modulated by the mechanical action. Strong experimental evidence indicates that surfactant is taken up by type II pneumocytes, and it is believed that the material taken up has previously been present in the surface film. This material may be catabolized or recycled (25,43).

When surfactant is recovered by BAL, the surface film is lost, and phospholipids are present in several forms. These include relatively dense large aggregates, which sediment at relatively low gravitational forces, and small less dense vesicular structures, which sediment at high gravitational forces (44,45). Of interest, the different forms have different apoprotein content and biophysical function, and conversion from heavy to light forms can be accomplished in vivo. Compared to the heavier aggregates, the lighter forms appear to contain little or no SP-A or SP-B and to have relatively poor biophysical function (44,46,47). Recent data suggest that conversion requires repetitive variations in surface area and the presence of a surfactant convertase which has been tentatively identified as a 75-kDa serine protease that is closely associated with surfactant phospholipid and may be a product of alveolar type II cells (48).

C. Biophysical Effects

Surfactant phospholipids and apoproteins combine to play an essential role in preventing alveolar collapse during normal tidal breathing. Traditionally, the law of Laplace ($P = 2\sigma/R$), which pertains to thin-walled spheres, is cited in introducing this role. This law relates surface tension (σ) and bubble radius (R) to trans-bubble pressure (P). If one takes the simplified view of the alveolus as a single bubble, expiration is accompanied by decreasing radius of that bubble. Were surface tension to stay constant as alveolar radius decreases, trans-bubble pressure would increase in a fashion that would promote alveolar collapse. However, in vitro observations support the idea that compression of the surfactant film on the contracting gas–liquid interface is associated with diminished surface tension. These observations may be made using a Wilhelmy balance to measure surface tension of a surfactant film that is cyclically compressed and expanded, or using a bubble surfactometer in which a small bubble is cycled between two known sizes and the trans-bubble pressure is measured at those extremes (49).

Certainly, the forces acting on individual alveoli are more complex than predicted by the law of Laplace, and are likely to be influenced by a variety of factors, including forces due to tissue elements and dynamic alteration of surface tension due to the behavior of lung surfactant. The critical elements of surfactant behavior, which are influenced by surfactant composition, include (a) lowering of surface tension during alveolar contraction, (2) respreading of surfactant on the fluid surface after the phospholipid monolayer has been compressed past the point of monolayer collapse, and (3) adsorption of lung surfactant to the monolayer from the aqueous subphase.

One of the ongoing challenges in understanding the physiology of lung surfactant has been to relate the behavior of the material in vitro to pressure–volume relationships of the intact lung. Excellent reviews of the physical chemistry (50) and influence on lung mechanics (51) have been published.

D. Cellular Effects

Phagocytes in the inflamed lung, including the alveolar macrophage and blood neutrophils that migrate into the alveolar space, are in intimate contact with lung surfactant. Surfactant may modulate one or more functions of these cells, perhaps by effecting production by phagocytes of superoxide radical (O_2^-), the product of the phagocyte membrane NADPH oxidase, by modulating release of cytokines, or by affecting degranulation. As evidence is accumulating that surfactant may have important effects on the production of O_2^- by phagocytes, it is important to understand the ways in which this production may be altered and how effects may be detected.

Superoxide is produced by a multicomponent oxidase unique to phagocytes that catalyzes the single electron reduction of O_2 to O_2^- using NADPH as an electron donor. Enzyme subunits exist in both the cytosolic and membrane fractions of the phagocyte and are assembled into a catalytically active oxidase only following exposure of the cell to soluble or particulate agonists. In the former group are phorbol myristate acetate (PMA), which directly stimulates protein kinase C. In ways that are not fully understood, this stimulation results in phosphorylation of cytosolic oxidase subunits with consequent conformational changes that lead to assembly of cytosolic components onto an integral membrane protein, cytochrome b_{558} (6,52,53). A second and perhaps biologically more relevant cell stimulus is the formylated tripeptide met-leu-phe (fMLP) that acts by binding to a distinct membrane receptor with subsequent G-protein-mediated activation of phospholipase C and production of diacylglycerol and polyphosphoinositols. The latter may act directly to mobilize stored intracellular Ca^{2+}, while the former may act on Ca^{2+}-dependent protein kinase C to effect assembly of the oxidase. Alternatively, parallel pathways that do not involve protein kinase C activation may link fMLP receptor occupancy to oxidase assembly and activation.

Phagocyte oxidase assembly and activation are frequently studied using the superoxide dismutase–dependent reduction of ferricytochrome c to quantify directly the production of O_2^-. Plots of O_2^- versus time following addition of an agonist (see Fig. 1) allow measurement of the interval between addition of agonist and initiation of oxidase activity. The times required for signal transduction and oxidase component assembly contribute to this interval. In addition, plots reveal the maximal rate of O_2^- production, which can be used as one measure of oxidase activity. Chemiluminigenic probes, including luminol and lucigenin, react with ROIs to produce light, and detection of this light following exposure of cells to agonist provides an alternative mechanism for studying cell activation (54). Phagocyte reactivity may be altered by an initial priming exposure of the cell to subthreshold levels of agonist that are insufficient to result in cell activation. Subsequent exposure to low levels of an alternative agonist result in responses that are much greater than those seen in the absence of priming. Although the events responsible for priming are not well understood, they may include partial assembly of the oxidase.

Several investigators have studied phagocytes exposed to lung surfactant, phospholipid mixtures, or surfactant apoprotein A. Hayakawa et al. exposed alveolar macrophages from infant or adult rabbits to PMA or opsonized zymosan (op-zym) and detected production of reactive oxygen species

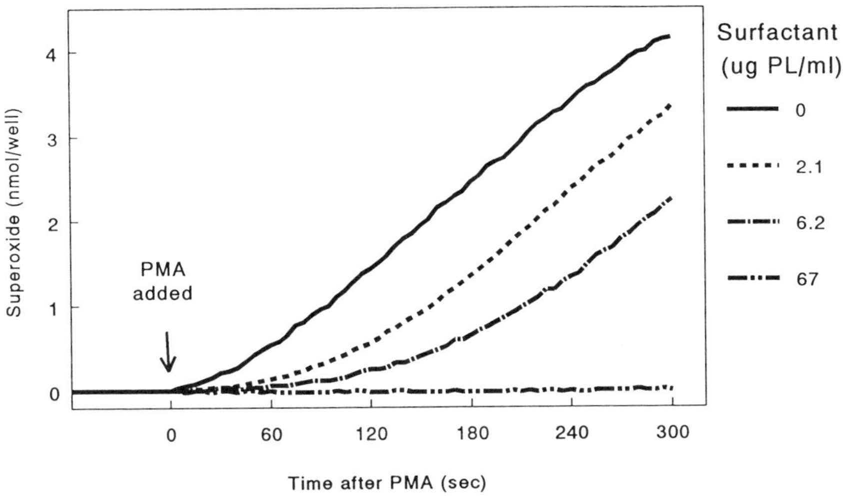

Figure 1 Effect of increasing concentrations of porcine lung surfactant on human neutrophil superoxide production elicited by phorbol myristate acetate (PMA) is shown. Neutrophils, 2×10^5 cells, were incubated for 2 min in the presence of the indicated concentrations of surfactant. At $t = 0$, PMA (80 n*M* final concentration) was added and superoxide production at 37°C measured as the superoxide dismutase inhibitable reduction of cytochrome *c*. At low concentrations of surfactant there is a slowing of the response to PMA but only a minimal decrease in maximal oxidase activity (as indicated by the slope of the line). In contrast, at higher concentrations of surfactant, there is complete inhibition of superoxide production.

by chemiluminescence (55). Exposure of macrophages to natural rabbit surfactant produced significant (55 to 70%) inhibition of both PMA- and op-zym-elicited chemiluminescence responses. This effect was seen in the presence of 5% fetal bovine serum, was observable at concentrations of surfactant as low as 121 μg/mL phospholipid, and was even more pronounced when previously primed cells were studied. To study the effects of artificial surfactant, DPPC, dioleoylphosphatidylcholine (DOPC), dipalmitoyl phosphatidylglycerol (DPPG), dioleoylphosphatidylglycerol (DOPG), and sphingomyelin were combined. This preparation was able to reduce PMA-elicited chemiluminescent responses by approximately 75%.

More recently, Hayakawa et al. have investigated the effect of individual phospholipids contained in pulmonary surfactant on the chemiluminescent response induced by PMA or op-zym (56). Primed macrophages incubated with natural surfactant exhibited 17% and 35%, respectively, of the PMA- or op-zym-elicited chemiluminescent response obtained in the absence of surfactant. Complete artificial surfactant reduced the PMA-elicited chemiluminescent responses to 32% control, and a similar reduction was caused by DOPC concentrations above 250 μM. DPPG and PI had less inhibitory activity, and DPPC, DOPG, phosphatidyl ethanolamine, and sphingomyelin had no significant effects. In contrast, primed alveolar macrophages incubated with natural surfactant, DPPC, or DOPC produced significantly greater chemiluminescence after exposure to op-zym than did identical cells that had not been exposed to lipid, indicating that the effect of natural surfactant is not due uniformly to the effect of a single lipid.

To investigate the effect of SP-A on phagocyte function, Weber et al. incubated dog alveolar macrophages or blood neutrophils with canine SP-A, exposed the cells to PMA, and measured the chemiluminescent response (57). SP-A, but none of nine other proteins tested, resulted in significant diminution of the chemiluminescent response. Treatment of SP-A with collagenase prior to incubation destroyed the effect. Thus both the lipid and apoprotein components of lung surfactant may modulate phagocyte superoxide production.

We have confirmed the inhibitory effect of complete porcine lung surfactant on human neutrophil superoxide production and have also demonstrated that a synthetic surfactant (KL_4-surfactant) (29) shares this property (Fig. 1). Human neutrophils were incubated with either preparation, exposed to PMA or fMLP, and respiratory burst activity was measured as the superoxide dismutase inhibitable reduction of ferricytochrome *c*. The presence of surfactant resulted in significant inhibition of fMLP-elicited superoxide production and was more pronounced for KL_4 surfactant (IC_{50} = 30 μg phosphorus/mL) than for porcine surfactant (IC_{50} = 200 μg/mL). PMA-elicited $O_2^{\bar{\cdot}}$ production was profoundly inhibited by both surfactant preparations over a range of PMA concentrations up to 10^{-6} M, and the IC_{50} for porcine surfactant (10 μg/mL) was, in this case, lower than that for KL_4 surfactant (70 to 100 μg/mL). Neither KL_4 surfactant nor porcine surfactant altered cytochrome *c* reduction, and production of ionized calcium within the cell in response to agonist appeared to be independent of the presence of surfactant. Thus these observations are consistent with an effect of surfactant on protein kinase C or on the assembly and activity of the respiratory burst oxidase.

Although these findings describe an inhibitory effect of lung surfactant on phagocyte function, conflicting reports exist (58,59). Confusion in the literature may be due to interspecies variability in phagocyte or surfactant function or differences in assay techniques.

Phagocyte functions other than superoxide production may also be altered by lung surfactant. Data obtained with human cells are scant, but preliminary reports suggest that an artificial surfactant preparation lacking apoproteins (Exosurf) causes dose-dependent inhibition of cytokine release by human alveolar macrophages stimulated by endotoxin, TNF-α, or IL-1 (60,61). Exposure of human PMNs to bovine surfactant results in a dose-dependent inhibition of adherence, directed migration, and ROI production (62), and exposure of human peripheral blood monocytes to porcine surfactant impairs phagocytosis and killing of staphylococci (63).

Thus multiple lines of investigation suggest that lung surfactant may alter important cellular functions of inflammatory cells. The mechanism of these effects is not known, although evidence suggests that the NADPH oxidase enzyme system responsible for ROI production during the respiratory burst may be sensitive to alterations in its phospholipid environment. The presence of long-chain unsaturated phosphatidylcholines partially inhibits the respiratory burst elicited by PMA in intact PMNs (53), while certain phospholipids (notably phosphatidylserine) appear to augment oxidase activity in solubilized enzyme preparations (64) and are required for activity under certain conditions (65). Whether different surfactant preparations act directly on the respiratory burst oxidase or act by incorporation into the PMN plasma membrane bilayer with secondary effects on membrane fluidity is unknown, although agents that alter membrane fluidity are known to alter respiratory burst activity (66,67). Whether a similar mechanism may also explain the effect of surfactant on other cell functions is unknown.

III. Surfactant Alterations Occurring in the Presence of Acute Lung Injury

Lung surfactant recovered in BAL from patients with ARDS or from animals with experimental acute lung injury demonstrate disturbances of the lung surfactant system. Since treatment of these animals with exogenous lung surfactant has therapeutic effects, it is reasonable to hypothesize that disturbances of the lung surfactant system contribute to the pathophysiology of the acute lung injury. Ashbaugh et al. provided initial observations on fluid recovered from minced lung tissue obtained postmortem from two patients with ARDS and suggested that surfactant dysfunction contributed to the clinical syndrome (13). Subsequently, Petty et al. showed that surfactant in BAL from patients with ARDS was both abnormally aggregated and functionally inactive (68). Hallman et al. studied the phospholipid composition and surface activity of surfactant recovered from BAL from adults with a variety of pulmonary diseases, including ARDS, pneumonia, malignancy, and chronic obstructive lung disease (COPD) and from normal controls. The recovery of phospholipid in all of these conditions was equivalent to that of controls, with the exception of COPD, in which recovery was 50% of control values. Surfactant from patients with ARDS had a significantly lower fractional content of phospholipid present as PC and phosphatidylglycerol (PG) and had little or no surface-tension-lowering activity (19).

Pison et al. have studied the composition and function of surfactant from patients with respiratory failure following multiple trauma (69,70). They found similar amounts of total phospholipid in lavage samples from traumatized patients with low respiratory failure scores, from patients with high respiratory failure scores, and from normal controls. However, the fractional contents of PC and PG were lower in traumatized patients than in controls; the fractional content of PC was significantly lower in patients with high respiratory failure scores than in patients with low scores. Study of surfactant function as reflected by hysteresis area on the Wilhelmy balance gave similar results. Hysteresis area declined over 1 to 2 weeks following trauma and was significantly less in patients with high as compared to patients with low respiratory failure scores.

A recent report by Gregory et al. provides information on surfactant recovered in lavage fluids from three groups of patients: those with established ARDS, those "at risk" for ARDS (who, in fact, had respiratory failure of insufficient magnitude to meet criteria for ARDS), and normals (71). The surfactant pellet recovered by centrifugation from patient BAL samples contained substantially less total phospholipid (median values of 1.33, 1.62, and 7.54 μmol/mL for the three groups, respectively) and less total protein (0.97, 1.01, and 2.40 mg/mL, respectively). These results are consistent with a decreased total lavagable surfactant pool but might also reflect increased difficulty capturing lung surfactant during the lavage procedure. Surfactant phospholipid composition was altered as reported previously, with decreased levels of PC and PG. Concentrations of SP-A and SP-B recovered in lavage fluid were also markedly decreased. Median values of SP-A were 6.18, 8.81, and 118.04 μg/mL, respectively, in the three groups. Values for SP-B were 0.05, 0.36, and 0.66 μg/mL, respectively. Taken together, these data provide strong evidence for decreased synthesis or altered metabolism of SP-A and SP-B in patients with ARDS. Finally, surfactant function was confirmed to be abnormal. Measured on the pulsating bubble surfactometer, median minimum surface tension was 24.0, 16.7, and 4.1 dyn/cm.

In all, seven investigations address the biochemical composition and/or function of surfactant recovered from the lungs of patients with ARDS (Table 2). The size of the lung surfactant pool is difficult to determine accurately, but might be reflected in the phospholipid content of lavage fluid. Recent studies suggest that such content is decreased. Studies by most authors of surface-tension-lowering ability suggest abnormal surfactant function, although minimum surface tension has not been found uniformly to be abnormal. Taken together, the results suggest strongly that abnormal surfactant composition and function are present in patients with ARDS, and

Table 2 Surfactant Abnormalities in Patients with ARDS[a]

Investigator	Year	Number of patients	Total PL	PC or DPPC	PG	Minimum surface tension
Ashbaugh (13)	1967	2				I
Petty (116)	1977	1	NC			NC
Petty (68)	1979	6				NC
Hallman (19)	1982	26	NC	D	D	I
Pison (70)	1989	17	D	D	D	NC
Pison (70)	1990	30	D	D	D	
Gregory (71)	1991	49	D	D	D	I

Source: Adapted from Ref. 115.
[a]PL, phospholipid; PC, phosphatidylcholine; DPPC, disaturated phosphatidylcholine; NC, no change; I, increased; D, decreased.

suggest the possibility that surfactant supplementation would be of benefit to those patients.

A variety of mechanisms might explain the alterations in surfactant phospholipid composition in the alveolar spaces of patients with acute respiratory failure. Phospholipid production by injured type II cells, or by the hypertrophic or hyperplastic type II cells found in patients with ARDS, may be abnormal. Care of patients with acute lung injury includes use of partial pressures of oxygen that may cause diffuse alveolar damage. Phospholipid biosynthesis by type II pneumocytes has been shown to be depressed in rabbits exposed to high levels of oxygen (72). The effect may be indirect, however, as exposure of type II cells to hyperoxia in vitro has only minor effects on lipid metabolism (73).

A second cause of abnormal phospholipid composition in BAL recovered from patients with ARDS might be plasma phospholipids that gain access to the alveolar space. To investigate this possibility, Hallman et al. compared BAL and plasma lipids. Several differences were noted: the ratio of DPPC to PC in BAL was seven times that of plasma; phosphatidylserine (PS) in BAL was 10 times that in plasma; BAL lyso-PC was approximately one-fifth that of plasma. They concluded that plasma lipid contamination of BAL could not explain the BAL phospholipid abnormalities (19). Finally, phospholipids from nonsurfactant sources (e.g., from necrotic cells in the alveolar space or from bacteria) might be recovered in BAL. Phospholipids from tissue sources should contribute a relatively higher fractional content of sphingomyelin. In fact, BAL surfactant phospholipid content of sphingomyelin is increased two to four times control values in the presence of acute lung injury, consistent with contribution of phospholipid from necrotic cells. In addition to quantitative and qualitative abnormalities of surfactant synthesis by injured or hyperplastic type II epithelial cells, biochemical changes of lung surfactant in the alveolar space may result from lipid peroxidation or apoprotein cleavage; inhibitors of surfactant function may be present; and lung surfactant clearance mechanisms may be perturbed.

Toxic oxygen products and macrophage proteases present in the lungs of patients with acute lung injury may react with lung surfactant. Oxidative attack is predicted to involve primarily di- or polyunsaturated acyl groups of the phospholipid molecule, which represent approximately 2% of such groups in bovine lung surfactant (21). Aldehydic products of peroxidation, including hydroxynonenal and hydroxyoctenol, have significant chemotactic or cytotoxic activity (74).

Seeger et al. have exposed rabbit lung lavage fluid to a variety of oxidizing stresses and found altered surface tension properties as measured on a Wilhelmy balance (75). The surfactant-containing lavage fluid exposed to one of several oxyradical generating systems demonstrated delayed reduction of surface tension during compression with reduction of hysteresis area, augmented monolayer collapse rate, and an increase in minimal compressibility. These alterations in surface tension properties were less marked or absent after exposure to lavage fluid to H_2O_2 or O_2.

We have recently observed that oxidation of lung surfactant phospholipids may be accompanied by loss of surface activity as determined both in vitro and in vivo (76). The hydrophobic components of porcine surfactant were exposed to Fe^{2+} + H_2O_2 or to Cu^{2+}. The former system results in lipid peroxidation secondary to formation of complex iron–oxygen radicals (77), while exposure to Cu^{2+} has been shown to result over hours in oxidation of plasma low-density lipoprotein (78). In both cases we detected production of conjugated dienes and thiobarbituric acid–reactive substances, consistent with formation of aldehydic products of lipid peroxidation. In addition, the surface-tension-lowering ability was impaired after exposure to oxidizing conditions. Minimum surface tension in the bubble surfactometer after 5 min of pulsation at 37°C remained above 10 mN/m for surfactant samples exposed to $FeCl_2$ + H_2O_2, but fell to < 3 mN/m within 5 min for surfactant alone or surfactant exposed only to Fe^{2+} or to H_2O_2. Similar impairment of function was seen after exposure of surfactant to Cu^{2+} at room temperature for 24 h. Surfactant to which Cu^{2+} was added but which was not incubated showed no loss of function (Fig. 2). These in vitro findings were confirmed by in vivo experiments in which surfactant-deficient rabbit fetuses failed to demonstrate the expected improvement in lung compliance when treated with surfactant that had been oxidized using either exposure to Fe^{2+} + H_2O_2 or to Cu^{2+} (Fig. 3).

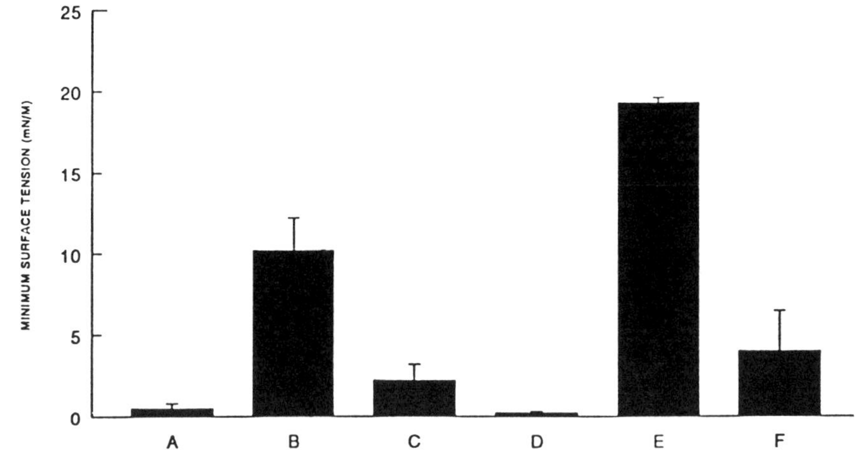

Figure 2 Samples of porcine lung surfactant were unexposed (A), exposed to $FeCl_2$ + H_2O_2 (B), to $FeCl_2$ only (C), to H_2O_2 only (D), to $CuCl_2$ for 24 h (E), or to $CuCl_2$ for 5 min (F). Minimum surface tension was measured after 5 min of pulsation in a bubble surfactometer. Samples exposed to $FeCl_2$ + H_2O_2 or $CuCl_2$ for 24 h had a significant increase in minimum surface tension.

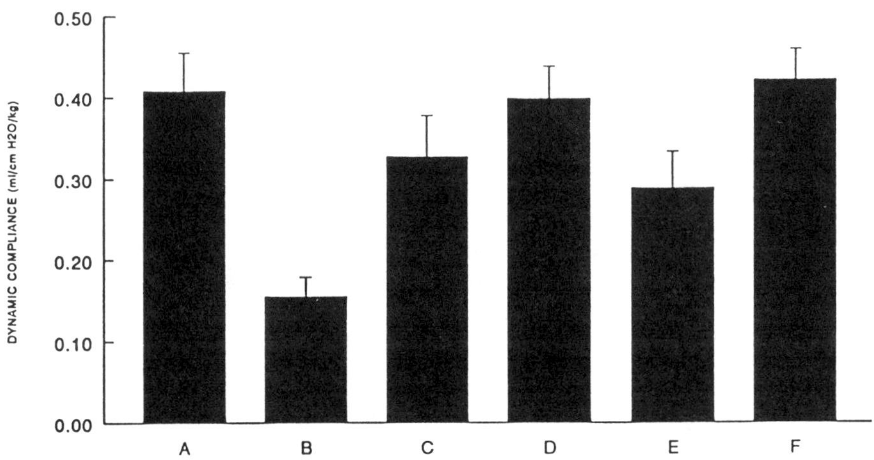

Figure 3 Samples of porcine lung surfactant were unexposed (A), exposed to $FeCl_2$ + H_2O_2 (B), to $FeCl_2$ only (C), to H_2O_2 only (D), to $CuCl_2$ for 24 h (E), or to $CuCl_2$ for 5 min (F). Physiologic activity was measured in vivo by determining the ability of samples to augment compliance of the lungs of surfactant-deficient fetal rabbits after 1 h. Samples exposed to $FeCl_2$ + H_2O_2 failed to augment compliance fully, and samples exposed to $CuCl_2$ tended to show similar results.

It should be stressed that while the functional alterations cause by exposure to oxidant correlate with evidence of lipid peroxidation, a variety of mechanisms may explain the results. These include formation of lysophospholipid, apoprotein degradation, or formation of malondialdehyde-apoprotein or hydroxynonenal-apoprotein ligands (79).

Finally, both collagenase and human neutrophil elastase have been shown to cleave surfactant apoprotein A with in vitro loss of surface activity after exposure of surfactant to elastase. Elastase may also attack hydrophobic apoproteins, with associated loss of function (80,81). Ryan et al. have investigated further the potential for PMN to alter lung surfactant function in the inflamed lung, and have reported decreased adsorption and increased minimum surface tension and compressibility of canine surfactant after exposure to products of neutrophils activated by exposure to PMA (82). Superoxide dismutase inhibited all effects except decreased adsorption, and oxidant-generating systems produced results similar to those obtained with activated cells. Analysis of SP-A disclosed evidence of cleavage, suggesting that neutrophil proteases had caused the degradation. These results suggest that ROIs are largely responsible for the loss of function associated with oxidant injury, but that proteolysis of SP-A may contribute to decreased adsorption.

While the surface-tension-lowering ability of lung surfactant is abnormal in surfactant purified from lavage fluid of patients with ARDS, that ability is markedly more abnormal when equimolar amounts of phospholipid are examined in the original lavage fluid. Figure 4 presents results of analysis of BAL fluids from three patients that demonstrated the presence of functional inhibitor. Surfactant function was determined either in the presence of BAL proteins or after extraction into chloroform/methanol. The marked difference in minimum surface tension between samples examined in the presence or absence of BAL proteins suggests removal of hydrophilic inhibitors of surfactant function. A variety of plasma proteins have been identified that demonstrate inhibitory function, including glycolipid, albumin, hemoglobin, fibrin monomer, and bilirubin (referenced in Ref. 83). These proteins are likely to gain access to alveolar lining fluid, as integrity of the alveolar-capillary membrane is disrupted during acute lung injury. Recently, Li et al. have reported the inhibitory activity of C-reactive protein, a plasma acute-phase reactant that may increase in concentration up to 1000-fold during illness related to trauma or inflammation. Patients with ARDS also have increased levels of C-reactive protein in BAL, and this protein has the ability to bind to liposomes containing DPPC or PG. Incubation of extracted calf lung surfactant with C-reactive protein impairs surface-tension-lowering function. Thus C-reactive protein in the alveolar space may also contribute to loss of surfactant function during acute lung injury (84).

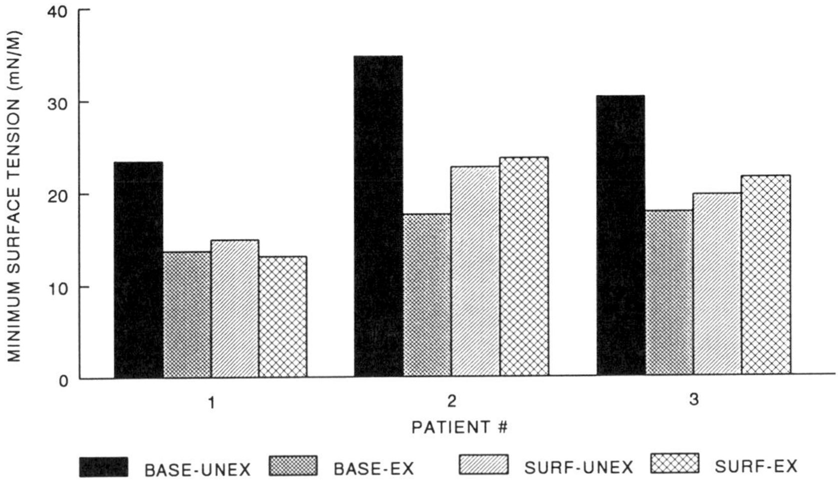

Figure 4 Pulmonary surfactant was recovered by centrifugation from bronchoalveolar lavage fluid from three patients with ARDS. The surfactant was dialyzed, lyophilized, and part of each sample was extracted into organic solvent to remove hydrophilic contaminants. Samples were then resuspended in 100 m*M* NaCl, 1.5 m*M* $CaCl_2$ at a concentration of 2 mg of phospholipid per milliliter, and surface-tension-lowering ability of samples was determined using the bubble surfactometer. Mean values (m*N*/m) are presented. The results demonstrate the presence of hydrophilic inhibitor in BAL prior to surfactant administration; after administration of exogenous surfactant inhibitor, activity is not detectable. BASE, baseline; SURF, after surfactant administration; UNEX, unextracted; EX, extracted. (From Ref. 115.)

The usefulness of surfactant supplementation in the setting of acute lung injury may depend, in part, on the ability of exogenous surfactant to overcome functional inhibitors. In a study of surfactant inhibition due to meconium, Moses et al. showed, using the pulsating bubble surfactometer, that at low surfactant concentrations (1.5 mg phospholipid/mL) meconium exhibited a dose-dependent inhibition of surface-tension-lowering function (over the range of 0 to 65 μg/mL). When surfactant was present in a concentration of 6 mg phospholipid/mL, no inhibition of function was seen over this range (85). Holm has demonstrated that surfactant inhibitors may have synergistic effects. Concentrations of phospholipase A_2 and albumin that, separately, cause minimal or no inhibition of bovine surfactant surface-tension-lowering function (+3 and +0 mN/m, respectively), when present together resulted in an increase of 14 mN/m (86).

Surfactant composition appears to partially modify the sensitivity to inhibition. The surface-tension-lowering ability of bovine surfactant devoid of SP-A is markedly inhibited by albumin; however, if the surfactant is supplemented with SP-A, this inhibition is not apparent (87). In summary, when considering surfactants to use in the treatment of acute lung injury, it will be important to determine the resistance of the material to inhibition as well as its biophysical and cellular effects.

IV. Surfactant Supplementation in the Setting of Acute Lung Injury

A. Available Preparations

Substantial experience has been gained in the treatment of infant RDS with exogenous surfactant. Nevertheless, direct clinical comparisons of the efficacy of various surfactant preparations are not yet available. Even were such comparisons at hand, it is quite possible that the requirements for an ideal surfactant for the treatment of ARDS might differ from those of a surfactant designed for the treatment of infant RDS. The latter condition requires relatively small absolute quantities of material, and surfactants available in limited supply may be useful. Such surfactants include that derived from human amniotic fluid or those requiring addition of protein that is not abundantly available. In contrast to the amount that may be required for treatment of a neonate, treatment of an adult with instilled doses of comparable amount (relative to body weight) may require up to 200 times the quantity of surfactant. Laboratory data suggest that the treatment of acute lung injury with surfactant that is aerosolized rather than instilled might require substantially less material. Therefore, additional considerations might include the ability to aerosolize a surfactant preparation and the resistance of the material to inactivation during aerosolization.

Finally, the requirement that surfactant used to treat ARDS be resistant to inhibition by protein provides a further consideration. While appearance of plasma proteins in alveolar fluid of infants with RDS may occur, formation of protein-rich alveolar edema is a hallmark of ARDS, and a surfactant of maximal efficacy might be predicted to be one with relative resistance to inhibition by such proteins. To date, the surfactants containing apoprotein have been found to be relatively less sensitive to the inhibitory effects of protein than surfactant formulated without apoprotein (88).

As displayed in Table 3, a variety of natural, modified, and synthetic preparations are potentially available for therapeutic use. The desirability

Table 3 Examples of Surfactant Preparations Available for Possible Therapeutic Use

Class	Origin	Composition[a]	Ref.
Natural	Human amniotic fluid	Phospholipids, neutral lipids, hydrophobic and hydrophilic apoproteins	117
Modified natural	Bovine, e.g., Survanta	Phospholipids and hydrophobic apoproteins	40
	Porcine, e.g., Curosurf	Phospholipids and hydrophobic apoproteins	118
Synthetic, without protein/peptide	Synthetic, e.g., Exosurf	DPPC + hexadecanol + tyloxapol	119
Synthetic, with protein/peptide	Synthetic	Arg(Leu-Leu-Leu-Leu-Arg)$_4$ + DPPC + PG	29
		Amphipathic α-helical 24-residue peptide + DPPC	

Source: Adapted from Ref. 115.
[a]DPPC, dipalmitoylphosphatidylcholine; PG, phosphatidylglycerol.

of a synthetic preparation cannot be overstated, as such a preparation should be available in large quantity, be inexpensive, and carry no risk of transmitted disease. Recent reports of effective synthetic preparations that incorporate peptide analogs of hydrophobic surfactant apoproteins are encouraging (89,90). However, it should be kept in mind that excellence of function and resistance to inhibition of function may prove to be separable, and in that case the optimal surfactant preparation for a lung rich in inhibitor may prove to be different from the preparation that is optimal for the immature lung. Comparative studies of efficacy in patients with ARDS will be of particular value, as they include information of comparative resistance to inhibitors found in the inflamed lung.

Several considerations bear on the amount of supplemental surfactant that might be rational to deliver. The size of the endogenous surfactant pool in babies beyond the neonatal period is approximately 10 mg/kg (91), approximately three times the amount required to coat the interior wet surface of the lung with a monolayer of phospholipid (92). Pool sizes have not been measured in the human. It is likely that problems with distribution particularly with overcoming inhibition due to plasma proteins that have gained access to the lower airway, will require use of doses much larger than this. Experience from treating patients with IRDS and from treating a very small number of patients with ARDS suggests that as a first approximation, the required dose of exogenous surfactant for a single treatment might be in the range 50 to 200 mg/kg or 3500 to 14,000 g/dose for a 70-kg adult. Survanta and Exosurf are currently available in sufficient quantity to allow clinical evaluation; the availability of synthetic surfactants will accelerate the pace of this activity.

B. Animal Experimentation

Models

A relatively small number of animal models of acute lung injury have been demonstrated to be responsive to surfactant supplementation. These have recently been reviewed by Robertson (93) and include acute lung injury induced by exposure to hyperoxia (94), *N*-nitroso-*N*-methylurethane (NNNMU) (95,96), saline lung lavage (97–99), hydrochloric acid instillation (100), anti-lung serum, oxygen radicals (101), or bilateral vagotomy (102).

Acute lung injury in the experimental animal is associated variably with decreased pulmonary compliance, atelectasis, consequent ventilation–perfusion imbalance, and intrapulmonary shunting with consequent hypoxemia. These findings, which tend to reverse with surfactant supplementation, are central features of the pathophysiology of the patient with ARDS. Although they are undoubtedly multidetermined, they are consistent with loss of surfactant function.

Administration Techniques

The methodology to be used for administration of exogenous surfactant to patients with ARDS is a complex issue, Dose, volume used for suspension and administration, and method of delivery and patient position are likely to be significant variables (103). Prior experience has been gained by administration in bolus form either directly down the endotracheal tube, or via a catheter or bronchoscope placed in the endotracheal tube. Use of a bronchoscope has the advantage that fractions of the total dose may be delivered to discrete portions of the lung. Administration via nebulization is of theoretical interest, as material might be delivered with maximum homogeneity. However, the rate at which surfactant may be delivered is significantly less than the bolus method. A dose of 100 mg/kg (suspended in 4 mL/kg) can be delivered by instillation over 15 to 60 min, whereas aerosol delivery of a similar dose may take 20 h or more. Nevertheless, recent reports suggest that small doses delivered by nebulization to surfactant deficient animals are of significant benefit. Treatment of surfactant-deficient preterm lambs with aerosolized surfactant (2 mg deposited phospholipid/kg) caused improvement in gas exchange and lung mechanics comparable to that resulting from surfactant instilled intratracheally (50 mg phospholipid/kg) (104).

Subsequent comparison of delivery methods in adult rabbits with acute lung injury induced by *N*-nitroso-*N*-methylurethane has shown significant improvement in lung function occurring with deposition of very small quantities of aerosolized surfactant (95). More recently, Lewis et al. (99) have studied the distribution and physiologic effect of surfactant administered by aerosol to adult sheep rendered surfactant deficient by saline lavage. Delivery by aerosol results in a more uniform distribution of the exogenous surfactant than that provided by the heterogeneously distributed tracheally instilled liquid surfactant. Significant physiological improvements are noted with very small quantities of aerosolized surfactant deposited in lung tissue (0.5 mg of phospholipid per kilogram). Recent measurement of distribution of aerosolized surfactant in surfactant-deficient adult sheep with one protected (surfactant sufficient) lobe disclosed preferential deposition in the normal lobe with predictable compromise of the physiological response compared to that of animals with a uniform surfactant deficiency (105). This observation may explain variability in clinical response in the face of heterogeneous lung injury and underscores the potential importance of ventilatory techniques that optimize homogeneous delivery of aerosol.

Recent clinical trials in humans with ARDS suggest beneficial effects of exogenous surfactant delivered as an aerosol (106). However, the optimal dose of aerosolized surfactant required for maximal physiological response, the pulmonary distribution of the aerosolized exogenous surfactant, and the total quantity of surfactant actually deposited in lung tissue are not yet known.

C. Clinical Observations

A fascinating history of surfactant research has been provided by Obladen (107). Although the clinical importance of lung surfactant was suggested by von Neergaard in 1929, Avery and Mead provided the first evidence in humans that surfactant deficiency—specifically in the lungs of premature neonates—was associated with development of respiratory distress. Subsequent carefully controlled multicenter investigations of the value of surfactant administration to infants at risk of IRDS or with established IRDS have documented improvement in survival, decrease in disease severity, and decrease in complications (cited in Ref. 108).

Experience with administration of exogenous surfactant to adults with acute lung injury is quite limited. However, the rationale for attempting this therapy rests on the results of animal experimentation and on the observation that surfactant administration to infants with established RDS is of benefit. While such infants have an absolute deficiency of lung surfactant rather than the functional deficiency described in patients with ARDS, they resemble patients with ARDS in that their lungs are undergoing an acute inflammatory process. Evidence for this includes presence of neutrophil proteases and oxidized protein in tracheal aspirate (14).

Reports of treating patients with ARDS with surfactant suggest the possibility of benefit. Lachmann administered a natural surfactant preparation (300 mg phospholipid/kg) to a terminally ill 4-year-old patient with sepsis and severe ARDS. Within 4 h, the p_aO_2 value is reported to have increased from 19 mmHg to 240 mmHg (109). Nosaka et al. have reported administration of surfactant-TA in the treatment of two adults. Surfactant was administered in a dose of 240 mg/day approximately 15 times over 38 days to a 66-year-old woman with severe burns and ARDS and three times to a 51-year-old man with postoperative pneumonitis. Modest increments in gas exchange were reported, and development of antibodies to surfactant-TA were not detected (110).

We have used a porcine surfactant preparation (Curosurf) (41) to treat six patients without preexisting lung disease who satisfied an operational definition of ARDS (panlobar infiltrates, pulmonary artery wedge pressure $\leqslant$ 18 torr, acute onset over $\leqslant$ 7 days). These patients had a baseline F_1O_2 requirement of 0.658 $\pm$ 0.7 (mean $\pm$ SEM) and were receiving 8 to 15 cm of PEEP to maintain a baseline p_aO_2 value of 78 $\pm$ 4.6 torr. Patients initially received either 4 g of porcine surfactant (Curosurf) in 50 mL (10 mL delivered via a fiber optic bronchoscope to each lobe of the lung) or 50 mL of placebo (saline in one patient, air in five) delivered in a similar fashion. Administration of 50 mL of exogenous porcine surfactant was well tolerated and was accompanied by a transient improvement in gas exchange. Compared to preinstillation values, the maximal improvement in p_aO_2 seen during the 3 h after treatment with surfactant averaged +48.8 $\pm$ 15.5 torr. In contrast, the maximal improvement in p_aO_2 after treatment with placebo was +8.8 $\pm$ 4.6 torr. Analysis of BAL disclosed increased phospholipid concentration after administration of exogenous surfactant that was sufficient to overcome the presence of BAL inhibitor activity. This effect was transient, persisting in the 24-h BAL sample in only one patient—the one who had sustained improvement in gas exchange. No alterations in pulmonary compliance, chest radiograph, hemodynamic parameters, or serum chemistries were associated with the use of exogenous surfactant. Finally, no evidence of immunologic sensitization occurred, as indicated by failure to develop evidence of immune complexes or anti-Curosurf antibodies.

More recently, surfactant has been delivered as an aerosol rather by instillation. Weg et al., in a prospective randomized placebo-controlled trial, administered aerosolized Exosurf to 52 ventilated patients with ARDS secondary to sepsis (106). They found improvement in shunt (−4.6 versus −1.5%) and static respiratory compliance (+1.9 versus −1.4 mL/cmH_2O and a trend toward decreased mortality (35 versus 47%) when comparing patients receiving surfactant continuously for 5 days versus patients receiving aerosolized saline. Subsequently, a sixfold concentration of Exosurf has been used under similar conditions and has resulted in trends suggesting improvement in alveolar–arterial oxygen gradient and reduction in mortality (111).

More extensive studies of the benefits of instilled Survanta or of aerosolized Exosurf are in progress. If these studies confirm the improvement in survival suggested by prior studies, it will be of interest to attempt to discern if this improvement is due entirely to improvement in gas exchange and compliance of the lung, or if down-regulation of the inflammatory response by exogenous surfactant contributes to the improved survivorship. As the majority of patients who die of ARDS do not die of primary gas exchange failure, improvement in survivorship due to surfactant treatment must be explained either by improved pulmonary function and secondary survival benefit (i.e., earlier discharge from the ICU, with consequent decreased likelihood of iatrogenic injury) or by a primary impact on the systemic disease process.

Initial considerations of surfactant treatment of premature infants with surfactant deficiency suggested that beneficial effects would occur as a consequence of reducing surface tension of alveolar lining fluid, improving lung compliance, reducing shunt and extent of low-VA/Q units, and improving gas exchange. The presumed association between decreased lung compliance and lack of native surfactant function was so powerful that it was quite surprising when several investigators reported a dissociation between improvement in gas exchange and alteration in compliance following administration of exogenous surfactant to neonates with respiratory distress syndrome (RDS) (112,113). These observations suggest that improvement in oxygenation in the absence of detectable change in lung compliance may be due to reduction in shunt, improved VA/Q, and/or decreased interstitial edema. Our own observations of the effects of treatment of adults with ARDS who each received 4 g of extracted porcine surfactant (Curosurf) (114) suggests that treatment is accompanied by a modest improvement in gas exchange that is not associated with a discernible improvement in lung compliance. Certainly, the reduction in lung compliance in patients with ARDS is multidetermined and is due not only to loss of surface activity in small airways and alveoli, but also to profound interstitial edema and, after the initial stages of disease, to proliferation of fibroblasts and myocytes and production of excessive amounts of extracellular matrix proteins.

Acknowledgments

This work was supported by grant HL-23584 from the National Institutes of Health, Bethesda, Maryland, and a grant from the California Tobacco Related Disease Research Program.

References

1. Weiland JE, Davis WB, Holter JF, Mohammed JR, Dorinsky PM, Gadek JE. Lung neutrophils in the adult respiratory distress syndrome: clinical and pathophysiologic significance. Am Rev Respir Dis 1986; 133:218–225.
2. Schlag G, Voigt W-H, Schnells G, Glatzl A. Vergleichende Untersuchungen der Ultrastrucktur von menschlicher Lunge und Skeletmuskulatur im Schock. II. Anaesthetist 1977; 26:612–622.
3. Bachofen M, Weibel ER. Structural alterations of lung parenchyma in the adult respiratory distress syndrome. Clin Chest Med 1982; 3:35–56.
4. Ognibene FP, Martin SE, Parker MM, et al. Adult respiratory distress syndrome in patients with severe neutropenia. N Engl J Med 1986; 315:547–551.
5. Spragg RG, Smith RM. Biology of acute lung injury. In: Crystal RG, West JB, eds. The Lung: Scientific Foundations. New York: Raven Press, 1990:2003–2017.
6. Smith RM, Woodman RC, Babior BM. The elusive oxidase: the respiratory burst oxidase of human phagocytes. In: Harris JR, ed. Subcellular Biochemistry. Vol 17. Blood Cells. New York: Plenum Press, 1991.
7. Weiss SJ. Tissue destruction by neutrophils. N Engl J Med 1989; 320:365–376.
8. Johnson D, Travis J. The oxidative inactivation of human alpha 1-proteinase inhibitor: further evidence for methionine at the reactive center. J Biol Chem 1979; 254:4022–4026.
9. McGuire WW, Spragg RG, Cohen AB, Cochrane CG. Studies on the pathogenesis of the adult respiratory distress syndrome. J Clin Invest 1982; 69:543–553.
10. Lee CT, Fein AM, Lippman M, Holtzmann H, Kimbel P, Wienbaum G. Elastolytic activity in pulmonary lavage fluid from patients with adult respiratory distress syndrome. N Engl J Med 1981; 304:192–196.
11. Wewers MD, Herzyk DJ, Gadek JE. Alveolar fluid neutrophil elastase activity in the adult respiratory distress syndrome is complexed to alpha-2-macroglobulin. J Clin Invest 1988; 82:1260–1267.
12. Pratt PC, Vollmer RT, Shelburne JD, Crapo JD. Pulmonary morphology in a multihospital collaborative extracorporeal membrane oxygenation project. Am J Pathol 1979; 95:191–214.
13. Ashbaugh DG, Bigelow DB, Petty TL, Levine BE. Acute respiratory distress in adults. Lancet 1967; 2:319–323.
14. Merritt TA, Cochrane CG, Holcomb K, et al. Elastase and alpha 1-proteinase inhibitor activity in tracheal aspirates during respiratory distress syndrome. Role of inflammation in the pathogenesis of bronchopulmonary dysplasia. J Clin Invest 1983; 72:656–666.
15. Collaborative European Multicenter Study Group. Factors influencing the clinical response to surfactant replacement therapy in babies with severe respiratory distress. Eur J Pediatr 1991; 150:433–439.
16. King RJ. Pulmonary surfactant. J Appl Physiol 1982; 53:1–8.
17. Baughman RP, Mangels DJ, Strohofer S, Corser BC. Enhancement of macrophage and monocyte cytotoxicity by the surface active material of the lung lining fluid. J Lab Clin Med 1987; 109:692–697.

18. Lewis J, Ikegami M, Jobe A. Surfactant metabolism and surfactant aggregate pool sizes and function in a lung injury model. Clin Res 1990; 38:204a.
19. Hallman M, Spragg RG, Harrell JH, Moser KM, Gluck L. Evidence of lung surfactant abnormality in respiratory failure: study of bronchoalveolar lavage phospholipids, surface activity, phospholipase activity, and plasma myoinositol. J Clin Invest 1982; 70:673–683.
20. Shelley SA, Balis JU, Paciga JE, Espinoza CG, Richman AV. Biochemical composition of adult human lung surfactant. Lung 1982; 160:195–206.
21. Yu S, Harding GR, Smith M, Possmayer F. Bovine pulmonary surfactant: chemical composition and physical properties. Lipids 1983; 18:522–529.
22. Suzuki Y, Fujita Y, Kogishi K. Reconstitution of tubular myelin from synthetic lipids and proteins associated with pig pulmonary surfactant. Am Rev Respir Dis 1989; 140:75–81.
23. Kuroki Y, Mason RJ, Voelker DR. Alveolar type II cells express a high-affinity receptor for pulmonary surfactant protein A. Proc Natl Acad Sci USA 1988; 85:5566–5570.
24. Dobbs LG, Wright JR, Hawgood S, Gonzalez R, Venstrom K, Nellenbogen J. Pulmonary surfactant and its components inhibit secretion of phosphatidylcholine from cultured rat alveolar type II cells. Proc Natl Acad Sci USA 1987; 84:1010–1014.
25. Wright JR, Wager RE, Hamilton RL, Huang M, Clements JA. Uptake of lung surfactant subfractions into lamellar bodies of adult rabbit lungs. J Appl Physiol 1986; 60:817–825.
26. Possmayer F. A proposed nomenclature for pulmonary surfactant-associated proteins. Am Rev Respir Dis 1988; 138:990–998.
27. Revak SD, Merritt TA, Hallman M, Cochrane CG. Reconstitution of surfactant activity using purified human apoprotein and phospholipids measured in vitro and in vivo. Am Rev Respir Dis 1986; 134:1258–1265.
28. Shiffer K, Hawgood S, Duzgunes N, Goerke J. Interactions of the low molecular weight group of surfactant-associated proteins (SP 5-18) with pulmonary surfactant lipids. Biochemistry 1988; 27:2689–2695.
29. Cochrane CG, Revak SD. Pulmonary surfactant protein (SP-B): structure-function relationships. Science 1992; 254:566–568.
30. Vincent JS, Revak SD, Cochrane CG, Levin IW. Raman spectroscopic studies of model human pulmonary surfactant systems: phospholipid interactions with peptide paradigms for the surfactant protein SP-B. Biochemistry 1991; 30:8395–8401.
31. Takahashi A, Waring AJ, Amirkhanian J, Fan B, Taeusch HW. Structure-function relationships of bovine pulmonary surfactant proteins, Sp-B and SP-C. Biochim Biophys Acta 1990; 1044:43–49.
32. Curstedt T, Johansson J, Persson P, et al. Hydrophobic surfactant-associated polypeptides: SP-C is a lipopeptide with two palmitoylated cysteine residues, whereas SP-B lacks covalently linked fatty acyl groups. Proc Natl Acad Sci USA 1990; 87:2985–2989.
33. Takahashi A, Fujiwara T. Proteolipid in bovine lung surfactant: its role in surfactant function. Biochem Biophys Res Commun 1986; 135:527–532.

34. Notter RH, Shapiro DL, Ohning B, Whitsett JA. Biophysical activity of synthetic phospholipids combined with purified lung surfactant 6000 dalton apoprotein. Chem Phys Lipids 1987; 44:1–17.
35. Whitsett JA, Ohning BL, Ross G, et al. Hydrophobic surfactant-associated protein in whole lung surfactant and its importance for biophysical activity in lung surfactant extracts used for replacement therapy. Pediatr Res 1986; 20: 460–467.
36. Hawgood S, Benson BJ, Schilling J, Damm D, Clements JA, White RT. Nucleotide and amino acid sequences of pulmonary surfactant protein SP 18 and evidence for cooperation between SP 18 and SP 28-36 in surfactant lipid adsorption. Proc Natl Acad Sci USA 1987; 84:66–70.
37. Revak SD, Merritt TA, Degryse E, et al. Use of human surfactant low molecular weight apoproteins in the reconstitution of surfactant biologic activity. J Clin Invest 1988; 81:826–833.
38. Hall SB, Venkitaraman AR, Whitsett JA, Holm BA, Notter RH. Importance of hydrophobic apoproteins as constituents of clinical exogenous surfactants. Am Rev Respir Dis 1992; 145:24–30.
39. Enhorning G, Shennan A, Possmayer F, Dunn M, Chen CP, Milligan J. Prevention of neonatal respiratory distress syndrome by tracheal instillation of surfactant: a randomized clinical trial. Pediatrics 1985; 76:145–153.
40. Fujiwara T, Chida S, Watabe Y, Maeta H, Morita T, Abe T. Artificial surfactant therapy in hyaline-membrane disease. Lancet 1980; i:55–59.
41. Collaborative European Multicenter Study Group. Surfactant replacement therapy in severe neonatal respiratory distress syndrome; an international randomized clinical trial. Pediatrics 1988; 82:683–691.
42. Crouch E, Rust K, Marienchek W, Parghi D, Chang D, Persson A. Developmental expression of pulmonary surfactant protein D (SP-D). Am J Respir Cell Mol Biol 1991; 5:13–18.
43. Baritussio A, Carraro R, Bellina J, et al. Turnover of phospholipids isolated from fractions of lung lavage fluid. J Appl Physiol 1985; 59:1055–1060.
44. Magoon MW, Wright JR, Baritussio A, et al. Subfractionation of lung surfactant: implications for metabolism and surface activity. Biochim Biophys Acta 1983; 750:18–31.
45. Gross NJ. Surfactant subtypes in experimental lung damage: radiation pneumonitis. Am J Physiol 1991; 260:L302–L310.
46. Baritussio A, Bellina L, Carraro R, et al. Heterogeneity of alveolar surfactant in the rabbit: composition, morphology and labelling of subfractions isolated by centrifugation of lung lavage. Eur J Clin Invest 1984; 14:24–29.
47. Yamada T, Ikegami M, Jobe AH. Effects of surfactant subfractions on preterm rabbit lung function. Pediatr Res 1990; 27:592–598.
48. Gross NJ, Schultz RM. Requirements for extracellular metabolism of pulmonary surfactant: tentative identification of serine protease. Am J Physiol 1992; 262:L446–L453.
49. Enhorning G. A pulsating bubble technique for evaluating pulmonary surfactant. J Appl Physiol 1977; 43:198–201.

50. Keough KMW. Physical chemistry of pulmonary surfactant in the terminal air spaces. In: Robertson B, Van Golde LMG, Batenburg JJ, eds. Pulmonary Surfactant: From Molecular Biology to Clinical Practice. 2nd ed. New York: Elsevier, 1992:109–164.
51. Goerke J. Surfactant and lung mechanics. In: Robertson B, Van Golde LMG, Batenburg JJ, eds. Pulmonary Surfactant: From Molecular Biology to Clinical Practice. 2nd ed. New York: Elsevier, 1992:165–192.
52. Park JW, Ma M, Reudi JM, Smith RM, Babior BM. The cytosolic components of the respiratory burst oxidase exist as a M(r) approximately 240,000 complex that acquires a membrane-binding site during activation of the oxidase in a cell-free system. J Biol Chem 1992; 267:17327–17332.
53. Yoshida K, Mohsenin V. Unsaturated phosphatidylcholines inhibit superoxide production in human neutrophils. Life Sci 1991; 49:1359–1365.
54. Schadelin J, Schadelin R, Mandell GL. Chemiluminescence of phagocytic cells. Crit Rev Clin Lab Sci 1981; 13:1–19.
55. Hayakawa H, Myrvik QN, St Clair RW. Pulmonary surfactant inhibits priming of rabbit alveolar macrophage. Evidence that surfactant suppresses the oxidative burst of alveolar macrophage in infant rabbits. Am Rev Respir Dis 1989; 140:1390–1397.
56. Hayakawa H, Giridhar G, Myrvik QN, Kucera L. Pulmonary surfactant phospholipids modulate priming of rabbit alveolar macrophages for oxidative responses. J Leukoc Biol 1992; 51:379–385.
57. Weber H, Heilmann P, Meyer B, Maier KL. Effect of canine surfactant protein (SP-A) on the respiratory burst of phagocytic cells [published erratum appears in FEBS Lett 1990 (Dec. 10); 276(1–2):232–233]. FEBS Lett 1990; 270:90–94.
58. Webb DSA, Jeska EI. Enhanced luminol-dependent chemiluminescence of stimulated rat alveolar macrophages by pretreatment with alveolar lining material. J Leukoc Biol 1986; 40:55–64.
59. van Iwaarden F, Welmers B, Verhoef J, Haagsman HP, van Golde LM. Pulmonary surfactant protein A enhances the host-defense mechanism of rat alveolar macrophages. Am J Respir Cell Mol Biol 1990; 2:91–98.
60. Thomassen MJ, Lee F, Meeker D, Antal J, Connors M, Wiedemann HP. Synthetic surfactant (Exodurf) inhibits endotoxin-stimulated cytokine secretion by human alveolar macrophages. Am Rev Respir Dis 1992; 145:A875.
61. Allen JN, Herzyk DJ, Wewers MD. Surfactant suppresses interleukin-1- and tumor necrosis factor-α release by human alveolar macrophages. Am Rev Respir Dis 1992; 145:A875.
62. Suziki H, Otake K, Yakuwa N, et al. Effects of artificial surfactant on human peripheral blood polymorphonuclear neutrophil function. Am Rev Respir Dis 1992; 145:A876.
63. Speer CP, Gotze B, Curstedt T, Robertson B. Phagocytic functions and tumor necrosis factor secretion of human monocytes exposed to natural porcine surfactant (Curosurf). Pediatr Res 1991; 30:69–74.
64. Tamura M, Tamura T, Tyagi SR, Lambeth JD. The superoxide-generating respiratory burst oxidase of human neutrophil plasma membrane. Phosphatidyl-

serine as an effector of the activated enzyme. J Biol Chem 1988; 263:17621–17626.
65. Gabig TG, Babior BM. The O_2^--forming oxidase responsible for the respiratory burst in human neutrophils. Properties of the solubilized enzyme. J Biol Chem 1979; 254:9070–9079.
66. Yuli I, Tomonaga A, Snyderman R. Chemoattrractant receptor functions in human polymorphonuclear leukocytes are divergently altered by membrane fluidizers. Proc Natl Acad Sci USA 1982; 79:5906–5910.
67. Newton JA, Ashwood ER, Yang KD, Augustine NH, Hill HR. Effect of pentoxifylline on developmental changes in neutrophil cell surface mobility and membrane fluidity. J Cell Physiol 1989; 140:427–31X.
68. Petty TL, Silvers WG, Paul GW, Stanford RE. Abnormalities in lung elastic properties and surfactant function in adult respiratory distress syndrome. Chest 1979; 75:571–574.
69. Pison U, Seeger W, Buchhorn R, et al. Surfactant abnormalities in patients with respiratory failure after multiple trauma. Am Rev Respir Dis 1989; 140:1033–1039.
70. Pison U, Overtacke U, Brand M, et al. Altered pulmonary surfactant in uncomplicated and septicemia-complicated courses of acute respiratory failure. J Trauma 1990; 30:19–26.
71. Gregory TJ, Longmore WJ, Moxley MA, et al. Surfactant chemical composition and biophysical activity in acute respiratory distress syndrome. J Clin Invest 1991; 88:1976–1981.
72. Holm BA, Matalon S, Finkelstein JN, Notter RH. Type II pneumocyte changes during hyperoxic lung injury and recovery. J Appl Physiol 1988; 65:2672–2678.
73. Haagsman HP, Schuurmans EA, Batenburg JJ, van Golde LM. Phospholipid synthesis in isolated alveolar type II cells exposed in vitro to paraquat and hyperoxia. Biochem J 1987; 245:119–126.
74. Esterbauer H. Lipid peroxidation products: formation, chemical properties and biological activities. In: Poli G, Cheeseman KH, Dianzani MU, Slater TF, eds. Free Radicals in Liver Injury. Oxford: IRL Press, 1985:29–47.
75. Seeger W, Lepper H, Hellmut RD, Neuhof H. Alteration of alveolar surfactant function after exposure to oxidative stress and to oxygenated and native arachidonic acid in vitro. Biochim Biophys Acta 1985; 835:58–67.
76. Gilliard N, Heldt G, Merritt TA, Pappert D, Spragg RG. Functional consequences of porcine lung surfactant oxidation. Am Rev Respir Dis 1990; 141: A635.
77. Minotti G, Aust SD. The requirement for the iron(III) in the initiation of lipid peroxidation by iron(II) and hydrogen peroxide. J Biol Chem 1987; 262:1098–1104.
78. Esterbauer H, Striegl G, Puhl H, Rotheneder M. Continuous monitoring of in vitro oxidation of human low density lipoprotein. Free Radic Res Commun 1989; 6:67–75.
79. Rosenfeld ME, Palinski W, Yla-Herttuala S, Butler S, Witztum JL. Distribution of oxidation specific lipid-protein adducts and apolipoprotein B in atherosclerotic lesions of varying severity from WHHL rabbits. Arteriosclerosis 1990; 10:336–349.

80. Pison U, Tam EK, Caughey GH, Hawgood S. Proteolytic inactivation of dog lung surfactant-associated proteins by neutrophil elastase. Biochim Biophys Acta 1989; 992:251–257.
81. Haagsman HP, Hawgood S, Sargeant T, et al. The major lung surfactant protein, SP 28-36, is a calcium-dependent, carbohydrate-binding protein. J Biol Chem 1987; 262:13877–13880.
82. Ryan SF, Ghassibi Y, Liau DF. Effects of activated polymorphonuclear leukocytes upon pulmonary surfactant in vitro. Am J Respir Cell Mol Biol 1991; 4: 33–41.
83. Spragg RG, Richman P, Gilliard N, Merritt TA. The future for surfactant therapy of the adult respiratory distress syndrome. In: Lachmann B, ed. Surfactant Replacement Therapy. New York: Springer-Verlag, 1987:203–211.
84. Li JJ, Sanders RL, McAdam KP, et al. Impact of C-reactive protein (CRP) on surfactant function. J Trauma 1989; 29:1690–1697.
85. Moses D, Holm BA, Spitale P, Liu MY, Enhorning G. Inhibition of pulmonary surfactant function by meconium. Am J Obstet Gynecol 1991; 164:477–481.
86. Holm BA, Keicher L, Liu MY, Sokolowski J, Enhorning G. Inhibition of pulmonary surfactant function by phospholipases. J Appl Physiol 1991; 71:317–321.
87. Venkitaraman AR, Hall SB, Whitsett JA, Notter RH. Enhancement of biophysical activity of lung surfactant extracts and phospholipid–epoprotein mixtures by surfactant protein A. Chem Phys Lipids 1990; 56:1–10.
88. Holm BA, Venkitaraman AR, Enhorning G, Notter RH. Biophysical inhibition of synthetic lung surfactants. Chem Phys Lipids 1990; 52:243–250.
89. Revak SD, Merritt TA, Hallman M, et al. The use of synthetic peptides in the formation of biophysically and biologically active pulmonary surfactants. Pediatr Res 1991; 29:460–465.
90. McLean LR, Krstenansky JL, Jackson RL, Hagaman KA, Olsen KA, Lewis JE. Mixtures of synthetic peptides and dipalmitoylphosphatidylcholine as lung surfactants. Am J Physiol 1992; 262:L292–L300.
91. Hallman M, Merritt TA, Pohjavuori M, Gluck L. Effect of surfactant substitution on lung effluent phospholipids in respiratory distress syndrome: evaluation of surfactant phospholipid turnover, pool size, and the relationship to severity of respiratory failure. Pediatr Res 1986; 20:1228–1235.
92. Marks LB, Notter RH, Oberdorster G, McBride JT. Ultrasonic and jet aerosolization of phospholipids and the effects on surface activity. Pediatr Res 1983; 17:742–747.
93. Robertson B. Surfactant inactivation and surfactant replacement in experimental models of ARDS. Acta Anaesthesiol Scand 1991; 35(suppl 95):22–28.
94. Matalon S, Holm BA, Notter RH. Mitigation of pulmonary hyperoxic injury by administration of exogenous surfactant. J Appl Physiol 1987; 62:756–761.
95. Lewis J, Ikegami M, Higuchi R, Jobe A, Absolom D. Nebulized vs. instilled exogenous surfactant in an adult lung injury model. J Appl Physiol 1991; 71: 1270–1276.
96. Harris JD, Jackson F Jr, Moxley MA, Longmore WJ. Effect of exogenous surfactant instillation on experimental acute lung injury. J Appl Physiol 1989; 66: 1846–1851.

97. Berggren P, Lachmann B, Curstedt T, Grossmann G, Robertson B. Gas exchange and lung morphology after surfactant replacement in experimental adult respiratory distress syndrome induced by repeated lung lavage. Acta Anaesthesiol Scand 1986; 30:321–328.
98. Kobayashi T, Kataoka H, Ueda T, Murakami S, Takada Y, Kokubo M. Effects of surfactant supplement and end-expiratory pressure in lung-lavaged rabbits. J Appl Physiol 1984; 57:995–1001.
99. Lewis JF, Tabor B, Ikegami M, Jobe A, Joseph M, Absalom D. Lung function and surfactant distribution in saline lavaged sheep given instilled vs nebulized surfactant. J Appl Physiol 1993; 74:1256–1264.
100. Kobayashi T, Ganzuka M, Taniguchi J, Nitta K, Murakami S. Lung lavage and surfactant replacement for hydrochloric acid aspiration in rabbits. Acta Anaesthesiol Scand 1990; 34:216–221.
101. Lachmann B. Animal models and clinical pilot studies of surfactant replacement in adult respiratory distress syndrome. Eur Respir J Suppl 1989; 3:98s–103s.
102. Berry D, Ikegami M, Jobe A. Respiratory distress and surfactant inhibition following vagotomy in rabbits. Am J Physiol 1986; 61:1741–1748.
103. Gilliard N, Richman PM, Merritt TA, Spragg RG. Effect of volume and dose on the pulmonary distribution of exogenous surfactant administered to normal rabbits or to rabbits with oleic acid lung injury. Am Rev Respir Dis 1990; 141:743–747.
104. Lewis JF, Ikegami M, Jobe AH, Tabor B. Aerosolized surfactant treatment of preterm lambs. J Appl Physioll 1991; 70:869–876.
105. Lewis J, Ikegami M, Tabor B, Jobe A, Absalom D. Aerosolized surfactant is preferentially deposited in normal versus injured regions of lung in a heterogenous lung injury model. Am Rev Respir Dis 1992; 145:A184.
106. Weg J, Reines H, Balk R, et al. Safety and efficacy of an aerosolized surfactant (Exosurf) in human sepsis-induced ARDS. Chest 1992; 100:137S.
107. Obladen M. History of surfactant research. In: Robertson B, Van Golde LMG, Batenburg JJ, eds. Pulmonary Surfactant: From Molecular Biology to Clinical Practice. 2nd ed. New York: Elsevier, 1992:1–18.
108. Smith BT. Clinical experience with modified natural surfactant. In: Robertson B, Van Golde LMG, Batenburg JJ, eds. Pulmonary Surfactant: From Molecular Biology to Clinical Practice. 2nd ed. New York: Elsevier, 1992: 593–604.
109. Lachmann B. Surfactant replacement in acute respiratory failure: animal studies and first clinical trials. In: Lachmann B, ed. Surfactant Replacement Therapy. New York: Springer-Verlag, 1987:212–223.
110. Nosaka S, Sakai T, Yonekura M, Yoshikawa K. Surfactant for adults with respiratory failure. Lancet 1990; 336:947–948.
111. Wiedemann H, Baughman R, deBoisblanc B, et al. A multicenter trial in human sepsis-induced ARDS of an aerosolized synthetic surfactant (Exosurf). Am Rev Respir Dis 1992; 145:A184.
112. Davis JM, Veness-Meehan K, Notter RH, Bhutani VK, Kendig JW, Shapiro DL. Changes in pulmonary mechanics after the administration of surfactant

to infants with respiratory distress syndrome. N Engl J Med 1988; 319:476–479.

113. Bhat R, Dziedzic K, Bhutani VK, Vidyasagar D. Effect of single dose surfactant on pulmonary function. Crit Care Med 1990; 18:590–595.
114. Richman PS, Spragg RG, Robertson B, Merritt TA, Curstedt T. The adult respiratory distress syndrome: first trials with surfactant replacement. Eur Respir J Suppl 1989; 3:109s–111s.
115. Spragg RG, Gilliard N, Richman P, et al. The adult respiratory distress syndrome: clinical aspects relevant to surfactant supplementation. In: Robertson B, Van Golde LMG, Batenburg JJ, eds. Pulmonary Surfactant: From Molecular Biology to Clinical Practice. 2nd ed. New York: Elsevier, 1992:685–703.
116. Petty TL, Reiss OK, Paul GW, Silvers GW, Elkins ND. Characteristics of pulmonary surfactant in adult respiratory distress syndrome associated with trauma and shock. Am Rev Respir Dis 1977; 115:531–536.
117. Hallman M, Merritt TA, Schneider H, et al. Isolation of human surfactant from amniotic fluid and a pilot study of its efficacy in respiratory distress syndrome. Pediatrics 1983; 71:473–482.
118. Robertson B, Berggren P, Curstedt T, et al. Clinical trials of natural surfactant phospholipids in the treatment of the neonatal respiratory distress syndrome (RDS). In: Lucey JF, ed. Ross Laboratories Special Conference: Clinical Trials of the Use of Surfactant in the Neonatal Respiratory Distress Syndrome. Columbus, OH: Ross Laboratories, 1984:145–156.
119. Durand DJ, Clyman RI, Heymann MA, et al. Effects of a protein-free, synthetic surfactant on survival and pulmonary function in preterm lambs. J Pediatr 1985; 107:775–780.

17

Highly Invasive Therapies for Acute Lung Injury

JESUS VILLAR

Mount Sinai Hospital
Toronto, Ontario, Canada

ARTHUR S. SLUTSKY

University of Toronto
and Mount Sinai Hospital
Toronto, Ontario, Canada

Air in the body we call breath;
outside we call wind.

Hippocrates (460–377 B.C.)

I. Introduction

Acute respiratory failure due to acute lung injury (ALI) is one of the most common problems in patients admitted into intensive care units (ICUs). Traditionally, mechanical ventilation has been a supportive, not a curative form of therapy, providing adequate oxygenation and carbon dioxide elimination. Mechanical ventilation is limited by (1) the alveolar exchange surface, which must be adequately ventilated and perfused; and (2) the detrimental effects of high inflation pressures and high concentration of inspired oxygen, which can damage the lung. These factors have been under intense scrutiny over the past decade. Despite substantial progress in the development of supportive therapy, the mortality rate of ALI remains at 50 to 70% (1), and even higher when prolonged mechanical ventilation is required. Much of the research effort directed to improving the prognosis of patients with ALI has been directed toward improving the means of supporting the injured lung. Mechanical support provides time for disease-specific therapy until remodeling of lung architecture can take place and eventual recovery of respiratory function. At the present time, there is no evidence that any specific ventilatory therapy has a prophylactic effect against the development of ALI (2).

Mechanical ventilation and hyperinflation, even at normal pressures, may cause severe lung injury in previously healthy lungs and augment lung injury in the diseased lung (3). The adverse effects of conventional mechanical ventilation are the direct consequence of pulmonary pressures and/or volume changes induced by cyclic mechanical insufflation of the diseased lungs. The most important disadvantages of this therapy include reduction in cardiac output, barotrauma, water retention, reduction in pulmonary lymph flow and organ blood flow, and ventilation–perfusion mismatch. It is now known that the lung in severe ALI is nonhomogeneous, with the normally aerated lung comprising as little as 20 to 30% of the normal lung volume (4). Since the total ventilation occurs in this small portion of functional lung, the ventilation of this "small" lung may be up to five times greater than normal. In addition to the high peak pressures and hyperventilation, the use of high concentrations of oxygen is another important damaging factor during mechanical ventilation.

The purpose of this chapter is to review a number of alternative techniques of oxygenation and ventilation that have been developed to minimize the adverse effects of conventional mechanical ventilation by resting the lung, reducing the mechanical stress, and maintaining effective blood volume. Although these new approaches might facilitate lung repair, they are in general more laborious and highly invasive. Some of them are currently in use (extracorporeal lung support), others have been indicated for the treatment of different diseases (lung transplant), and others represent promising experimental techniques that may become feasible therapeutic modalities in the future (intravascular oxygenation, liquid respiration).

II. Extracorporeal Life Support

Interest in resting the lungs has produced an increase in the number of strategies for achieving extrapulmonary gas exchange. These techniques of extracorporeal lung support were originally designed to support patients with acute respiratory failure of such severity that it was impossible to provide adequate oxygenation by conventional mechanical ventilation.

Extracorporeal life support with an artificial lung was first introduced in 1885 as a pump oxygenator in patients with acute respiratory failure (5). Since 1951, extracorporeal oxygenation has become a routine procedure during cardiopulmonary bypass in patients undergoing open heart surgery (6). The use of early oxygenators was limited to only a few hours of support since the oxygenator itself caused injury to blood cells, resulting in hemolysis and platelet destruction. The use of artificial lungs for longer periods of time was not possible until the introduction of the membrane oxygenator, which avoided direct contact between gas and blood.

There are two components of extracorporeal gas exchange: the transfer of oxygen into, and carbon dioxide out of, the extracorporeal blood. The extent to which these two largely independent components contribute to the success of the technique depends on its indications. These techniques are limited to a few centers since they are extremely labor intensive and require special expertise.

A. Extracorporeal Membrane Oxygenation

The main objective of extracorporeal membrane oxygenation (ECMO) is to provide immediate improvement in the oxygenation of the arterial blood (7). Unfortunately, since at any given hemoglobin concentration, the oxygen content of the blood leaving an artificial membrane lung is limited by the oxyhemoglobin disociation curve and the dissolved oxygen, adequate oxygenation requires extracorporeal blood flow (ECBF) rates exceeding 60% of normal cardiac output. Thus, to oxygenate the blood adequately without using the gas-exchanging surface of the lungs, most of the cardiac output must go through the ECMO circuit. During ECMO some CO_2 is removed by the extracorporeal circuit, but this CO_2 removal is usually supplemented with conventional mechanical ventilation at ventilatory rates between 8 and 12 breaths per minute, FiO_2 less than 0.5, and with low tidal volumes to maintain a peak airway pressure below 35 cmH_2O.

The first long-term bypasses were performed in the venoarterial mode (which partially bypasses the pulmonary vascular bed) (8). Blood is drained from a catheter advanced into the femoral vein up to the level of the vena cava. Arterialized blood is then returned to the patient's arterial system. Because venoarterial bypass reduces pulmonary blood flow, which may lead to stasis and thrombosis, this approach may result in lung tissue ischemia and hence compromise lung repair (9). Most long-term bypasses since the original ECMO study (8) have been performed in the venovenous mode (which preserves blood flow through the pulmonary vasculature), using silicone spiral-coiled membrane lungs, using up to 9 m^2 of total surface area when applied to adult patients (10,11) (Fig. 1). Several kinds of vessel cannulations are possible. Venovenous bypass uses the femoral, saphenous, or jugular veins for access, and may also be performed percutaneously. The inner diameter of the drainage catheters, which varies from 6 to 11 mm, is the main determinant of the maximal ECBF obtainable. The drainage line delivers blood from the patient, where it passes through an oximeter, to a

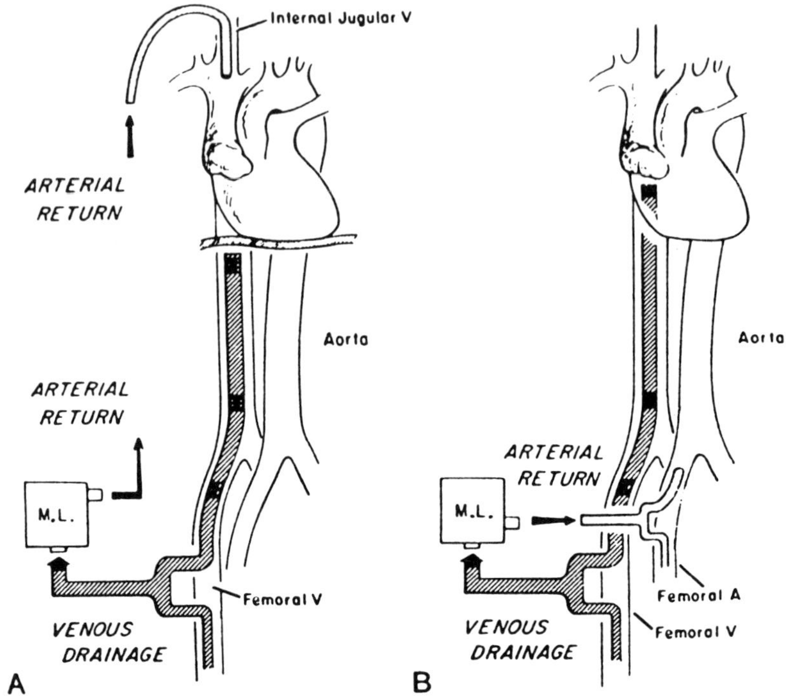

Figure 1 Various strategies for ECMO: (A) venovenous perfusion route; (B) venoarterial perfusion route with femoral artery cannulation and distal aortic return. ML, membrane lung. (From Ref. 11.)

roller pump that propels blood through the membrane oxygenator. In addition, a silicone rubber bubble trap and an outflow oximeter complete the circuit (7).

Bleeding due to anticoagulation is the major complication of extracorporeal support. A single vein bypass using a double-lumen catheter inserted into the femoral vein may reduce the incidence of bleeding (the outer catheter used for drainage, the inner catheter for returning the blood) (12). The need for anticoagulation with standard management using the spiral-coil membrane lung results in a substantial risk of bleeding from any surgical procedure (cannulation sites, chest tubes, etc.). The development of new heparinized lungs will hopefully reduce the extent of bleeding (13). Other major causes of morbidity and mortality associated with the use of ECMO include disseminated intravascular coagulopathy, sepsis, and brain hemorrhage.

ECMO in Adults with Respiratory Failure

ECMO was first used in the late 1950s and early 1960s in adults undergoing cardiac surgery. With the development of the membrane oxygenator, patients could be supported for prolonged periods with cardiopulmonary bypass, and several investigators applied this technique to treat patients with severe ALI. The rationale for the use of an artificial lung in the treatment of patients with severe ALI was multifactorial: (1) to correct the hypoxia and hypercapnia, from which patients with severe ALI often died 20 years ago (14); (2) to reduce ventilatory pressures; (3) to decrease inspired oxygen concentrations; and (4) to enable the lung to heal while other tissues were being perfused (11). Hill and colleagues (15,16) were the first to report the successful use of ECMO in patients with severe acute respiratory failure. Following this initial report of success, ECMO was used in some centers for the treatment of patients with ALI that were impossible to oxygenate by conventional mechanical ventilatory support. In 1976, Gille and Bagniewski (17) reviewed the worldwide experience with ECMO for severe ALI in 233 patients and reported a 15% overall survival rate.

In view of the large number of anecdotal reports, from 1974 to 1976, the National Heart, Lung, and Blood Institute in the United States supported a multicenter trial in patients with severe ALI to compare extracorporeal venoarterial support plus conventional mechanical ventilation with conventional mechanical ventilation alone (8,18,19). This prospective randomized trial involved nine centers and enrolled 90 patients with respiratory failure of various etiologies (e.g., pneumonia, pulmonary embolism, multiple trauma) during a 2½-year period. Forty-eight patients were treated with mechanical ventilation alone, and 42 had their treatment augmented with venoarterial ECMO employing a high flow system (approximately 90% of the cardiac output). Conventional ventilatory support included application of positive end-expiratory pressure (PEEP), enriched oxygen, and tidal volumes of at least 0.5 L at a rate of 15 breaths per minute. The entry criteria for the study were based on gas exchange during standard ventilatory conditions (18). The "fast" entry criteria included patients with $PaO_2 < 50$ mmHg for more than 2 h when ventilated with an FiO_2 of 1 and a PEEP ≥ 5 cmH_2O; the "slow" entry criteria included patients with $PaO_2 < 50$ mmHg for more than 12 h when ventilated with $FiO_2 \geq 0.6$ and PEEP ≥ 5 cmH_2O. Patients were excluded if the pulmonary insult was longer than 21 days in duration, if the wedge pressure was greater than 25 mmHg, or if there was chronic or irreversible disease of other organs. In this study, ECMO was associated with frequent complications (e.g., bleeding, sepsis) and was very labor intensive. Survival rates for the two groups were poor and virtually the same (8.3 versus 9.5%).

Despite the excitement that had been generated by earlier reports of success, the results of the ECMO trial led to a loss of enthusiasm for the use of ECMO support in acute respiratory failure. However, a number of issues have been raised regarding the applicability of these results to current practice. The study design was such that ECMO was employed not as an alternative to more traditional means of support but as a rescue therapy when there was essentially no alternative. In addition, the limits of conventional treatment defined in the study, especially the level of PEEP that was used, are very different from those that are acceptable today, when it is not unusual to use levels of PEEP up to 15 to 20 cmH_2O. In addition, there have been many significant advances in our knowledge about the pathophysiology of ALI, improvements in life-support technology, and better nursing care in the last 20 years. On the other hand, currently, most patients with severe ALI die not from hypoxemia but from nonpulmonary organ failure, in particular, sepsis (20).

Despite the dissapointing results of the ECMO trial, a few centers in North America continue to use it in selected circumstances. Some investigators believe that there exists a role for long-term ECMO in patients who are deemed to have potentially reversible pulmonary dysfunction. Egan and colleagues from Toronto have demonstrated that ECMO is a feasible technique for candidates undergoing lung transplantation (21). The same group of investigators reported their experience for the period 1978–1987 in 17 patients with respiratory failure from diverse causes (7). Four patients were treated with venoarterial perfusion and 13 with the venovenous approach. The reported survival rate (24%) was better than the survival rates in the original ECMO report.

By contrast, the development of extracorporeal techniques for ALI continued in Europe, where venovenous ECMO proved to be an alternative form of treatment (22–26). In the past, this invasive procedure required almost complete systemic anticoagulation using heparin; however, recently extracorporeal systems whose blood-contacting surfaces are coated with covalently bound heparin are used, reducing the need for systemic anticoagulation (24). Lewandowski and colleagues (26) from Berlin have reported a 54% survival rate in 24 patients with severe ALI treated with venovenous ECMO. The patients had the same entry criteria as patients in the U.S. ECMO trial. Although this was a nonrandomized study, the reported high survival rate represents an impressive outcome that must be examined in a randomized fashion.

ECMO in Neonates and Children

Far more impressive than its performance in adults has been the success of ECMO in newborns and infants with respiratory failure. Neonatologists, pediatric intensivists, and surgeons in United States were less discouraged by the results of the adult ECMO study. They continued to treat neonates with severe acute respiratory failure, and currently it is an accepted technique, albeit untested in a randomized study, that at the time of writing (1992) has been performed in more than 6000 pediatric patients (in over 80 centers in North America alone) as a mode of cardiopulmonary support (27–31). By contrast, only a few neonates or infants have so far been treated in Europe (32).

The pioneering work in this area was performed by White et al. (27) and Bartlett (33), who in the early 1970s began using ECMO in term infants with respiratory failure who had a predicted mortality greater than 80%. They obtained a 70% survival rate (33). Since then, more than 5000 neonates with respiratory failure unresponsive to conventional mechanical ventilation have been treated with ECMO, resulting in an overall survival rate of 83% (30).

Respiratory failure remains one of the most common causes of morbidity and mortality in neonates. The majority of patients develop respiratory failure from hyaline membrane disease, persistent pulmonary hypertension, pneumonia, meconium aspiration syndrome, sepsis, or congenital diaphragmatic hernia (34). The rationale of neonatal ECMO is to provide temporary cardiorespiratory support, allowing the lungs to heal without the damaging effects of high airway pressures and high inspired oxygen tensions (35). Typical ventilatory settings while on ECMO (36) are rate of 5 to 10 breaths/minute, peak pressures between 20 and 25 cmH_2O, PEEP 8 to 14 mmHg, and FiO_2 of 0.21 to 0.30. Criteria for the use of neonatal ECMO are well established. Infants must have a gestational age > 35 weeks; a birth weight > 2 kg; absence of, or grade I intraventricular hemorrhage; absence of severe and uncorrectable coagulopathy; and absence of congenital anomalies incompatible with life (37). Although these criteria exclude most preterm infants with the respiratory distress syndrome, they select a group of patients with a predicted mortality rate greater than 80%. The two measures most commonly used to select neonates for ECMO are the alveolar–arterial difference of oxygen ($A\text{-}aDO_2$) and the oxygenation index [(mean airway pressure × FiO_2)/PaO_2]. An $A\text{-}aDO_2$ value > 610 mmHg for more than 8 h or an oxigenation index >40 have been associated with 80% mortality (37,38).

Most neonatal ECMO uses a venoarterial approach, as shown in Fig. 2. A catheter is inserted into the right internal jugular vein and its tip is advanced to the right atrium for venous drainage. Desaturated blood flows by gravity to a roller-head pump, which delivers blood in a nonpulsatile fashion to the membrane oxygenator (0.6 to 0.8 m^2), where gas exchange takes place. Oxygenated blood is perfused back into the aortic arch via a catheter inserted into the right common carotid artery. A venoarterial circuit provides both

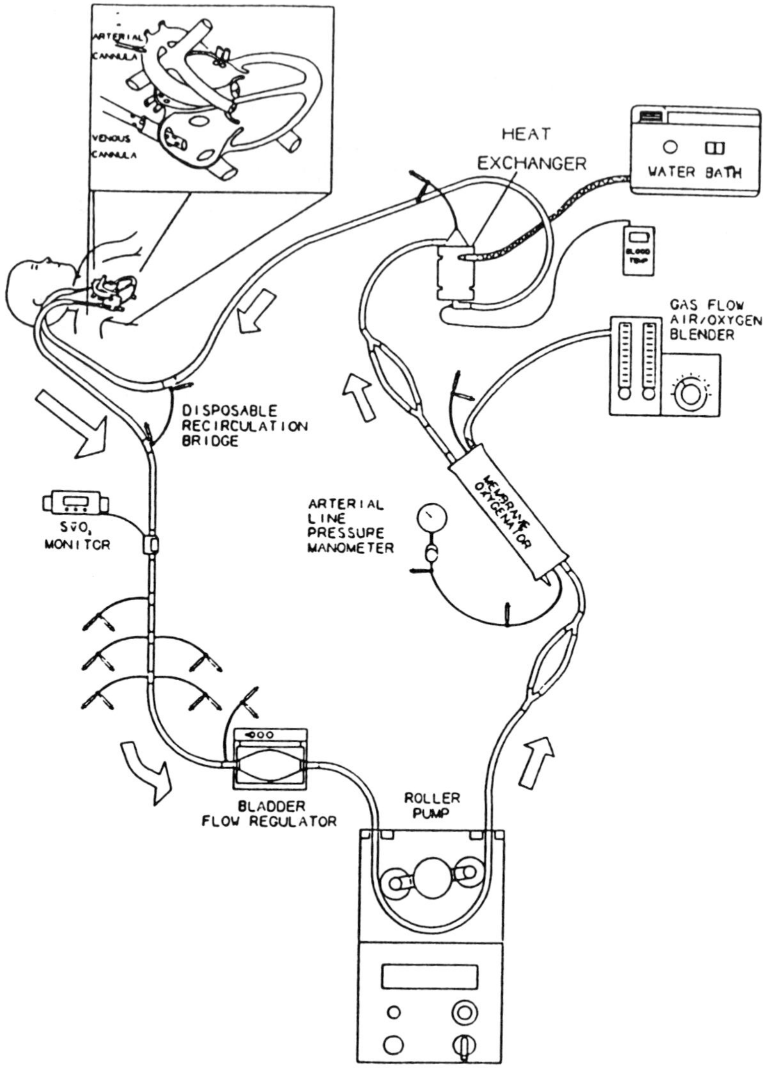

Figure 2 Venoarterial ECMO circuit for neonates. (From Ref. 37.)

cardiac and pulmonary support. The venovenous approach provides gas exchange but no cardiac support. In this case, desaturated blood is removed from the right atrium and the saturated blood is returned to the femoral vein after transit through the ECMO circuit. The use of venoarterial or venovenous depends primarily on the need for cardiac support as well as vascular accessibility. ECMO is discontinued when significant recovery of pulmonary function occurs and the infant can be treated with mechanical ventilation only. The duration of ECMO support for lung failure can be as short as 3 to 4 days for an uncomplicated meconium aspiration syndrome, and as long as 3 to 4 weeks for a severe congenital diaphragmatic hernia (31). Koumbourlis and associates have found that improvement in the clinical condition and oxygenation permitting successful weaning from ECMO are achieved before clinically important improvements in lung mechanics (39).

Systemic heparinization for anticoagulation places all patients on ECMO at risk for bleeding. Other major complications include intracranial hemorrhage due an increase in cerebral blood flow velocity and vasodilation of the cerebral vessels (40); a decrease in ionized calcium due to inadequate calcium priming, which is probably responsible for the occurrence of hypotensive episodes commonly seen during ECMO (41); and neurologic sequelae. With the increasing survival rate of infants with severe acute lung disease, a number of these patients treated with ECMO have been identified as having chronic lung disease (42). Factors contributing to the development of chronic lung injury are probably related to the underlying disease and duration of the illness. Although approximately 15 to 25% of ECMO survivors suffer clinical neurologic sequelae, these complication rates do not differ significantly from those associated with conventional therapy (43). Certainly, carotid artery ligation may not be a benign event for the newborn, but repair is now feasible and is being performed in some centers, despite lack of evidence of any clear benefit (44).

Although extracorporeal life support has achieved acceptance in neonatology, improvements in our understanding of the pathophysiology of respiratory failure in the newborn will make these techniques less necessary in the future. Different ventilatory techniques, particularly high-frequency oscillatory ventilation (45) and surfactant replacement therapy (46), have shown promise as new treatment modalities in the management of term and preterm neonates with severe respiratory failure.

The ability to support neonates with respiratory failure using ECMO has prompted its use for respiratory and cardiac support in older children with severe myocardial dysfunction following cardiac surgery or congenital heart disease (47), in anticipation of heart transplantation (37), and in severe

ALI caused by diverse etiologies, including pneumonia and sepsis (31,48, 49). The techniques used for pediatric ECMO are essentially the same as those used for neonatal ECMO, except that in many instances, the venovenous approach is preferred. The reported survival rates in the pediatric population treated with ECMO are between 40 and 50%, lower than those found in neonatal respiratory failure. Extracorporeal life support for pediatric respiratory and cardiac failure must be studied rigorously since little is known about factors predicting death in these patients suffering from a multitude of disease processes.

B. Extracorporeal Carbon Dioxide Removal

As described above, one of the major problems with the use of ECMO is that more than 60% of the cardiac output has to flow through the membrane to achieve adequate arterial oxygenation. Since many of the detrimental effects of ECMO are related to the residence time of the blood in the oxygenator (reduction in pulmonary blood flow, increased tendency for bleeding, activation of neutrophils), a technique in which less of the cardiac output flows through the oxygenator would be useful. Such a technique was developed by Kolobow and co-workers at the National Heart, Lung, and Blood Institute, working in conjunction with Gattinoni from Milan (50–52). With this technique, termed extracorporeal carbon dioxide removal ($ECCO_2R$) (52), it was possible to decrease mechanical ventilation to zero while augmentation of oxygenation was achieved via the lungs by intratracheal insufflation of oxygen, apneic oxygenation, or low levels of conventional ventilation (53).

With $ECCO_2R$, total elimination of metabolic carbon dioxide can be accomplished with an ECBF of about 20 to 30% of the cardiac output. By contrast to ECMO, adequate carbon dioxide removal can occur with this small blood flow because of the shape and position of the carbon dioxide dissociation curve. This avoids the severe reduction in pulmonary blood flow associated with ECMO that may result in lung ischemia (9,54).

$ECCO_2R$ refers only to the venovenous bypass mode. Most long-term bypasses in adults have been performed with a roller pump; only occasionally have centrifugal pumps been used (10). The alveolar gas composition is determined by the composition and flow of gas ventilating the oxygenator, and is identical in all ventilated alveoli, ventilation-to-perfusion ratio inequality thus being eliminated (55). The objective in $ECCO_2R$ is to reduce the need for alveolar ventilation, thereby disminishing the potential for further pulmonary injury due to barotrauma. Partial $ECCO_2R$ indicates partial removal of the CO_2 produced with a venovenous circulation, with blood flow at as little as 0.8 to 1 L/min and only one gas exchange membrane. Total $ECCO_2R$ requires higher flow rates (1 to 2 L/min) and the use of two membranes in series (13).

In 1978, the same group of investigators found that it was advantageous to use low-frequency positive-pressure ventilation (LFPPV) at 3 to 4 breaths per minute at limited peak pressure to avoid the progressive atelectasis observed during apnea (56,57). These authors postulated that the application of $ECCO_2R$ with LFPPV (LFPPV-$ECCO_2R$) could promote healing of the lungs by avoiding the deleterious effects of high volumes and pressures used in patients with severe ALI. In this mode, the patients's lungs are inflated with PEEP ranging from 15 to 25 cmH_2O and ventilated 3 to 5 times per minute, with the peak airway pressure limited to 35 to 40 cmH_2O in order to provide intermittent stretching of the lung (58). With these conditions, mean airway pressure approximated PEEP. During $ECCO_2R$, some oxygen is provided by the artificial lung, but to provide for oxygen consumed during the long end-expiratory pause, a continuous flow of 1 to 2 L/min of O_2 is directed into the trachea through a small catheter advanced through a side port in the tracheal tube connector (Fig. 3).

Although $ECCO_2R$ is currently performed in more than 20 centers worldwide, there are no uniform entry criteria for $ECCO_2R$. The entry criteria are based mainly on gas exchange values during standard ventilatory conditions, and in most cases, it is applied only as a last resort, when conventional mechanical ventilation fails.

$ECCO_2R$ in Adults

Originally, $ECCO_2R$ was designed as an extracorporeal technique for the treatment of adult patients with severe ALI. At the time of writing, almost 300 patients have been treated with $ECCO_2R$ worldwide, with an impressive survival rate between 45 and 55% (59) in patients with a predicted mortality rate greater than 90% based on the ECMO study (8). Although an European registry is not yet available, the most experience comes from Europe. Since published studies are generally uncontrolled trials, it is important to point out a number of limitations in interpretation of the results. The major issue relates to the use of a historic control group, since simple differences in patient selection or changes in patient care could explain the differences in mortality. Second, patients from the Milanese study were treated with a sequential three-step therapeutic program, employing pressure-controlled inverse ratio ventilation, continuous positive airway pressure (CPAP), and LFPPV-$ECCO_2R$, a protocol markedly different from that used in the original ECMO trial. Third, the initiation of LFPPV-$ECCO_2R$ was always ac-

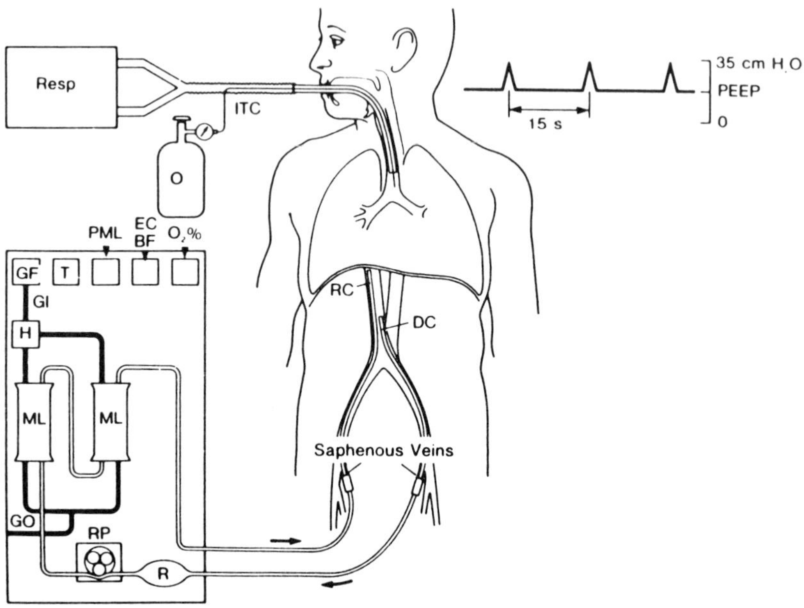

Figure 3 Cannulation and perfusion circuit for extracorporeal CO_2 removal. DC, blood drainage catheter; ECBF, extracorporeal blood flow; GF, gas flow monitor; GI, gas inlet; GO, gas outlet; H, humidifier; ITC, intratracheal catheter; ML, membrane lung; O_2%, venous drainage blood oxygen monitor; PEEP, positive end-expiratory pressure; PML, membrane lung pressure, in–out; R, venous reservoir; RC, blood return catheter; Resp, respiratory; RP, roller pump; T, ambient temperature control. (From Ref. 57.)

companied by furosemide administration; in the ECMO study, extracellular fluid volume overexpansion was treated aggressively prior to reaching a decision on whether to proceed with therapy using ECMO. Thus it is possible that the actual entry criteria between the two studies may have been somewhat different.

Two kinds of artificial lungs have been used in most of the long-term $ECCO_2R$ systems in adults: some centers use only the surface-heparinized artificial lung (Carmeda, Stockholm, Sweden) (60), some use the silicone spiral-coil membrane lung, and others have experience with both devices. Over the past 15 years, different techniques of cannulation have been used. Percutaneous cannulation using the femoral–femoral or femoral–jugular approach is currently the technique of choice, since the catheters are easily inserted or removed and the risk of bleeding is minimized.

Using animal models, LFPPV-$ECCO_2R$ prevents the hemodynamic and renal complications reported during mechanical ventilation (61) and does not cause lung damage in normal lungs (62). Borelli and co-workers have shown that in an animal model of high-pressure ventilation-induced lung injury, all animals with severe acute respiratory failure died with either conventional mechanical ventilation or CPAP with $ECCO_2R$, whereas $ECCO_2R$ was superior to mechanical ventilation in animals with mild respiratory failure (63).

To evaluate the utility of extracorporeal strategies in managing patients with severe ALI, Morris and colleagues from Salt Lake City, Utah, initiated a randomized controlled trial in 1987 comparing the efficacy of a three-step program (which included $ECCO_2R$) with that of a conventional mechanical ventilation strategy (64). This was a highly standardized, protocol-controlled approach to ensure uniformity of care with equal intensity and frequency of monitoring, and consistent decision-making logic for all randomized patients from the time of randomization to extubation or death, regardless of the support mode used. The trial was completed in 1992 and has recently been published (65). From a population of 249 patients with severe ALI, 40 patients were enrolled in the clinical trial (65). Nineteen patients were assigned to conventional mechanical ventilation and 21 to LFPPV-$ECCO_2R$. Although the overall (38%) and group survival were statistically different from those of the previous ECMO trial, survival between the two groups (42% versus 33%) was not statistically different. Therefore, the role of extracorporeal support for acute respiratory failure is uncertain.

$ECCO_2R$ in Neonates and Children

Newborns who develop hyaline membrane disease (HMD) usually do not have severe pulmonary dysfunction at the time of birth. Similarly, the pathology of HMD is never found in utero, but only following birth after pulmonary ventilation has been established, suggesting that pulmonary ventilation may play a major role in the development of HMD. Maximum therapy varies among centers, but it may include pressure-controlled ventilation, permissive hypercapnia, high-frequency ventilation, topical surfactants, and ECMO.

Pesenti et al. (66) hypothesized that by keeping the lungs motionless and inflated it should be possible to avoid the sequence that almost always results in respiratory failure and death in premature animals and neonates. These investigators used $ECCO_2R$ in the management of HMD in fetal lambs, and showed that $ECCO_2R$ combined with tracheal insufflation of oxygen could prevent the development of HMD. After 6 to 66 h of extracorporeal support, the lungs had cleared sufficiently to allow normal pulmonary ventilation in 10 of 11 lambs. In a control group treated with mechanical ventilation alone, 5 of 7 lambs died of severe HMD within the first 24 h (66). As indicated in their paper, $ECCO_2R$ was able "to buy time" for the immature lungs to reach a state compatible with adequate pulmonary ventilation, since the fetal lungs have sufficient vascularity and airspaces to sustain potential effective pulmonary gas exchange before the fetus is considered viable (67). These investigators (66) chose a venovenous bypass using the external jugular and umbilical veins. This technique caused minimal hemodynamic derangement and allowed the lungs to be perfused with oxygenated blood at normal PCO_2, resulting in a decrease in pulmonary vascular resistances and a rise in pulmonary blood flow.

These observations were confirmed by Dorrington et al. (68) a few years later in a randomized trial in a surfactant-depleted rabbit model of HMD. By permitting oxygenation to take place via the animal's own lungs rather than via the extracorporeal membrane lung, they observed that $ECCO_2R$ combined with tracheal insufflation of oxygen prevented the deterioration of lung function and prevented the development of hyaline membranes, which are often markers of alveolar mechanical disruption (69). This study provides further evidence that repeated mechanical inflation of surfactant-deficient lungs is one step in the process leading to alveolar exudates (68).

Chevalier et al. (70,71) from Paris have recently published their experience with $ECCO_2R$ in 91 neonates and 53 children with severe respiratory failure, reporting a survival rate of 85% in neonates and 55% in children with severe ALI. The venovenous extracorporeal perfusion circuit consists of a nonocclusive roller pump, a heat exchanger, a liquid heater, and a membrane lung with a cross-sectional area of 0.8 to 4.5 m^2. The novel feature of this circuit is that the same cannula inserted through the right internal jugular vein is used for drainage and injection, and the flow is generated by an alternating clamp and by the roller pump, which turns at a constant rotor speed (70). The flow is started at 10% of cardiac output and gradually increased to 30 to 40% while oxygenation is performed by tracheal insufflation of oxygen through a catheter positioned at the level of the carina. To prevent the lungs from alveolar collapse, lungs were mechanically inflated 4 to 6 times/min at a maximum pressure of 35 cmH_2O. Independent of the outcome, satisfactory gas exchange was achieved rapidly in all patients.

Other Devices for Extracorporeal CO_2 Removal

Extracorporeal Lung Assist

The term *extracorporeal lung assist* (ECLA), introduced by Japanese investigators (72,73), emphasizes neither oxygen nor CO_2 transfer. With this technique of partial ventilatory support, spontaneous ventilation is assisted with an extracorporeal artificial lung using low ECBF rates and without endotracheal intubation and mechanical ventilation. The membrane lung is ventilated with a humidified gas mixture of about 40% oxygen in nitrogen through an oxygen blender. Terasaki et al. (72) used this technique in two patients with CO_2 narcosis after spontaneous pneumothoraces to decrease the patient's respiratory effort, in the hope that pulmonary collapse and quiet respiration would allow the air leak to close spontaneously. ECLA was instituted using a venovenous jugulofemoral circuit with a membrane with a cross-sectional area of 2.5 m^2 at a flow rate of 20 to 25 mL/kg per minute. Gas exchange was adequate and respiratory rate and minute ventilation decreased by more than 50%.

Extracorporeal CO_2 Removal by Hemodialysis

Excretion of carbon dioxide without a membrane lung was first reported in the 1970s (74). This approach involves CO_2 removal in the form of bicarbonate by hemodialysis against a bicarbonate-free solution. Removal of bicarbonate by dialysis decreases bicarbonate concentration in body fluids. This reduces the amount of dissolved CO_2 and consequently, the $PaCO_2$.

Using this technique, Gille et al. (75) were able to remove 46% of the total metabolic CO_2 production in paralyzed, anesthetized dogs. The extracorporeal circuit included a double roller pump and two cellulose membrane dialyzers in parallel. The cross-sectional area of the hemodialysis membrane is relatively high (2.7 m^2 of regenerated cellulose), but unlike ECMO or $ECCO_2R$, the ECBF rate is low (less than 7 mL/kg per minute, which would correspond to less than 1 L/min in humans). With mass bicarbonate elimination, the key issue becomes that of maintaining the acid-base equilibrium in body fluids. The body acid-base balance was controlled by continuous infusion of Tham, an organic buffer. Throughout the period of hemodialysis, ventilation of the lungs was kept at half the control respiratory minute volume. This work proved the effectiveness of this procedure regarding mass CO_2 transfer. It confirmed that blood processing by hemodialysis (CO_2 removal as bicarbonate) is inherently more efficient than CO_2 removal in a blood-gas exchange unit such as a membrane lung (i.e., removal of dissolved CO_2).

Extracorporeal CO_2 Removal via a Pumpless Arteriovenous Hollow Fiber Membrane System

Animal studies have demonstrated than pumpless arteriovenous extracorporeal membrane oxygenation can improve arterial oxygenation and carbon dioxide elimination if blood flow through the oxygenator is approximately 20 to 30% of the cardiac output (76,77). Chapman and colleagues (77) studied the respiratory and hemodynamic response of this technique in dogs with oleic acid–induced ALI. Using a Bentley CM40 hollow-fiber oxygenator, significant improvement in blood gases was achieved, although adverse hemodynamic effects were noted. However, these changes were ameliorated by volume expansion and/or dopamine infusion.

A new hollow-fiber oxygenator with low resistance to blood flow and high gas transfer characteristics has been developed (78). This device was evaluated in pigs using an arteriovenous configuration in a pumpless system driven by aortic pressure. Animals were anesthetized and mechanically ventilated. The experimental conditions were established to produce normal conditions, moderate respiratory failure, and severe respiratory failure. The results from this study showed that (1) up to 25% of CO_2 production can be removed with this pumpless $ECCO_2R$ device at an extracorporeal shunt fraction of 10% or less; and (2) oxygen exchange is limited to states of moderate to marked hypoxemia.

C. Artificial Endocrine Lungs

The technique of continuous arteriovenous or venovenous hemofiltration (CAVH, CVVH) is applied widely to the management of patients with a variety of renal, metabolic, and cardiorespiratory disturbances. It uses membranes made of polyacrylonitrile or polycarbonate which differ from the standard cellophane membrane used in hemodialysis because of their increased biocompatibility and hydraulic permeability. CAVH requires only an arterial and venous cannula and the hemofilter. The vascular resistance of the hemofilter is very low, so that a systolic pressure of greater than 80 mmHg is adequate for spontaneous hemofiltration, avoiding some of the risks associated with a blood pump (79).

The application of hemofiltration has been examined in patients with acute pulmonary edema, acute renal failure, sepsis, ALI, and multiple-organ failure. The rationale for its indication includes fluid removal, to facilitate parenteral feeding and administration of vasoactive drugs, urea and creatinine removal, and removal of complement-activating fractions and endotoxin-associated peptides. Most of these vasoactive, endotoxin-induced peptides (i.e., serotonin, histamine, bradykinin, endorphin, prostaglandins, thromboxane, leukotrienes, etc.) cause peripheral vasodilatation, and in the lungs cause a significant rise in pulmonary vascular resistance and microvascular permeability. Approximately 80% of most of these peptides are cleared from the blood during one passage through the pulmonary circulation (80), but this clearance property of the pulmonary endothelium is severely impaired during sepsis and hypoxia (80).

The use of CAVH in the management of critically ill patients was first conceived by Kramer et al. in 1980 (81). Since then, there have been several studies demonstrating the effectiveness of CAVH in cardiovascular and respiratory function along with the manipulation of fluid balance during acute renal failure (82,83). Gotloib and Barzilay (82) reported a very high survival rate of 92% in 24 patients with sepsis-induced ALI and a relatively low period of mechanical ventilation (5.1 ± 1.2 days). In a retrospective study, Bagshaw et al. (83) reported a significant improvement in respiratory function in 13 patients with multiple-organ-system failure who had undergone a prolonged period of intense CAVH. However, the overall mortality of this group was 77%. It may be that hemofiltration is more effective in treating isolated, acute pulmonary dysfunction, and less capable of influencing the eventual outcome when the more chronic, multisystemic changes associated with end-stage ALI are present. Further prospective trials of CAVH are needed. If it shortens the period of mechanical ventilation and the stay in the ICU, this relatively easy-to-use technique has important therapeutic and economic implications.

In summary, it is difficult to make predictions regarding the potential utility of extracorporeal life-support techniques. Although several reports have shown encouraging results, these techniques are costly and associated with complications. Development of more effective and less invasive therapies for pulmonary failure may make extracorporeal lung support obsolete.

III. Intracorporeal Life Support

A. Intravascular Oxygenation

In an attempt to provide adequate gas exchange without excessive mechanical ventilation and without an extracorporeal membrane, Mortensen (84) pioneered the use of an intravascular oxygenator (IVOX) which can be implanted surgically into the inferior vena cava. IVOX is perhaps the most innovative approach for providing adequate gas exchange, and the objective of this device is to provide partial respiratory support during conventional mechanical ventilation. In essence, the device consists of a miniaturised, thromboresistant hollow-fiber membrane oxygenator (up to 1200 fibers) designed to provide exchange of oxygen and carbon dioxide. It is placed into the venous system percutaneously and is advanced through the inferior vena cava and right atrium to the superior vena cava (Fig. 4).

Gas exchange using the IVOX occurs through a double-lumen gas conduit. The gas inlet limb is attached to an oxygen source. A flow of 1 to 3 L/min of oxygen is drawn through the fibers under subatmospheric pressure

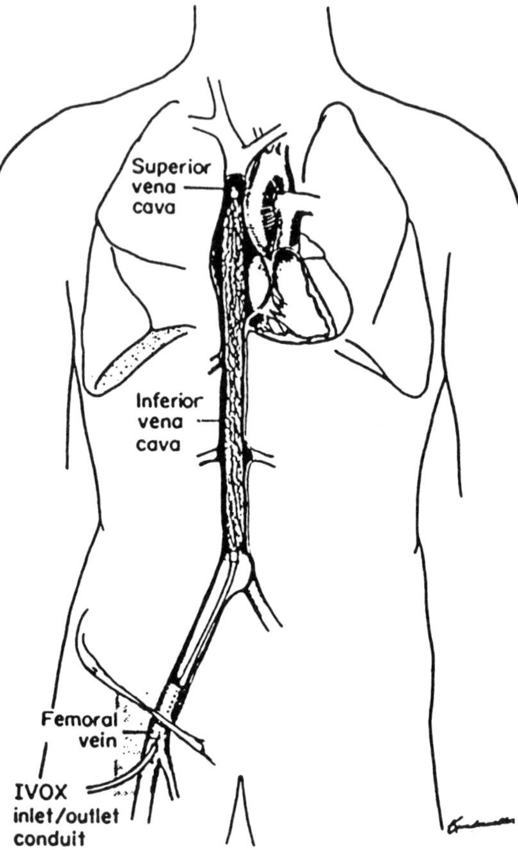

Figure 4 IVOX fiber oxygenator in situ within the caval venous system. (From Ref. 86.)

to prevent air emboli while anticoagulated venous blood flows between them. Oxygen diffuses passively down its concentration gradient into the blood, while carbon dioxide diffuses out. The gas outlet limb connects to a vacuum source which pulls the gas through the device, exhausting the O_2–CO_2 mixture (Fig. 5). The internal diameter of each IVOX hollow fiber is 200 μm, the microporous hollow fiber wall is 30 μm thick, and the continuous siloxane gas transfer membrane is less than 1 μm thick and is treated with covalently bonded heparin. The size of the device when it expands in vivo is 10 to 14 mm in diameter, with a gas transfer membrane surface area of 0.2 to 0.5 m^2 (84–86). The hollow fibers have been arranged to produce a turbulent blood flow pattern over/around each hollow fiber. Current models

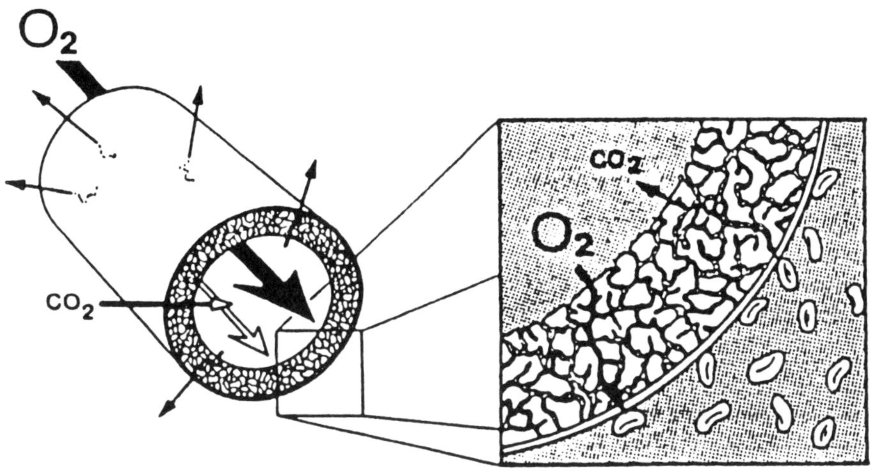

Figure 5 IVOX hollow-fiber gas-transfer membrane. Oxygen and carbon dioxide permeate through the gas-permeable but fluid-impermeable siloxane membrane in response to partial pressure gradients between the gas inside the hollow fiber and blood outside the hollow fiber, as shown in the inset drawing. The similarities between the IVOX fiber and the natural pulmonary alveolar–capillary membrane complex is apparent. (From Ref. 87.)

of IVOX provide only one-third to one-half of the metabolic oxygen and CO_2 gas transfer requirements of most adult animals or patients with ALI (50 to 200 mL/min of O_2 and CO_2) (86).

Compared with extracorporeal gas exchangers with surface areas in excess of 10 m^2 and higher gas flows, an intracorporeal device will necessarily be limited in performance by size. Because of these limitations, it is unlikely that current designs of the IVOX could offer the same support as with an extracorporeal membrane oxygenator. By contrast, this relatively simple technique may play a role in augmenting ventilation by providing partial pulmonary support with much less danger of barotrauma and toxic effects of high inspired oxygen concentrations. In addition, IVOX will reduce the incidence of complications associated with the use of extracorporeal life assistance (hemorrhage, sepsis, heat loss). The device appears to function effectively for up to 18 days with no major risks, hazards, or complications and with only minor, incidental adverse effects at the surgical venotomy site (87).

To date, several acute and chronic implantations in normal and hypoxemic/hypercapnic animals have been evaluated and produced favorable safety and efficacy data (86–88). IVOX does not produce recognizable effects on hemodynamics, including cardiac output, pulmonary artery pressure, or cause significant hemolysis, and there has been no gross or microscopic evidence of particulate thromboemboli to lungs or systemic target organs. In addition, IVOX allows a reduction in the minute ventilation by removing CO_2 from the venous blood.

IVOX appears to be suitable as a means for temporarily augmenting blood gas transfer in selected patients in severe, acute, potentially reversible respiratory failure. In 1990, the first IVOX was implanted in a human being. The procedure was undertaken on a 16-year-old woman in moribund condition with end-stage pulmonary disease. To date, more than 50 patients with advanced ALI have been treated worldwide with an IVOX implantation (87,89,90). Phase II of a clinical trial supervised by the U.S. Food and Drug Administration is now under way to assess the efficacy of this new, promising technique.

B. Implantable Artificial Lung

In 1977, Morin and colleagues from Laval University, Quebec, reported the design of a totally implantable artificial lung (91). The geometry proposed attempts to recreate the spongelike texture of the normal lung. The prototypes were constructed with silicone rubber capillary tubing with an inner diameter of 0.3 mm and an outer diameter of 0.6 mm. The capillary tubing was woven by hand into a tight rectangular weave with six strands in one direction and one strand in the other. This artificial lung can occupy the space of the removed lung in one hemithorax. The oxygenator and CO_2 removal device is connected to the pulmonary artery and left atrium for circulation of blood. The lung is perfused by blood driven by the native heart and ventilated by the respiratory muscles of the recipient. Air entry and exit occur through a chest wall opening equipped with an inspiratory and expiratory valve. The maximum oxygen transfer rate was 48 mL/min. Carbon dioxide transfer rate was 23 mL/min per square meter. Although a very interesting device, artificial lungs will probably not become available until far into the future.

IV. Liquid Respiration

Contemporary terrestrial mammals may breathe fluid for considerable periods of time, much like their fish ancestors, if the liquid respiratory medium contains sufficient dissolved oxygen. In 1962, Kylstra et al. (92) performed the first liquid respiration study using various salt solutions as the respiratory medium. They were able to maintain a nonanesthetised mice alive for up to 18 h while breathing a balanced and buffered salt solution at 20°C and in equilibrium with oxygen at a tension of 8 atm. These mice made respiratory movements while subjected to uniform compression of up to 160 atm, similar to the pressure existing in the oceans at a depth of approximately 1 mile. These authors speculated that if humans possess a similar tolerance to uniform compression, escape from submarines might be possible at depths far exceeding current estimates based on the concept that compressible gases are toxic or narcotic at increased densities.

In 1966, Clark and Gollan (93) were the first to ventilate small animals under normobaric conditions using perfluorocarbon (PFC), an inert liquid with a very low surface tension (10 to 15 dyn/cm), an extraordinary ability to dissolve oxygen and carbon dioxide, and evenly distributed throughout the lung at low insufflation pressures (94). Since then, several groups of investigators have examined the physiologic implications and explored possible therapeutic uses of liquid respiration in animal models of ALI (94–97). These studies suggest that liquid breathing using PFC may be a very effective means for achieving gas exchange in ALI.

PFC is a liquid with negligible pulmonary absorption that is removed by vaporization through the lungs and does not cause adverse histological or biological effects. PFC acts as an exogenous surfactant by coating the alveolar lining or flooding the alveoli. Consequently, it reduces surface tension and eliminates the alveolar air–liquid interface, resulting in a significant reduction in lung recoil and improvement in lung compliance (96) (Fig. 6).

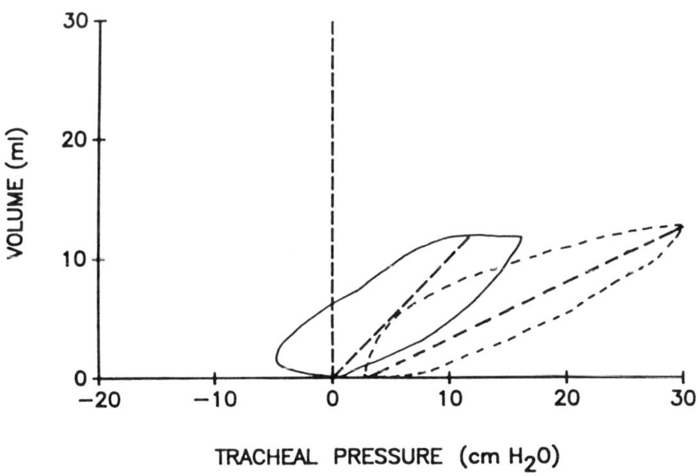

Figure 6 Dynamic pressure–volume relationship of tidal breaths in two lambs: gas ventilated (dashed line) and liquid ventilated (solid line). Dashed line connects points of zero flow slope, which graphically illustrates respiratory compliance. (From Ref. 96.)

It is this property of liquid ventilation that makes it attractive as a technique for ventilation in various pulmonary diseases. Normal cardiac output can be maintained during liquid respiration if there is an adequate intravascular volume (94).

Wolfson and colleagues (96) studied the effects during 3 h of gas and liquid ventilation on cardiopulmonary function in 16 preterm lambs. Warmed, oxygenated PFC was delivered by a mechanically assisted liquid ventilation system (V_T = 10 to 20 mL/kg of PFC, FiO_2 = 1, rate = 3 to 8 breaths/min, FRC = 20 to 40 mL/kg). Mortality rate of gas-ventilated animals was much higher than those ventilated with liquid (67% versus 14%). Their data on respiratory compliance and histological analysis indicated that liquid ventilation during early development provides more effective gas exchange with less barotrauma than does gas ventilation.

Fuhrman et al. (95,97) have developed a new way to assist ventilation by using a hybrid of liquid and gas techniques termed perfluorocarbon-associated gas exchange (PAGE). With PAGE, conventional gas ventilation is superimposed on PFC-filled lungs. A volume of PFC equivalent to the normal pulmonary FRC (30 mL/kg) was instilled into the trachea and "bubble-oxygenated" in vivo, within the lungs, where it participated directly in gas exchange. This approach is a major simplification of conventional liquid respiration techniques since it was accomplished using a conventional ventilator. Three patterns of interfacial surfaces are possible: gas/PFC, alveolar lining/PFC, and alveolar lining/gas (Fig. 7). Inspiration might be expected to distend surfaces so as to minimize the total increase in surface forces that occurs with these possible interfaces. Fuhrman et al. have performed these experiments in piglets with normal lungs (95) and with premature surfactant deficient lungs (97). Values of peak airway pressure during PAGE and during mechanical ventilation were nearly identical. Blood gases were maintained normal, and in surfactant-deficient lungs, compliance increased dramatically and immediately after liquid ventilation.

Based on these reports, liquid respiration appears to be a promising, alternative therapeutic modality for the management of neonates and adults with pulmonary diseases affecting the gas–liquid interface. A recent preliminary clinical trial showed that it is possible to maintain gas exchange in very immature neonates (98). It is likely that liquid ventilators could be developed and would find their place in the ICUs of the twenty-first century, as a form of ventilatory support.

V. Lung Transplantation

Human organ transplantation is an extraordinary medical advance that has benefited more than 100,000 patients in North America over the last 20 years.

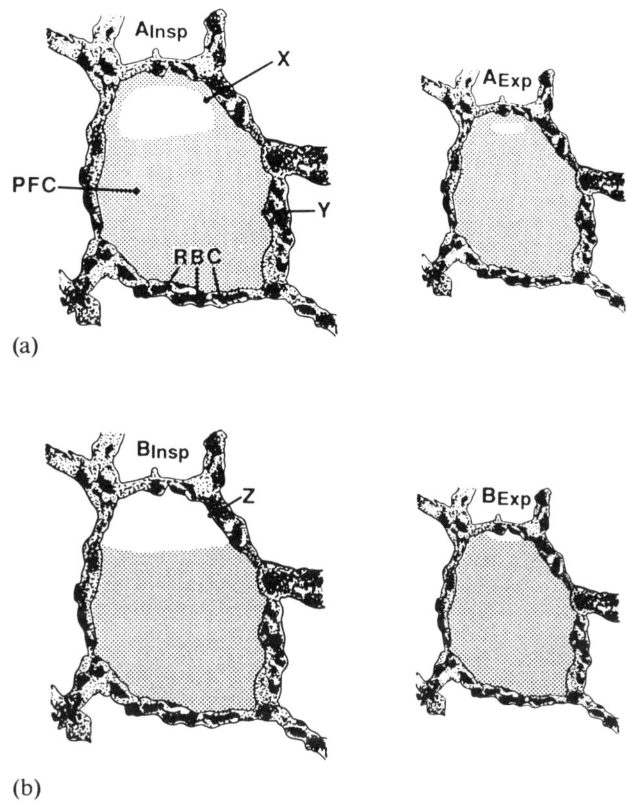

Figure 7 Two possible patterns of bubble growth within a fluid-filled alveolus during perfluorocarbon-associated gas exchange (PAGE). Bubble growth within the fluid (a) presents two interfacial surfaces: gas–PFC (X) and alveolar lining–PFC (Y). Bubble growth against the alveolar lining (b) presents an additional interface, alveolar lining–gas (Z). Insp, inspiration; Exp, expiration; PFC, perfluorocarbon; RBC, red blood cells. (From Ref. 95.)

In recent years, heart–lung transplantation and double or unilateral lung transplantation have emerged as acceptable therapeutic options for patients with end-stage lung disease, such as cystic fibrosis, advanced chronic obstructive lung disease, and pulmonary fibrosis (99,100). Long-term follow-up studies of combined heart and lung transplants now indicate approximately an 80% 2-year survival (101). However, lungs suitable for transplantation are particularly scarce among donor organs since reliable preservation times are only about 6 h.

Under circumstances where no other therapeutic option is available, and in the absence of evidence of multiorgan damage, lung transplantation may be a justifiable procedure for severe, irreversible ALI. In this respect, the use of extracorporeal support as a bridge to transplantation deserves special comment. Although the multicenter ECMO study (8) did not show any differences in overall survival between those treated with mechanical ventilation and those treated with ECMO, survival between days 2 and 11 was greater in the group having ECMO. Therefore, extracorporeal life support may increase the time available to search for suitable organs. This approach has been used by some investigators. Egan and colleagues from Toronto (7,21) placed two patients with Paraquat-induced ALI on ECMO before they performed three lung transplants. Yacoub's group in London performed heart-lung transplants in six patients with severe respiratory failure (25,102). Three of these patients left the hospital after transplantation.

Although lung transplantation may prove to be a therapeutic option for patients with ALI, most are not suitable in the acute phase because of the high incidence of sepsis. One of the major difficulties with this approach in ALI is the fact that patients with severe respiratory failure who die often develop multisystem organ failure (MSOF) as part of their clinical course (20). It is hard to justify lung transplantation early in their clinical course because many patients survive without heroic measures. Later in their course, when they start to develop MSOF, transplantation would be much more difficult and dangerous, especially due to the fact that potent immunosuppresors are required to prevent rejection. The second major problem alluded to earlier is the difficulty in organ procurement, with patients often waiting months for a suitable lung to become available. Patients with ALI requiring transplantation would not be expected to survive for prolonged periods of time until a suitable organ is obtained. As a consequence, lung transplantation as a therapeutic modality in ALI is not expected to be a feasible option for large numbers of patients until xenograft transplantation becomes a reality (i.e., use of animal organs for transplantation) (103).

A novel approach has recently been reported by Pennock and colleagues in Pittsburgh (104). These investigators demonstrated in dogs that a heterotopic lung transplant into the abdominal space, ventilated with a volume ventilator, can provide augmented systemic oxygenation to an animal whose native lungs do not provide adequate oxygenation. The rationale for the application of this technique is to allow the native lungs to recover without using an external pump or membrane oxygenator. In this system, the blood flow through the transplanted lung was provided by anastomoses from the transplanted pulmonary artery and left atrial cuff to the host aorta and inferior vena cava, respectively. A mathematical model predicted that this transplant procedure has the potential for providing oxygenation in patients whose native lungs have a venous admixture greater than 50%.

VI. Summary

Over the past 20 years, our understanding of the pathophysiology of acute lung injury has increased tremendously. Perhaps, one of the most important discoveries has been our understanding of the detrimental effects of mechanical ventilation on lung structure and function. It is for this reason that many investigators have begun examining therapeutic modalities that limit airway pressures or lung stretch. Unfortunately, many of these techniques are highly invasive and have other detrimental side effects that may limit their utility. Further studies are required to determine the clinical indications for these highly invasive techniques.

References

1. Villar J, Slutsky AS. The incidence of the adult respiratory distress syndrome. Am Rev Respir Dis 1989; 140:814–816.
2. Swami A, Keogh BF. The injured lung: conventional and novel respiratory therapy. Thorax 1992; 47:555–562.
3. Hickling KG. Ventilatory management of ARDS: can it affect the outcome? Intensive Care Med 1990; 16:219–226.
4. Gattinoni L, Pesenti A, Torresin A, et al. Adult respiratory distress syndrome profiles by computed tomography. J Thorac Imag 1986; 1:25–30.
5. von Frey M, Gruber M. Untersuchungen uber den Stoffwechsel isolirter Organe. Ein Respirationsapparat fur isolirte Organe. Arch Anat Phys Phys Abt 1985; 9:519–532.
6. Hill JD, Gibbon JH Jr. The development of the first successful heart-lung machine. Ann Thorac Surg 1982; 34:337–341.
7. Egan TM, Duffin J, Glynn MF, et al. Ten-year experience with extracorporeal membrane oxygenation for severe respiratory failure. Chest 1988; 94:681–687.
8. Zapol WM, Snider MT, Hill JD, et al. Extracorporeal membrane oxygenation in severe acute respiratory failure: a randomized prospective study. JAMA 1979; 242:2193–2196.
9. Kolobow T, Spragg RG, Pierce JE. Massive pulmonary infarction during total cardiopulmonary bypass in unanaesthetised spontaneously breathing lambs. Int J Artif Organs 1981; 4:76–81.
10. Gattinoni L, Pesenti A, Marcolin R, Damia G. Extracorporeal support in acute respiratory failure. Intensive Care World 1988; 5:42–45.
11. Zapol WM, Qvist J, Pontopiddan H, et al. Extracorporeal perfusion for acute respiratory failure: recent experience with the spiral coil membrane lung. J Thorac Cardiovasc Surg 1975; 69:439–449.

12. Pesenti A, Kolobow T, Riboni A, et al. Single vein cannulation for extracorporeal respiratory support. Life Support Syst 1982; 1(S):165–167.
13. Evans TW, Keogh BF. Extracorporeal membrane oxygenation: a breath of fresh air or yesterday's treatment? Thorax 1991; 46:692–694.
14. Villar J, Slutsky AS. The adult respiratory distress syndrome: two decades later. Intensive Crit Care Digest 1988; 7(1):18–20.
15. Hill JD, Fallat R, Cohn K, et al. Clinical cardiopulmonary dynamics during prolonged extracorporeal circulation for acute respiratory insufficiency. Trans Am Soc Artif Intern Organs 1971; 17:355–361.
16. Hill JD, O'Brien TG, Murray JJ, et al. Prolonged extracorporeal oxygenation for acute post-traumatic respiratory failure (shock-lung syndrome). N Engl J Med 1972; 286:629–634.
17. Gille JP, Bagniewski AM. Ten years of use of extracorporeal membrane oxygenation (ECMO) in the treatment of acute respiratory insufficiency (ARI). Trans Am Soc Artif Intern Organs 1976; 22:102–109.
18. National Heart, Lung, and Blood Institute. Protocol for extracorporeal support for respiratory insufficiency. Bethesda, MD: NHLBI, 1974.
19. National Heart, Lung, and Blood Institute. Extracorporeal support for respiratory insufficiency (RFP-NHLI-73-20). Bethesda, MD: NHLBI, 1979.
20. Villar J, Manzano JJ, Blazquez MA, Quintana J, Lubillo S. Multiple system organ failure in acute respiratory failure. J Crit Care 1991; 6:75–80.
21. Egan TM, Saunders NR, Murphy E, Cooper JD. Succesful extracorporeal membrane oxygenation for 24 days in a patient who underwent 2 lung transplants for Paraquat poisoning: a case report. Proc Am Acad Cardiovasc Perf 1983; 4: 154–158.
22. Dawids S, Agostoni A. Extracorporeal oxygenation. A European concerted action. Int J Artif Organs 1979; 2:233.
23. Guillett DS, Gunning KEJ, Sawicka EH, Bellingham AJ, Ware RJ. Life threatening sickle chest syndrome treated with extracorporeal membrane oxygenation. Br Med J 1987; 294:81–82.
24. Bindslev L. Adult ECMO performed with surface-heparinized equipment. ASAIO Trans 1988; 34:1009–1013.
25. Mudalier MY, Hunter DN, Morgan CJ, Evans TW. Extracorporeal gas exchange for respiratory failure. Hosp Update 1991; 17:410–415.
26. Lewandowski K, Falke KJ, Rossaint R, Slama K, Pappert D, Kuhlen R. Low mortality associated with advanced treatment including v-v ECMO for severe ARDS (abstr). Intensive Care Med 1992; 18:S42.
27. White JJ, Andrews HG, Risemberg H, Mazur D, Haller JA Jr. Prolonged respiratory support in newborn infants with a membrane oxygenator. Surgery 1971; 70:288–296.
28. Barlett RH, Toomasian J, Roloff D, et al. Extracorporeal membrane oxygenation (ECMO) in neonatal respiratory failure. Ann Surg. 1986; 204:236–245.
29. Stork E. Extracorporeal membrane oxygenation in the newborn and beyond. Clin Perinatol 1988; 15:815–829.
30. ECMO Registry. Report of the Extracorporeal Life Support Organization, July 1991. Ann Arbor, MI: ELSO, 1991.

31. Bohn D. the current role of ECMO in paediatric practice. Intensive Care World 1991; 8:162–166.
32. Chavalier JY, Durandi Y, Lecompte Y, et al. Assistance respiratoire extracorporelle en pediatrie. Rean Soins Intens Med Urg 1987; 3:269.
33. Bartlett RH, Gazzaniga AB, Jeffries MR, et al. Extracorporeal membrane oxygenation (ECMO) cardiopulmonary support in infancy. Trans Am Soc Artif Intern Organs 1976; 22:80–93.
34. Pfenninger J, Gerber A, Tschappeler H, Zimmermann A. Adult respiratory distress syndrome in children. J Pediatr 1982; 101:352–357.
35. Van Meurs KP, Frankel LR, Pearl RG. Extracorporeal life support: issues of who, when, why, and how. Crit Care Med 1992; 20:1200–1202.
36. Keszler M, Subramanian KNS, Smith YA, et al. Pulmonary management during extracorporeal membrane oxygenation. Crit Care Med 1989; 17:495–500.
37. Dalton HJ, Thompson E. Extracorporeal membrane oxygenation. In: Grenvik A, Downs J, Rasanen J, Smith R, eds. Mechanical Ventilation and Assisted Respiration. New York: Churchill Livingstone, 1991:115–137.
38. Beck R, Anderson KD, Pearson GD, Cronin J, Miller MK, Short BL. Criteria for extracorporeal membrane oxygenation in a population of infants with persistent pulmonary hypertension of the newborn. J Pediatr Surg 1986; 21:297–302.
39. Koumbourlis AC, Motoyama EK, Mutich RL, Nakayama DK, Thompson AE. Lung mechanics during and after extracorporeal membrane oxygenation for meconium aspiration syndrome. Crit Care Med 1992; 20:751–756.
40. van de Bor M, Walther FJ, Gangitano ES, Snyder JR. Extracorporeal membrane oxygenation and cerebral blood flow velocity in newborn infants. Crit Care Med 1990; 18:10–13.
41. Meliones JN, Moler FW, Custer JR, et al. Hemodynamic instability after the initiation of extracorporeal membrane oxygenation: role of ionized calcium. Crit Care Med 1991; 19:1247–1251.
42. Schwendeman CA, Clark RH, Yoder BA, Null BM, Gerstmann DR, deLemos RA. Frequency of chronic lung disease in infants with severe respiratory failure treated with high-frequency ventilation and/or extracorporeal membrane oxygenation. Crit Care Med 1992; 20:372–377.
43. Rosenberg AA, Kinsella JP. Effect of extracorporeal membrane oxygenation on cerebral hemodynamics in newborn lambs. Crit Care Med 1992; 20:1575–1581.
44. Dacar D, Zobel G, Kuttnig-Haim M, et al. Extracorporeal membrane oxygenation without ligation of the carotid artery: use in two neonates. Crit Care Med 1992; 20:436–438.
45. deLemos R, Yoder B, McCurnin D, Kinsella J, Clark R, Null D. The use of high-frequency oscillatory ventilation (HFOV) and extracorporeal membrane oxygenation (ECMO) in the management of the term/near term infant with respiratory failure. Early Human Dev 1992; 29:299–303.
46. Shapiro DL. Surfactant replacement therapy for respiratory distress syndrome. In: Brigham KL, Stahlman MT, eds. Respiratory Distress Syndromes. Molecules to Man. Nashville TN: Vanderbilt University Press, 1990:236–249.

47. Delius RE, Bove EL, Meliones JN, et al. Use of extracorporeal life support in patients with congenital heart disease. Crit Care Med 1992; 20:1216–1222.
48. Butt W, Karl T, Horton A, Shann R, Mullaly R. Experience with extracorporeal membrane oxygenation in children more than one month old. Anaesth Intensive Care 1992; 20:308–310.
49. Moler FW, Custer JR, Bartlett RH, et al. Extracorporeal life support for pediatric respiratory failure. Crit Care Med 1992; 20:1112–1118.
50. Kolobow T, Gattinoni L, Tomlinson TA, White D, Pierce JE, Iapichino G. The carbon dioxide membrane lung (CDML): a new concept. Trans Am Soc Artif Intern Organs 1977; 23:17–21.
51. Kolobow T, Gattinoni L, Tomlinson TA, Pierce JE. An alternative to breathing. J Thorac Cardiovasc Surg 1978; 75:261–266.
52. Gattinoni L, Kolobow T, Damia G, et al. Extracorporeal carbon dioxide removal ($ECCO_2R$): a new form of respiratory assistance. Int J Artif Organs 1979; 2:183–185.
53. Villar J, Winston B, Slutsky AS. Non-conventional techniques of ventilatory support. Crit Care Clin 1990; 6:579–603.
54. Ratliff JL, Hill D, Fallat RJ, Parrot J, Tucker HJ. Complications associated with membrane lung support by veno-arterial perfusion. Ann Thorac Surg 1975; 19:537–539.
55. Hickling KG. Extracorporeal CO_2 removal in severe adult respiratory distress syndrome. Anaesth Intensive Care 1986; 14:46–53.
56. Gattinoni L, Kolobow T, Tomlinson T, et al. Low frequency positive pressure ventilation with extracorporeal carbon dioxide removel (LFPPV-$ECCO_2R$): an experimental study. Anaesth Analg 1978; 57:470–477.
57. Gattinoni L, Pesenti A, Mascheroni D, et al. Low-frequency positive pressure ventilation with extracorporeal CO_2 removal in severe acute respiratory failure. JAMA 1986; 256:881–886.
58. Gattinoni L, Pesenti A, Caspani ML, et al. The role of total static lung compliance in the management of severe ARDS unresponsive to conventional treatment. Intensive Care Med 1984; 10:121–126.
59. Gattinoni L. Extracorporeal respiratory support for acute respiratory failure. Curr Opin Anesthesiol 1991; 4:261–265.
60. Bindslev L, Eklund J, Norlander O, et al. Treatment of acute respiratory failure by extracorporeal carbon dioxide elimination performed with a surface heparinized artificial lung. Anesthesiology 1987; 67:117–120.
61. Gattinoni L, Agostoni A, Damia G, et al. Hemodynamics and renal function during low frequency positive pressure ventilation with extracorporeal CO_2 removal. A comparison with continuous positive pressure ventilation. Intensive Care Med 1980; 6:155–161.
62. Peters J, Radermacher P, Lenhsen U, Lohe B, Rosen P, Falke KJ. Lung extravascular volume during venovenous bypass with extracorporeal CO_2 removal in dogs. Intensive Care Med 1988; 14:123–130.
63. Borelli M, Kolobow T, Spatola R, Prato P, Tsuno K. Severe acute respiratory failure managed with continuous positive airway pressure and partial extracorporeal carbon dioxide removal by an artificial membrane lung. Am Rev Respir Dis 1988; 138:1480–1487.

64. Morris AH, Menlove RL, Rollins RJ, Wallace CJ, Beck E. A controlled clinical trial of a new 3-step therapy that includes extracorporeal CO_2 removal for ARDS. J Am Soc Artif Intern Organs 1988, 11:48–53.
65. Morris AH, Wallace CJ, Menlove RL, et al. Randomized clinical trial of pressure-controlled inverse ratio ventilation and extracorporeal CO_2 removal for adult respiratory distress syndrome. Am J Respir Crit Care Med 1994; 149:295–305.
66. Pesenti A, Kolobow T, Buckhold DK, Pierce JE, Huang H, Chen V. Prevention of hyaline membrane disease in premature lambs by apneic oxygenation and extracorporeal carbon dioxide removal. Intensive Care Med 1982; 11–17.
67. Born GVR, Dawes GS, Mott JC. The viability of premature lambs. J Physiol (London) 1955; 130:191–212.
68. Dorrington KL, McRae KM, Gardaz JP, Dunnill MS, Sykes MK, Wilkinson AR. A randomized comparison of total extracorporeal CO_2 removal with conventional mechanical ventilation in experimental hyaline membrane disease. Intensive Care Med 1989; 15:184–191.
69. Hamilton PP, Onayemi A, Smyth JA, et al. Comparison of conventional and high-frequency ventilation: oxygenation and lung pathology. J Appl Physiol 1983; 55:131–138.
70. Chevalier JY, Durandy Y, Batisse A, Mathe JC, Costil J. Preliminary report: extracorporeal lung support for neonatal acute respiratory failure. Lancet 1990; 335:1364–1366.
71. Chevalier JY. Extracorporeal lung support (ECLS) for neonatal and pediatric acute respiratory failure. In: Artigas A. ed. Proceedings of the 6th European Congress of Intensive Care Medicine, Barcelona, 1992:80–82.
72. Terasaki H, Nogami T, Saito Y, Otsu T, Morioka T. Extracorporeal lung assist without endotracheal intubation and mechanical ventilation. Crit Care Med 1987; 15:84–85.
73. Terasaki H. Extracorporeal lung assist. Respir Res 1985; 4:742–745.
74. Updike SJ, Shults MC. Excretion of carbon dioxide using an artificial kidney. J Appl Physiol 1973; 34:274–276.
75. Gille JP, Saunier C, Baluais G, Battistella P. Extracorporeal CO_2 removal by hemodialysis. ASAIO 1984; 7:80–84.
76. Barthelemy R, Galleti PM, Trudell LA, et al. Total extracorporeal CO_2 removal in a pumpless artery-to vein shunt. Trans Am Soc Artif Intern Organs 1982; 28:354–358.
77. Chapman J, Adams M, Geha AS. Hemodynamic response to pumpless extracorporeal membrane oxygenation. J Thorac Cardiovasc Surg 1990; 99:741–750.
78. Grier LR, Blount J, Conrad SA, Brown EG, Mortensen JD. Extracorporeal CO_2 removal via a pumpless arteriovenous hollow fiber membrane sytem (abstr). Crit Care Med 1992; 20:S60.
79. Simonds AK, Mansell MA. Hemodialysis or hemofiltration in the ITU. Intensive Crit Care Digest 1986; 5(1):5–7.
80. Gillis CN, Pitt BR. The fate of circulatory amines within the pulmonary circulation. Annu Rev Physiol 1982; 44:269–281.
81. Kramer P, Kaufhold G, Grone HJ, et al. Management of anuric intensive care patients with arteriovenous hemofiltration. Int J Artif Organs 1980; 3:225–230.

82. Gotloib L, Barzilay E. Hemofiltration in septic ARDS—the artificial kidney as an artificial endocrine lung. Resuscitation 1986; 13:123–132.
83. Bagshaw ONT, Anaes FRC, Hutchinson A. Continuous arteriovenous haemofiltration and respiratory function in multiple organ systems failure. Intensive Care Med 1992; 18:334–338.
84. Mortensen JD. An intravenacaval blood gas exchange (IVCBGE) device: a preliminary report. Trans Am Soc Artif Intern Organs 1987; 33:570–573.
85. Mortensen JD, Berry G. Conceptual and design features of a practical, clinically effective intravenous mechanical blood oxygen/carbon dioxide exchange device (IVOX). Int J Artif Organs 1989; 12:384–389.
86. Cockroft S, Kuo J, Colvin MP, Lewis CT, Innis RF, Withinton PS. Initial evaluation of an intracorporeal oxygenation device. Anaesthesia 1992; 47:48–51.
87. Mortensen JD. Augmentation of blood gas transfer by means of an intravascular blood gas exchanger (IVOX). In: Marini JJ, Roussos C, eds. Ventilatory Failure. London: Springer-Verlag, 1991:318–346.
88. Cox CS, Zwischenberg JB, Traber DL, Traber LD, Herndon DN. Intravascular oxygen/carbon dioxide removal allows reduction of ventilatory support in an ovine model of acute lung injury. Crit Care Med 1992; 20:S60.
89. Kallis P, Al-Saady NM, Bennet D, Treasure R. Clinical use of intravascular oxygenation (letter). Lancet 1991; 337:549.
90. Mira JP, Belghith M, Renaud B, et al. Can IVOX modify mechanical ventilation of patients with ARDS? (abstr). Intensive Care Med 1992; 18:S222.
91. Morin PJ, Gosselin C, Picard R, Vincent M, Guidoin R, Nicholl CIH. Implantable artificial lung. Preliminary report. J Thorac Cardiovasc Surg 1977; 74:130–136.
92. Kylstra JA, Tissing MO, van der Maen A. Of mice as fish. Trans Am Soc Artif Intern Organs 1962; 8:378–383.
93. Clark LC Jr, Gollan F. Survival of mammals breathing organic liquids equilibrated with oxygen at atmospheric pressure. Science 1966; 152:1755–1756.
94. Curtis S. Perfluorocarbon-associated gas exchange: a hybrid approach to mechanical ventilation. Crit Care Med 1991; 19:600–601.
95. Fuhrman BP, Paczan PR, DeFrancisis M. Perfluorocarbon-associated gas exchange. Crit Care Med 1991; 19:712–722.
96. Wolfson MR, Greenspan JS, Deoras KS, Rubenstein SD, Shaffer TH. Comparison of gas and liquid ventilation: clinical, physiological, and histological correlates. J Appl Physiol 1992; 72:1024–1031.
97. Leach CL, Fuhrman BP. Perfluorocarbon associated gas exchange in surfactant deficiency (abstr). Am Rev Respir Dis 1992; 145:A454.
98. Greenspan JS, Wolfson MR, Rubenstein SD, Shaffer TH. Liquid ventilation of human preterm neonates. J Pediatr 1990; 117:106–111.
99. Toronto Lung Transplant Group. Unilateral lung transplantation for pulmonary fibrosis. N Engl J Med 1986; 314:1140–1145.
100. Cooper JD, Patterson GA, Grossman R, Maurer J, and the Toronto Lung Transplant Group. Double-lung transplant for advanced chronic obstructive lung disease. Am Rev Respir Dis 1989; 139:303–307.

101. McCarthy PM, Starnes VA, Theodore J, Stinson EB, Oyer PE, Shumway NE. Improved survival after heart-lung transplantation. J Thorac Cardiovasc Surg 1990; 99:54–60.
102. Tsang V, Evans TW, Morgan C, Yacoub M. Heart-lung transplantation for adult respiratory distress syndrome. Crit Care Med 1991; 19:286–287.
103. Auchincloss H Jr. Xenogenic transplantation. A review. Transplantation 1988; 46:1–20.
104. Pennock BE, Lazzara RR, Kaplan PD, Magovern JA. Auxiliary oxygenation with a heterotopic lung transplant (abstr). Am Rev Respir Dis 1992; 145:A301.

18

Prospects for Gene Therapy for Endotoxin-Induced Lung Injury

KENNETH L. BRIGHAM, ANGELO E. CANONICO,
BARBARA O. MEYRICK, JON T. CONARY, and HANS SCHREIER

Vanderbilt University School of Medicine
Nashville, Tennessee

I. Introduction

With the development of technologies for manipulating nucleic acids and using these techniques to instruct cellular synthetic machinery, the prospect of gene therapy is rapidly emerging as a clinical reality. In the lungs, the principal focus has been on the inherited diseases, α-1 antitrypsin deficiency and cystic fibrosis (1–3). As in other inherited diseases, the concept is that if a normal gene could be supplied to the host in a way that the gene is expressed in vivo, the disease could be cured.

Although initial studies in humans with cystic fibrosis have been approved by the reviewing bodies and are under way, several difficulties remain. Strategies for delivering DNA in vivo exist, but whether they are efficient enough to have the physiologic effects necessary to cure the clinical disease is yet to be proven. A fundamental difference in cystic fibrosis and α-1 antitrypsin deficiency is that the defect in the latter case is in a gene encoding a secreted protein, while in the former case the defect is in a gene encoding a protein integral to the cell membrane. Thus it is theoretically possible to cure α-1 antitrypsin deficiency without supplying the normal DNA to the actual cells that normally produce the protein, whereas with cystic fibrosis the physiologic defect will be corrected only in the cells that generate the functional protein. Effective delivery strategies may be very different for the two cases.

As far as is known, most diseases of the lungs are not a consequence of a genetic abnormality. For example, it may well be that biologic variability based on genetic makeup affects the host response to endotoxin, but even if it were possible, it would not seem prudent to propose to alter the host genome in order to alter that response. Thus the usual concept of gene therapy does not apply very readily to the majority of lung diseases, particularly those that are acute.

II. Concept of Transient Gene Therapy (Gene Therapeutics)

It is possible that the therapeutic potential of in vivo gene transfer techniques is not limited to diseases requiring permanent correction of a genetic defect. If DNA could be introduced into the cells of living organisms in a form that assured transient expression of selected transgenes, either increased (sense constructs) or decreased (antisense constructs) generation of specific proteins could be effected. By choosing transgenes encoding proteins that protect against or contribute to the pathogenesis of cell and organ injury, it is theoretically possible to treat a large number of diseases. This idea has been called transient gene therapy or gene therapeutics (4,5). In concept this approach is very similar to traditional pharmacology, using DNA as a drug and developing formulations that deliver the drug effectively. This approach would require that foreign DNA not be incorporated into the host genome and thus might be safer than technologies designed to permanently alter gene expression by alteration of host DNA (6). Numerous genes encoding potentially therapeutic proteins have been cloned. Expression vectors are available which incorporate into the host genome at very low frequency and exuberantly express transgenes in mammalian cells. The principal obstacle to clinical application of gene therapeutics to acute lung injury of the type seen following endotoxemia is development of delivery systems which are safe and sufficiently efficient to generate enough product in vivo to have a physiological effect. Rapid advances are being made in this area. Some of these advances will be discussed below, as they may apply to the lungs.

III. Strategies for Gene Delivery

Transfection of mammalian cells in culture with functioning foreign DNA can be accomplished by several different techniques; this approach has become a routine part of the laboratory technology for investigating regulation of gene expression. Based on that experience, several experimental techniques have been developed for in vivo delivery of functioning DNA, with the usual goal of treating diseases that are a consequence of a genetic deficiency.

A. Introduction of Genetically Engineered Cells into the Host

For treatment of diseases requiring continuous production of a secreted protein, it is possible to modify cells in culture genetically so that they produce and secrete increased amounts of the protein and introduce the cells into a host where the transplanted cells continue to function. Garver and colleagues (7) used a retroviral vector to transform mouse fibroblasts in culture permanently so that the cells produced human α_1-antitrypsin. They implanted the transformed cells into the peritoneum of nude mice and detected the human protein in serum and bronchoalveolar lavage fluid 4 weeks later, but the amount of human α_1-antitrypsin detected was far below therapeutic levels. This idea, the implantation of genetically modified cells, has been explored by several investigators. Vascular grafts have been seeded with genetically modified endothelial cells and the graft implanted into experimental animals where the foreign cells continue to function (8). So-called "organoids" have been constructed where matrices are impregnated with genetically modified cells and with angiogenesis factors, so that upon implantation into the peritoneum of experimental animals, the gel becomes vascularized and the genetically modified cells continue to function (1–3). Recently, Culver and associates (9) injected murine fibroblasts that were producing a retroviral vector in which the herpes simplex thymidine kinase gene had been inserted directly into cerebral gliomas in rats. The animals were then treated with the antiherpes drug, gancyclovir, and the tumors regressed, presumably because of selective in vivo transformation of the rapidly dividing tumor cells with the retrovirus vector.

B. Direct in Vivo Gene Transfer

For foreign DNA to be expressed, it must access the interior of cells in an appropriate form. Since DNA is a large lipophobic molecule, it will not easily traverse cell membranes, and most strategies for delivering DNA are designed to get the DNA across cell membranes. There is some evidence that DNA can be injected directly into tissues, or even intravenously, and be expressed in vivo, although the mechanism by which the DNA accesses the cell interior is not clear.

Direct injection of expression vectors containing reporter genes into mouse muscle resulted in demonstrable gene expression in the injected sites (10). Intravenous administration of recombinant adenovirus vectors expressing reporter genes into mice resulted in long-term gene transfer in skeletal and cardiac muscle (11). Buttrick and associates injected DNA in the form of a plasmid containing a chloramphenicol acetyltransferase reporter gene into adult rat hearts in vivo and showed reporter gene expression in the injected cardiac muscle (12). Barr and Leiden (13) injected genetically modified murine myoblasts intravenously into mice and showed transgene expression in muscle which persisted for at least 3 weeks. Histological studies of muscle from injected animals showed that the myoblasts appeared to fuse with native muscle cells to form multinucleated myotubes.

In summary, it is clear that the phenomenology of transgene expression following direct injection of DNA in several different forms can be demonstrated in experimental animals. Whether the amount of gene product achievable by this technique is sufficient to have therapeutic potential is not yet clear, and it is unlikely that such approaches would be feasible in the lungs.

C. Delivery of DNA via Cell Surface Receptors

Attachment of ligand to a variety of cell surface receptors results in internalization of the ligand into the cell interior. This process has been used by several investigators to effectively delivery DNA, particularly in the form of plasmids. Zenke and colleagues (14) achieved transfection of cells by receptor-mediated endocytosis using DNA complexed to transferrin–polycation conjugates. Wu and Wu (15) used a similar technique to target liver cells by covalently linking an asialoglycoprotein to a polycation. Since liver cells possess unique receptors that internalize asialoglycoproteins, and DNA is anionic and thus will readily associate with a polycation by charge–charge interaction, this provides a hepatocyte-targeted delivery system. Using reporter gene constructs, they showed specific transgene expression in the livers of rats following intravenous injection. Similarly, Wilson and associates (16) complexed DNA to a protein specific for hepatocyte uptake and delivered the gene encoding the hepatic low-density lipoprotein (LDL) receptor to rabbits that were hypercholesterolemic secondary to deficiency of LDL receptors. Intravenous infusion of the DNA–protein complex caused hepatocyte specific gene expression and amelioration of hypercholesterolemia. Curiel and associates (17) used the cell entry mechanisms of viruses to deliver DNA. They constructed adenovirus–polysine–DNA complexes using an antibody bridge with an antibody specific for the virion and showed efficient DNA uptake in cultured cell systems. Curiel and associates (18) also effected gene transfer to respiratory epithelial cells using DNA–polylysine conjugates.

Utilization of receptors on the surfaces of cells to target DNA delivery to specific cell populations has particular attractiveness for the lungs. The lungs are readily accessible through the intravenous route since all of the peripheral venous blood obligatorily traverses the large pulmonary capillary bed prior to entering the systemic circulation. Obviously, respiratory epithelial cells and alveolar macrophages are readily accessible through the airways. Thus, choosing molecular escorts for DNA which have specific affinity for subpopulations of lung cells might be more practicable than in other organs.

IV. Liposome Delivery Systems

The use of liposomes as a drug delivery system has had varying degrees of popularity over the past several years. Conceptually, liposomes are attractive for delivery of any substance that needs to access the cell interior to be optimally effective. An important study by Freeman and colleagues (19,20) showed that liposome-encapsulated antioxidant enzymes administered intravenously could protect the lungs of rats from the toxic effects of breathing high concentrations of ambient oxygen more effectively than could administration of unencapsulated enzymes. Pitt and Huang (21) have recently reviewed current approaches to the use of liposomes for deliverying drugs to the lungs.

Liposome delivery systems are particularly interesting in the lungs because liposomes can be designed and constructed which have particular affinities for subpopulations of cells either because of their lipid composition or by inserting specific targeting proteins into the lipid membrane (21). Since the geography of the lungs makes it possible to deliver liposomes selectively to endothelium (intravenous injection) or to respiratory epithelium and alveolar macrophages (intratracheal instillation or aerosol administration), opportunities for targeted delivery of drugs (including DNA) to selected subpopulations of lung cells are readily available. Schreier and colleagues have investigated extensively the effects of liposome lipid composition and particle size on the cellular distribution of liposomes delivered by aerosol in experimental animals (22,23). The same investigators have shown the safety of aerosol delivery of liposomes to experimental animals and humans (24).

Liposomes are attractive as an in vivo delivery technology for DNA because DNA must access the interior of cells in order to function, a requirement not easily satisfied because of the size and lipophobicity of nucleic acid molecules. In some of the earliest in vivo experiments in this area, Nicolau and colleagues encapsulated a recombinant plasmid encoding rat preproinsulin I into large liposomes and injected the plasmid–liposome complexes intravenously into rats (25). They found uptake of the liposomes principally in the spleen and liver and generation of transgene product primarily in those organs as well. In later studies, the same workers could target liposomes to different subpopulations of liver cells by altering lipid composition (26).

Huang and co-workers have designed liposomes that target specific cell types by incorporating an antibody to cell surface proteins into the liposome membrane. In addition, they have rendered the liposomes pH sensitive by altering lipid composition in an effort to prevent degeneration of administered DNA in lysozomes (27). These "immunoliposomes" can deliver encapsulated DNA efficiently in cell culture systems (27).

In 1987, Felgner and associates (28,29) described synthesis of a different liposome preparation with particular advantages for nucleic acid delivery. They incorporated positively charged groups into the lipid membranes, producing liposomes that would simply associate with DNA by charge–charge interaction upon incubation at room temperature. This obviated the necessity of incorporation into the liposome interior, simplifying generation of liposome–DNA complexes. These cationic liposomes were shown to deliver DNA efficiently to cells in culture (28), and since the original description, a number of cationic liposome preparations have become available commercially.

We reported the first use of cationic liposomes to deliver DNA in vivo in 1989 (31,32) (see below). Since that time, a number of investigators have used this technology to delivery foreign DNA to different cell types and organs. These include rabbit and human secretory epithelial cells in culture (33), endothelial cells (30), and the endothelium of arteries in intact animals (34).

V. Delivery of Genes to the Lungs in Vivo

Two general technologies have emerged as potentially useful in delivering transgenes to the lungs: replication deficient viruses as vectors (adenovirus

Table 1 CAT Activity in Lungs, Liver, and Kidneys 72 H Following DNA–Liposome Injection into Mice

	% Chloramphenicol acetylation/h/μg protein × 10^{-2}		
Route	Lungs	Liver	Kidneys
Intravenous			
30 μg DNA per mouse	24.5	0	0
15 μg DNA per mouse	10.1	0	0
Intratracheal			
30 μg DNA per mouse	35.7	0	0
Intraperitoneal			
30 μg DNA per mouse	0	0	0

or adeno-associated virus) and liposome technology. Clinical trials in patients with cystic fibrosis are currently under way using the adenovirus technology.

Initially, we used cationic liposomes and a plasmid construct containing a reporter gene to demonstrate that functioning foreign DNA could be delivered to the lungs of mice by delivering cationic liposome–DNA complexes either intravenously or by injection into the trachea (31). Table 1 shows data from that study illustrating that specific transgene expression was seen in the lungs; the time course of gene expression for the particular plasmid construct that we used is shown in Fig. 1. There was a dose-related magnitude of transgene expression, and greater expression was seen in the lungs following intratracheal injection than after intravenous delivery (Table 1). In a more recent publication, we confirmed that relatively specific expression of a human growth hormone transgene was seen in the lungs after intravenous injection of cationic liposome plasmid complexes (32). The organ localization of transgene expression following intravenous injection of DNA–cationic liposome complexes is different from that following intravenous injection of liposome-encapsulated DNA. That observation has been confirmed (35).

We have demonstrated expression of several other transgenes in the lungs. Using a plasmid construct in which a CMV promoter drives a human

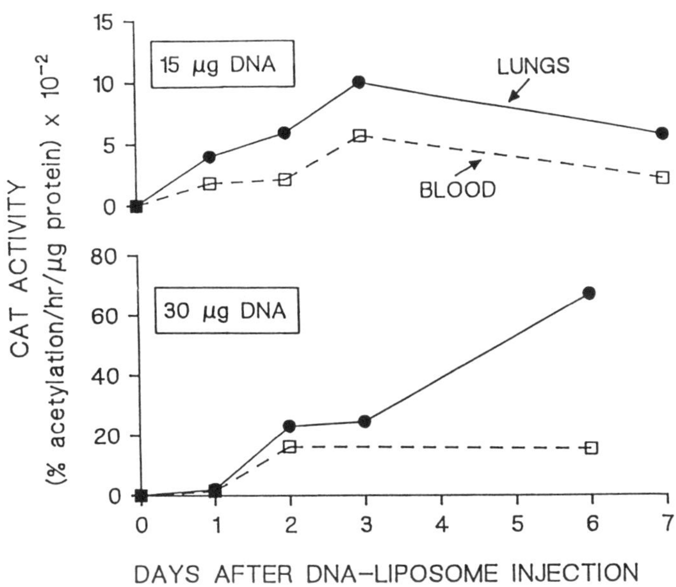

Figure 1 Time course of CAT reporter gene expression following intravenous delivery of DNA-liposome in mice (Reference no. 31).

α_1-antitrypsin cDNA, Canonico and co-workers showed that delivery of plasmid–cationic liposome complexes, by either intravenous or aerosol routes to rabbits, resulted in expression of the α_1-antitrypsin gene in the lungs (36–38). Immunohistochemical studies localized the human protein to both lung vascular endothelium and airway epithelium following intravenous delivery of the plasmid and to airway epithelium following aerosol delivery. Expression of the transgene was found in the liver as well as the lungs, especially following intravenous delivery. In similar studies, Conary and associates demonstrated expression of an ovine prostaglandin G/H synthase gene in the lungs of rabbits following intravenous injection of plasmid–cationic liposome complexes (39,40).

Additional work suggesting that cationic liposomes may be a useful delivery technology has been published. In mice, Stribling and associates showed expression of a chloramphenicol acetyltransferase reporter gene in the lungs following aerosol delivery of plasmid–cationic liposome complexes (41). Following intratracheal injection of plasmid–cationic liposome complexes in mice, Yoshimura et al. found expression of both a luciferase reporter gene and a cystic fibrosis transmembrane conductance regulator transgene (42). In the first demonstration of an in vivo physiologic effect of a CFTR transgene, Hyde and associates showed correction of the ion transport defect in cystic fibrosis transgenic mice following intratracheal administration of a plasmid containing the CFTR complexed to cationic liposomes (43).

Crystal and co-workers used a replication deficient adenovirus injected into the trachea of cotton rats to deliver the human α_1-antitrypsin gene (44) and the human cystic fibrosis transmembrane conductance regulator gene (45) to the lungs and showed expression of the transgene; in situ hybridization localized transgene expression in airway epithelium. Whether the efficiency of this delivery system is sufficient to make it practicable as a therapeutic strategy is not yet known. In addition, there may be generation of some viral proteins from the delivery vector which could limit the practical use of the method. Some of these problems may be addressed using adeno-associated viral vectors or other viral vectors.

Curiel has pioneered the use of adenovirus–polylysine–DNA complexes as a system for delivering DNA to cells in culture (46,47). This approach allows utilization of receptor-mediated gene delivery. Recently, Gao and associates, in Curiel's laboratory, showed in vivo delivery of reporter genes to airway epithelium of rodents using this delivery technology (48).

It is clear that functioning foreign DNA can be delivered to the lungs in vivo, both epithelium and endothelium, by several different techniques. Few studies have addressed the physiologic consequences of transgene expression, and there is reason to believe that the magnitude of transgene product necessary to produce the desired physiologic effect may be difficult to achieve.

VI. Transgenes Possibly Therapeutic for Endotoxin Injury

Several genes have been cloned and expressed in experimental systems that have the potential for protecting lungs from endotoxin-induced cellular injury or hastening recovery from such injury. There is abundant evidence that at least part of the injury resulting from endotoxin (and other injurious processes as well) involves oxidative stress. Hyperexpression of the genes encoding protective antioxidant enzymes could therefore be therapeutic. Such possibilities include superoxide dismutase, catalase, and enzymes involved in glutathione metabolism. Part of the injury resulting from inflammation may involve release of proteolytic enzymes from inflammatory cells. Hyperexpression of antiproteases such as α_1-antitrypsin might therefore be therapeutic in these conditions. Interestingly, it is even possible to design a transgene that encodes a biologically active α_1-antitrypsin which is resistant to inactivation by aggressive oxygen species.

Conary and co-workers (49) reported protection of rabbit lungs from the effects of intravenous endotoxin following in vivo transfection with a plasmid expressing the cDNA for prostaglandin G/H synthase. They reasoned that intravenous delivery of the DNA would primarily transfect endothelial cells, and that the resulting increased production of the principal endothelial prostanoids, prostacyclin and PGE_2, would protect the lungs from endotoxin. Since production of the physiologically active substances was catalyzed by the transgene product, they reasoned that the physiologic consequences of in vivo transfection would be amplified and therefore that the magnitude of expression necessary for an effect would be minimized. They injected rabbits intravenously with a construct containing the ovine PGG/H cDNA driven by a CMV promoter complexed to cationic liposomes. Twenty-four hours later, they showed increased generation of prostacyclin and PGE_2 by the lungs and increased amounts of prostacyclin in the urine. In an in situ perfused lung preparation, they showed marked attenuation of the pulmonary hypertensive response to endotoxin and a decrease in endotoxin-induced pulmonary edema.

VII. Regulation of Transgene Expression

Several strong promiscuous promoters have been used in a variety of preparations to drive selected genes. We have used SV40, RSV, and CMV promoters and have demonstrated that each of these constructs functions in the lungs following in vivo delivery of plasmid constructs complexed to cationic liposomes.

If DNA constructs could be designed so that regulatory sequences which respond to injurious substances would drive genes that encode a protein which protects against the injury, transgene expression could be limited to cells exposed to injury, making this approach more effective and safer. Using DNA constructs that assure organ-specific expression of transgenes could also permit a more precisely targeted manipulation of mediator generation.

Regulatory regions of the tumor necrosis factor gene confer endotoxin responsiveness (50). Plasmid constructs containing both promoter and 3′UTR regulatory sequences for tumor necrosis factor (TNF) driving a CAT reporter gene respond to endotoxin [lipopolysaccharide (LPS)] when transfected into macrophages (50). The LPS responsiveness appears to depend on both the 5′ promoter sequence (which contains an NFκB site) and 3′UTR sequences which normally confer translational repression but are derepressed in the presence of LPS (50,51). Maximum endotoxin effect requires the presence of both the 5′ and 3′UTR sequences. RAW 264.7 cells transfected with a plasmid with these sequences and a CAT reporter gene increase CAT activity more than 500-fold in the presence of endotoxin (50).

Several endothelial cell adhesion molecules are expressed selectively in areas of inflammation (52). Studies of regulation of expression of some of these molecules (e.g., VCAM-1) demonstrate responsiveness of the promoter regions to cytokines, especially tumor necrosis factor (49). Use of regulatory regions for VCAM-1, an adhesion molecule that is expressed with some specificity in lung endothelial cells, could permit specific transgene expression in areas of inflammation.

In addition to proximal promoter sequences that regulate initiation of transcription, there are several other regulatory elements located farther upstream. These include enhancer elements that modulate transcription independent of distance and orientation (53,54), and these elements sometimes regulate transcription in a tissue-specific manner (55,56). There is also a class of upstream functional elements encompassing what is defined as a dominant control region (DCR). In experiments with transgenic mice, the DCR upstream regions from several genes facilitate chromatin-position-independent high-level expression of both homologous and heterologous promoters in a tissue-specific manner (57). In the liver, a specific factor with an enhancer 4.8 kb upstream of the phosphoenolpyruvate carboxykinase (PEPCK) gene (55) with enhancing activity specific for the liver has been identified (58,59). Such regulatory regions might prove useful for targeting transgene expression to specific organs.

VIII. Summary and Conclusions

Gene therapy for inherited diseases has become a clinical reality, and the application of this technology is developing rapidly. The concept can be expanded to include diseases not due primarily to a genetic defect by designing delivery systems and expression vectors that express transgenes transiently. This transient gene therapy, or gene therapeutics, uses DNA in a manner similar to that of any other pharmacologic agent. Delivery technologies have been developed that work in the lungs. The lungs may be particularly appropriate for this approach because they can be accessed by either the circulatory or airway route, and it may even be possible to target subpopulations of cells in this manner. A number of genes have been cloned that might protect or cure the lungs from endotoxin injury. Early studies in animals show that the pathophysiology of the endotoxin response can be altered using this approach. As information concerning safety of DNA delivery to humans accumulates, the delivery technologies mature, and animal data for efficacy are obtained, application of this approach to humans with endotoxin-induced lung injury will occur. If successful, this will provide an entirely new category of therapies for this often fatal clinical syndrome.

Acknowledgment Supported by Grants No. HL-19153, HL-07123, HL-45151, HL-34208 from the National Institutes of Health; The American Lung Association; and the Cystic Fibrosis Foundation. Dr. Brigham is the Joe and Morris Werthan Professor of Investigative Medicine.

References

1. Friedman T. Progress toward human gene therapy. Science 1989; 244:1275–1281.
2. Anderson WF. Editorial: the NIH human gene therapy symposium. Hum Gene Ther 1992; 3:127–128.
3. Verma IM. Gene therapy. Sci Am 1990; 261:68–84.
4. Brigham KL, Canonico AE, Conary JT. Mediator gene expression in sepsis: implications for therapy. In: Lamy M, Thijs LG, eds. Update in Intensive Care and Emergency Medicine. Vol 16. Mediators of Sepsis. New York: Springer-Verlag, 1992:393–405.
5. Brigham KL, Schreier H. Cationic liposomes and DNA delivery. J Liposome Res 1993; 3:31–49.
6. Hazinski TA. Prospects for gene therapy in acute lung injury. Am J Med Sci 1992; 304:131–135.
7. Garver RI Jr, Chytil A, Courtney M, Crystal RG. Clonal gene therapy: transplanted mouse fibroblast clones express human α1-antitrypsin gene in vivo. Science 1987; 237:762–764.
8. Wilson JM, Birinya LK, Salomon RN, Libby P, Callow AD, Mulligan RC. Implantation of vascular grafts lines with genetically modified endothelial cells. Science 1989; 244:1344–1346.

9. Culver KW, Ram Z, Wallbridge S, Ishii H, Oldfield EH, Blaese RM. In vivo transfer with retroviral vector-producer cells for treatment of experimental brain tumors. Science 1992; 256:1550–1552.
10. Wolff JA, Malone RW, Williams P, et al. Direct gene transfer into mouse muscle in vivo. Science 1990; 247:1465–1468.
11. Stratford-Perricaudet LD, Makeh I, Perricaudet M, Briand P. Widespread long-term gene transfer to mouse skeletal muscles and heart. J Clin Invest 1992; 90:626–630.
12. Buttrick PM, Kaas A, Kitis RN, Kaplan ML, Leinwand LA. Behavior of genes directly injected into the rat heart in vivo. Circ Res 1992; 70:193–198.
13. Barr E, Leiden JM. Systemic deliver of recombinant proteins by genetically modified myoblasts. Science 1991; 254:1507–1509.
14. Zenke M, Steinlein P, Wagner E, Cotten M, Beug H, Birnstiel ML. Receptor-mediated endocytosis of transferrin-polycation conjugates: an efficient way to introduce DNA into hematopoietic cells. Proc Natl Acad Sci USA 1990; 87: 3655–3659.
15. Wu GY, Wu CH. Delivery systems for gene therapy. Biotherapy 1991; 3:87–95.
16. Wilson JM, Grossman M, Wu CH, Chowdhury NR, Wu GY, Chowdhury JR. Hepatocyte-directed gene transfer in vivo leads to transient improvement of hypercholesterolemia in low density lipoprotein receptor–deficient rabbits. J Biol Chem 1992; 267:963–967.
17. Curiel DT, Wagner E, Cotten M, et al. Hum Gene Ther 1992; 3:147–154.
18. Curiel DT, Agarwal S, Romer MU, et al. Gene transfer to respiratory epithelial cells via the receptor-mediated endocytosis pathway. Rapid Commun 1992; 6: 247–252.
19. Freeman B, Young S, Crapo J. Liposome-mediated augmentation of superoxide dismutase in endothelial cells prevents oxygen injury. J Biol Chem 1983; 258:12534–12542.
20. Randall BR, Czopf L, Jilling T, Freeman BA, Kirk KL, Matalon S. Quantitation of alveolar distribution of liposome-entrapped antioxidant enzymes. Am J Physiol 1992; 263:L585–L594.
21. Pitt BR, Huang L. Targeted drug delivery to the lung. Contemp Intern Med 1992 (Sept):81–92.
22. Niven RW, Schreier H. Nebulization of liposomes. I. Effects of lipid composition. Pharm Res 1990; 7:1127–1133.
23. Niven RW, Speer M, Schreier H. Nebulization of liposomes. II. The effects of size and modeling of solute release profiles. Pharm Res 1991; 8:217–221.
24. Gonzales-Rothi RJ, Straub L, Cacace JL, Schreier H. Liposomes and pulmonary alveolar macrophages: function and morphologic interactions. Exp Lung Res 1991; 17:687–705.
25. Nicolau C, Pape AL, Soriano P, Fargette F, Juhel M-F. In vivo expression of rat insulin after intravenous administration of the liposome-entrapped gene for rat insulin I. Proc Natl Acad Sci USA 1983; 80:1068–1072.
26. Soriano P, Dukstra J, Legrand A, et al. Targeted and nontargeted liposomes for in vivo transfer to rat liver cells of a plasmid containing the preproinsulin I gene. Proc Natl Acad Sci USA 1983; 80:7128–7131.

27. Wang C-Y, Huang L. Highly efficient DNA delivery mediated by pH-sensitive immunoliposomes. Biochemistry 1989; 9508-9514.
28. Felgner PL, Gadek TR, Holm M, et al. Lipofection: a highly efficient, lipid-mediated DNA-transfection procedure. Proc Natl Acad Sci USA 1987; 84:7413-7417.
29. Felgner PL, Ringold GM. Cationic liposome-mediated transfection. Nature 1989; 337:387-388.
30. Brigham KL, Meyrick B, Christman B, Berry LC Jr, King G. Expression of a prokaryotic gene in cultured lung endothelial cells following lipofection with a plasmid vector. Am J Respir Cell Mol Biol 1989; 1:95-100.
31. Brigham KL, Meyrick B, Christman B, Magnuson M, King G, Berry LC Jr. Rapid communication: in vivo transfection of murine lungs with a functioning prokaryotic gene using a liposome vehicle. Am J Med Sci 1989; 298:278-281.
32. Brigham K, Meyrick B, Christman B, et al. Expression of human growth hormone fusion genes in cultured lung endothelial cells and in the lungs of mice. Am J Respir Cell Mol Biol 1993; 8:209-213.
33. Lu L, Zeitlin PL, Guggino WB, Craig RW. Gene transfer by lipofection in rabbit and human secretary epithelial cells. Eur J Physiol 1989; 415:198-203.
34. Nabel EG, Plautz G, Nabel DJ. Site-specific gene expression in vivo by direct gene transfer into the arterial wall. Science 1990; 249:1285-1288.
35. Stewart M, Plautz G, del Buono F, et al. Gene transfer in vivo with DNA-liposome complexes: safety and acute toxicity in mice. Hum Gene Ther 1992; 3: 267-275.
36. Canonico AE, Conary JT, Meyrick BO, Brigham KL. Aerosol and intravenous transfection of human α-1 antitrypsin gene to lungs of rabbits. Am J Respir Cell Mol Biol 1994; 10:24-29.
37. Canonico AE, Conary JT, Christman BW, Meyrick BO, Brigham KL. Expression of a CMV promoter driven human α-1 antitrypsin gene in cultured lung endothelial cells and in the lungs of rabbits. Clin Res 1991; 39:219A.
38. Canonico AE, Conary JT, Meyrick BO, Brigham KL. In vivo expression of a CMV promoter driven human alpha-1 antitrypsin (alpha-1AT) gene after intravenous or airway administration of DNA/liposome complex. Am Rev Respir Dis 1992; 145:A200.
39. Conary JT, Canonico AE, Parker RE, Christman BW, King G, Brigham KL. Expression of a CMV promoter driven ovine prostaglandin G/H synthase gene in the lungs of rabbits. Am Rev Respir Dis 1992; 145:A850.
40. Conary JT, Brown DL, Shepherd VE, Christman BW, King G, Brigham KL. Expression of a CVM promoter driven ovine prostaglandin G/H synthase gene in cultured lung endothelial cells. Am Rev Respir Dis 1992; 145:A839.
41. Stribling R, Brunette E, Liggitt D, Gaensler K, Debs R. Aeroxol gene delivery in vivo. Proc Natl Acad Sci USA 1992; 89:11277-11281.
42. Yoshimura K, Rosenfeld MA, Nakamura H, et al. Expression of human cystic fibrosis transmembrane conductance regulator gene in the mouse lung after in vivo intratracheal plasmid-mediated gene transfer. Nucleic Acids Res 1992; 12: 3233-3240.
43. Hyde SC, Gill DR, Higgins CF, et al. Correction of the ion transport defect in cystic fibrosis transgenic mice by gene therapy. Nature 1993; 362:250-255.

44. Rosenfeld MA, Siegfried W, Yoshimura K, et al. Adenovirus-mediated transfer of a recombinant alpha-1 antitrypsin gene to the lung epithelium in vivo. Science 1991; 252:431–434.
45. Rosenfield MA, Yoshimura K, Trapnell BC, et al. In vivo transfer of the human cystic fibrosis transmembrane conductance regulator gene to the airway epithelium. Cell 1992; 68:143–155.
46. Curiel DT, Agarwal S, Wagner E, Cotten M. Adenovirus enhancement of transferrin-polylysine-mediated gene delivery. Proc Natl Acad Sci USA 1991; 88: 8850–8854.
47. Curiel DT, Wagner E, Cotten M, et al. High-efficiency gene transfer mediated by adenovirus coupled to DNA–polylysine complexes. Hum Gene Ther 1992; 3:147–154.
48. Gao L, Wagner E, Cotten M, et al. Direct in vivo gene transfer to airway epithelium employing adenovirus–polylysine–DNA complexes. Hum Gene Ther 1993; 4:17–24.
49. Conary JT, Parker RE, Christman BW, Faulks RD, King G, Meyrick B, Brigham KL. Protection of rabbit lungs from endotoxin injury by *in vivo* hyperexpression of the prostaglandin G/H synthase gene. J Clin Invest 1994; 93:1834–1 1840.
50. Han J, Huez G, Beutler B. Interactive effects of the tumor necrosis factor promoter and 3 untranslated regions. J Immunol 1991; 146:1843–1848.
51. Kruys V, Kemmer K, Shakov A, Jongeneel V, Beutler B. Constitutive activity of the tumor necrosis factor promoter is cancelled by the 3′ untranslated region in nonmacrophage cell lines: a trans-dominant factor overcomes this suppressive effect. Proc Natl Acad Sci USA 1992; 89:673–677.
52. Carlos T, Harlan J. Membrane proteins involved in phagocyte adherence to endothelium. Immunol Rev 1990; 114:5–28.
53. Cullen B, Raymond K, Ju G. Transcriptional activity of avlan retroviral long terminal repeats directly correlates with enhancer activity. J Virol 1985; 53:515–521.
54. Queen C, Baltimore D. Immunoglobulin gene transcription is activated by downstream sequence elements. Cell 1983; 33:741–748.
55. Ott MO, Sperling L, Herbonnel P, Yaniv M, Weiss M. Tissue-specific expression is conferred by a sequence from the 5′ end of the rat albumin gene. EMBO J 1984; 3:2505–2510.
56. Theisen M, Stief A, Sippel M. The iysozyme enhancer: cell specific activation of the chicken lysozyme gene by a far upstream DNA element. EMBO J 1986; 5:719–724.
57. Grosveld F, vanAssendelft G, Greaves D, Kolloas G. Position-independent, high level expression of the human B-globin gene in transgenic mice. Cell 1987; 51:975–985.
58. Ip T, Poon D, Stone D, Granner G, Chalkley R. Interaction of a liver-specific factor with an enhancer 4.8 kilobases upstream of the phosphoenolpyruvate carboxykinase gene. Mol Cell Biol 1990; 10:3770–3781.
59. Ip T, Fournier K, Chalkley R. Extinction of phosphoenolpyruvate carboxykinase gene expression is associated with loss of a specific chromatin-binding protein from a far upstream domain. Mol Cell Biol 1990; 10:3782–3787.

AUTHOR INDEX

Italic numbers give the page on which the complete reference is listed.

A

B

C

D

H

I

L

M

O

Q

R

T

W

X

Y

SUBJECT INDEX

A

B

C

D

F

G

H

I

M

N

R

T

U

V